Network Theory and Filter Design

Network Theory and Filter Design

(THIRD EDITION)

Vasudev K. Aatre
Naval Physical and Oceanographic Laboratory
Ministry of Defence, India
Formerly with Nova Scotia Technical College
Halifax, Canada
Presently Scientific Adviser to Defence Minister
& Secretary, Department of Defence R & D
Ministry of Defence, India

Publishing Globally
NEW AGE INTERNATIONAL (P) LIMITED, PUBLISHERS
New Delhi • Bangalore • Chennai • Cochin • Guwahati
Hyderabad • Kolkata • Lucknow • Mumbai
Visit us at **www.newagepublishers.com**

Published by New Age International (P) Ltd., Publishers
Second Edition: 1980
Third Edition: 2014

BRANCHES

- **Bangalore** 37/10, 8th Cross (Near Hanuman Temple), Azad Nagar, Chamarajpet Bangalore-560 018, Tel.: (080) 26756823, Telefax: 26756820 **E-mail: bangalore@newagepublishers.com**
- **Chennai** 26, Damodaran Street, T. Nagar, Chennai-600 017, Tel.: (044) 24353401 Telefax: 24351463, **E-mail: chennai@newagepublishers.com**
- **Cochin** CC-39/1016, Carrier Station Road, Ernakulam South, Cochin-682 016 Tel.: (0484) 2377004, Telefax: 4051303 **E-mail: cochin@newagepublishers.com**
- **Guwahati** Hemsen Complex, Mohd. Shah Road, Paltan Bazar, Near Starline Hotel Guwahati-781 008, Tel.: (0361) 2513881, Telefax: 2543669 **E-mail: guwahati@newagepublishers.com**
- **Hyderabad** 105, 1st Floor, Madhiray Kaveri Tower, 3-2-19, Azam Jahi Road Nimboliadda, Hyderabad-500 027, Tel.: (040) 24652456 Telefax: 24652457, **E-mail: hyderabad@newagepublishers.com**
- **Kolkata** RDB Chambers (Formerly Lotus Cinema) 106A, 1st Floor S.N. Banerjee Road, Kolkata-700 014, Tel.: (033) 22273773 Telefax: 22275247, **E-mail: kolkata@newagepublishers.com**
- **Lucknow** 16-A, Jopling Road, Lucknow-226 001, Tel.: (0522) 2209578, 4045297 Telefax: 2204098, **E-mail: lucknow@newagepublishers.com**
- **Mumbai** 142C, Victor House, Ground Floor, N.M. Joshi Marg, Lower Parel Mumbai-400 013, Tel.: (022) 24927869, Telefax: 24915415 **E-mail: mumbai@newagepublishers.com**
- **New Delhi** 22, Golden House, Daryaganj, New Delhi-110 002 Tel.: (011) 23262368, 23262370, Telefax: 43551305 **E-mail: sales@newagepublishers.com**

ISBN: 978-81-224-3409-5

₹ 299.00

C-13-06-7001

Printed in India at Ajit Printers, Delhi.
Typeset at Pagiteck Graphics, Delhi.

PUBLISHING FOR ONE WORLD
NEW AGE INTERNATIONAL (P) LIMITED, PUBLISHERS
7/30 A, Daryaganj, New Delhi - 110002
Visit us at **www.newagepublishers.com**

To

Prof. K. Ramakrishna

a friend, philosopher
and guide par excellence

Preface to the Third Edition

Since the publication of the second edition of this book, teachers of several universities have sent their comments and criticisms about the contents of the book. On study of these suggestions, it was decided to modify the book incorporating them for its third edition.

Though the philosophy of the book has remained the same, it has been sufficiently modified to cater for specific comments. Several sections have been deleted, new sections added and the presentations in some of the chapters have been changed. In Chapter 3, an additional section on networks with controlled sources has been added. Most electrical engineering curriculums cover the sinusoidal steady state in the first course on circuits. Hence, the chapter on sinusoidal steady state has been deleted and relevant parts of the chapter merged with Chapter 4. The application of Telegens theorem to evaluate the sensitivity of networks through Adjoint-Networks has been included in the chapter on Network theorems. The chapters on Two-port and Indefinite admittance matrix have been merged to a single chapter on multi-terminal networks. The detailed formulation of state equations from graphtheoretic methods has been introduced. The chapter on Image parameter filter design has been relegated to an appendix with the section on frequency transformation being appended to the chapter on approximation. The concept of energy functions has been introduced as a prelude to the PR character of driving point immittances. The realisation of ladder networks using zero shifting technique and of higher order active filters by simulated ladders has also been included.

With the preceding changes, the book has now 12 chapters and 6 appendices. These changes are incorporated with a view to improve the usefulness of the book as a textbook for undergraduate electrical engineering students. The third edition of the book is also aimed at getting rid of some of the notational inconsistancies and 'printers devil' mistakes.

Prof. VGK Murthy of IIT Madras and KVV Murthy of IIT Bombay reviewed the book thoroughly and made recommendations for the third edition of the book.

The author is deeply indebted to them. Dr. K. Soundararajan of ADE, Bangalore has spent hours in recasting the chapter on Active Filters. I am gratefully acknowledge his help.

The guiding force behind the reviewed edition, as of the second edition, was late Prof. K. Ramakrishna of Indian Institute of Science. No amount of my acknowledgement can express my indebtedness to him. Indeed the book is dedicated to him.

V.K. AATRE

Preface to the First Edition

The basis for most electrical engineering courses depends on two fundamental theoretical concepts, viz., the lumped-parameter circuit theory based on laws like those of Kirchhoff's and Ohm's, and field theory based on Maxwell's equations.

Circuit and network theory courses may be considered as prerequisites to most other courses in a normal electrical engineering curricula as many concepts developed in such courses form a basis for teaching electronics and signal processing, control and linear systems. The two main branches of network theory are network analysis and network synthesis, the former being a prelude to the latter. Network synthesis naturally leads to filter design. The importance of the design and the application of filters cannot be over-emphasized as most signal acquisition and processing systems utilize filters-passive or active inductorless filters and more recently digital filters.

Network analysis, network synthesis and filter design are well developed subjects, and are endowed with many excellent textbooks, though most of them are devoted to the discussion of one of these topics. The stringent requirements placed by most electrical engineering curricula have forced network analysis and synthesis to a sequence of two courses. This book has attempted a unification of these two fundamental topics by discussing the required analytical methods and linking the essentials of approximation, synthesis and filter design.

Following a brief review of basic circuit elements, fundamental concepts of graph theory and concept of the generalized network element are unified to yield both a formal and the inspection method of writing various forms of network equations. Solution of network equations based on Laplace transforms, sinusoidal steady-state analysis and resonance, network theorems including Tellegen's theorem, concepts of natural frequency and poles-zeros of network functions, two-port parameters including analysis of the ubiquitous ladder network, indefinite admittance matrix analysis of multi-terminal networks and equivalent source method of writing state equations are all discussed in sufficient detail to form the analysis part of the book. Essentials of approximation and synthesis leading to practical filter design are presented in the synthesis part of the book. Image parameter filter design is introduced for historical reasons and to bring out the concept of matching and cascading in filter design. Foster and Cauer form synthesis of two-element kind networks and realization of all-pole transfer function are discussed as prelude to Darlington's insertion loss synthesis. Comprehensive introduction to active filter design methods for single and multi-amplifier filters, cascaded and GIC based structures, is also presented.

This book presupposes a knowledge of differential equations and transform calculus, elementary circuit analysis covering Kirchhoff's laws, resistive networks, transients and three-phase AC circuits. It is designed for use as a text for undergraduate and postgraduate courses in network analysis and fundamentals of filter design. Contents of the book can be covered in a sequence of two courses, each with 30 hours of lectures aided by 20 hours of laboratory or problem sessions. The first course can cover preliminary analysis (chapters 1–8) and the second one advanced methods of analysis (Chapters 9–10) and synthesis/design (chapters 11–15). In this form it has been used for several years at Nova Scotia Technical College, Halifax.

A project of writing a book like this can only materialize with the help of many. It is a great pleasure to thank my undergraduate students who suffered through the earlier versions of this book and graduate students, especially Mr. W.D. Eckford and Mr. A.K. Mitra, who helped me in formulating many of my thoughts. I am indebted to my colleagues, Drs. V. Bapeswara Rao of Indian Institute of Technology, Madras, R. Hastings James and O.K. Gashus of Nova Scotia Technical College, for many valuable criticisms. I am especially grateful to Dr. O.K. Gashus for providing assistance in the preparation of the earlier versions of the manuscript.

For permission to use copyright material, I am grateful to: Prentice Hall Inc., USA, for Figures 3.9, 3.10, 3.11 and 3.14 from D.E. Johnson: *Introduction to Filter Theory* and Addison-Wesley Inc., USA, for Problems 3.3 and 4.3 to 4.5 from S.P. Chan, S.Y. Chan and S.G. Chan: *Analysis of Linear Networks and Systems.*

The final draft of the book was completed while the author spent his sabbatical year at the Indian Institute of Science, Bangalore, and the author thanks the authorities of the Institute for many facilities. The assistance offered by Dr. R. Vittal Rao in proof reading and manuscript, Mr. N. Govinda Raju in typing the manuscript and late Mr. V.P. Naidu in drafting the diagrams is acknowledged. This book would not have materialized but for the invaluable guidance of Dr. K. Ramakrishna through all phases of the book and I remain ever so grateful to him. Finally I would like to express my gratitude to my family.

Halifax, Nova Scotia

V.K. AATRE

Contents

Preface to the Third edition vii

Preface to the First edition ix

1. Introduction **1–16**

1.1 Lumped Circuit Elements *2*
- 1.1.1 Resistors *2*
- 1.1.2 Independent Sources *4*
- 1.1.3 Capacitors *5*
- 1.1.4 Inductors *7*

1.2 Kirchhoff's Laws *9*
1.3 Classification of Networks *10*
1.4 Nullators and Norators *13*
Problems 14
Notes and References 16

2. Network Topology **17–40**

2.1 Basic Definitions and Properties *18*
2.2 Matrices of Graphs *25*
- 2.2.1 Incidence Matrix *26*
- 2.2.2 Fundamental Cut-Set Matrix *28*
- 2.2.3 Fundamental Circuit Matrix *31*
- 2.2.4 Inter-Relationships Among the Matrices *35*

Problems 37
Notes and References 40

3. Network Equations **41–79**

3.1 Node and Mesh Transformations *41*
3.2 Generalized Element *45*
3.3 Source Transformation *49*

3.4 Formulation of Network Equations *50*
3.4.1 Node Equations *50*
3.4.2 Mesh Equation *54*
3.4.3 Cut-Set Equations *57*
3.4.4 Loop Equations *59*
3.5 Network Equations by Inspection *61*
3.5.1 Node Equations *61*
3.5.2 Loop Equations *62*
3.6 Mutual Inductance *63*
3.7 Networks with Controlled Sources *68*
3.8 Transform Networks *71*
3.8.1 Resistors *72*
3.8.2 Inductors *72*
3.8.3 Capacitors *73*
3.8.4 Coupled Coils *73*
Problems 76
Notes and References 79

4. Solution of Network Equations and Sinusoidal Response 80–121

4.1 Properties of Network Matrices *80*
4.2 Solution of Equations *82*
4.3 Partial-Fraction Expansion *83*
4.4 Impulse Response and Convolution *89*
4.5 Linear Time-Invariant Networks *97*
4.6 Evaluation of Initial Conditions *98*
4.6.1 Network with Degenerate Circuits *100*
4.6.2 Network with Degenerate Cut-Set *102*
4.7 Sinusoidal Steady State *104*
4.8 Resonance *107*
4.8.1 Parallel Resonance *107*
4.8.2 Series Resonance *112*
4.9 Frequency and Impedance Scaling *113*
Problems 115
Notes and References 121

5. Network Theorems 122–145

5.1 Superposition Theorem *122*
5.2 Substitution Theorem *123*
5.3 Thevenin-Norton Theorem *127*
5.4 Tellegen's Theorem *133*
5.5 Reciprocity Theorem *136*
5.6 Adjoint Networks *140*
5.7 Maximum Power Transfer Theorem *141*

Problems *143*
Notes and References *145*

6. Natural Frequencies and Network Functions **146–169**

6.1 Natural Frequency *146*
6.2 Network Functions *150*
6.3 Poles and Zeros *152*
6.4 Time Response *159*
6.5 Frequency Response *162*
Problems *164*
Notes and References *169*

7. Multiterminal Networks **170–207**

7.1 Two-Port Parameters *171*
7.1.1 Impedance Parameters *171*
7.1.2 Admittance Parameters *173*
7.1.3 Transmission Parameters *174*
7.1.4 Hybrid Parameters *176*
7.1.5 Interrelationship among Parameters *177*
7.1.6 Ladder Networks *181*
7.2 Interconnection of Two-Ports *184*
7.2.1 Series Connection *184*
7.2.2 Parallel Connection *185*
7.2.3 Cascade Connections *186*
7.3 Equivalent Networks *186*
7.4 Indefinite Admittance Matrix *189*
7.5 Properties of IAM *190*
7.6 Network Functions *195*
7.7 Operational Amplifier Networks *197*
Problems *201*
Notes and References *206*

8. State Equations **208–226**

8.1 Concept of State *208*
8.2 State Equations *209*
8.3 Formulation of State Equation *212*
8.3.1 Equivalent Source Method *213*
8.3.2 Graph Theoretic Method *218*
8.4 State Equations and Transfer Function *222*
Problems *224*
Notes and References *226*

9. Elements of Network Synthesis **227–275**

9.1 Positive Real Function *227*

9.2 Reactance Functions *235*
9.3 RC Functions *244*
9.4 RL Network *254*
9.5 Two-Port Functions *255*
9.5.1 RC Ladders *256*
9.5.2 LC Ladders *266*
9.6 Minimum-Phase Networks *272*
Problems 273
Notes and References 275

10. Approximation **276–310**

10.1 Filter Specifications *277*
10.2 Butterworth Approximation *279*
10.3 Chebyshev Approximation *285*
10.4 Comparison between Butterworth and Chebyshev Transfer Functions *291*
10.5 Frequency Transformation *294*
10.5.1 Low-Pass to High-Pass *296*
10.5.2 Low-Pass to Band-Pass *297*
10.5.3 Low-Pass to Band Elimination *299*
10.6 Band-Pass Filter Approximation *300*
10.7 Other Methods of Approximation *303*
10.7.1 Inverse Chebyshev Approximation *303*
10.7.2 Elliptic Filter Approximation *304*
10.7.3 Bessel Approximation *305*
Problems 309
Notes and References 310

11. Insertion Loss Synthesis **311–328**

11.1 Coefficient Matching Technique *311*
11.2 Darlington's Method *313*
11.2.1 Insertion Loss and Reflection Coefficient *314*
11.2.2 Realization Procedure *318*
11.3 Discussions *325*
Problems 326
Notes and References 328

12. Active Networks and Filters **329–379**

12.1 Active Elements *329*
12.1.1 Operational Amplifier *331*
12.1.2 Negative Impedance Converter *332*
12.1.3 Gyrators *334*
12.1.4 Generalised Impedance Converter *336*
12.2 Nullator-Norator Models *337*

12.3 Sensitivity *343*
12.4 Single Amplifier Filters *349*
12.4.1 Low-Pass Filter *350*
12.4.2 HP, BP and BE Filters *354*
12.5 State-Variable Realization *363*
12.6 All-Pass and Notch-Filter *366*
12.7 Higher Order Filters *367*
12.7.1 FDNR Transformation *371*
12.7.2 Coupled Biquad Structure *373*
12.8 Non-Ideal Op. Amp *375*
12.9 Discussions *377*
Notes and References *378*

Appendices **380–422**

Appendix A. **Matrix Algebra** *380*
Appendix B. **Laplace Transforms** *389*
Appendix C. **Natural Frequencies** *394*
Appendix D. **Bode Plots** *397*
Appendix E. **Scattering Parameters** *402*
Appendix F. **Image Parameter Filter Design** *410*

Index **423–427**

CHAPTER 1

Introduction

A network or a system may be described as a collection of objects or elements governed by certain laws. In electric *network theory* we concern ourselves with four basic quantities: voltage, current, charge and flux. The study of interrelationship among these variables and the physical laws governing their behaviour is the subject of network theory. Network theory, in general, may be divided into three components:

1. The mathematical *modelling* of electromagnetic phenomena.
2. The mathematical *analysis* of models of individual elements and systems, and their signal processing capabilities.
3. The synthesis and *design* of systems and devices.

One of the most useful approaches in scientific analysis is that of modelling. The primary reason for constructing models is that the physical systems are usually too complex to be amenable to practical analysis. The complexity arises due to the presence of non-essential factors and, hence, a model should extract only the essential factors. In essence, we can state that the model should adequately describe the physical behaviour of a system. The more detailed the model the more accurate the predicted solutions. But then the analysis of such a model also becomes more complicated. Hence, a model is always a compromise between the reality and simplicity. Consider, for example, an ideal and the corresponding non-ideal model of a coil shown in Fig. 1.1. The latter has three basic circuit elements, whereas the former has only one element.

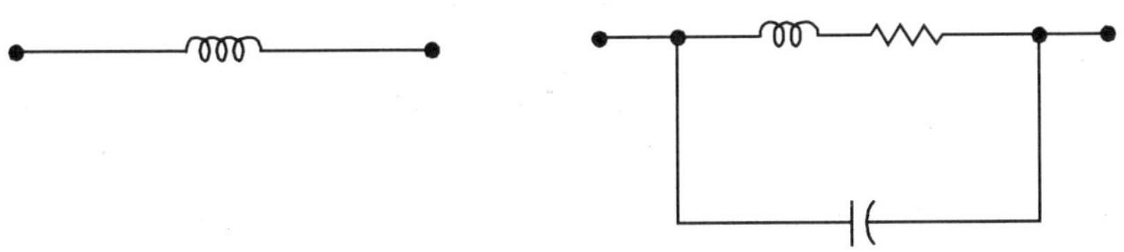

Fig. 1.1

An electric network (amplifier, filter or equalizer circuits and the like) can be modelled using certain lumped circuit elements. A basic problem faced by an engineer is to design a network model to satisfy certain signal processing specifications and then fabricate physical components which approximate the idealized elements. A prelude to design (and synthesis) is analysis, which is mainly an algebraic problem. A major part of this book deals with it.

In this chapter, we discuss the characterization and modelling of *lumped circuit elements.* A network, consisting of lumped elements, exhibits certain basic properties depending on the type of elements used which themselves are classified according to these properties.

1.1 LUMPED CIRCUIT ELEMENTS

A large part of the network theory deals with the study of lumped circuits. Lumped circuit is a result of interconnection of lumped elements. An element is classified as a *lumped element* if its physical dimensions are small compared to the wavelength corresponding to the highest frequency of operation of that element. For example, a frequency of operation of 10 kHz (audio frequency) corresponds to a wavelength of 30 km and any circuit built in a laboratory has negligible physical dimension compared to this wavelength. If the frequency of operation is 1 GHz or 10^9 Hz (microwave frequency), the wavelength is 0.3 m and a circuit's dimension becomes comparable to the wavelength. In the latter case, the "lumped circuit" approximation is not valid.

For a lumped element the instantaneous current entering one terminal is equal to the instantaneous current leaving the other terminal. A television antenna is not a lumped element as the current at the foot of the antenna is not the same as that at the tip of the antenna. If the two were the same, there would be no radiation. The current through and the voltage across a lumped element are well defined quantities and satisfy Kirchhoff's laws (Section 1.2). Typical lumped circuit elements are resistors inductors, capacitors, voltage and current sources.

1.1.1 Resistors

An element which can be characterized by a curve in the v–i plane is called a resistor (v represents voltage and i represents current). A resistor is linear if it is characterized by a straight line passing through the origin of the v–i plane [Fig. 1.2(*a*)]. A linear resistor is represented by the symbol shown in Fig. 1.2(*b*) and is described mathematically by

$$i = Gv \text{ or } v = Ri \qquad \ldots(1.1)$$

The constant G, called the conductance, represents the slope of the line in Fig. 1.2(*a*). The constant R is the reciprocal of the slope and is called the resistance. The practical unit of G is mho (℧) and R is ohm (Ω). If $R = 0$, a linear resistor is said to be a *short circuit*; and if $R = \infty$, it is said to be an *open circuit.*

If a resistor is characterized by a *v–i curve* other than a straight line through the origin, it is called a *non-linear resistor.* Figure 1.3 shows the characteristics (v–i curves) of two non-linear resistors. Figure 1.3(*a*) corresponds to a tunnel diode and it can be seen that the current is a single valued function of the voltage.

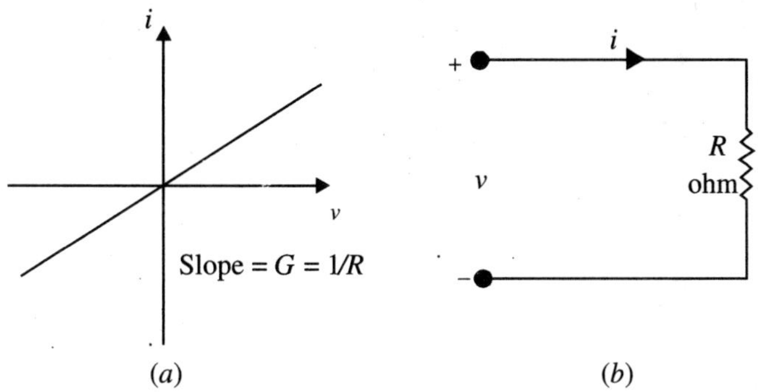

Fig. 1.2

Such a resistor is said to be *voltage-controlled.* On the other hand, in the characteristics of Fig. 1.3(*b*) (silicon-controlled rectifier) the voltage is a single valued function of the current. Such a resistor is said to be *current-controlled.* These non-linear resistors have the unique property that the slope of the *v–i* characteristic is negative in some range of voltage and current. Such devices are often used as *negative-resistance devices* for these ranges. Evidently, non-linear

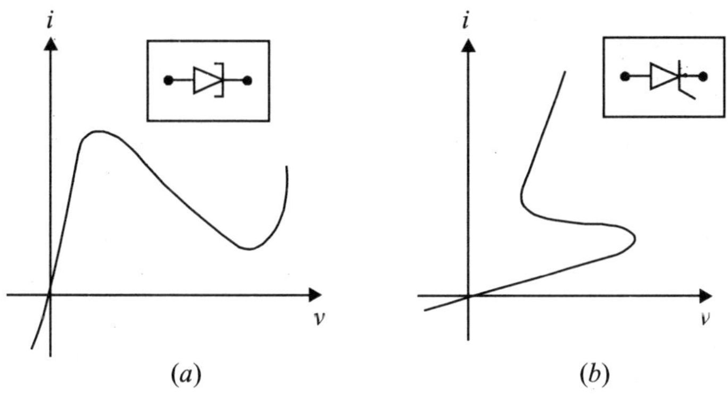

Fig. 1.3

resistors cannot be characterized by a single value of resistance *R,* as in the case of linear resistors. The value of the non-linear resistor depends on the operating point and hence is characterized by *incremental resistance* which is defined as the slope of the *v–i* characteristic at the particular operating point. Non-linear resistors find use in rectification, frequency multiplication, current and voltage limiting, and many other electronic applications. Figure 1.4 shows the symbol and the *v–i* characteristic of a semiconductor diode (continuous curve). In analysing a network with a non-linear element, such as a diode, one often resorts to piece-wise linearization. The piece-wise linear model of a diode (broken line) is also shown in Fig. 1.4.

A distinct property of a linear resistor not usually possessed by a non-linear resistor is that the *v–i* characteristic is symmetric with respect to the origin.

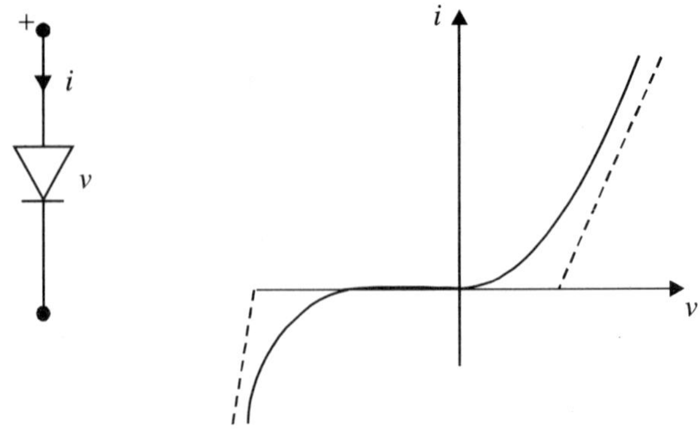

Fig. 1.4

An element exhibiting such a symmetric property is called a *'bilateral element'*. A physical consequence of this property is that for a bilateral element it is not important to keep track of the two terminals of the element. All linear circuit elements are bilateral. Not all non-linear elements are bilateral. The semiconductor diode of Fig. 1.4 is not bilateral.

1.1.2 Independent Sources

Figure 1.5 shows an arbitrary network N excited by a source of electric energy. If we change the network N, then v and i, in general, also change. If the prescribed voltage is maintained, the source of energy is called a *voltage source.* On the other hand, if the prescribed current is maintained it is called a *current source.*

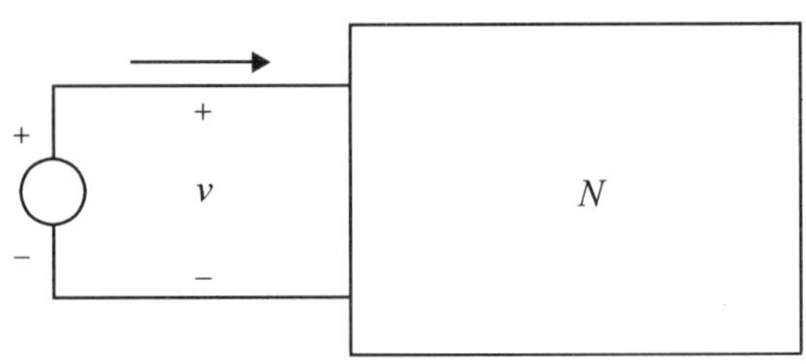

Fig. 1.5

A two-terminal element is called an ideal *independent voltage source,* if it is capable of supplying any current at the same prescribed voltage, *i.e.* the voltage across the source is independent of the current drawn from the source Figure 1.6 shows the symbol and the i–v characteristic of a voltage source. If the voltage of a voltage source is identically zero, the voltage source is effectively a short circuit. The internal resistance of an ideal voltage source is considered to be zero. In the physical world, there is no ideal independent voltage source.

A two-terminal element is called an *ideal independent current source,* if it is capable of supplying a prescribed current at any voltage, *i.e.* the current drawn from a current source is independent of the voltage across the source. Figure 1.7

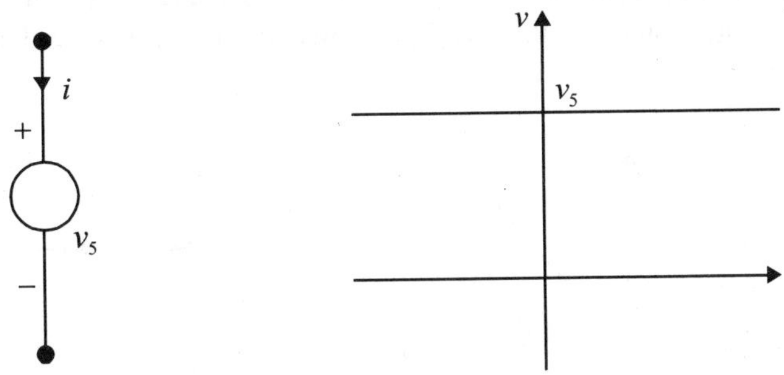

Fig. 1.6

shows the symbol and the i–v characteristic of a current source. If the current drawn from a current source is identically zero, the current source is effectively an open circuit. The internal shunt resistance of an ideal current source is considered to be infinite. In the physical world, there is no ideal independent current source.

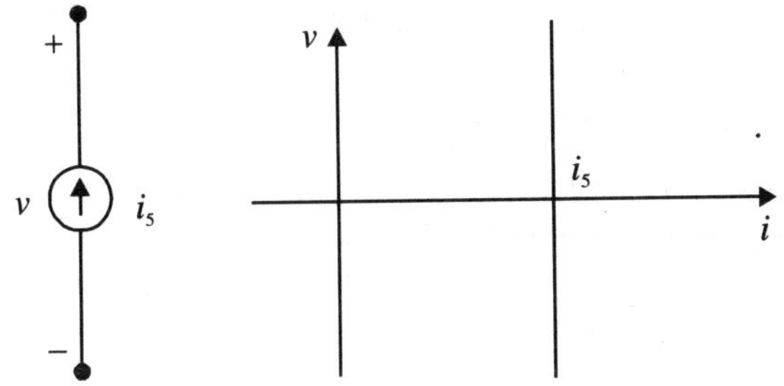

Fig. 1.7

A *dependent* voltage (or current) source is a source the voltage (or current) of which depends on another voltage (or current). A dependent or *controlled* source is said to be voltage- (or current-) controlled if its terminal behaviour is controlled by another voltage (or current). This leads to four different controlled sources which are:

(*a*) Voltage-controlled voltage source.

(*b*) Voltage-controlled current source.

(*c*) Current-controlled voltage source.

(*d*) Current-controlled current source.

Some physical devices operate almost like ideal dependent sources. For example, an operational amplifier is a voltage-controlled voltage source, a field-effect transistor a voltage-controlled current source, and a common base transistor a current-controlled current source.

1.1.3 Capacitors

An element which can be characterized by a curve in the v–q plane is called a capacitor (q represents the charge). A capacitor is linear if its characteristic is a straight line passing through the origin of the v–q plane [Fig. 1.8(a)]. A linear capacitor is represented by the symbol shown in Fig. 1.8(b) and is described analytically by

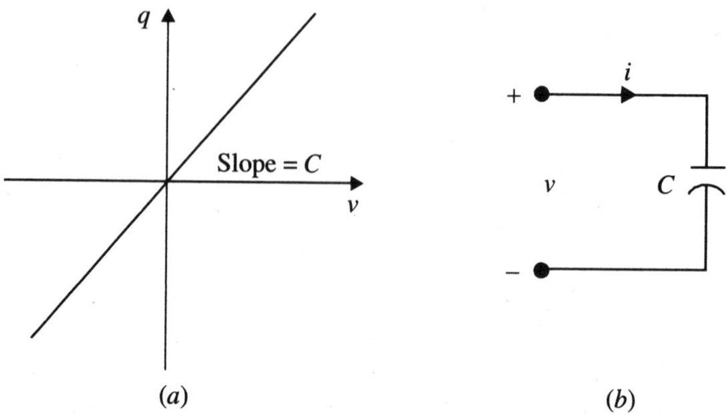

Fig. 1.8

$$q = Cv \qquad ...(1.2)$$

where the constant C is the slope of the straight line, and is called the *capacitance.* The unit of capacitance is farad. (A farad is a very big unit for measuring capacitance; a more practical unit in a network being μF—microfarads, or pF—picofarads). The current-charge relation being $i = dq/dt$, for a linear capacitor, we have

$$i = C\frac{dv(t)}{dt} \qquad ...(1.3)$$

If a capacitor is characterized by a v–q curve other than a straight line through the origin, it is called a non-linear capacitor. Figure 1.9 shows the characteristic of a non-linear capacitor (metal-oxide semiconductor capacitor). A non-linear capacitor is characterized by

$q = f(v)$ *(for a voltage-controlled capacitor)*

and by $v = \psi(q)$ *(for a charge-controlled capacitor)*

For a voltage controlled capacitor the current through the capacitor is

$$i(t) = \frac{dq(t)}{dt} = \frac{df(v)}{dv} \cdot \frac{dv(t)}{dt}$$

or

$$i(t) = C(v)\frac{dv(t)}{dt}$$

where $C(v) = df(v)/dv$ is called the *incremental capacitance* of the capacitor.

The instantaneous power entering an element is given by

$$p(t) = v(t)\,i(t)$$

where v is the voltage in volts and i the current in amperes. Hence, the energy in the element from t_0 to t is

$$W(t_0, t) = \int_{t_0}^{t} v(\tau)i(\tau)d\tau \qquad \ldots(1.4)$$

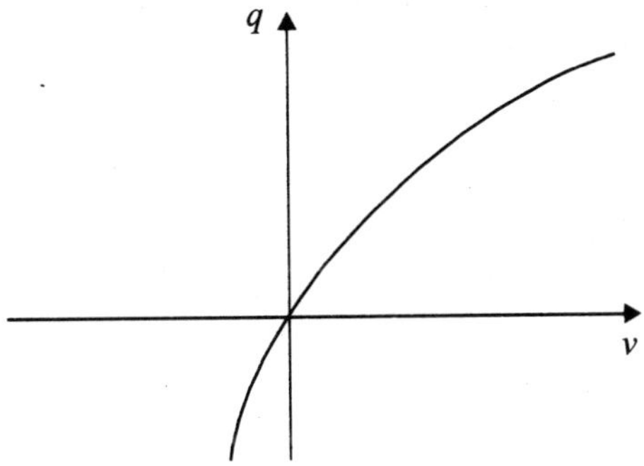

Fig. 1.9

Let a capacitance be characterized by

$$v = \hat{v}(q) \qquad \ldots(1.5)$$

The energy in the capacitor, from eqns. 1.4 and 1.5, is

$$W(t_0, t) = \int_{t_0}^{t} v(\tau)i(\tau)d\tau$$

$$= \int_{q(t_0)}^{q(t)} \hat{v}(q)dq$$

If the initial charge on the capacitors, $q(t_0)$, is zero, then the energy stored in the capacitor is

$$W(t) = \int_{0}^{q(t)} \hat{v}(q)(dq)$$

For a linear capacitor ($q = Cv$) it reduces to

$$W(t) = \frac{1}{2}\frac{q^2(t)}{C} = \frac{1}{2}Cv^2(t) \qquad \ldots(1.6)$$

1.1.4 Inductors

An element which can be characterized by a curve in the $i - \phi$ plane is called an inductor (ϕ represents the flux). An inductor is called a linear inductor if its characteristic is a straight line passing through the origin of the i–ϕ plane (Fig. 1.10). A linear inductor is represented by the symbol shown in Fig. 1.10(b)

and is described analytically by

$$\phi = Li \quad \text{...(1.7)}$$

where the constant L is the slope of the straight line and is called the *inductance.* The unit of inductance is henry. (Since an inductance of one henry is quite large, typical values of inductors are given is mH—millihenrys.) The voltage flux linkage relation being $v = d\phi/dt$ for a linear inductor, we have

$$v = L\frac{di(t)}{dt} \quad \text{...(1.8)}$$

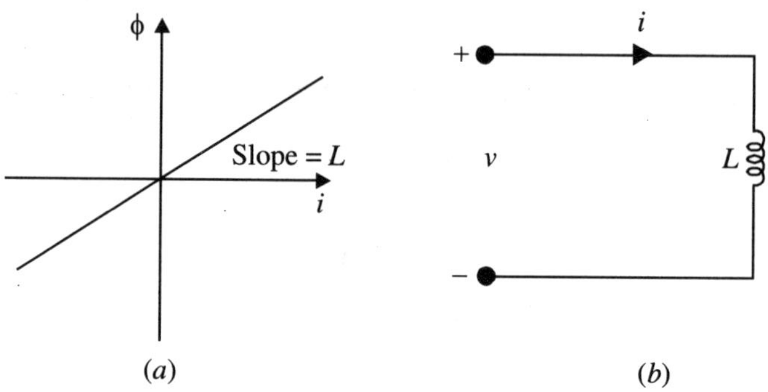

Fig. 1.10

If an inductor is characterized by a $i - \phi$ curve other than a straight line through the origin, it is called a non-linear inductor. A non-linear inductor is characterized by

$$\phi = f(i) \text{ (for a current-controlled inductor)}$$

and by $$i = \psi(\phi) \text{ (for a flux-controlled inductor)}$$

For a current-controlled inductor the voltage across the inductor is

$$v(t) = \frac{d\phi(t)}{dt} = \frac{df(i)}{dt} \cdot \frac{di(t)}{dt}$$

or $$v(t) = L(i)\frac{di(t)}{dt}$$

where $L(i) = \dfrac{df(i)}{di}$ is called the *incremental inductance.*

Non-linear inductors can be used in frequency conversion, memory and storage.

A special type of nonlinear inductor, such as a ferromagnetic-core inductor, has a characteristic that exhibits the phenomenon of *hysteresis* (Fig. 1.11).

Let an inductor be characterized by

$$i = \hat{i}(\phi) \quad \text{...(1.9)}$$

The energy in the inductor, from eqns. 1.4 and 1.9, is

$$W(t_0, t) = \int_{t_0}^{t} v(\tau)i(\tau)d\tau$$

$$= \int_{\phi(t_0)}^{\phi(t)} \hat{i}(\phi)d\phi$$

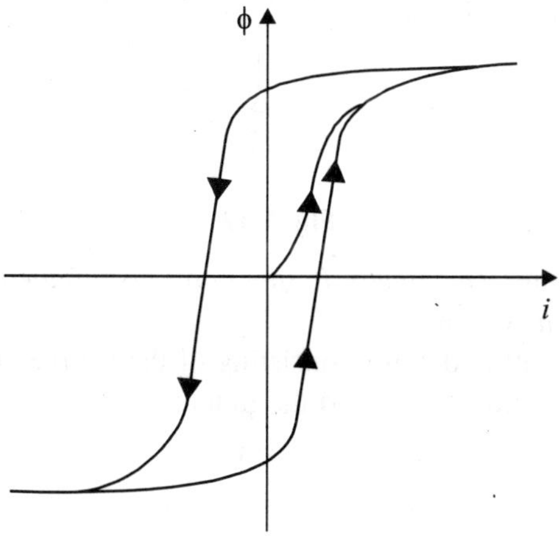

Fig. 1.11

If the initial flux, ϕ (t_0), is zero, then the energy stored in the inductor is

$$W(t) = \int_{0}^{\phi(t)} \hat{i}(\phi)d\phi$$

For a linear inductor ($\phi = Li$) this reduces to

$$W(t) = \frac{1}{2}\frac{\phi^2(t)}{L} = \frac{1}{2}Li^2(t) \qquad \ldots(1.10)$$

1.2 KIRCHHOFF'S LAWS

Figure 1.12 shows the model of an electric network (from here on we use the word 'network' for its model). A basic problem in network theory is to analyse the network and find the voltages across and the currents through all the elements of the network. Certain physical laws are fundamental to the algebraic formulation of such a problem. Two of the fundamental physical laws in network theory are the 'Kirchhoff's current and voltage laws'. If we consider the junction points A and B of the network we can write

at $\qquad A : \; i = i_{C_1} + i_{L_2}$

at $\qquad B : \; i_{L_2} = i_{C_2}$

that is, the current flowing towards a junction is equal to the total current flowing away from the junction. This is *Kirchhoff's current law* (KCL). More formally, the

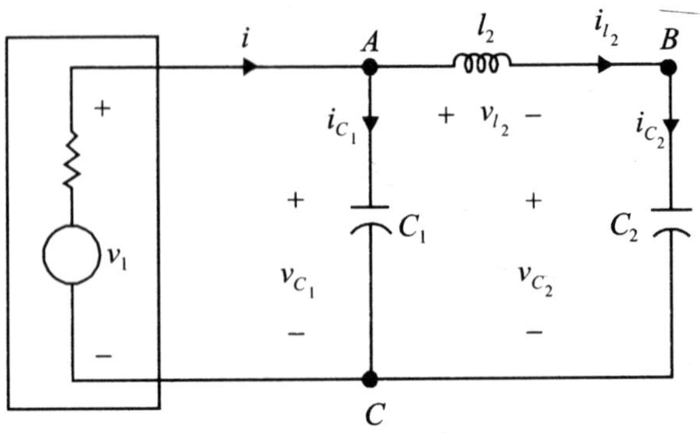

Fig. 1.12

KCL can be stated thus: *The algebraic sum of all instantaneous currents entering or leaving a junction is zero.*

If we consider the closed paths consisting of the source and capacitor C_1 (say path 1) and the capacitors C_1, C_2 and the inductor L_2 (say path 2) we can write

$$\text{path 1:} \quad v_1 = v_{C_1}$$

$$\text{path 2:} \quad v_{C_1} = v_{l_2} + v_{C_2}$$

where v_1, v_{C_1}, v_{C_2} and v_{l_2} are all voltage drops, *i.e.* the voltage drop between any two junctions is independent of the path between the junctions. This is *Kirchhoff's voltage law* (KVL). More formally, KVL states: *The algebraic sum of all voltage drops or rises in a closed path is zero.*

The KCL imposes linear constraints on the currents through the elements whereas the KVL imposes linear constraints on the voltage drops across the elements. Together, these two laws imply conservation of energy (Sec. 5.4).

1.3 CLASSIFICATION OF NETWORKS

The overall behaviour of an electrical network can be predicted by the constituent elements and their interconnection. The behaviour of the network, considered as a black box, leads to a number of classifications like linear, nonlinear; time-invariant and time-variant; passive, active.

(*i*) **Linear and Non-linear networks:** In a linear network, the relationship between the voltage and current is described by a linear equation. Consider two networks N_1 and N_2 as shown in Fig. 1.13(*a*) and (*b*) respectively. Network (*a*) is made up of a linear resistor R, while (*b*) is made up of a semi-conductor diode and linear resistor R. Let the cut-in voltage of the diode be 0.6 volt. In network (*a*) current I_1 is given by V/R and exists for all values of V. In network (*b*) if V is less than 0.6 volt the current I_2 is zero and for voltages higher than 0.6 volts I_2 is given by $(V - 0.6)/(R + R_F)$, where R_F is the forward resistance of the diode. Obviously, in network (*a*) the current response is linear in contrast to that in (*b*).

Let both the networks be excited by two serially connected voltage sources V_1 and V_2. Then it can be seen that

$$I_1(V_1 + V_2) = I_1(V_1) + I_1(V_2)$$

and

$$I_2(V_1 + V_2) \neq I_2(V_1) + I_2(V_2)$$

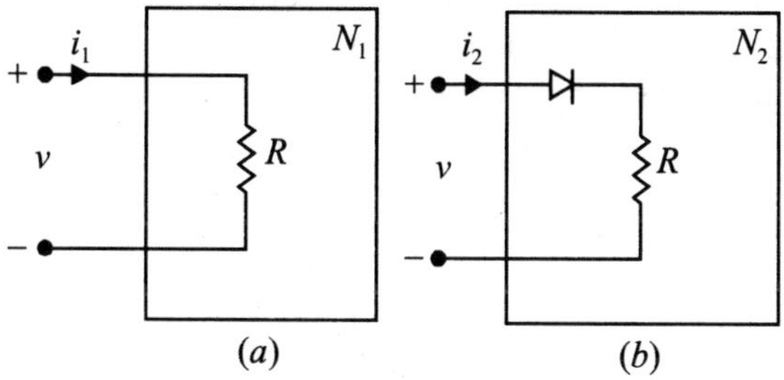

Fig. 1.13

where I_i (V_j) is the current into the terminals of network N_i when excited by a voltage source V_j. We say that network N_1 is linear and N_2 is non-linear as the principle of superposition holds for N_1 but not for N_2.

Linearity of a network can be defined as follows: Let a network be characterized by

$$F(x_l) = y_i$$

where x_i is the input and y_i the output, and $F(\cdot)$ denotes some function. Then the network is linear if, and only if,

$$F(\alpha_1 x_1 + \alpha_2 x_2) = \alpha_1 F(x_1) + \alpha_2 F(x_2) = \alpha_1 y_1 + \alpha_1 y_2$$

where α_1 and α_2 are arbitrary constants, and x_1 and x_2 are any two allowable inputs. The principle expressed by this equation is called the *principle of superposition and homogeneity*. Hence, we conclude that a network is linear if it satisfies this principle; otherwise, it is non-linear.

(ii) **Time-invariant and Time-variant networks:** Let a linear resistor be characterized by

$$v(t) = R(t)i(t)$$

where R (t) is a prescribed time function. This can be achieved, for example, by the sliding contact of a potentiometer being moved back and forth by a motor. Such a resistor is called a *time-varying* resistor. Similarly, it is possible to build time-varying capacitors and inductors. The elements we considered in Sec. 1.1 were all time-invariant in that they were characterized by parameters which were not dependent on time.

A time-invariant network is characterized by a constant coefficient equation whereas the time-variant one by a time-variant coefficient equation. Mathematically, we can describe a time-invariant network by

$$F\left[x(t - t_0)\right] = y(t - t_0)$$

when the network is characterized by

$$F[x(t)] = y(t)$$

i.e. the response (output) depends on the shape of the excitation (input) but not on the time of application. A network composed of time-invariant elements is necessarily time-invariant whereas network composed of time-variant elements may exhibit time-invariant terminal behaviour.

(*iii*) **Passive and Active networks:** Consider a network made up of a single linear resistor. The energy supplied to (or dissipated by) the resistor, from eqn. 1.4, is

$$W(t_0, t) = \int_{t_0}^{t} i^2(t) R\, dt$$

If the resistor has to deliver power to the external world, R has to be negative. As long as R is positive the resistor will consume power, and such a resistor is called a passive resistor.

Let $v(t)$ and $i(t)$ be the voltage and current at the terminals of a network. Then the energy delivered to the network is given by

$$W(t_0, t) = \int_{t_0}^{t} v(\tau) i(\tau) d\tau$$

A network is said to be passive if

$$W(t_0, t) + E(t_0) \geq 0$$

for all t and t_0 and for all $v(t)$ and $i(t)$, where $E(t_0)$ is the energy in the network at $t = t_0$. Otherwise, the network is said to be active. In other words, if the energy delivered to the network is non-negative for all time and input, the network is said to be passive. The conditions for activity and passivity of an element can also be obtained by a study of its characteristics. For example, we can state that *a non-linear resistor is passive if its characteristic, for all time, is in the first and third quadrants of v–i plane.* Similarly, a capacitor (inductor) is passive if, its characteristic is in the first and third quadrants of the v–q (ϕ–i) plane.

The preceding classification of networks was based on the basic behaviour of the response of the overall network. Often, networks are grouped without any reference to their response. Such a grouping includes *n-terminal* and *n-port* networks. If, for example, a voltage source is connected across the terminals k and k' of a network, the current i_k flowing into the network at terminal k must be necessarily the same as the current i_k', flowing out of the terminal k'. Under these conditions, the pair of the terminal $k - k'$ constitute a *port.* A two-terminal or a one-port network has only one accessible terminal pair (Fig. 1.14) and a two-port network has two accessible terminal pairs (Fig. 1.15). This definition can be extended to n-port networks.

Networks can be classified in a number of other ways—lumped and distributed networks; reciprocal networks; large scale networks and power distribution networks.

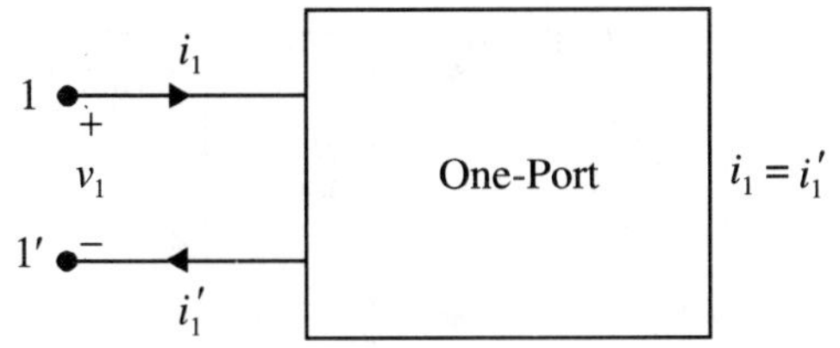

Fig. 1.14

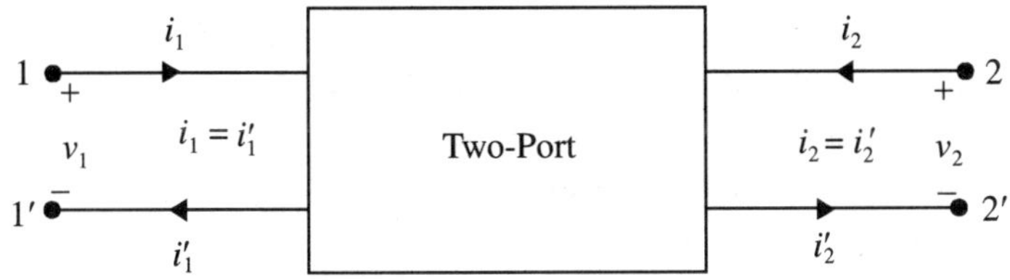

Fig. 1.15

1.4 NULLATORS AND NORATORS

The lumped circuit elements introduced in Sec. 1.1 are sufficient to model most of the networks. (Sometimes we may need coupled coils to model a network. We postpone the introduction of coupled coils till Sec. 3.4). But the modelling and analysis of active networks (*i.e.* networks with active components like transistors and operational amplifiers) is facilitated by two singular (or pathological) network elements—the 'Nullator' and 'Norator'. A nullator is a two-terminal device which is simultaneously open and short, *i.e.* the current through and the voltage across a nullator are zero. A nullator is represented by the symbol shown in Fig. 1.16(*a*). A norator is a two-terminal device whose current and voltage are arbitrary, *i.e.* the current through and the voltage across a norator are determined by the external circuitry. The symbol for a norator is shown in Fig. 1.16(*b*). It is obvious that

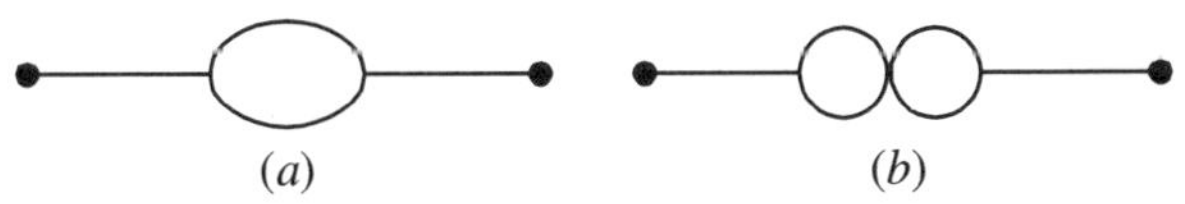

Fig. 1.16

nullators and norators cannot exist in the real world. But the ideal behaviour of certain active components can be represented by a combination of nullators and norators. For example, an ideal common emitter transistor can be represented by the nullor (nullator-norator) model shown in Fig. 1.17. In Chapter 12, we discuss the application of nullor models in analysing and in synthesizing active networks.

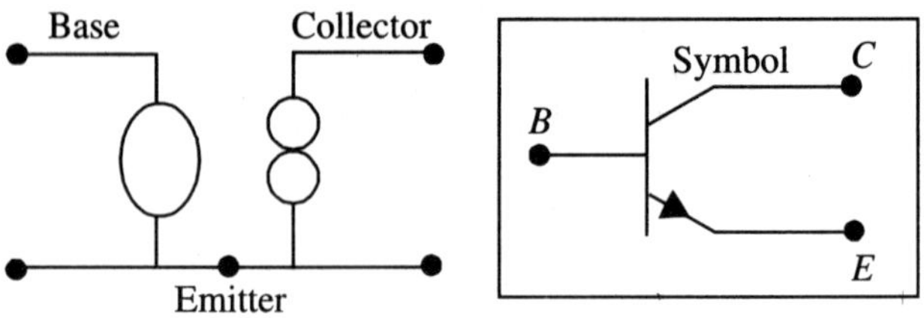

Fig. 1.17

As indicated earlier, modelling of the physical devices is basic to the analysis and the synthesis of signal processing networks and devices. In this chapter, we have learnt the basic models and the properties of certain circuit elements. The interconnection of these elements and the geometry of the network is the topic of the next chapter.

PROBLEMS

1.1 The characteristic of a non-linear resistor is described by the i–v relationship

$$i = \alpha_0 + \alpha_1 v + \alpha_2 v_2$$

If the resistor excited by a signal $v = A \cos \omega t$, find an expression for the current i. If we are to generate a third harmonic signal what should be the characteristic of the resistor.

1.2 Find the current through the resistors R_1, R_2 and R_3 in the network of Fig. P. 1.2.

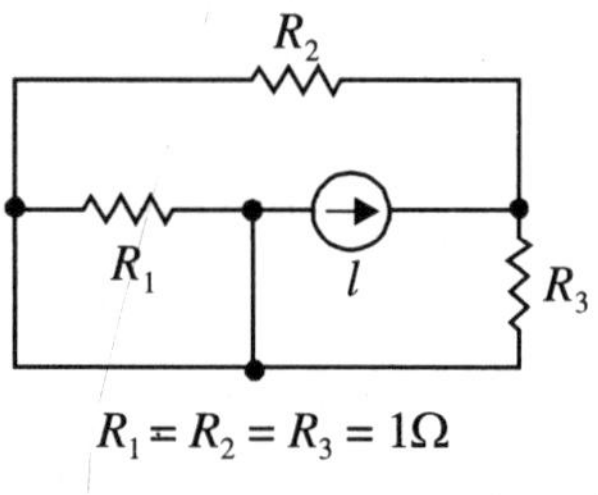

Fig. P. 1.2

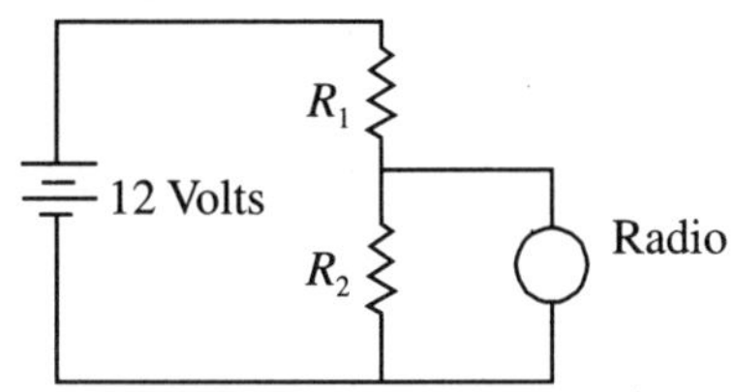

Fig. P. 1.3

1.3 A 9-volt transistor radio is to be run on a 12-volt battery (Fig. P. 1.3). The radio draws a current of 6 mA. Find R_1 and R_2.

1.4 The open-circuit voltage of a source was found to be 10 volts. A variable resistor R is connected across the terminals of the source (Fig. P. 1.4) and the resistor R is adjusted so that voltage V_0 is 4 volts. If the value of the resistor is 400 Ω what is the internal impedance of the source.

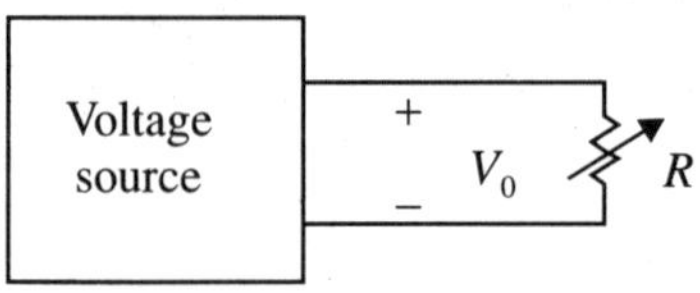

Fig. P. 1.4

1.5 Find the currents I and I_i, $i = 1, 2, \ldots 5$ in the network of Fig. P. 1.5.

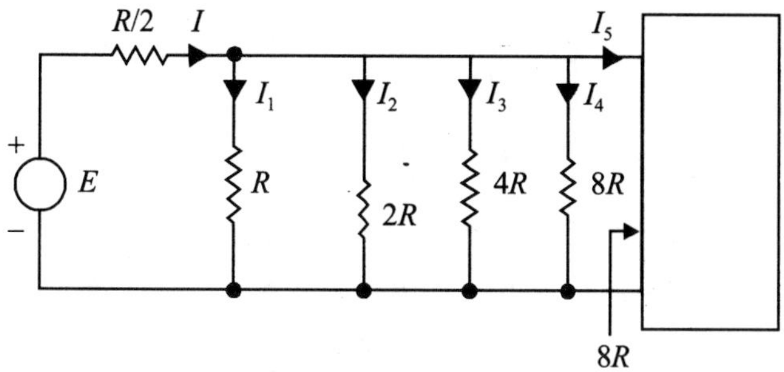

Fig. P. 1.5

1.6 Find the conditions for the passivity of time-varying capacitor.

1.7 Classify the following elements as passive or active:

(*a*) Resistor: $i = v^2$ (*b*) Resistor: $i = v^3$

(*c*) Inductor: $i = -\phi + \phi^3$ (*d*) Inductor: $L(t) = a + b \cos \omega t$

(*e*) Capacitor: $q = V + V^2$ (*f*) Capacitor: $C(t) = a \cos \omega t + b \sin \omega t$

1.8 The voltage across an inductor of 100 mH is shown in Fig. P. 1.8. Plot the current through the inductor assuming $i_1(0) = 0$. If $i_1(0) = 0.25$ amps how would you modify the plot?

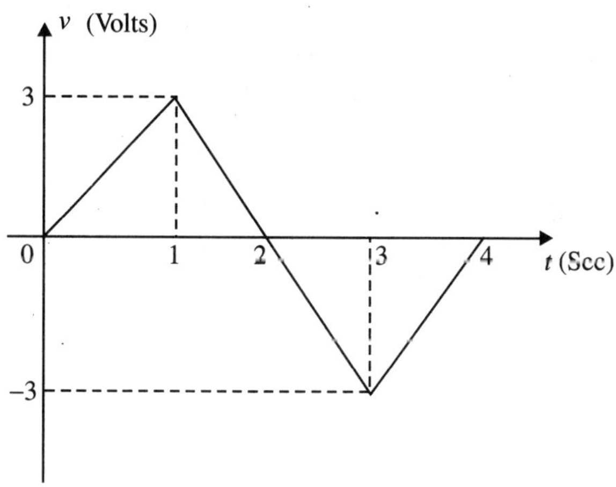

Fig. P. 1.8

1.9 A nonlinear capacitor is characterized by the equation $q = 1 - e^{-v}$. Find the small signal capacitance at $v = v_1$.

1.10 Figure P. 1.10 shows a 'Darlington Pair'. Draw the nullator-norator equivalent diagram and find the voltages across and currents through R_1 and R_2.

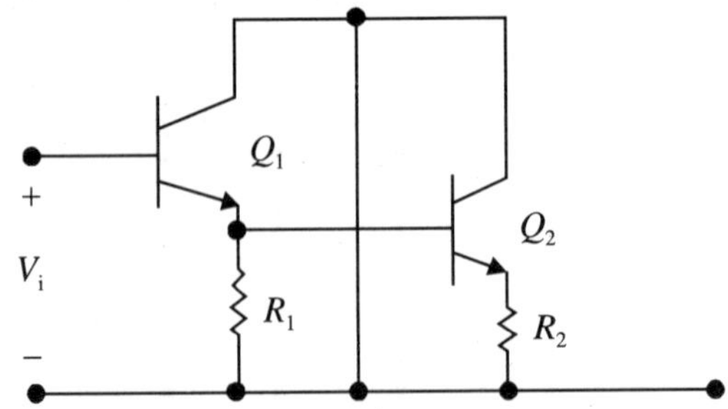

Fig. P. 1.10

NOTES AND REFERENCES

Study of circuit elements and their models is fundamental to the design of electronic devices and circuits. Here we have introduced a few passive circuits elements and the classification of Networks has been briefly reviewed. The coverage, by intent, is neither elaborate nor exhaustive. Several textbooks discuss the details of modelling and of the range of application of these models.

Desoer and Kuh's book[1] (Chaps. 1 & 2) discusses in detail, the lumped circuit elements and their models. Chua's book[2] (Chaps. 1 – 3) on non-linear networks introduces many non-linear elements, their models and signal processing abilities. The concepts and models of *R, L* and *C* can also be developed from the field point of view and this is well discussed in Rohrer's book[3] (Chaps. 2 – 3). The concept of Nullators and Norators was introduced by Carlin[4].

1. C.A. Desoer and E.S. Kuh: *Basic Circuit Theory*, McGraw-Hill, 1969.
2. L.O. Chua: *Introduction to Nonlinear Network Theory*, McGraw-Hill, 1969.
3. R.A. Rohrer: *Circuit Theory: An Introduction to State-variable Approach*, McGraw-Hill, 1970.
4. H.J. Carlin: *Singular Network Elements, IEEE Trans. on Circuit Theory*, CT-11, March 1964, pp. 67-72.

CHAPTER

2

Network Topology

Topology formalizes the formulation of the network equilibrium equations (loop equations, node equations and the like). Most of the computer aided analysis and design methods utilize topological formulation. The derivation of the state equations of a network (Chapter 10) inherently depends on the topological matrices of the network.

Consider the network shown in Fig. 2.1. The equations describing this network are:

$$(a)\quad v_2 = R_2 i_2\ ;\ v_3 = R_3 i_3\ ;\ i_4 = C_4 \frac{dv_4}{dt}$$

$$(b)\quad i_1 = i_2 + i_3,\ i_3 = i_4,\ i_1 = i_2 + i_4,\ v_2 = v_3 + v_4$$

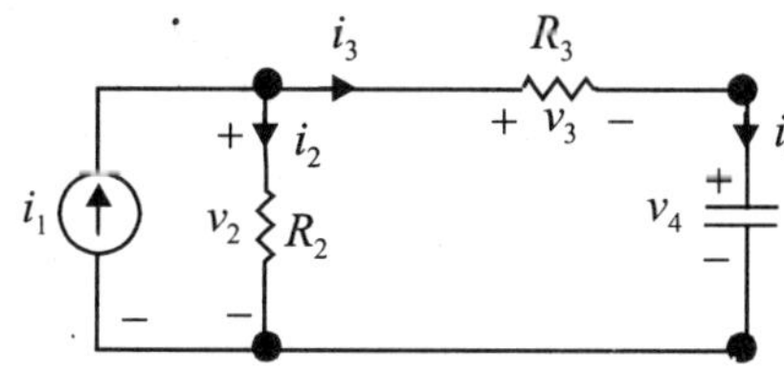

Fig. 2.1

The set of equations in (a) are the terminal relations of the various elements and these do not depend on the interconnection of the elements. The set of equations in (b) are the constraints imposed by the two Kirchhoff's laws and, as such, do not depend on the elements but only on the geometry or the *topology* of the network. Also, there are seven equations in six unknowns. Of course, in this particular case, it is easy to recognize the redundant equation in (b). But in large networks this becomes a complex problem and a study of topology proves helpful in solving such a problem. Topology, or geometry of a network, is concerned with the interconnections of the elements in a network. The network is represented by a *linear graph*, and in this chapter we discuss the properties and applications of such a graph.

2.1 BASIC DEFINITIONS AND PROPERTIES

The linear graph of a network is obtained by replacing every element by a line-segment and the junction points by nodes. For example, the linear graph of the network of Fig. 2.1 is shown in Fig. 2.2. By definition a linear graph is a collection of line segments and their end points. A line-segment with its end points is called an *edge* of the graph and the end points are called *vertices* (or *nodes*) of the graph. A linear graph or a graph of a network carries all the information about the inter-connection of the elements of the network. The length, thickness and formation of the edges are of no consequence. Further, we impose the restriction that the end points of an edge are distinct. An edge is said to be incident at a vertex if that vertex is an end point of the edge. For example, edges 1, 2 and 3 are incident at vertex v_1 in Fig. 2.2.

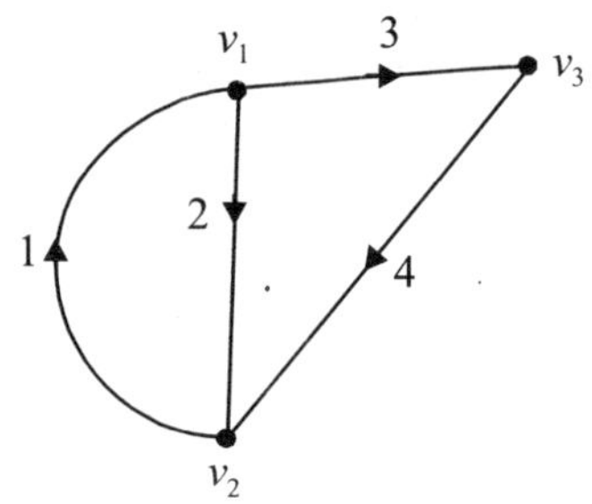

Fig. 2.2

Sub-graph of a graph G is a graph whose edges and vertices form a subset of the edges and the vertices of G. In particular, G is itself a sub-graph of G. If a sub-graph of G does not contain the entire graph G, it is called a *proper sub-graph* of G. If a sub-graph contains all the vertices of G, it is called a *spanning sub-graph* of G. A *complementary sub-graph* of a sub-graph of the graph G is the graph formed by the edges not in the sub-graph of G. Consider a graph G shown in Fig. 2.3. (The edges of the graph do not carry any arrows as in Fig. 2.2. The arrows on the edges indicate the orientation of the edges; oriented edges and oriented graphs are discussed at a later stage.) This graph has six vertices v_i; i = 1 to 6 and

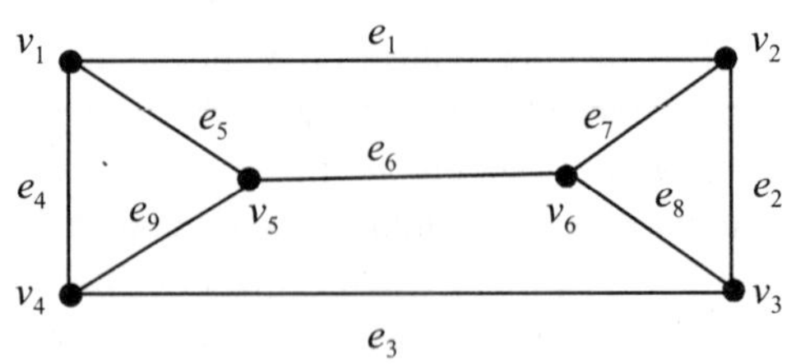

Fig. 2.3

nine edges e_1, $i = 1$ to 9. The edge sets (e_1, e_2, e_3, e_4) and (e_1, e_7, e_6, e_9) constitute two sub-graphs of G. Both the sub-graphs are proper sub-graphs of G but not spanning sub-graphs. On the other hand, the sub-graph formed by the edges e_5, e_6, e_7, e_8 and e_9 constitute a spanning sub-graph of G. Of course, there are many more sub-graphs and spanning sub-graphs of G. Consider two sub-graphs G_1 and G_2 of G formed by the edges (e_1, e_2, e_3, e_7) and (e_4, e_5, e_6, e_8, e_9) respectively. Then G_2 is the complementary sub-graph of G_1 and vice-versa.

The *degree of a vertex* v_i, $\delta(v_i)$, is the number of edges and incidents at the vertex v_i (the degree of the vertex v_1 in the graph of Fig. 2.3 is three and that of v_6 is also three). It can be easily seen that for any graph

$$\sum_{i=1}^{v} \delta(v_i) = 2e$$

where G has v vertices and e edges.

A path between two vertices, v_i and v_j, is a sequence of edges between v_i and v_j so that the vertices and the edges appearing in the path are all distinct. In this path, the vertices v_i and v_j, *terminal vertices*, are of degree one and all other vertices, *internal vertices*, are of degree two. A closed path, a path where the terminal vertices v_i and v_j are the same, is called a circuit. In a circuit, all vertices are of degree two. The edge sets (e_1, e_2) and (e_1, e_7, e_6, e_9, e_3) are two paths between vertices v_1 and v_3 of the graph G of Fig. 2.3 and the edge sets (e_1, e_2, e_3, e_4) and (e_1, e_7, e_6, e_9, e_4) are two circuits of G.

A graph G is called a *connected graph* if there exists at least one path between every pair of vertices. Graphs shown in Fig. 2.2 and 2.3 are connected graphs. If a graph G is not connected, then it has *separate parts*. A graph in which there is one and only one edge between every pair of vertices is called a *complete graph*. Obviously, a complete graph is a connected graph, but the converse is not, in general, true. The graph G of Fig. 2.3 is not a complete graph (the edges v_1v_3, v_1v_6, v_2v_4, v_2v_5, v_3v_5 and v_4v_6 are missing). Figure 2.4 shows a complete graph on five vertices. It can be easily shown that the number of edges in a complete graph of v vertices is $v(v - 1)/2$.

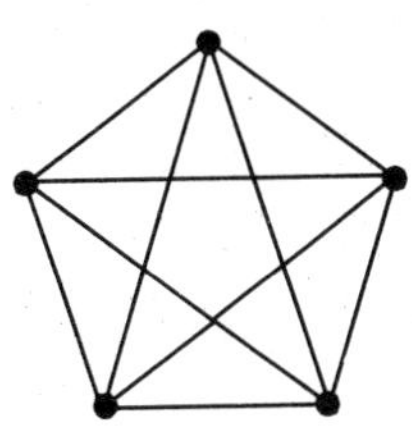

Fig. 2.4

A *tree* of a connected graph is a spanning sub-graph of G which is connected but circuitless. Connectedness implies that there is at least one path between any two vertices and circuitless implies that there is only one path between any two vertices, *i.e.*, in a tree of a graph there is one and only one path between every pair of vertices. A connected graph has at least one tree. If the graph is not connected, then every separate part of the graph has a tree and the collection of these trees is called a *forest* of the non-connected graph. Obviously, a forest of a graph with p parts has p trees. The edges of a graph G, which are the edges of a tree, are called the *branches* of the tree. The complementary sub-graphs of the tree, of a graph is called the *co-tree* of the tree and the edges of the co-tree are called *links* or *chords*. For example, the edge set (e_5, e_6, e_7, e_8, e_9) constitutes a tree of the graph of Fig. 2.3. With respect to this tree, the edges e_5, e_6, e_7, e_8, e_9 are the branches, the edges e_1, e_2, e_3, e_4 are the chords and the edge set (e_1, e_2, e_3, e_4) is the co-tree. It can be easily seen that a given graph may have more than one tree (the edges e_1, e_5, e_7, e_8, e_9 form another tree for Fig. 2.3). But all the trees of a given graph have the same number of edges or branches.

Theorem 2.1: The number of branches in a tree of a connected graph G with v vertices is $(v - 1)$.

Corollary 2.1: The number of chords of a tree of a connected graph G of v vertices and e edges is $(e - v + 1)$.

Let T be a tree of graph G, and L be the corresponding co-tree. If we add a chord from L to the tree T then the resulting graph will have exactly one circuit. This is reasoned as follows: suppose the chord under consideration is the edge between the vertices v_a and v_b. In the tree T, there is one and only one path between the vertices v_a and v_b. Hence, with the addition of the chord to the tree we have created two paths only between the vertices v_a and v_b and thus there is exactly one circuit formed. This circuit, which contains one chord and the unique tree-path between the end vertices of the chord, is called a *fundamental circuit* or *f-circuit*. We say that the *f*-circuit is defined by the chord. From Cor. 1.1 it is clear that the number of *f*-circuits in a graph is $(e - v + 1)$. This is the maximum number of independent circuits that can be formed in a graph of v vertices and e edges.

Theorem 2.2: For a connected graph of e edges and v vertices a set of $(e - v + 1)$ *f*-circuits with respect to a tree constitutes a complete set of independent circuits of the graph.

This theorem implies that all the other circuits of the graph necessarily depend on the $(e - v + 1)$ *f*-circuits. We will not try to prove the theorem here (see notes and references at the end of the Chapter). In the next chapter, we use this theorem to write a set of independent KVL equations and then show that a set of loop or mesh currents constitutes a set of independent currents for a network.

Example 2.1. Figure 2.5 shows a graph with 6 vertices and 8 edges. For this graph, the edge set (1, 2, 3, 4, 5) constitutes a tree (solid lines) and the edge set (6, 7, 8) constitutes the corresponding co-tree (broken lines). Every tree of this graph has 6 – 1 = 5 branches and every co-tree has 8 – 5 = 3 chords. Hence, the number of *f*-circuits is three and with respect to the chosen tree the *f*-circuits are:

f-circuit of chord 6 $\rightarrow$ (1, 2, 4, 6)

f-circuit of chord 7 $\rightarrow$ (2, 4, 5, 7)

f-circuit of chord 8 $\rightarrow$ (2, 3, 5, 8)

A *cut-set* of a connected graph is a minimal set of edges whose removal from the graph results in two separate parts. One of the separate parts may be an isolated vertex (vertex without any edges incident on it). It is important to note that when we remove the edges we do not remove the end vertices of the edges. The word 'minimal' in the definition of a cut-sets implies that no proper subset of this set has the property of separating the graph into two pieces. We can extend the definition of cut-sets to non-connected graphs. A cut-set of a graph with p parts is a minimal set of edges whose removal results in a graph with $(p + 1)$ parts. Some

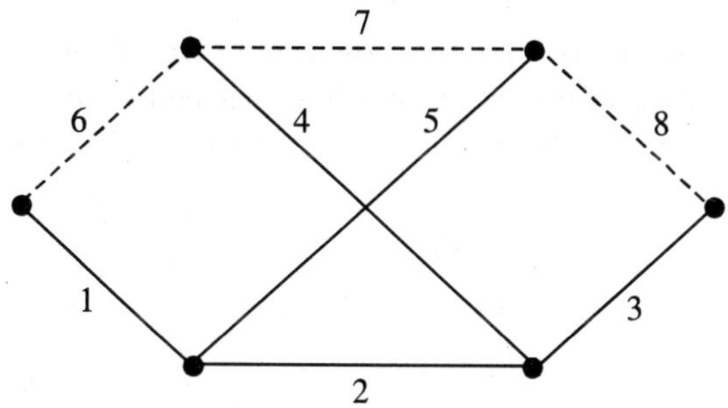

Fig. 2.5

of the cut-sets for the graph of Fig. 2.5 are shown in Fig. 2.6 by drawing lines on the graph. The edge sets (1, 6), (4, 6, 7), (2, 3, 4) and (2, 4, 8) are some of the cut-sets of the graph. The edge set (4, 5, 6, 7) does separate the graph into two parts.

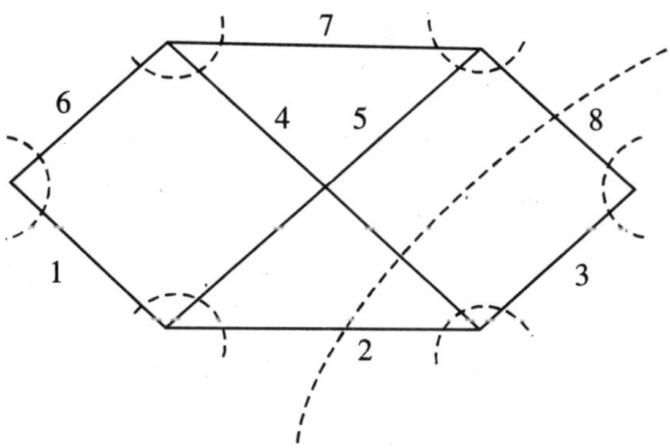

Fig. 2.6

Yet, it does not constitute a cut-set for its proper subset, the edge set (4, 6, 7), is a cut-set of the graph. For this particular graph all the edges incident at a vertex constitute a cut-set. The fact that this need not be true is demonstrated in the graph of Fig. 2.7. The edge set (1, 2, 3, 4) at the vertex v is not a cut-set of the graph as a subset (1, 4) is a cut-set of the graph.

In a tree, because of exactly one path between every pair of vertices, the removal of any branch separates the tree into two connected parts and, hence, every branch of a tree is a cut-set of that tree. Let b be a branch of a tree T of a graph G. Let the

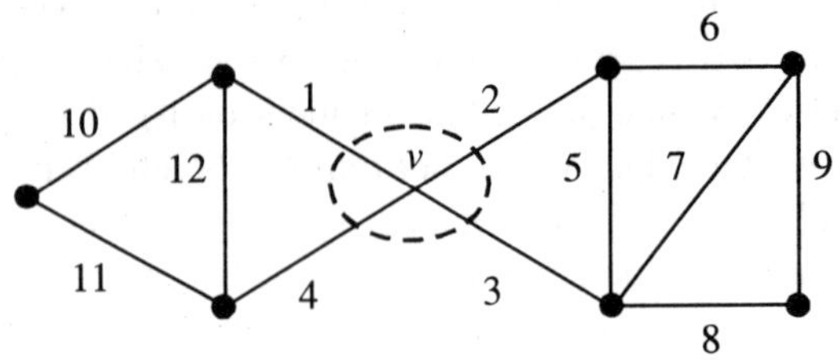

Fig. 2.7

removal of b separate the tree into two connected parts, T_1 and T_2. If we replace all the chords of T, some of the chords may have one end vertex in T_1 and the other in T_2 (Fig. 2.8). In Fig. 2.8, these chords are shown by broken lines. The set of all

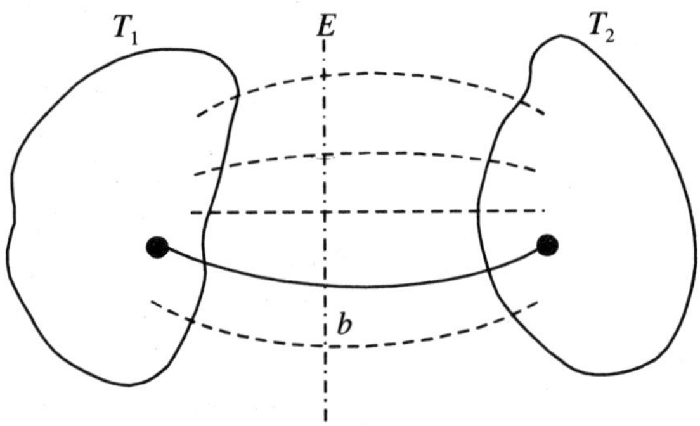

Fig. 2.8

such chords and the branch b obviously constitute a cut-set of G. This cut-set has only one branch (b) of the tree T. A cut-set, which has only one branch of a tree T, is called *fundamental* or *f-cut-set* with respect to the tree T. Every branch of every tree defines its own f-cut-set. The total number of f-cut-sets from Theorem 2.1 is $(v - 1)$, where v is the number of vertices of the graph. This is the maximum number of independent cut-sets that can be formed in a graph. Hence,

Theorem 2.3: For a connected graph of v vertices a set of $(v - 1)$ f-cut-sets with respect to a tree constitutes a complete set of independent cut-sets of the graph.

This theorem implies that all the other cut-sets of the graph necessarily depend on the $(v - 1)$ f-cut-sets. In the next chapter, we use this theorem to write a set of independent KCL equations and then show that a set of branch voltages constitutes a set of independent voltages for a network.

Example 2.2. Figure 2.9 shows a few cut-set of the graph of Fig. 2.5. The tree branches are shown by solid lines and the chord by broken lines. The number of f-cut-sets is five. They are:

f-cut-set of branch 1 $\rightarrow$ (1, 6)

f-cut-set of branch 2 $\rightarrow$ (2, 6, 7, 8)

f-cut-set of branch 3 → (3, 8)

f-cut-set of branch 4 → (4, 6, 7)

f-cut-set of branch 5 → (5, 7, 8)

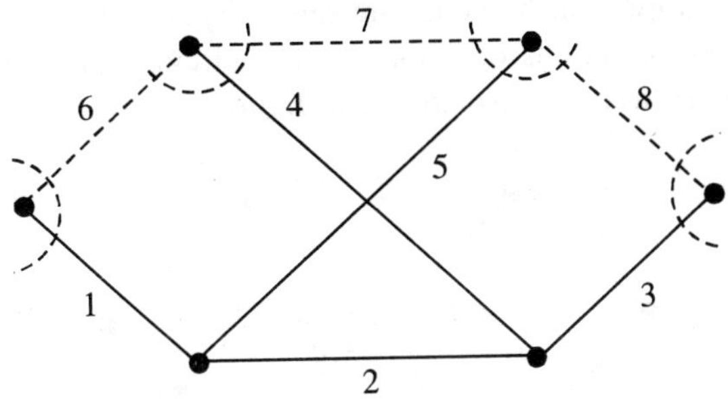

Fig. 2.9

Notice that it is not possible to draw a line to show the cut-set of branch 2. To do this the graph has to be redrawn, as shown in Fig. 2.10.

So far we have considered only a non-oriented graph, *i.e.*, a graph in which the end vertices of an edge are not an ordered pair. But in the study of networks we are, in general, interested in the direction of flow of current or of voltage drop in an element. Hence, the edges of a linear graph of an electrical network carry an orientation or an arrow. The general convention is to orient an edge corresponding

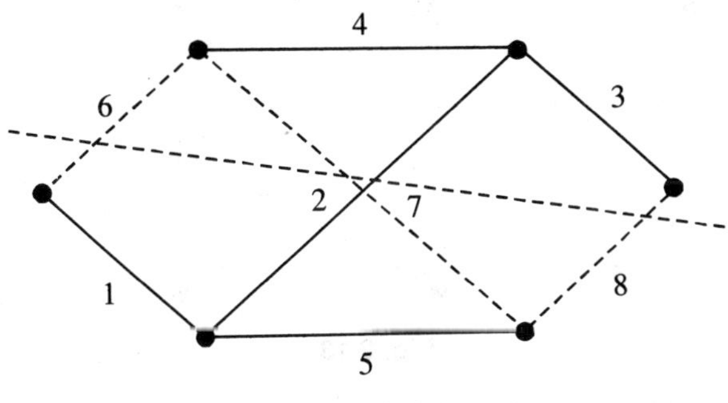

Fig. 2.10

to an element such that the orientation coincides with the direction of current flow or voltage drop. Figure 2.11 shows the orientation of edges corresponding to the sources.

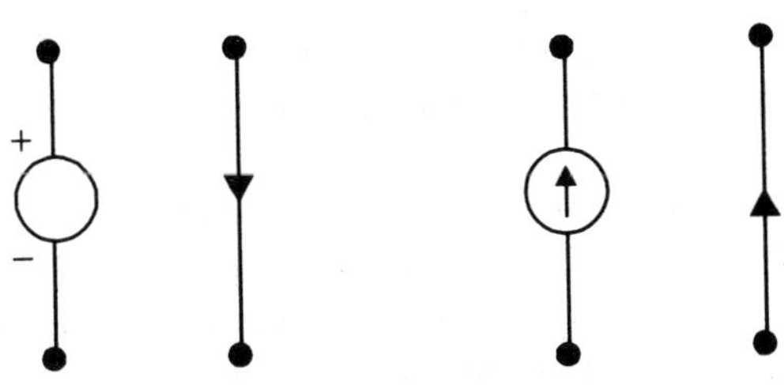

Fig. 2.11

An *oriented graph* is a graph in which all the edges of the graph have an assigned orientation. In an oriented graph the *f*-cut-sets and *f*-circuits are all oriented. The orientation of an *f*-circuit coincides with the orientation of the defining chord and that of an *f*-cut-set with the orientation of the defining branch. Figure 2.12 shows an oriented graph. The branches are shown by solid lines and the chords by broken lines. The orientations of some of the *f*-circuits and the *f*-cut-sets with respect to the chosen tree are also shown in figure.

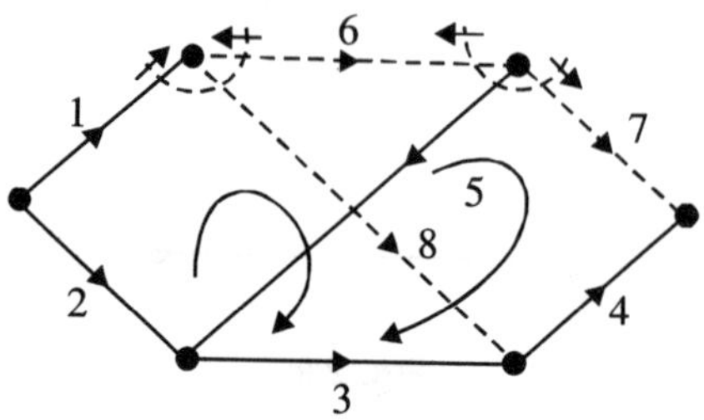

Fig. 2.12

An important concept involved in developing the mesh equations for a network is the planarity of a graph. A graph is said to be planar if it can be mapped on to a plane so that no two edges intersect at a point which is not a vertex of the graph. Otherwise, the graph is nonplanar. Figure 2.13 shows two (basic) nonplanar graphs. Tests for planarity of a graph are rather complicated.

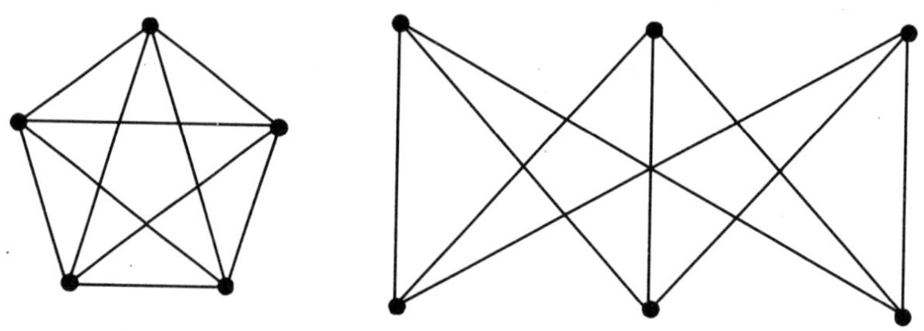

Fig. 2.13

Planar graphs can also be characterized by the concept of duality. Two graphs G_1 and G_2 are said to be *duals* of each other if, there is one to one correspondence between the circuits of one to the cut-sets of the other. The algebraic characterization of planarity is not simple. We state a theorem, without proof, relating planar graphs and duals.

Theorem 2.4: A graph G is planar if, it has dual.

The dual of graph, if it exists, can be found by the so-called *window-dot method* as follows:

When a planar graph is mapped on to a plane without any intersecting edges, the plane is divided into m inside regions or windows (meshes) and one outside region. The dual of such a graph can be constructed as follows:

(*i*) Place a dot in each window, including the external or outer region. There will be $(m + 1)$ dots.

(*ii*) Connect these $(m + 1)$ dots by line segments so that for each edge j between two windows there exists a line segment j' which crosses edge j and connects the two dots corresponding to the two windows. Thus, if the original graph has e edges, the dual has also e edges.

The resulting planar graph with $(m + 1)$ vertices and e edges is the dual of the original graph.

Figure 2.14 shows an example for the construction of the dual of a graph G and the dual G'. We can extend the concept of duality to networks. Two networks N_1 and N_2 are said to be duals of each other, if graphs of these networks G_1 and G_2 are duals and, if in addition, the elements of N_2 are duals of the elements of N_1. The topological and electrical duality relations can be summarized as

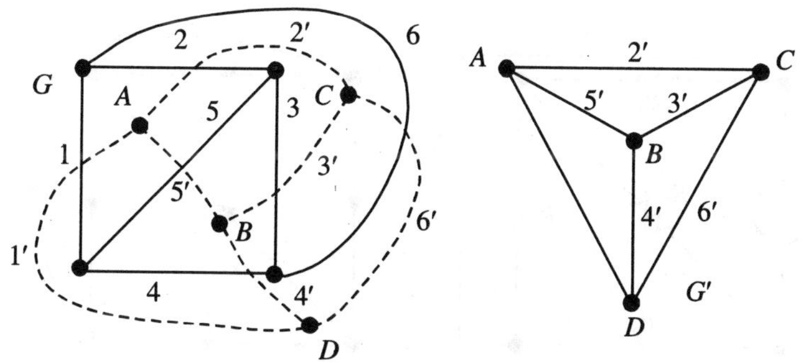

Fig. 2.14

(*i*) Topological duality:

Graph G_1 ⇔ Graph G_2
Circuits ⇔ Cut-sets
Series ⇔ Parallel

(*ii*) Electrical duality:

Current ⇔ Voltage
Resistor ⇔ Resistor (of value $1/R$)
Capacitor ⇔ Inductor
Impedance ⇔ Admittance

where ⇔ 'implies the dual of'.

Of course, it is easy to see that a planar network has no dual if it contains mutual inductances.

2.2 MATRICES OF GRAPHS

The Kirchhoff's current and voltage laws for a lumped network form two sets of linear equations. A method of writing these equations in compact form is to resort to matrix notation. To this end we define and study the properties of incidence, *f*-cut-set and *f*-circuit matrices of a network graph.

2.2.1 Incidence Matrix

Consider the resistive network shown in Fig. 2.15(*a*). Fig. 2.15(*b*) shows its oriented linear graph. The orientations of the edges corresponding to the resistors are arbitrarily chosen. There is no loss of generality in selecting a resistive network, as the Kirchhoff's laws do not depend on the elements. There are four vertices in the graph and writing the KCL at the four vertices we have:

$$\text{At vertex } v_1 \qquad -i_1 + i_2 - i_6 = 0$$

$$\text{At vertex } v_2 \qquad -i_2 - i_3 + i_4 = 0$$

$$\text{At vertex } v_3 \qquad -i_4 + i_5 + i_6 = 0$$

$$\text{At vertex } v_4 \qquad -i_1 + i_3 - i_5 = 0$$

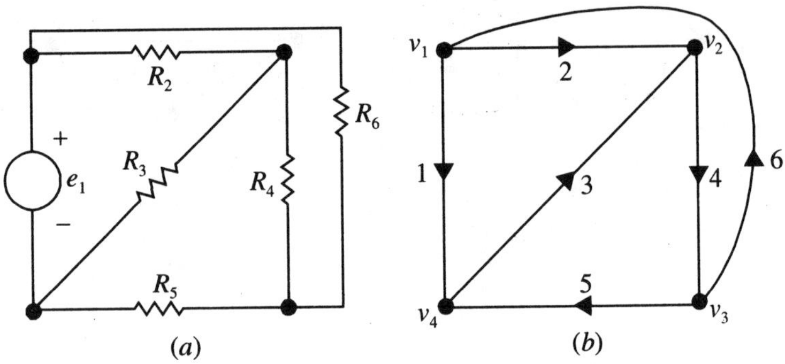

Fig. 2.15

These equations can be written in matrix form as

$$\begin{bmatrix} 1 & 1 & 0 & 0 & 0 & -1 \\ 0 & -1 & -1 & 1 & 0 & 0 \\ 0 & 0 & 0 & -1 & 1 & 1 \\ -1 & 0 & 1 & 0 & -1 & 0 \end{bmatrix} \begin{bmatrix} i_1 \\ i_2 \\ i_3 \\ i_4 \\ i_5 \\ i_6 \end{bmatrix} = 0$$

or $$A_a I_e = 0$$

where $\mathbf{I}_c = [i_1 \;\; i_2 \;\; i_3 \;\; i_4 \;\; i_5 \;\; i_6]^t$ is the column vector of the edge currents and A_a is the incidence matrix of the graph.

Let G be a graph with v vertices and e edges. Then the incidence matrix of G is defined by

$$A_a = [a_{ij}]_{v \times e} \qquad \text{...(2.1)}$$

where a_{ij} = 1, if the edge j is incident at the vertex i and oriented away from it.
= −1, if the edge j is incident at the vertex i and oriented towards it.
= 0, otherwise.

The edge orientation is considered as positive if it is away from vertex and negative if it is towards the vertex. Of course, there is no loss of generality in this.

The KCL for a network can now be written as

$$A_a \mathbf{I}_e = 0 \qquad ...(2.2)$$

where $\mathbf{I}_e$ is the column vector of the edge currents and A_a the incidence matrix of the graph of the network.

Several properties of the matrix A_a can be noted from its structure:

(*i*) Every column of A_a has exactly two non-zero entries, one +1 and the other –1. This is evident as every edge has two distinct ends.

(*ii*) As a consequence of property (*i*), any row of A_a can be deleted without loss of information. In the sequel, we only consider a matrix in which one of the rows is deleted and designate it by A. A has $(v - 1)$ rows and e columns and is called the *reduced* incidence matrix. The vertex corresponding to the row deleted is identified as the *reference* or *datum* vertex.

(*iii*) Another consequence of property (*i*) is that the sum of all rows of A_a results in a row of zeros. Hence, the rank of A_a is at most $(v - 1)$. In fact, we can prove that the rank of A_a, for a connected graph is at least $(v - 1)$ indicating that the rank of A_a, and that of A, is $(v - 1)$.

To prove this, we show that $(v - 1)$ KCL equations for a network are linearly independent. If it were not, then there exists some k KCL equations where $k \leq v - 1$, so that

$$\sum_{i=1}^{k} \alpha_i \zeta_i = 0 \qquad ...(2.3)$$

where α_1 is a constant and ζ_i is the KCL equation corresponding to vertex i. Equation 2.3 is equivalent to adding the corresponding rows of A_a after multiplying it by appropriate α_1. Consider the set of all vertices 1, 2, ...k. Split the graph G of the network into two parts, G_1 and G_2, so that G_1 has the k vertices and their incident edges and G_2 has the remaining $(v - k)$ vertices and their corresponding edges (Fig. 2.16). Since $k \leq v - 1$, there

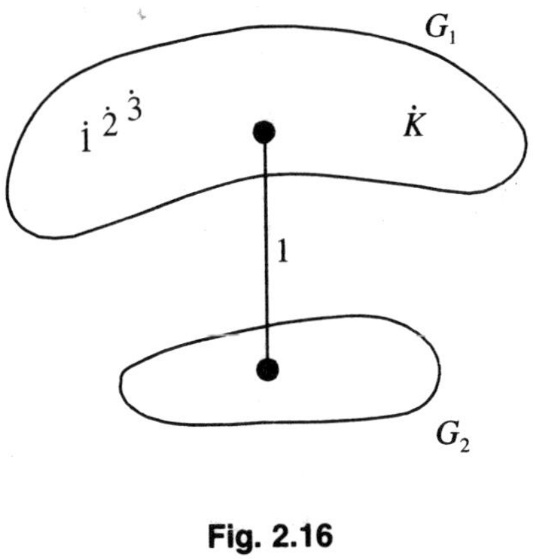

Fig. 2.16

is at least one vertex in G_2. There must be at least one edge, say 1, between G_1 and G_2 (G is a connected graph). Thus, i_1 appears in only one of the k equations ζ_1, ζ_2, ... and hence cannot cancel out in the sum

$$\sum_{i=1}^{m} \alpha_i \zeta_i$$

This contradicts the original assumption that k KCL equations satisfy eqn. 2.3. Hence, we can conclude: *There are* $(v - 1)$ *independent KCL equations for a network whose graph G is connected and has v vertices.*

The arguments can be easily verified by the KCL equations of the network in Fig. 2.15. The sum of any three or less rows of the corresponding incidence matrix will not add to a zero row.

The number $(v - 1)$ is also called the *rank of the graph* and this is the same as the number of branches in a tree. If the graph has p parts, it can be shown that the rank of A_a is $v - p$, the number of edges in a forest of the graph.

(*iv*) The determinant of any sub-matrix of A is ±1 or zero.

(*v*) The determinant of a sub-matrix of order $(v - 1)$ of A is ±1 if, the edges corresponding to the columns of this sub-matrix form a tree of the graph. This property indicates the one to one correspondence between the non-singular majors of A and the trees of G.

We will neither prove the last two properties nor use them. They are stated just for the sake of completeness. In passing, we state that the property (*v*) is the cornerstone of the topological analysis of networks.

2.2.2 Fundamental Cut-Set Matrix

In selecting the *f*-cut-set, the first step is to select a tree. Let the edges 1, 4 and 5 constitute a tree of the graph of Fig. 2.15(*b*). The graph is redrawn in Fig. 2.17

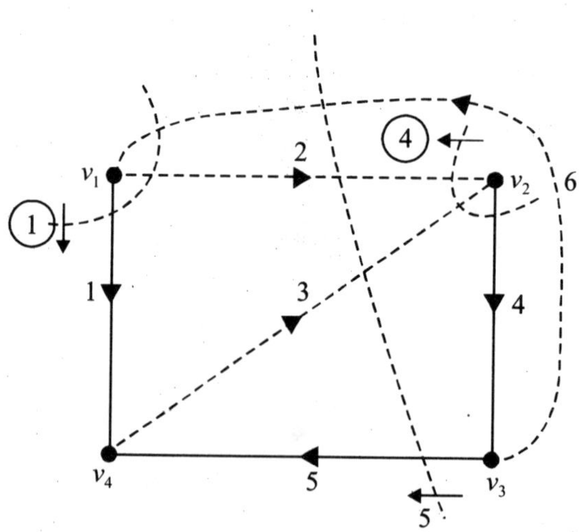

Fig. 2.17

with solid lines representing branches and broken lines the chords. There are three f-cut-sets and these are defined by the branches 1, 4 and 5 and are marked on the figure. The KCL applied to the cut-sets (*i.e.*, the algebraic sum of currents in the edges in a cut-set is zero) yields

$$i_1 + i_2 - i_6 = 0$$

$$-i_2 - i_3 + i_4 = 0$$

$$-i_2 - i_3 + i_5 + i_6 = 0$$

The cut-sets defined by branches 1 and 4 are *node cut-sets,* whereas the cut-set defined by the branch 5 is not a node cut-set. It is a linear combination of two node cut-sets corresponding to nodes v_2 and v_3. In a graph, every cut-set is either a node cut-set or a linear combination of node cut-sets (see eqn. 2.15). In the former case, KCL is as stated in section 1.2. In the latter case, the KCL equation corresponding to the nonnode cut-set is obtained as a linear combination of KCL equation at certain nodes. Hence, we see that KCL is applicable to cut-sets of graph of a network and in particular to f-cut-sets. The preceding equations can be written in the matrix form as

$$\begin{bmatrix} 1 & 1 & 0 & 0 & 0 & -1 \\ 0 & -1 & -1 & 1 & 0 & 0 \\ 0 & -1 & -1 & 0 & 1 & 1 \end{bmatrix} \begin{bmatrix} i_1 \\ i_2 \\ i_3 \\ i_4 \\ i_5 \\ i_6 \end{bmatrix} = 0$$

or $$Q_f \mathbf{I}_e = 0 \qquad \text{...(2.4)}$$

where $\mathbf{I}_e$, as before, is the edge current vector and Q_f is the f-cut-set matrix.

Let G be a connected graph with v vertices and e edges. Then the f-cut-set matrix with respect to the selected tree of G is defined by

$$Q_f = [Q_{ij}]_{(v-1) \times e}$$

q_{ij} = 1, if the chord j appears in the f-cut-set defined by the branch i and oriented in the same direction as that of i.

= –1, if the chord j appears in the f-cut-set defined by the branch i and oriented opposite to that of i.

= 0, otherwise.

It is more convenient to write the f-cut-set matrix with the edges of the graph numbered in a particular order. After selecting a tree, the chords are numbered 1, 2, ...$e - v + 1$ and the branches $e - v + 2$, $e - v + 3$, ...e. The columns of the f-cut-set matrix are also arranged in the same order, *i.e.*, the first $(e - v + 1)$ columns correspond to chords and the next $(v - 1)$ columns to branches. The rows of the matrix are arranged so that the first row corresponds to the f-cut-set defined by the branch numbered $(e - v + 2)$ and the second row to that defined by branch numbered $(e - v + 3)$, and so on. Then the f-cut-set matrix can be written as

$$Q_f = \begin{array}{c|c} \text{Chords} & \text{Br} \\ [Q_{11} & U] \end{array} \qquad \text{...(2.5)}$$

where U is a unit matrix of order $(v - 1)$. For example, the matrix for the graph of Fig. 2.17 is

$$Q_f = \begin{array}{c} \begin{matrix} 1' & 2' & 3' & 4' & 5' & 6' \end{matrix} \\ \begin{bmatrix} 1 & 0 & -1 & 1 & 0 & 0 \\ -1 & -1 & 0 & 0 & 1 & 0 \\ -1 & -1 & 1 & 0 & 0 & 1 \end{bmatrix} \end{array}$$

and the corresponding edge current vector is

$$\mathbf{I}_e = [i_2 \quad i_3 \quad i_6 \quad i_1 \quad i_4 \quad i_5]^t$$

(Here, we have re-numbered the edges of the graph. $1 \to 4', 2 - 1', 3 - 2', 4 \to 5', 5 \to 6', 6 \to 3'$. $4'\ 5'\ 6'$ constitutes the tree and tree and $1'\ 2'\ 3'$ the co-tree.) From now on when we refer to f-cut-set matrix we assume that the matrix is in the form of eqn. 2.5.

Some of the properties of Q_f are:

(*i*) Rank of $Q_f = v - 1$. This is evidenced by the unit matrix of order $(v - 1)$. This indicates that there are $(v - 1)$ independent cut-sets (Theorem 3) and hence KCL equations.

(*ii*) There is one to one correspondence between the non-singular majors of Q_f and the trees of graph.

(Compare the properties of the incidence and the f-cut-set matrices)

If we partition the edge current vector to conform to that of the f-cut-set matrix, we can write

$$\mathbf{I}_e = \begin{bmatrix} \mathbf{I}_c \\ \mathbf{I}_b \end{bmatrix}$$

where $\mathbf{I}_c$ is the set of chord currents and $\mathbf{I}_b$ the set of branch currents with respect to the selected tree. The KCL can now be written as

$$[Q_{11} \quad U] \begin{bmatrix} \mathbf{I}_c \\ \mathbf{I}_b \end{bmatrix} = 0$$

From which we have

$$\mathbf{I}_b = -Q_{11} \mathbf{I}_c \qquad \text{...(2.6)}$$

This equation shows that the branch currents can be expressed as linear combination of the chord currents. The converse is generally not true, since Q_{11}^{-1} may not exist. Hence, we conclude that the branch currents are uniquely determined by the chord currents. We use this fact in the next chapter to write the network equations.

2.2.3 Fundamental Circuit Matrix

Let, as before, the edges 1, 4 and 5 constitute the tree of the graph of Fig. 2.15(*b*). There are three *f*-circuits which are defined by the chords 2, 3 and 6. These circuits

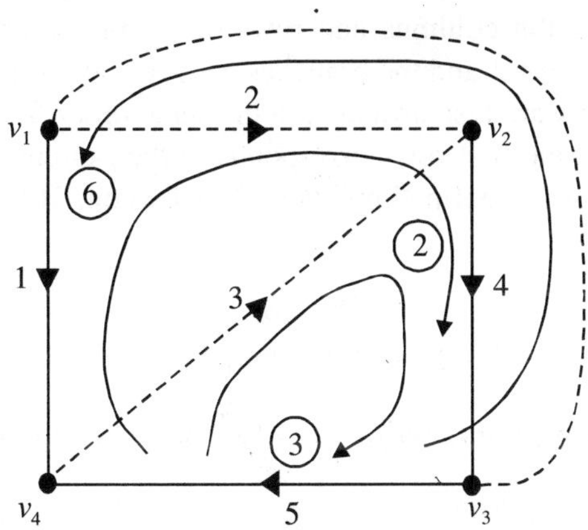

Fig. 2.18

are marked on Fig. 2.18. The KVL applied to the *f*-circuits yields:

$$-v_1 + v_2 + v_4 + v_5 = 0$$

$$v_3 + v_4 + v_5 = 0$$

$$v_1 - v_5 + v_6 = 0$$

where v_1 is the voltage of the edge *i*. Writing these equations in matrix form we have

$$\begin{bmatrix} -1 & 0 & 0 & 1 & 1 & 0 \\ 0 & 0 & 1 & 1 & 1 & 0 \\ 1 & 0 & 0 & 0 & -1 & 1 \end{bmatrix} \begin{bmatrix} v_1 \\ v_2 \\ v_3 \\ v_4 \\ v_5 \\ v_6 \end{bmatrix} = 0$$

or $$B_f \mathbf{V}_e = 0 \qquad \text{...(2.7)}$$

where V_e is the edge voltage vector and B_f the *f*-circuit matrix.

Let *G* be a connected graph with *v* vertices and *e* edges. Then the *f*-circuit matrix with respect to the selected tree is defined by

$$B_f = \left[b_{ij}\right]_{(e-v+1)\times e}$$

where $b_{ij} = 1$, if the branch *j* is in the *f*-circuit defined by the chord *i* and its orientation is the same as that of the *f*-circuit.

= −1, if the branch j is in the f-circuit defined by the chord i and its orientation is opposite to that of the f-circuit.

= 0, otherwise.

As in the case of f-cut-set matrix, we will number the edges of the graph in order and rearrange the columns and rows of B_f. In the graph, the chords are numbered 1, 2, ...$e - v + 1$ and the branches $e - v + 2$, $e - v + 3$, ...e. The columns of the f-circuit matrix are also arranged in the same order. The rows of the matrix are so arranged that the first row corresponds to the f-circuit defined by chord 1, the second row to that is defined by chord 2 and so on. Then the f-circuit matrix can be written as

$$B_f = \begin{array}{c|c} \text{Ch} & \text{Br} \\ [U & B_{12}] \end{array} \qquad \text{...(2.8)}$$

where U is a unit matrix of order $(e - v + 1)$. The f-circuit matrix for the graph of Fig. 2.18 is

$$B_f = \begin{array}{c} \begin{matrix} 1' & 2' & 3' & 4' & 5' & 6' \end{matrix} \\ \begin{bmatrix} 1 & 0 & 0 & -1 & 1 & 1 \\ 0 & 1 & 0 & 0 & 1 & 1 \\ 0 & 0 & 1 & 1 & 0 & -1 \end{bmatrix} \end{array}$$

and the corresponding edge voltage vector is

$$v_a = \begin{bmatrix} v_2 & v_3 & v_6 & v_1 & v_4 & v_5 \end{bmatrix}'$$

(observe the renumbering of edges of the graph).

From now on, whenever we refer to f-circuit matrix we assume that the matrix is in the form of eqn. 2.8.

Some of the properties of B_f are:

(*i*) Rank of $B_f = e - v + 1$. This is evidenced by the unit matrix of order $(e - v + 1)$. This indicates that there are $(e - v + 1)$ independent circuits (Theorem 2.2) and hence KVL equations.

(*ii*) There is one to one correspondence between the non-singular majors of B_f and the trees of the network.

If we partition the edge voltage vector to conform with that of the circuit matrix, we can write

$$\mathbf{V}_e = \begin{bmatrix} \mathbf{V}_c \\ \mathbf{V}_b \end{bmatrix}$$

where V_c is the set of chord voltages, and V_b the set of branch voltages. The KVL can now be written as

$$\begin{bmatrix} U & B_{12} \end{bmatrix} \begin{bmatrix} \mathbf{V}_c \\ \mathbf{V}_b \end{bmatrix} = 0$$

which yields

$$V_c = -B_{12} V_b \qquad ...(2.9)$$

This equation shows that the chord voltages can be expressed as a linear combination of the branch voltages. The converse is generally not true since B_{12}^{-1} may not exist. Hence, we conclude that the chord voltages are uniquely determined by the branch voltages, a fact which will be used in the formulation of network equations.

If the graph is planar, there is an alternate way of selecting $(e - v + 1)$ independent circuits. This is achieved by selecting meshes instead of *f*-circuits. By definition, a *mesh* is a circuit in which there are no edges in its interior. Obviously, meshes can be defined only for a planar graph. If the graph is non-planar, we have no choice but to select a set of *f*-circuits. But for a planar graph we can either choose a set of *f*-circuits or the set of meshes for the graph in order to write a set of linearly independent KVL equations. It may be possible to select a tree so that the meshes themselves are *f*-circuits. But this need not, in general, be true. The graph of Fig. 2.15(*a*) has three meshes, and these are shown in Fig. 2.19. It is a

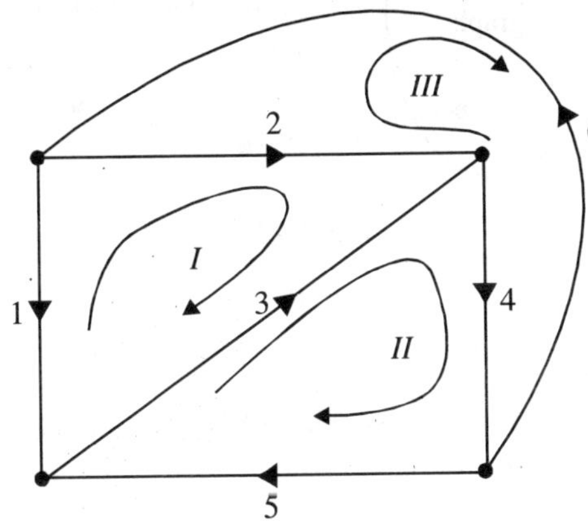

Fig. 2.19

general practice to orient all the meshes in either clockwise or anti-clockwise direction. It can be noticed that the three meshes (I, II and III) are also the *f*-circuits with respect to the tree 2, 3, 4. Of course, the *f*-circuit orientations and the mesh orientations do not coincide. It can be seen from the graph shown in Fig. 2.20, we cannot always select a tree so that all its *f*-circuits are also meshes.

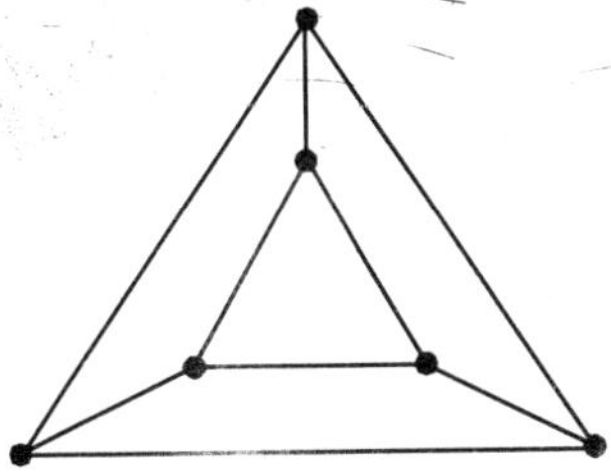

Fig. 2.20

Theorem 2.5: For a connected planar graph of v vertices and e edges the number of meshes is $(e - v + 1)$.

This can be proved by mathematical induction. Consider a graph which has only one mesh. Since all the vertices are of the second degree, $e - v + 1 = 1$ $(\Sigma\delta_y = 2v = 2e)$. Hence, the theorem is true for a single mesh graph. Let us assume that it is true for a l-mesh graph with e-edges and v-nodes.

i.e., $$e_l - v_l + 1 = l$$

Let us add an extra mesh to the graph. This can be achieved in two ways. We can add an edge between the two existing nodes. Then the total number of edges is increased by 1 and so is the number of meshes. We can also add a path between the two existing vertices by adding, say, m, edges (Fig. 2.21). The number of additional vertices added is obviously $m - 1$. Hence, we have

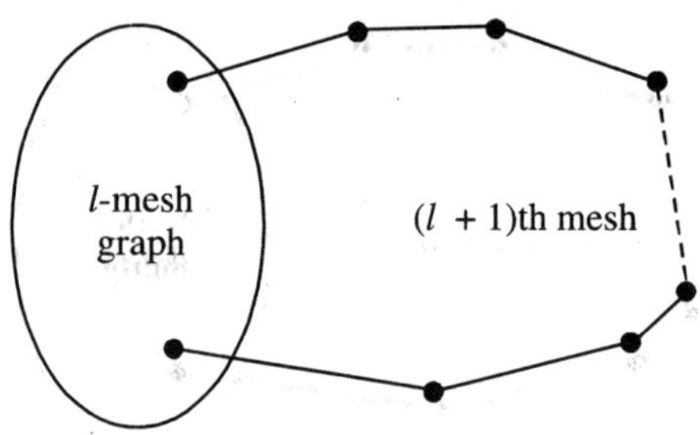

Fig. 2.21

$$e_{l+1} = e_l + m,\ v_{l+1} = v_l + m - 1$$

and $$e_{l+1} - v_{l+1} + 1 = (e_l - v_l + 1) + 1 = l + 1,$$

and hence the theorem.

KVL can be written for the meshes of a planar graph. The KVL for the graph of Fig. 2.19, in matrix form, is

$$\begin{bmatrix} -1 & 1 & -1 & 0 & 0 & 0 \\ 0 & 0 & 1 & 1 & 1 & 0 \\ 0 & 0 & 0 & -1 & 1 & -1 \end{bmatrix} \begin{bmatrix} v_1 \\ v_2 \\ v_3 \\ v_4 \\ v_5 \\ v_6 \end{bmatrix} = 0$$

or $$B_m \mathbf{V}_e = 0 \qquad \text{...(2.10)}$$

where B_m is the *mesh matrix*.

For a connected planar graph G with v vertices and e edges we can define the mesh matrix as

$$B_m = \left[b_{m_{ij}}\right]_{(e-v+1)\times e}$$

where $b_{m_{ij}} = 1$, if the edge j appears in the ith mesh and orientations are coincident.

$= -1$, if the edge j appears in the ith mesh and orientations are opposing.

$= 0$, otherwise.

We make use of the mesh matrix when we formulate the mesh equations for a network.

2.2.4 Inter-Relationships Among the Matrices

Though we have written three different matrices for the same graph, these matrices are not independent of one another. Given the incidence matrix it is possible to find the other two matrices without actually drawing the graph. Of course, the converse problem is rather complicated. Here we obtain certain inter-relationships among these matrices.

The KCL for a given network graph can be written in two ways, as

$$A\mathbf{I}_e = 0 \quad ...(a)$$

$$Q_f\mathbf{I}_e = 0 \quad ...(b) \qquad ...(2.11)$$

If we select a tree and partition the edge current vector $\mathbf{I}_e$ as

$$\mathbf{I}_e = \begin{bmatrix} \mathbf{I}_c \\ \mathbf{I}_b \end{bmatrix}$$

where $\mathbf{I}_c$ is the set of chord currents and $\mathbf{I}_b$ is the set of branch currents with respect to the selected tree, we can rewrite eqn. 2.11 as

$$\begin{bmatrix} A_{11} & A_{12} \end{bmatrix} \begin{bmatrix} \mathbf{I}_c \\ \mathbf{I}_b \end{bmatrix} = 0 \quad ...(a)$$

$$\begin{bmatrix} Q_{11} & U \end{bmatrix} \begin{bmatrix} \mathbf{I}_c \\ \mathbf{I}_b \end{bmatrix} = 0 \quad ...(b) \qquad ...(2.12)$$

In eqn. 2.12(a), the sub-matrix A_{11} is of the order $(v - 1) \times (e - v + 1)$ and the columns correspond to the chords. The sub-matrix A_{12} is of the order $(v - 1) \times (v - 1)$ and its columns correspond to a set of branches. From property 5 of the incidence matrix, A_{12} is a non-singular matrix. Hence, we can rewrite eqn. 2.12(a) as

$$\mathbf{I}_b = -A_{12}^{-1}A_{11}\mathbf{I}_c \qquad ...(2.13)$$

Equation 2.12(b) yields

$$\mathbf{I}_b = -U^{-1}Q_{11}\mathbf{I}_c \qquad ...(2.14)$$

From eqns. 2.13 and 2.14 we have

$$Q_{11} = A_{12}^{-1}A_{11} \qquad ...(2.15)$$

This equation implies that the rows of f-cut-set matrix are a linear combination of the rows of incidence matrix. We can obtain an f-cut-set matrix from an incidence matrix by elementary row transformations.

Similarly, a relationship between the incidence matrix A and the f-circuit matrix B_f can be obtained. The relationship is contained in the following theorem.

Theorem 2.6: If the columns of A and B_f are arranged in the same order (first chords and then branches), then

$$AB_f^T = 0. \quad ...(2.16)$$

Let $$AB_f^t = \left[p_{ij}\right]_{(y-1)\times(e-v+1)}$$

The element p_{ij} is obtained from the ith row of A (corresponding to ith vertex) and jth column of B_f^T, *i.e.*, jth row of B_f (corresponding to jth f-circuit). There exist two possibilities:

(*i*) The ith vertex is not in the jth circuit. In this case, edges incident at the ith vertex are not in the jth circuit. Hence, the non-zero entries in the ith row of A and jth row of B are not in the same columns and hence $p_{ij} = 0$.

(*ii*) The ith vertex is in the jth circuit. As every vertex of a circuit is of degree 2, the jth circuit will have exactly two edges of those incident at the ith vertex. Hence, in this case the ith row of A and the jth row of B_f have common non-zero entries in only two columns. The possible orientations are shown in Fig. 2.22.

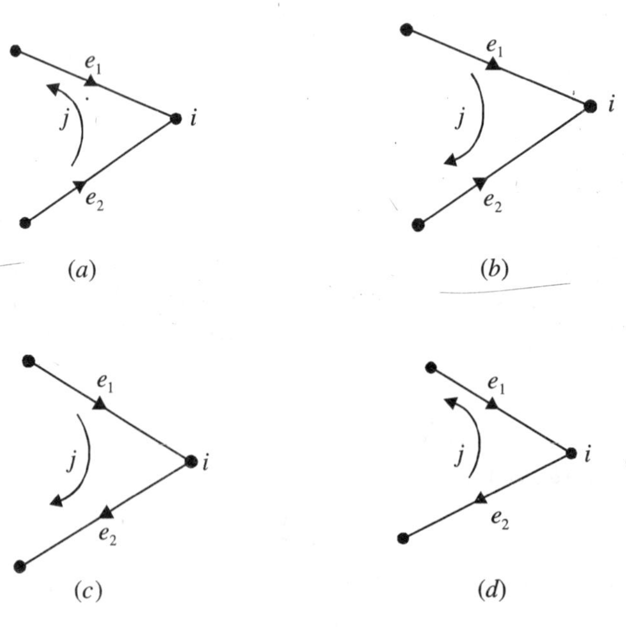

Fig. 2.22

				e_1	e_2
(*a*)	*i*th row *A*	→	...	–1	–1
	*j*th row B_f	→	...	–1	1
(*b*)	*i*th row A	→	...	–1	–1
	*j*th row B_f	→	...	1	–1
(*c*)	*i*th row *A*	→	...	–1	1
	*j*th row B_f	→	...	1	1
(*d*)	*i*th row *A*	→	...	–1	1
	*j*th row B_f	→	...	–1	–1

Hence, $p_{ij} = (-1)(-1) + (1)(-1) = 0$

or $\quad (-1)(+1) + (+1)(+1) = 0$

Hence, the theorem

If A and B_f are confirmingly partitioned, then eqn. 2.16 can be written as

$$\begin{bmatrix} A_{11} & A_{12} \end{bmatrix} \begin{bmatrix} U \\ B_{12}^T \end{bmatrix} = 0$$

which yields

$$B_{12}^T = -A_{12}^{-1} A_{12} \qquad \text{...(2.17)}$$

where columns of A_{12} correspond to a set of tree branches. Equation 2.17 shows that the columns of B_{12} are linear combinations of the rows of A_{11}.

The following relationship between the *f*-circuit and the *f*-cut-set matrices can be obtained from eqns. 2.15 and 2.17.

$$Q_{11} = -B_{12}^T \qquad \text{...(2.18)}$$

Hence, the *f*-circuit matrix, in terms of the *f*-cut-set matrix is

$$B_f = [U \quad -Q_{11}] \qquad \text{...(2.19)}$$

and the *f*-cut-set matrix, in terms of the *f*-circuit matrix, is

$$Q_f = \begin{bmatrix} -B_{12}^T & U \end{bmatrix} \qquad \text{...(2.20)}$$

Hence, finally we have

$$B_f Q_f^t = 0 \quad \text{and} \quad Q_f B_f^t = 0 \qquad \text{...(2.21)}$$

Equations 2.15, 2.17, 2.18 and 2.19 give the relationships between the three matrices of a network graph.

PROBLEMS

2.1 Give the proof of Theorem 2.1.

2.2 The graph of Fig. P. 2.2 has 16 trees. Mark all these trees.

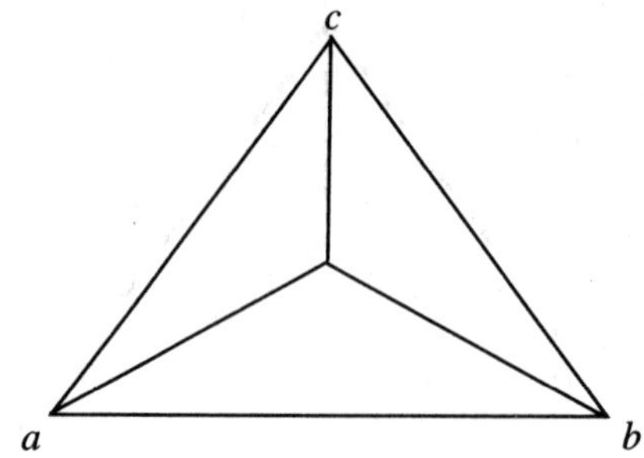

Fig. P. 2.2

2.3 The linear graph of a network is shown in Fig. P. 2.3. Select a tree for the graph and mark all the *f*-circuits and *f*-cut-sets for the graph.

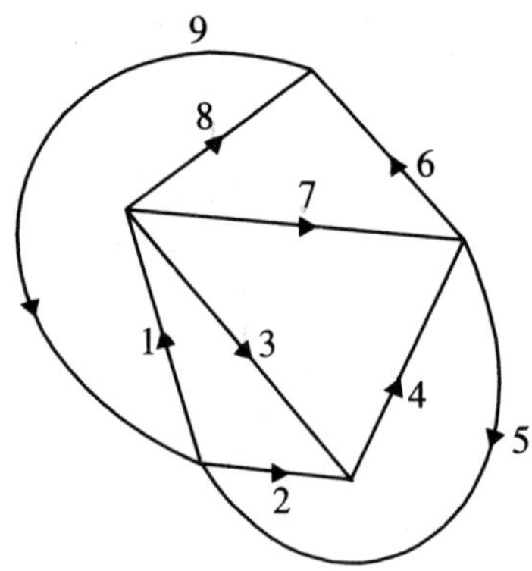

Fig. P. 2.3

2.4 For the graph shown in Fig. P. 2.4 select a set of *f*-circuits and *f*-cut-set. Show that all the other circuits and cut-sets of the graph can be obtained as a linear combination of the *f*-circuits and *f*-cut-sets.

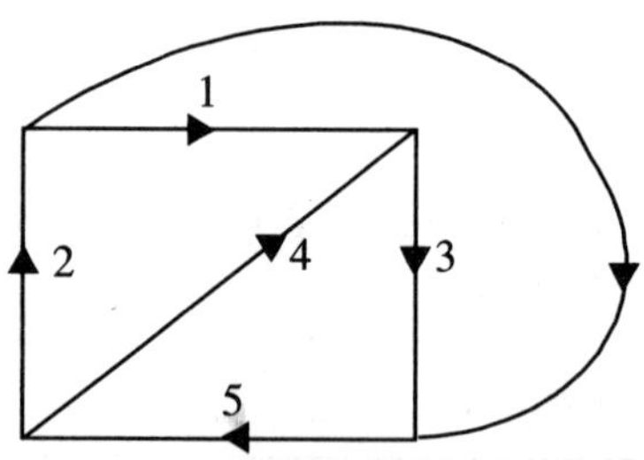

Fig. P. 2.4

2.5 Draw the dual graph of the graph shown in Fig. P. 2.5.

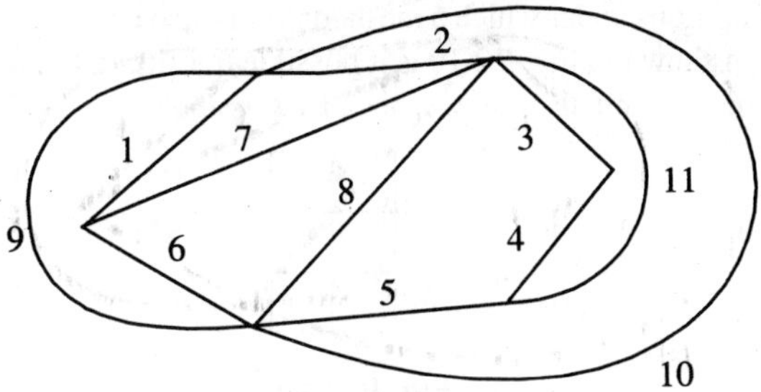

Fig. P. 2.5

2.6 Find the dual of the network shown in Fig. P. 2.6.

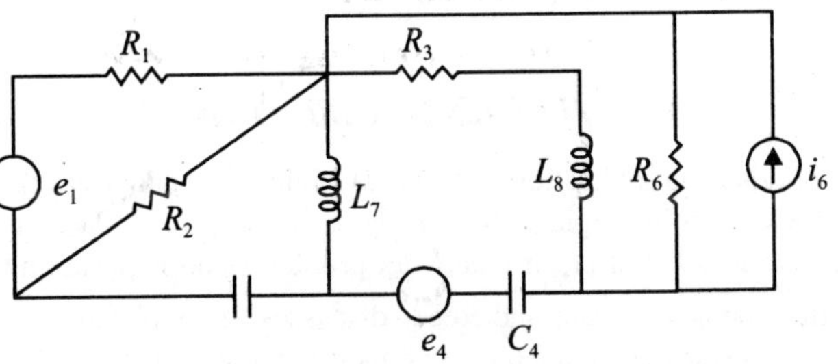

Fig. P. 2.6

2.7 For the graph of Fig. P. 2.3 write the incidence, *f*-circuit and *f*-cut-set matrices and verify eqns. 2.15, 2.17 and 2.18.

2.8 Indicate which of the following statements are true or false. Justify your answers.

(*i*) If a connected graph has but one tree, it has no circuits.

(*ii*) If a connected graph has but one circuit, it has only one tree.

(*iii*) The number of *f*-cut-sets is always less than the number of *f*-circuits.

(*iv*) The number of edges that are common to any *f*-circuit and any *f*-cut-set is either 0 or two.

(*v*) The dual of a planar graph is also planar.

(*vi*) Any $(v - 1)$ independent current equations for a v vertex connected network are sufficient to describe the current constraints in the network.

(*vii*) The set of $(e - v + 1)$ meshes of a v node e edge connected graph can always be made the *f*-circuits with respect to some tree of the graph.

(*viii*) The *f*-circuit matrix and the set of chord currents with respect to a tree determine all the currents of the network uniquely.

(*ix*) A v vertex circuitless connected graph has $(v - 1)$ cut-sets.

2.9 The *f*-circuit matrix of a graph is given by

$$\begin{array}{cccccccccccc} 7 & 8 & 9 & 10 & 11 & 12 & 1 & 2 & 3 & 4 & 5 & 6 \end{array}$$

$$B_f = \begin{bmatrix} 1 & 0 & 0 & 0 & 0 & 0 & -1 & 0 & 0 & 1 & 0 & 0 \\ 0 & 1 & 0 & 0 & 0 & 0 & 0 & -1 & 0 & 0 & 1 & 0 \\ 0 & 0 & 1 & 0 & 0 & 0 & 0 & 0 & -1 & 0 & 0 & 1 \\ 0 & 0 & 0 & 1 & 0 & 0 & 0 & 1 & 0 & -1 & 0 & 0 \\ 0 & 0 & 0 & 0 & 1 & 0 & 0 & 0 & 1 & 0 & -1 & 0 \\ 0 & 0 & 0 & 0 & 0 & 1 & -1 & 0 & 0 & 0 & 0 & 1 \end{bmatrix}$$

Draw the graph.

2.10 State the conditions under which a mesh matrix is also a *f*-circuit matrix of the graph. For the graph shown in Fig. P.2.10 is it possible to select a tree so that *f*-circuits with respect to this tree are also the meshes of the graph?

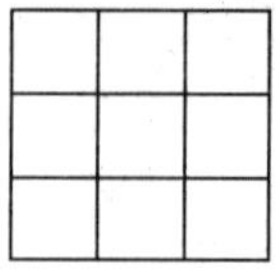

Fig. P. 2.10

2.11 From the first principles (*i.e.*, without using Theorem 2.5) show that

$$B_f Q_f^t = 0.$$

NOTES AND REFERENCES

Graph Theory finds wide application in the fields of electrical engineering, computer science and mathematics. With its origin in Konnigsberg bridge problem attributed to Euler, it is being used in the solution of electric network problems and in designing algorithms.

The excellent book of Seshu and Reed[1] discusses the application of graph theory to electric networks besides providing proofs of many theorems cited in this chapter. Additional information and problems can also be found in the books by Seshu and Balbanian[2] (Chapter 3) Desoer and Kuh[3] (Chapter 5, 9–11).

Graph theoretic concepts and methods provide a formal and algorithmic approach to the derivation of network equilibrium equations (Chapter 3) and are fundamental to the modelling of large scale networks and systems. Such models find extensive use in communication and transportation networks[4]. Many non-circuit theory applications are discussed in Deo's book[5].

Serious students of graph theory may find books of Harary[6], Swamy and Thulasiraman[7] to be of interest. The former gives a mathematician's point of view while the latter is concerned with engineering.

1. S. Seshu and M.B. Reed, *Linear Graphs and Electrical Networks*, Addison-Wesley, 1961.
2. S. Seshu and N. Balbanian, *Linear Network Analysis*, John Wiley, 1959.
3. C.A. Desoer and E.S. Kuh, *Basic Circuit Theory*, McGraw Hill, 1969.
4. H. Frank and I.T. Frisch, *Communication, Transmission and Transportation Networks*, Addison-Wesley, 1971.
5. N. Deo, *Graph Theory with Application to Engineering and Computer Science*, Prentice Hall, 1974.
6. F. Harary, *Graph Theory,* Addison-Wesley, 1969.
7. M.N.S. Swamy and K, Thulasiraman, *Graphs, Networks and Algorithms*, John Wiley, 1981.

CHAPTER

3 Network Equations

A linear network can be described by a set of linear equations. In this chapter, we present methods of writing such an equations. The basic questions to be answered before we have a set of equations are: How many equations do we need? How do we write these equations? When are we sure that we have enough equations? Some of these questions are answered by "Topology" of the network.

In order to completely solve a network of n elements, we need to find n voltages and n currents. In all, there are $2n$ variables. Hence, we need $2n$ linearly independent equations in these $2n$ variables. The terminal relations (*i.e.* $v - i$ characteristics) of the elements provide n independent equations. The second set of n equations is provided by the KCL and the KVL. We have already shown that there are $(v - 1)$ independent KCL equations and $(n - v + 1)$ independent KVL equations and, thus, in total we have n more equations. We have, thus, divided the $2n$ equations needed to solve the network, *i.e.* to find all the currents and voltages, into two groups: a group of n equations which depend on the elements but not on their interconnection, the second group of n equations which depend on the interconnection of the elements but not on the elements themselves. Different methods of combining these two groups of equations lead to different sets of equations, and this forms the subject of this chapter.

3.1 NODE AND MESH TRANSFORMATIONS

The Kirchhoff's current and voltage laws for any general network can be written as

$$Q_f \mathbf{I}_e = 0 \qquad \text{...(3.1)}$$

$$B_f \mathbf{V}_e = 0 \qquad \text{...(3.2)}$$

where $Q_f\,[(v-1)\times e]$ and $B_f\,[(e-v+1)\times e]$ are the f-cut-set and f-circuit matrices respectively of the network graph of v-vertices and e edges. If the edges of the network graph and the columns of these matrices Q_f and B_f are re-ordered, we can write

$$[Q_{11} \ \ U]\begin{bmatrix} \mathbf{I}_c \\ \mathbf{I}_b \end{bmatrix} = 0 \qquad \text{...(3.3)}$$

$$[U \ \ B_{12}]\begin{bmatrix} \mathbf{V}_c \\ \mathbf{V}_b \end{bmatrix} = 0 \qquad \text{...(3.4)}$$

where the subscript c indicates chord variables and b the branch variables. Equation 3.4 yields

$$\mathbf{V}_c = -B_{12} \, \mathbf{V}_b$$

On using eqn. 2.18, we get

$$\mathbf{V}_c = Q_{11}^t \mathbf{V}_b \qquad \text{...(3.5)}$$

By combining this with the trivial equation

$$\mathbf{V}_b = U \, \mathbf{V}_b,$$

we have

$$\mathbf{V}_e = \begin{bmatrix} Q_{11}^t \\ U \end{bmatrix} \mathbf{V}_b$$

or

$$\mathbf{V}_e = Q_f^t \ \mathbf{V}_b \qquad \text{...(3.6)}$$

This equation indicates that the edge voltages are expressed as a linear combination of $(v - 1)$ branch voltages, where v is the number of vertices in the network graph. Equation 3.6 is called the *branch transformation.* To verify the consistency of this equation we have only to see that

$$B_f \, \mathbf{V}_e = B_f(Q_f^t \ \mathbf{V}_b) = (B_f \, Q_f^t) \ \mathbf{V}_b = 0$$

Instead of using the branch voltages as a set of independent voltages, we can also use a set of node voltages as the independent set. If in a graph we select a node as the reference node, we can write

$$\mathbf{V}_e = A^t \, \mathbf{V}_n \qquad \text{...(3.7)}$$

where V_n is the column vector of $(v - 1)$ node voltages with respect to the reference node and A is the reduced incidence matrix of the graph. (The row corresponding to the reference node is deleted from A_a.)

To prove this result, we recognize two kinds of edges: those which are incident with the reference node and those which are not. For the former, the edge voltage is the same as the node voltage or its negative depending upon the orientation of the edge. For the latter, the edge voltage forms a circuit with two node voltages and hence the edge voltage can be expressed as a linear combination of node voltages. In both the cases, the edge voltages are linear combination of node voltages.

Let v_k be the edge voltage of the kth edge and e_i the node voltage of the node i. If the kth edge connects the ith node to the reference node then,

$$v_k = e_i, \text{ if edge } k \text{ leaves node } i.$$

$$= -e_i, \text{ if edge } k \text{ enters node } i.$$

On the other hand, if the kth branch leaves node i and enters node j, then,

$$v_k = e_i - e_j$$

Hence, we can write

$$\begin{bmatrix} v_1 \\ v_2 \\ v_3 \\ \vdots \\ v_e \end{bmatrix} = \begin{bmatrix} C_{11} & C_{1,\,v-1} \\ C_{12} & \vdots \\ \vdots & \vdots \\ C_{e1} & C_{e,\,v-1} \end{bmatrix} \begin{bmatrix} e_1 \\ e_2 \\ \vdots \\ e_{v-1} \end{bmatrix}$$

where

$C_{ij} = 1$, if edge i leaves node j.

$= -1$, if edge i enters node j.

$= 0$, if edge i is not incident with the node j.

From a definition of C_{ij}, it can be easily noticed that the matrix $C = A^t$ and hence

$$\mathbf{V}_e = A^t\,\mathbf{V}_n$$

This is known as *node transformation.*

We have expressed the edge voltages as linear combination of branch or node voltages. We can similarly express the edge currents as linear combinations of chord or mesh currents.

Equation 3.3 yields

$$\mathbf{I}_b = Q_{11}\,\mathbf{I}_c$$

In view of eqn. 2.18, we have

$$\mathbf{I}_b = B_{12}^t\,\mathbf{I}_c \qquad \ldots(3.8)$$

By combining this with the trivial equation

$$\mathbf{I}_e - U\mathbf{I}_c$$

we have

$$\mathbf{I}_e = \begin{bmatrix} U \\ B_{12}^t \end{bmatrix} \mathbf{I}_c$$

or

$$\mathbf{I}_e = B_f^t\mathbf{I}_e \qquad \ldots(3.9)$$

This equation indicates that the edge currents are expressed as linear combination of $(e - v + 1)$ chord currents, where the network graph has e edges and v vertices. Equation 3.9 represents the *chord transformation.*

If we are considering a planar network graph, we can select a set of mesh currents instead of chord currents. If $\mathbf{I}_m$ is the vector of mesh currents, we can easily show that

$$\mathbf{I}_e = B_m^t\,\mathbf{I}_m \qquad \ldots(3.10)$$

where B_m is the mesh matrix, indicating that the edge currents can also be expressed as linear combination of the set of mesh currents. This is known as *mesh transformation.*

Example 3.1. Consider the graph shown in Fig. 3.1. The branches are shown as solid lines and chords as broken lines. Vertex v_4 is selected as the reference node. There are three *f*-cut-sets, three *f*-circuits and three meshes, and these are marked on the graph. For this graph,

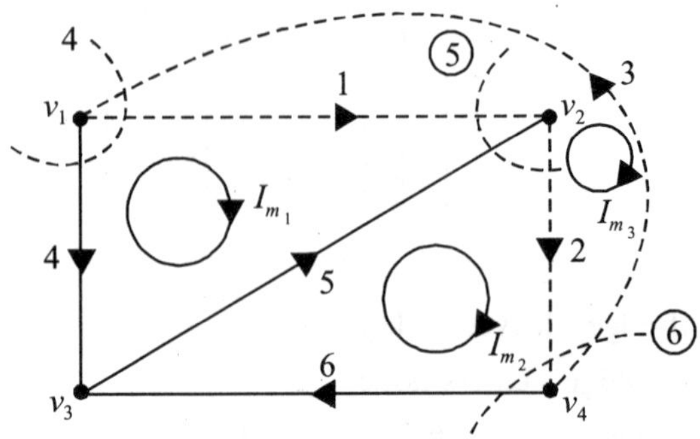

Fig. 3.1

$$A = \begin{matrix} & \begin{matrix} 1 & 2 & 3 & 4 & 5 & 6 \end{matrix} \\ \begin{matrix} v_1 \\ v_2 \\ v_3 \end{matrix} & \begin{bmatrix} 1 & 0 & -1 & 1 & 0 & 0 \\ -1 & 1 & 0 & 0 & -1 & 0 \\ 0 & 0 & 0 & -1 & 1 & -1 \end{bmatrix} \end{matrix}$$

$$Q_f = \begin{matrix} & \begin{matrix} 1 & 2 & 3 & 4 & 5 & 6 \end{matrix} \\ \begin{matrix} 4 \\ 5 \\ 6 \end{matrix} & \begin{bmatrix} 1 & 0 & -1 & 1 & 0 & 0 \\ 1 & -1 & 0 & 0 & 1 & 0 \\ 0 & -1 & 1 & 0 & 0 & 1 \end{bmatrix} \end{matrix}$$

$$B_f = \begin{matrix} & \begin{matrix} 1 & 2 & 3 & 4 & 5 & 6 \end{matrix} \\ \begin{matrix} 1 \\ 2 \\ 3 \end{matrix} & \begin{bmatrix} 1 & 0 & 0 & -1 & -1 & 0 \\ 0 & 1 & 0 & 0 & 1 & 1 \\ 0 & 0 & 1 & 1 & 0 & -1 \end{bmatrix} \end{matrix}$$

$$B_m = \begin{matrix} & \begin{matrix} 1 & 2 & 3 & 4 & 5 & 6 \end{matrix} \\ \begin{matrix} 1 \\ 2 \\ 3 \end{matrix} & \begin{bmatrix} 1 & 0 & 0 & -1 & -1 & 0 \\ 0 & 1 & 0 & 0 & 1 & 1 \\ -1 & -1 & -1 & 0 & 0 & 0 \end{bmatrix} \end{matrix}$$

$$\mathbf{I}_e = \begin{bmatrix} i_1 \\ i_2 \\ i_3 \\ i_4 \\ i_5 \\ i_6 \end{bmatrix}, \mathbf{V}_e = \begin{bmatrix} v_1 \\ v_2 \\ v_3 \\ v_4 \\ v_5 \\ v_6 \end{bmatrix}$$

$$\mathbf{V}_m = \begin{bmatrix} v_{v_1} \\ v_{v_2} \\ v_{v_3} \end{bmatrix}, \mathbf{V}_b = \begin{bmatrix} v_4 \\ v_5 \\ v_6 \end{bmatrix}, \mathbf{I}_m = \begin{bmatrix} i_{m_1} \\ i_{m_2} \\ i_{m_3} \end{bmatrix} \text{ and } \mathbf{I}_c = \begin{bmatrix} i_1 \\ i_2 \\ i_3 \end{bmatrix}$$

It can be seen that

$$\mathbf{V}_e = A^t\, \mathbf{V}_n,\ \mathbf{V}_e = Q_f^t\, \mathbf{V}_b$$

$$\mathbf{I}_e = B_m^t\, \mathbf{I}_m,\ \mathbf{I}_e = B_f^t\, \mathbf{I}_c$$

3.2 GENERALIZED ELEMENT

The terminal equations for the elements of a network can be written as

$$v_k = R_k i_k \quad \text{or} \quad i_k = G_k v_k \qquad \ldots(3.11)$$

$$v_k = L_k \frac{di_k}{dt} \quad \text{or} \quad i_k = \frac{1}{L_k}\int v_k dt \qquad \ldots(3.12)$$

$$v_k = \frac{1}{C_k}\int i_k dt \quad \text{or} \quad i_k = C_k \frac{dv_k}{dt} \qquad \ldots(3.13)$$

In a particular network these elements may appear in series or parallel as a group. In order to formulate the network equations, we introduce the concept of a *generalised element.* A generalized element k has an *operational impedance* z_k or *admittance* y_k in series with a voltage source e_k and in parallel with a current source j_k, as shown in Fig. 3.2. Let the operational impedance z_k be a series of combination of *R, L* and *C* and the admittance y_k a parallel combination of *R, L* and *C,* as shown in Fig. 3.3(*a*) and (*b*). (In writing the operational impedance all the elements in series are grouped together to form a single impedance. In writing the operational admittance all elements in parallel are grouped together to form a single admittance. If we want to write the operational admittance for the series of

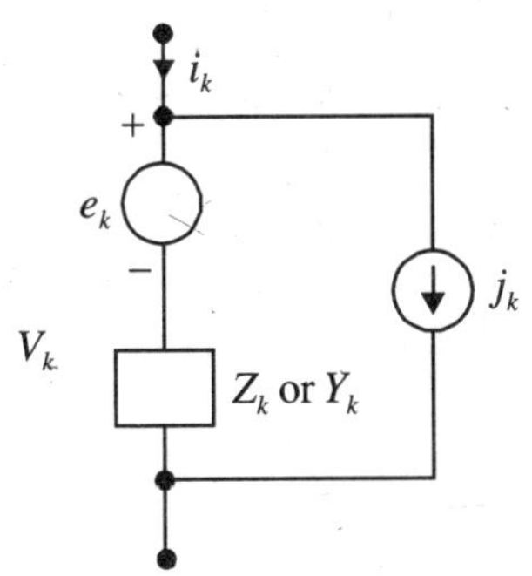

Fig. 3.2

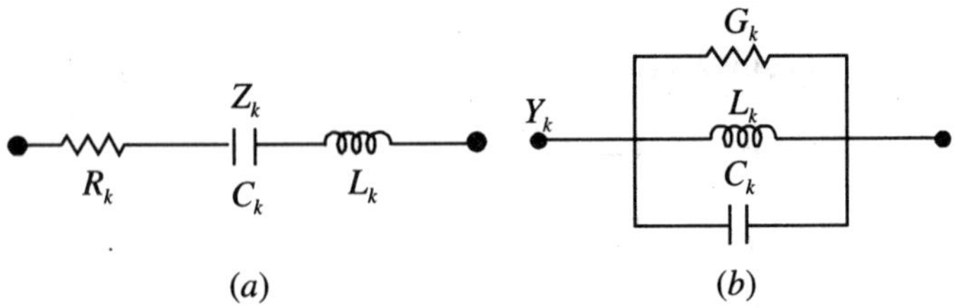

Fig. 3.3

combination of Fig. 3.3(*a*), each element is considered as a separate element while for writing the impedance for Fig. 3.3(*b*) each element is considered separately.) From Kirchhoff's laws v_k, the voltage across the edge k with an operational impedance, is

$$v_k(t) = e_k(t) + R_k\{i_k(t) - j_k(t)\} + L_k\frac{d}{dt}\{i_k(t) - j_k(t)\}$$

$$+\frac{1}{C_k}\int\{i_k(t) - j_k(t)\}dt \qquad \ldots(3.14)$$

The last three terms constitute the voltage drop across z_k, where $\{i_k(t) - j_k(t)\}$ is the current through z_k, and $i_k(t)$ is the edge current.

Equation 3.14 can be rewritten as

$$v_k(t) = e_k(t) + \left\{R_k i_k(t) + L_k\frac{d}{dt}i_k(t) + \frac{1}{C_k}\int i_k(t)\,dt\right\}$$

$$-\left\{R_k j_k(t) + L_k\frac{d}{dt}j_k(t) + \frac{1}{C_k}\int j_k(t)\,dt\right\} \qquad \ldots(3.15)$$

This integro-differential equation can be transformed to an algebraic equation by Laplace transformation (Appendix B). Consider the equation.

$$f(t) = R_k i_k(t) + L_k\frac{di_k(t)}{dt} + \frac{1}{C_k}\int i_k(t)\,dt$$

Taking the Laplace transform, we have

$$F(s) = R_k I_k(s) + L_k\{sI_k(s) - i_{l_k}(0)\} + \frac{1}{C_k}\left\{\frac{I_k(s)}{s} + \frac{q_{ck}(0)}{s}\right\}$$

where $i_{lk}(0)$ is the initial current through the inductor and $q_{ck}(0)$ the initial charge on the capacitor. Regrouping, we have

$$F(s) = \left(R_k + sL_k + \frac{1}{s}S_k\right)I_k(s) - L_k i_{lk}(0) + \frac{1}{s}v_{ck}(0)$$

where $1/C_k = S_k$ is the *elastance* of the capacitor and $v_{ck}(0)$ is the initial voltage across the capacitor. Further, this equation can be written as

$$F(s) = Z_k(s)I_k(s) - L_k i_{lk}(0) + \frac{1}{s}v_{ck}(0) \qquad ...(3.16)$$

where $Z_k(s) = R_k + sL_k + \frac{1}{s}S_k$ ...(3.17)

is the *impedance* of the edge k.

On carrying out the Laplace transformation of eqn. 3.15 and using eqns. 3.16 and 3.17, we have

$$V_k(s) = Z_k(s)I_k(s) + \left\{E_k(s) - Z_k(s)J_k(s) - L_k i_{lk}(0) + \frac{1}{s}V_{ck}(0)\right\} \qquad ...(3.18)$$

Equation 3.18 given the terminal relationship of a generalized element in Laplace domain. One such equation can be written for every element of the network, considering every element to be a generalized element. Combining all these elements we can write the overall terminal relationship in matrix form as:

$$\mathbf{V}_e(\mathbf{s}) = Z_e(s)\,\mathbf{I}_e(\mathbf{s}) + \mathbf{E}_q(\mathbf{s}) \qquad ...(3.19)$$

where $Z_e(s) = R_e + sL_e + \frac{1}{s}S_e$, ...(3.20)

and $\mathbf{E}_q(s) = \mathbf{E}_e(s) - Z_e(s)\mathbf{J}_e(s) - L_e\mathbf{i}_{ie}(0) + \frac{1}{s}\mathbf{V}_{ce}(0)$. ...(3.21)

$\mathbf{V}_e(s)$ is the edge voltage vector, $\mathbf{I}_e(s)$ the edge current vector, $\mathbf{E}_e(s)$ the voltage source vector, $\mathbf{J}_e(s)$ the current source vector, $\mathbf{i}_{le}(0)$ the initial current vector and $\mathbf{V}_{ce}(0)$ is the initial voltage vector defined by

$$\mathbf{V}_e(s) = \begin{bmatrix} V_1(s) \\ V_2(s) \\ \cdot \\ \cdot \\ \cdot \\ V_e(s) \end{bmatrix}, \mathbf{I}_e(s) = \begin{bmatrix} I_1(s) \\ I_2(s) \\ \cdot \\ \cdot \\ \cdot \\ I_e(s) \end{bmatrix}, \mathbf{E}_e(s) = \begin{bmatrix} E_1(s) \\ E_2(s) \\ \cdot \\ \cdot \\ \cdot \\ E_e(s) \end{bmatrix} \qquad ...(3.22)$$

$$\mathbf{J}_e(s) = \begin{bmatrix} J_1(s) \\ J_2(s) \\ \cdot \\ \cdot \\ \cdot \\ J_e(s) \end{bmatrix}, \mathbf{i}_{le}(0) = \begin{bmatrix} i_{l1}(0) \\ i_{l2}(0) \\ \cdot \\ \cdot \\ \cdot \\ i_{le}(0) \end{bmatrix}, \mathbf{V}_{ce}(0) = \begin{bmatrix} V_{c1}(0) \\ V_{c2}(0) \\ \cdot \\ \cdot \\ \cdot \\ V_{ce}(0) \end{bmatrix} \qquad ...(3.22a)$$

In eqn. 3.20, $Z_e(s)$ is the edge impedance matrix, R_e is the edge resistance matrix, L_e the edge inductance matrix and S_e the edge elastance matrix. The last three matrices are defined by

$$R_e = \text{diag } [R_1, R_2, ...]$$
$$L_e = \text{diag } [L_1, L_2, ...]$$
$$S_e = \text{diag } [S_1, S_2, ...]$$

In eqn. 3.19, $\mathbf{E}_q(\mathbf{s})$ can be thought of as the equivalent voltage source vector.

In an analogous manner, an equation similar to eqn. 3.20 can be developed when the element is described by an operational admittance. From Kirchhoff's laws i_k, the current through the generalized element is

$$i_k(t) = j_k(t) + G_k\{v_k(t) - e_k(t)\} + C_k\frac{d}{dt}\{v_k(t) - e_k(t)\} + \frac{1}{L_k}\int\{v_k(t) - e_k(t)\}dt$$

Regrouping, we have

$$i_k(t) = j_k(t) + \left\{G_k v_k(t) + C_k\frac{d}{dt}v_k(t) + \frac{1}{L_k}\int v_k(t)dt\right\}$$
$$-\left\{G_k e_k(t) + C_k\frac{d}{dt}e_k(t) + \frac{1}{L_k}\int v_k(t)dt\right\}$$

Carrying out the *Laplace transformation* and rearranging, we have

$$I_k(s) = Y_k(s) + V_k(s) + \left\{J_k(s) - Y_k(s)E_k(s) - C_k V_{ck}(0) + \frac{1}{s}i_{lk}(0)\right\} \quad ...(3.23)$$

where $$Y_k(s) = G_k + sC_k + \frac{1}{s}\Gamma_k \text{ and } \Gamma_k = \frac{1}{L_k}.$$

If we write one such equation for every element of the network, the overall equation can be written as:

$$\mathbf{I}_e(s) = Y_e(s)\,\mathbf{V}_e(s) + \mathbf{J}_q(s) \quad ...(3.24)$$

where $$Y_e(s) = G_e + sC_e + \frac{1}{s}\Gamma_e \quad ...(3.25)$$

and $$G_e = \text{diag } (G_1, R_2, ...)$$
$$C_e = \text{diag } (C_1, C_2, ...)$$
$$\Gamma_e = \text{diag } (\Gamma_1, \Gamma_2, ...).$$

Here

$$\mathbf{J}_q(s) = \mathbf{J}_e(s) - Y_e(s)\mathbf{E}_e(s) - C_e\mathbf{V}_{ce}(0) + \frac{1}{s}\mathbf{i}_{le}(0) \quad ...(3.26)$$

can be thought of as the equivalent current source vector.

Equations 3.19 and 3.24 are the required terminal relationships of the generalized elements of the network. We have not included mutual inductances in the network as yet. If we do include mutual inductances, then the matrices L_e and Γ_e will be no longer diagonal (see Sec. 3.6).

3.3 SOURCE TRANSFORMATION

It is often convenient if the network has only current sources or voltage sources. If both types of sources are present we can apply source transformation and arrive at a network containing only one type of source.

Consider the network (or an element of a network) shown in Fig. 3.4 which has a voltage source in series with an impedance. The voltage equation for this is

$$V(s) = V_g(s) - I(s)\,Z(s)$$

where $Z(s)$ is the impedance and $I(s)$ is the current. We can rewrite this equation as

$$\frac{V(s)}{Z(s)} = \frac{V_g(s)}{Z(s)} - I(s)$$

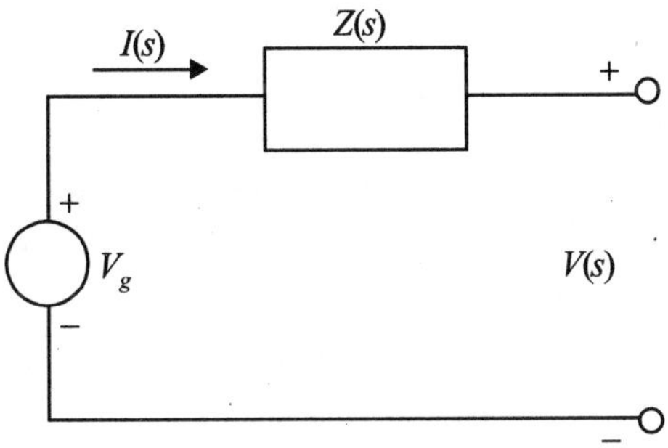

Fig. 3.4

i.e. $$V(s)\,Y(s) = V_g(s)/Z(s) - I(s)$$

Let $$I_g(s) = \frac{V_g(s)}{Z(s)}$$

than we have

$$V(s)\,Y(s) = I_g(s) - I(s)$$

This is the current equation for the network shown in Fig. 3.5. This is the *equivalent current source model* of the network in Fig. 3.4. Similarly, the first

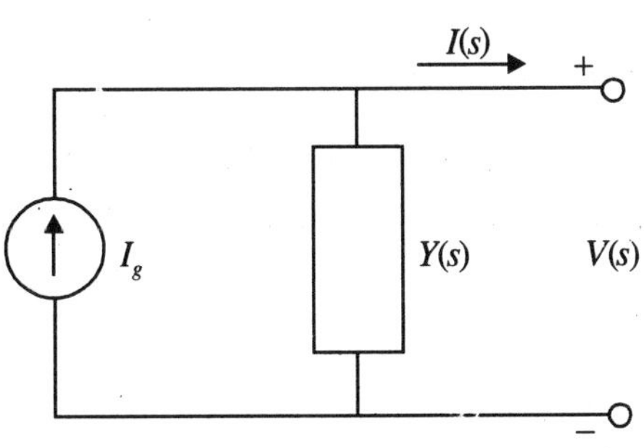

Fig. 3.5

network is the *equivalent voltage source model* of the second network. (The equivalent voltage source model is referred to as the *Thevenin equivalent* and the equivalent current source model the *Norton equivalent.* We shall discuss Thevenin's and Norton's equivalents in Chapter 5.) Hence, we see that a voltage (current) source in series (parallel) with an impedance (admittance) can be converted to a current (voltage) source in parallel (series) with an admittance (impedance). If, however, there exists a voltage source without a series impedance or a current source connected across several branches, the source transformation cannot be directly applied. In defining a generalized element we have assumed that every element has a voltage source in series and a current source in parallel with it. In Figs. 3.6(*a*) and 3.7(*a*), however, this is not the case. The *E*-shift and *I*-shift are employed to circumvent such situations.

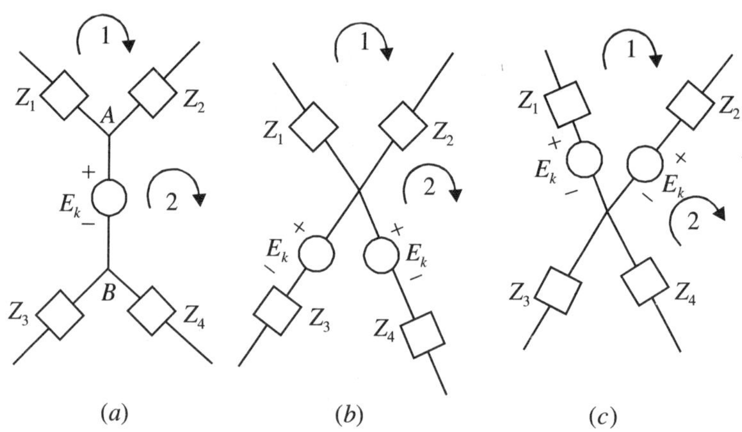

Fig. 3.6

Consider the situation shown in Fig. 3.6(*a*). We claim that we can use either of the networks in Fig. 3.6(*b*) or 3.6(*c*) as these are equivalent to (*a*). What we have done is shorted the nodes *A* and *B* and pushed the voltage source through the node into the edges as *A* or *B*. It is easy to show that the networks (*a*), (*b*) and (*c*) are equivalent insofar as the KVL are concerned. This method is called *E-shift.* The method of *I-shift* is demonstrated in Fig. 3.7. There is no element in parallel with the current source. We have introduced additional identical current sources as shown in (*b*) and (*c*) after removing the original source. It is easy to show that these are equivalent insofar as KCL is concerned.

3.4 FORMULATION OF NETWORK EQUATIONS

The generalized terminal relation equations obtained in Sec. 3.2 are combined with Kirchhoff's laws and transformations obtained in Sec. 3.1 to yield the final form of equations. We consider four methods with different sets of equations.

3.4.1 Node Equations

Node equation method is convenient for networks containing no voltage sources. If there are voltage sources in the network, they can be converted to equivalent

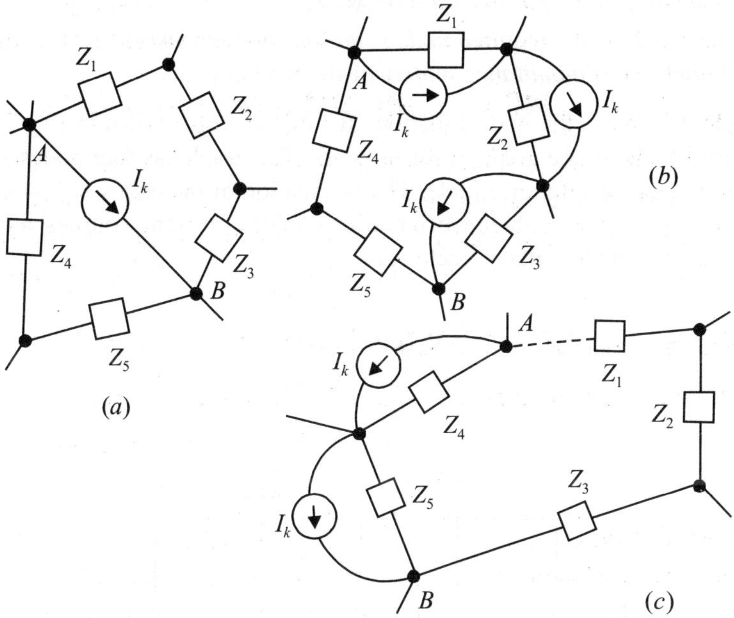

Fig. 3.7

current sources and the nodal equations can be easily written. In eqn. 3.23, the term $Y_k(s)\ E_k(s)$ is the equivalent current source corresponding to the voltage source $E_k(s)$. The terms $C_k\ V_{ck}\,(0), \frac{1}{s}\cdot i_{lk}\,(0)$ are the equivalent current sources due to the initial conditions in the network. For this reason the term $\mathbf{J}_q(\mathbf{s})$ in eqn. 3.24 is referred to as the 'equivalent current source' vector.

The starting point for writing the node equations is to write the terminal relations for the elements of the network in the form of eqn. 3.24, which is

$$\mathbf{I}_e(s) = Y_e(s)\ \mathbf{V}_e(s) + \mathbf{J}_q(s)$$

From Kirchhoff's current law, we have

$$A\mathbf{I}_e(s) = 0$$

where A is the reduced incidence matrix of the graph. Hence,

$$A\mathbf{I}_e(s) = Ay_e(s)\ \mathbf{V}_e(s) + A\mathbf{J}_q(s) = 0$$

i.e. $$AY_e(s)\ \mathbf{V}_e(s) = -A\mathbf{J}_q(s)$$

But from node-transformation, we have

$$\mathbf{V}_e(s) = A^t\mathbf{V}_n(s) \qquad \ldots(3.7)$$

where $\mathbf{V}_n(s)$ is a set of node voltages with respect to a pre-selected reference node. The last two equations together yield

$$AY_e\ (s)\ A^t\ \mathbf{V}_n(s) = -A\mathbf{J}_q(s)$$

or $$Y_n(s)\ \mathbf{V}_n(s) = -A\mathbf{J}_q\ (s) \qquad \ldots(3.27)$$

where $$Y_n(s) = AY_e(s)A^t \qquad \ldots(3.28)$$

Equation 3.27 is the required *node equation* for the network. $Y_n(s)$ is called the *node (datum-node) admittance matrix* of the network.

Example 3.2. Write the node equations for the network shown in Fig. 3.8.

Figure 3.9 shows the graph of the network. The graph has four nodes and node 4 is selected as the reference node. The orientation of the edges 1, 2, 4 and 5 are selected to coincide with the orientations of the current sources and initial conditions. The node voltage vector is

$$\mathbf{V}_n(\mathbf{s}) = \begin{bmatrix} V_1(s) \\ V_2(s) \\ V_3(s) \end{bmatrix}$$

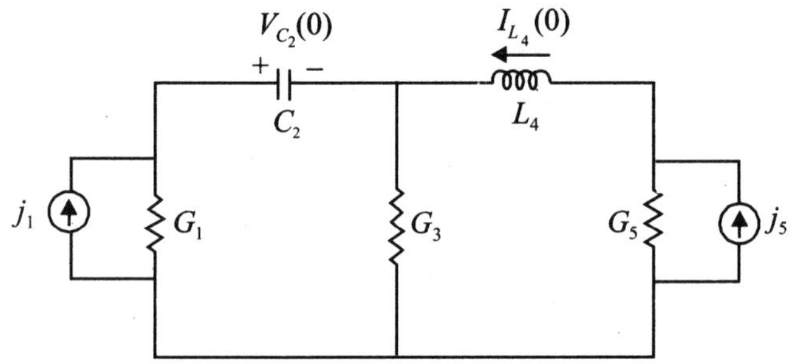

Fig. 3.8

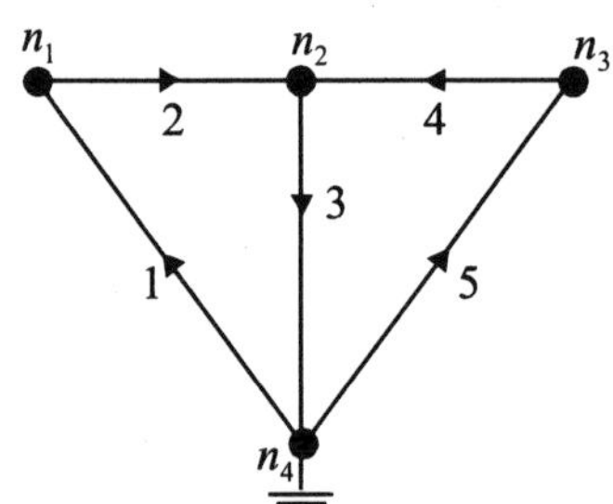

Fig. 3.9

where $V_i(s)$ is the voltage of the node n_i, with respect to the reference node.

The incidence matrix for this graph is

$$A = \begin{matrix} \\ n_1 \\ n_2 \\ n_3 \end{matrix} \begin{matrix} \begin{matrix} 1 & 2 & 3 & 4 & 5 \end{matrix} \\ \begin{bmatrix} -1 & 1 & 0 & 0 & 0 \\ 0 & -1 & 1 & -1 & 0 \\ 0 & 0 & 0 & 1 & -1 \end{bmatrix} \end{matrix}$$

The edge admittance matrix is

$$Y_e(s) = \begin{bmatrix} G_1 & 0 & 0 & 0 & 0 \\ 0 & 0 & 0 & 0 & 0 \\ 0 & 0 & G_3 & 0 & 0 \\ 0 & 0 & 0 & 0 & 0 \\ 0 & 0 & 0 & 0 & G_5 \end{bmatrix} + s \begin{bmatrix} 0 & 0 & 0 & 0 & 0 \\ 0 & C_2 & 0 & 0 & 0 \\ 0 & 0 & 0 & 0 & 0 \\ 0 & 0 & 0 & 0 & 0 \\ 0 & 0 & 0 & 0 & 0 \end{bmatrix}$$

$$+ \frac{1}{s} \begin{bmatrix} 0 & 0 & 0 & 0 & 0 \\ 0 & 0 & 0 & 0 & 0 \\ 0 & 0 & 0 & 0 & 0 \\ 0 & 0 & 0 & \Gamma_4 & 0 \\ 0 & 0 & 0 & 0 & 0 \end{bmatrix}$$

i.e.

$$Y_e(s) = \begin{bmatrix} G_1 & 0 & 0 & 0 & 0 \\ 0 & sC_2 & 0 & 0 & 0 \\ 0 & 0 & G_3 & 0 & 0 \\ 0 & 0 & 0 & \frac{\Gamma_4}{s} & 0 \\ 0 & 0 & 0 & 0 & G_5 \end{bmatrix}$$

The equivalent current source vector is

$$\mathbf{J}_q(\mathbf{s}) = \begin{bmatrix} J_1(s) \\ 0 \\ 0 \\ 0 \\ J_5(s) \end{bmatrix} - \begin{bmatrix} 0 & 0 & 0 & 0 & 0 \\ 0 & C_2 & 0 & 0 & 0 \\ 0 & 0 & 0 & 0 & 0 \\ 0 & 0 & 0 & 0 & 0 \\ 0 & 0 & 0 & 0 & 0 \end{bmatrix} \begin{bmatrix} 0 \\ V_{C2}(0) \\ 0 \\ 0 \\ 0 \end{bmatrix}$$

$$+ \frac{1}{s} \begin{bmatrix} 0 \\ 0 \\ 0 \\ i_{l4}(0) \\ 0 \end{bmatrix} \quad i.e., \quad \mathbf{J}_q(\mathbf{s}) = \begin{bmatrix} J_1(s) \\ -C_2 V_{C2}(0) \\ 0 \\ \frac{i_{l4}(0)}{s} \\ J_s(s) \end{bmatrix}$$

The terminal relationship of the elements is (eqn. 3.24)

$$\begin{bmatrix} I_1(s) \\ I_2(s) \\ I_3(s) \\ I_4(s) \\ I_5(s) \end{bmatrix} = \begin{bmatrix} G_1 & 0 & 0 & 0 & 0 \\ 0 & sC_2 & 0 & 0 & 0 \\ 0 & 0 & G_3 & 0 & 0 \\ 0 & 0 & 0 & \dfrac{\Gamma_4}{s} & 0 \\ 0 & 0 & 0 & 0 & C_5 \end{bmatrix}$$

$$\times \begin{bmatrix} V_1(s) \\ V_2(s) \\ V_3(s) \\ V_4(s) \\ V_5(s) \end{bmatrix} + \begin{bmatrix} J_1(s) \\ -C_2 V_{C2}(0) \\ 0 \\ \dfrac{i_{l4}(0)}{s} \\ J_s(s) \end{bmatrix}$$

The nodal admittance matrix (eqn. 3.28) is

$$AY_eA^t = \begin{bmatrix} G_1 + sC_2 & -sC_2 & 0 \\ -sC_2 & G_3 + sC_3 + \dfrac{\Gamma_4}{s} & -\dfrac{\Gamma_4}{s} \\ 0 & -\dfrac{\Gamma_4}{s} & G_5 + \dfrac{\Gamma_4}{s} \end{bmatrix}$$

and

$$-AJ_q(s) = \begin{bmatrix} J_1(s) + C_2 V_{C_2}(0) \\ -C_2 V_{C_2}(0) + \dfrac{1}{s} i_{l4}(0) \\ J_5(s) - \dfrac{1}{s} i_{l4}(0) \end{bmatrix}$$

The node equation for the network of Fig. 3.8 is $A\, Y_e\, A^t\, \mathbf{V}_n = -A\mathbf{J}_q$.

It can be observed from these equations that the diagonal elements of $Y_n(s)$ are always positive and the off-diagonal elements, if they exist, are negative. It is not necessary to go through the step-by-step formulation of the node-equations. They can be written by a mere inspection of the network (Sec. 3.5).

3.4.2 Mesh Equation

Mesh equations method is applicable to networks with planar graphs. Also, it is convenient if the network does not contain any current sources. In the eqn. 3.18 the term $Z_k(s)\, J_k(s)$ is the equivalent voltage source corresponding to the current source $J_k(s)$. The terms $L_k i_{lk}(0)$, $V_{ck}(0)/s$ are the equivalent voltage sources due to the initial conditions in the network. This is the reason that the term $\mathbf{E}_q(\mathbf{s})$ in the Eqn. 3.19 is referred to as the "equivalent voltage source" vector.

To formulate the mesh equations the terminal relationship for the elements of the network are written in the form of eqn. 3.19 which is

$$\mathbf{V}_e(s) = Z_e\,(s)\,\mathbf{I}_e(s) + \mathbf{E}_q(s)$$

From Kirchhoff's voltage law, we have

$$B_m\,\mathbf{V}_e(s) = 0$$

where B_m is the mesh matrix. (The orientations of all the meshes are chosen either clockwise or anti-clockwise.)

Hence

$$B_m\,\mathbf{V}_e(s) = B_m\,Z_e\,(s)\,\mathbf{I}_e(s) + B_m\,\mathbf{E}_q(s) = 0$$

i.e.

$$B_m\,\mathbf{V}_e(s)\,\mathbf{I}_e(s) = -B_m\,\mathbf{E}_q(s)$$

But, from mesh transformation we have

$$\mathbf{I}_e(s) = B_m^t\,\mathbf{I}_m(s) \qquad \text{...(3.10)}$$

where $\mathbf{I}_m(s)$ is the set of mesh currents. The last two equations together yield

$$B_m\,Z_e\,(s)\,B_m^t\,\mathbf{I}_m\,(s) = -B_m\,\mathbf{E}_q(s)$$

or

$$Z_m\,(s)\,\mathbf{I}_m\,(s) = -B_m\,\mathbf{E}_q(s) \qquad \text{...(3.29)}$$

where $Z_m(s) = B_m\,Z_e\,(s)\,B_m^t$ is called the *mesh impedance matrix.* Equation (3.29) is the required *mesh equation.*

Example 3.3. Write the mesh equations for the network shown in Fig. 3.10.

Figure 3.11 shows the graph of the network. There are three meshes and the mesh currents are marked on the graph. The orientation of the edges are selected to coincide with the orientation of the voltage sources and initial conditions. The mesh current vector is

$$\mathbf{I}_m(\mathbf{s}) = \begin{bmatrix} I_{m_1}(s) \\ I_{m_2}(s) \\ I_{m_3}(s) \end{bmatrix}$$

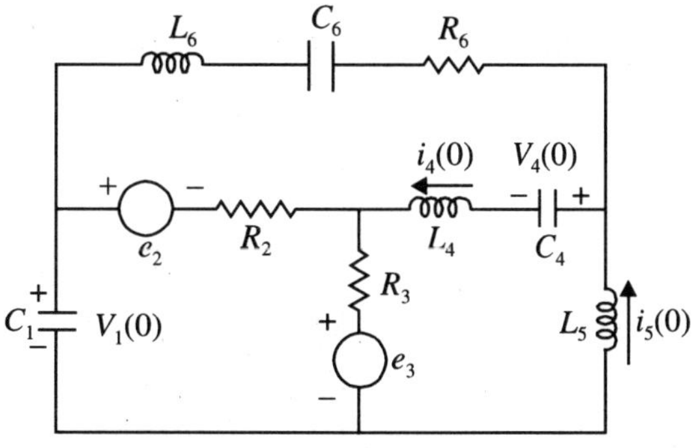

Fig. 3.10

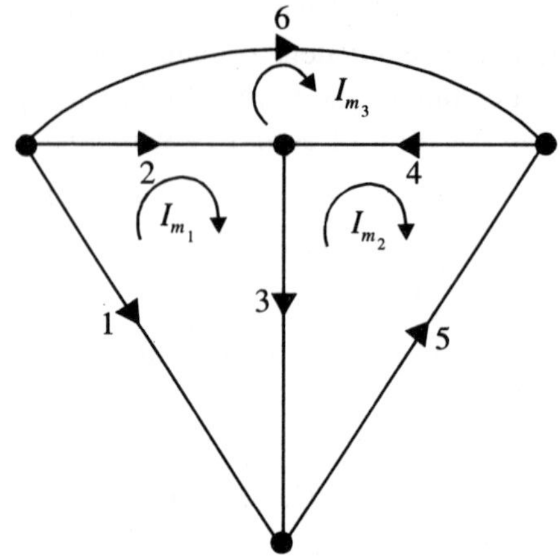

Fig. 3.11

The mesh matrix for this graph is

$$B_m = \begin{matrix} \\ m_1 \\ m_2 \\ m_3 \end{matrix} \begin{matrix} \begin{matrix} 1 & 2 & 3 & 4 & 5 & 6 \end{matrix} \\ \begin{bmatrix} -1 & 1 & 1 & 0 & 0 & 0 \\ 0 & 0 & -1 & -1 & -1 & 0 \\ 0 & -1 & 0 & 1 & 0 & 1 \end{bmatrix} \end{matrix}$$

The edge impedance matrix is

$$z_e(s) = \begin{bmatrix} \frac{S_1}{s} & 0 & 0 & 0 & 0 & 0 \\ 0 & R_2 & 0 & 0 & 0 & 0 \\ 0 & 0 & R_3 & 0 & 0 & 0 \\ 0 & 0 & 0 & sL_4 + \frac{S_4}{s} & 0 & 0 \\ 0 & 0 & 0 & 0 & sL_5 & 0 \\ 0 & 0 & 0 & 0 & 0 & R_6 + sL_6 + \frac{S_6}{s} \end{bmatrix}$$

and the equivalent voltage source vector is

$$\mathbf{E}_q(\mathbf{s}) = \begin{bmatrix} \frac{1}{s} V_{c1}(0) \\ E_2(s) \\ E_3(s) \\ \frac{1}{s} V_{c4}(0) - L_4 i_{l4}(0) \\ -L_5 i_{l5}(0) \\ 0 \end{bmatrix}$$

The mesh impedance matrix is

$$Z_m(s) = B_m\, Z_e\,(s)\, B_m^t$$

$$= \begin{bmatrix} R_2 + R_3 + \dfrac{S_1}{s} & -R_3 & -R_2 \\ -R_3 & R_3 + s\left(L_4 + L_5\right) + \dfrac{S_4}{s} & -\left(sL_4 + \dfrac{S_4}{s}\right) \\ -R_3 & -\left(sL_4 + \dfrac{S_4}{s}\right) & R_2 + R_6 + s\left(L_4 + L_6\right) + \dfrac{1}{s}\left(S_4 + S_6\right) \end{bmatrix}$$

and

$$-B_m \mathbf{E}_q(s) = \begin{bmatrix} \dfrac{1}{s} V_{c1}(0) - E_2(s) - E_3(s) \\ E_3(s) + \dfrac{1}{s} V_{c4}(0) - L_4 i_{l4}(0) - L_5 i_{l5}(0) \\ E_2(s) + L_4 i_{l4}(0) - \dfrac{1}{s} V_{c4}(0) \end{bmatrix}$$

The mesh equation for the network of Fig. 3.10 is

$$B_m\, Z_e B_m^t\, \mathbf{I}_m = -\, B_m\, \mathbf{E}_q$$

All the diagonal elements of $Z_m(s)$ for this example are positive and off-diagonal elements are negative. The mesh equations can also be written by inspection of the network (Sec. 3.5).

3.4.3 Cut-Set Equations

In the node equations method a set of node voltages were selected as the set of independent voltages. If, instead, we select a set of branch voltages with respect to a tree as the set of independent voltages we get *cut-set equations*.

The terminal relationship of elements is written in the form suitable for node equations.

i.e. $$\mathbf{I}_e(s) = Y_e\,(s)\,\mathbf{V}_e(s) + \mathbf{J}_q(s) \qquad \ldots(3.24)$$

By invoking the KCL in the form

$$Q_f\, \mathbf{I}_e(\mathbf{s}) = 0$$

we have

$$Q_f\, Y_e\,(s)\,\mathbf{V}_e(s) = -Q_f\, \mathbf{J}_q(s)$$

But from branch transformation, we have

$$\mathbf{V}_e(s) = Q_f^t\, \mathbf{V}_b(s) \qquad \ldots(3.6)$$

and finally

$$Q_f\, Y_e\,(s)\, Q_f^t\, \mathbf{V}_b(s) = -Q_f\, \mathbf{J}_q(s)$$

The cut-set equations then are

$$Y_e(s)\,\mathbf{V}_b(s) = -Q_f\,\mathbf{J}_q(s) \qquad \ldots(3.30)$$

where $\mathbf{V}_b(s)$ is a set of branch voltages and $Y_e(s) = Q_f\,Y_e(s)\,Q_f^t$ is the *cut-set admittance matrix.*

It can be easily seen that if the tree is selected in such a way that all the branches converge towards a particular node (Fig. 3.12), then the cut-set equations are identical to the node equations with that specific node as the reference node. (Such a tree is called a *Lagrangian tree or star tree*).

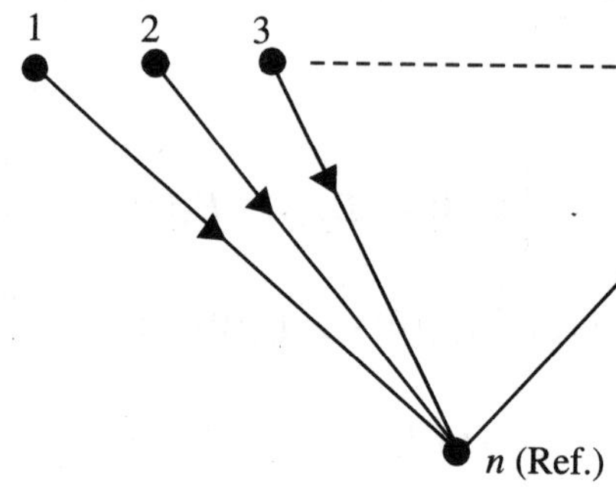

Fig. 3.12

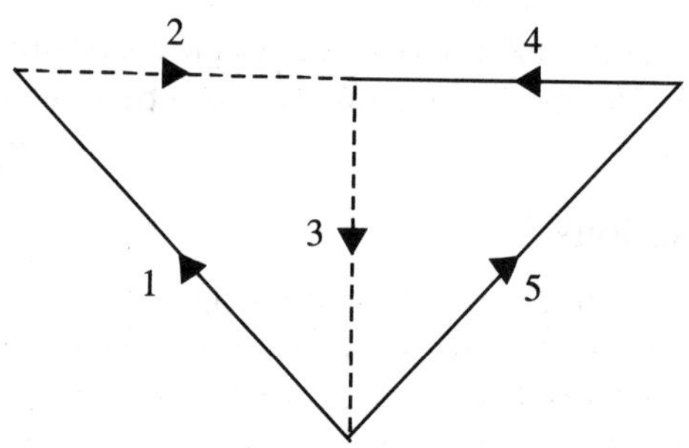

Fig. 3.13

Example 3.4. Write a set of cut-set equations for the network of Fig. 3.8.

Figure 3.13 shows the graph of the network. Edges 1, 5 and 4 constitute the tree (solid lines) and edges 2, 3 the co-tree (broken lines). The branch voltage vector is

$$\mathbf{V}_b = \begin{bmatrix} V_1(s) \\ V_4(s) \\ V_5(s) \end{bmatrix}$$

The f-cut-set matrix is

$$Q_f = \begin{matrix} \\ 1 \\ 4 \\ 5 \end{matrix} \begin{matrix} \begin{matrix} 1 & 2 & 3 & 4 & 5 \end{matrix} \\ \begin{bmatrix} 1 & -1 & 0 & 0 & 0 \\ 0 & 1 & -1 & 1 & 0 \\ 0 & 1 & -1 & 0 & 1 \end{bmatrix} \end{matrix}$$

Here we have not arranged the columns of Q_f in the order of the chords and the branches.

The terminal relationships of the elements are the same as those of example 3.1. Hence, the cut-set admittance matrix is

$$Y_c(s) = Q_f Y_e(s) Q^f = \begin{bmatrix} G_1 + sC_2 & -sC_2 & -sC_2 \\ -sC_2 & G_3 + sC_2 + \dfrac{\Gamma_4}{s} & sC_2 + G_3 \\ -sC_2 & sC_2 + G_3 & G_3 + G_5 + sC_2 \end{bmatrix}$$

and

$$-Q_f \mathbf{J}_q(s) = \begin{bmatrix} -J_1(s) - C_2 V_{c2}(0) \\ C_2 V_{c2}(0) - \dfrac{i_{l4}(0)}{s} \\ C_2 V_{c2}(0) - J_5(s) \end{bmatrix}$$

The cut-set equations are

$$Q_f\, Y_e\, Q_f^t\, \mathbf{V}_b = -Q_f\, \mathbf{J}_q$$

The diagonal elements of $Y_e(s)$ are always positive, but the off-diagonal elements may be positive or negative. The cut-set equations can also be written by inspection (Sec. 3.5).

3.4.4 Loop Equations

If the graph of a network is non-planar, we cannot write mesh equations for the network. But we can always write a set of equations if we consider a set of f-circuits instead of meshes. Such equations are generally called *loop equations.*

The terminal relationship of elements are written in the form of equation 3.19, *i.e.*

$$\mathbf{V}_e(s) = Z_e\,(s)\,\mathbf{I}_e(s) + \mathbf{E}_q(s)$$

The KVL for a set of f-circuits can be written as

$$B_f\,\mathbf{V}_e(s) = 0$$

Hence, we have

$$B_f\,Z_e\,(s)\,\mathbf{I}_e(s) = -B_f\,\mathbf{E}_q(s)$$

But chord transformation yields

$$\mathbf{I}_e(s) = B_f^t\,\mathbf{I}_c(s) \qquad \text{...(3.9)}$$

and finally

$$B_f Z_e(s) B_f^t \mathbf{I}_e(s) = -B_f \mathbf{E}_q(s)$$

The loop equations then are

$$Z_l(s) \mathbf{I}_c(s) = -B_f \mathbf{E}_q(s) \qquad \ldots(3.31)$$

where $\mathbf{I}_c(s)$ is a set of chord currents and $Z_l(s) = B_f Z_e(s) B_f^t$ is the *loop impedance matrix.*

Example 3.5. Write a set of loop equations for the network of Fig. 3.10.

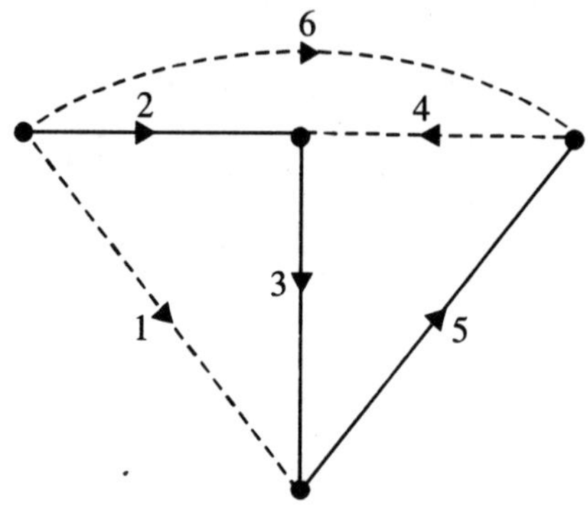

Fig. 3.14

Figure 3.14 shows the graph of the network. Edges 2, 3 and 5 constitute a tree (solid lines) and the edges 1, 4 and 6 the co-tree (broken lines). The chord current vector is

$$\mathbf{I}_e(s) = \begin{bmatrix} I_1(s) \\ I_4(s) \\ I_6(s) \end{bmatrix}$$

The f-circuit matrix is

$$B_f = \begin{array}{c} \\ 1 \\ 4 \\ 6 \end{array} \begin{array}{c} \begin{array}{cccccc} 1 & 2 & 3 & 4 & 5 & 6 \end{array} \\ \begin{bmatrix} 1 & -1 & -1 & 0 & 0 & 0 \\ 0 & 0 & 1 & 1 & 1 & 0 \\ 0 & -1 & -1 & 0 & -1 & 1 \end{bmatrix} \end{array}$$

It may be noted here that we have not arranged the columns of B_f in the order of the chords and the branches. The terminal relationships of the element are the same as those of example 3.2. Hence, the loop impedance matrix is

$$Z_l(s) = B_f Z_e(s) B_f^t$$

$$= \begin{bmatrix} R_2 + R_3 + \frac{S_1}{s} - R_3 & R_2 + R_3 & \\ -R_3 & R_3 + s(L_4 + L_5) + \frac{S_4}{s} & -(R_3 + sL_5) \\ R_2 + R_3 & -(R_3 + sL_5) & R_2 + R_3 + R_6 + s(L_5 + L_6) + \frac{S_6}{s} \end{bmatrix}.$$

and
$$-B_f \mathbf{E}_q(s) = \begin{bmatrix} E_2(s) + E_3(s) - \frac{1}{s} V_{c1}(0) \\ L_4 i_{l4}(0) + L_5 i_{l5}(0) - E_3(s) - \frac{1}{s} V_{c4}(s) \\ E_2(s) + E_3(s) - L_5 i_{l5}(0) \end{bmatrix}$$

The loop equations are

$$B_f Z_c B_f^t \mathbf{I}_c = -B_f \mathbf{E}_q$$

The diagonal elements of $Z_l(s)$ are always positive, but the off-diagonal elements may be either be positive or negative. Loop equations can also be written by inspection (Sec. 3.5).

3.5 NETWORK EQUATIONS BY INSPECTION

In Sec. 3.4, we discussed methods of formulating the network equations. In forming these equations one has to write all the matrices involved and perform the matrix triple product. If the network does not contain any controlled sources or mutual inductance, it is easy to write these equations by a mere inspection of the network.

3.5.1 Node Equations

In writing the node equations the first task is to select a datum or reference node. Let the network have n-nodes and they be numbered 1, 2, 3…n – 1, and n. The node numbered n is selected as the reference node. In the sequel, we assume that the network does not have any voltage source. (If it has voltage source, a source transformation, Sec. 3.3, can be effected.) Then the node equations in Laplace domain are written by inspection as follows:

(*i*) The node voltage vector is given by

$$\mathbf{V}_n = [V_1 \ V_2 \ \ldots \ V_{n-1}]^t$$

These voltages are defined with respect to the reference node.

(*ii*) The diagonal entries of Y_n are the sum of the admittances at these particular nodes, *i.e.*, $Y_{n_{ii}}$ is the sum of all the admittances connected to node i. (The diagonal entries are all positive for a passive RLC network.)

(*iii*) The off-diagonal entries of Y_n are the negative of the admittances common to the two nodes, *i.e.*, $Y_{n_{ij}}$ is the negative of the sum of the admittances connected between nodes i and j. (Off-diagonal entries are all negative for a passive RLC network).

(*iv*) The ith row of $A_q\mathbf{J}_q(s)$ (right-hand side of eqn. 3.27) corresponds to the ith node equivalent current source. This has three components: current sources, initial inductor currents and currents due to initial capacitor voltages.

(*a*) The current source component is the algebraic sum of all the current sources connected to node i. The sign of the sources is positive if it is directed away from node i and negative if it is towards node i. (The current source term should be written in Laplace domain.)

(*b*) The initial inductor current component is the algebraic sum of the initial current of the inductors (the entries are of the form $i_k(0)/(s)$ connected to node i. The sign conventions are the same as those of the current sources.

(*c*) The initial capacitor voltage component is the algebraic sum of the currents due to initial capacitor voltages the entries are of the form $C_kV_k(0)$) of the capacitors connected to node i. The sign convention is opposite to that of the current source, *i.e.*, if the positive of the initial voltage is at node i, the entry is negative, otherwise, positive.

3.5.2 Loop Equations

The first task is to select a tree for the network graph. Let the chords of the tree be numbered 1, 2, ... n_c ($n_c = e - v + 1$). We assume that there are no current sources of the network (Sec. 3.3). The loop equations (eqn. 3.31), in Laplace domain, can be written by inspection as follows:

(*i*) The chord current vector is given by

$$\mathbf{I}_c = [I_1\ I_2\ \ldots\ I_{n_c}]^t$$

where I_i is the current through chord i.

(*ii*) The diagonal entries of Z_l are the sum of the impedances of the elements appearing in the *f*-circuit of the particular chord, *i.e.*, z_{lii} is the sum of the impedance of the elements appearing in *f*-circuit of chord i. The diagonal entries are all positive for a passive RLC network.

(*iii*) The off-diagonal entries of Z_l are the impedances common to the two *f*-circuits, *i.e.* z_{lij} is the algebraic sum of the impedances common to the *f*-circuits defined by the chords i and j. The sign of the impedance is determined by the orientations of the edge corresponding to this impedance in the two *f*-circuits. If the particular edge is positively or negatively oriented in both the *f*-circuits (*f*-circuit orientation is defined by the orientation of the defining chord), the sign of the impedance is positive; otherwise it is negative. Obviously the off-diagonal entries can be either positive or negative.

(*iv*) The ith row of $B_f\mathbf{E}_q(s)$ corresponds to the equivalent voltage source of the *f*-circuit of the chord i and this has three components.

(*a*) The voltage source component is the algebraic sum of all the voltage sources in the ith f-circuit. The sign of the voltage source is positive if its direction coincides with the orientation of the ith f-circuit; negative, otherwise.

(*b*) The initial capacitor voltage component is the algebraic sum of the initial voltages of the capacitors in the ith f-circuit (the entries are of the form $V_k(0)/s$). The sign conventions are the same as those of the voltage sources.

(*c*) The initial inductor current component is the algebraic sum of the voltages due to the initial inductor currents of the inductors in the ith f-circuit (the entries are of the form $L_k i_k(0)$). The sign convention is opposite to that of the voltage sources, *i.e.*, if the initial inductor current is oriented in opposition to that of the f-circuit, the entry is positive; otherwise, negative.

Similar rules can be formulated to write the mesh and cut-set equations by inspection (Problems 3.9 and 3.10). Table 3.1 summarizes the formulation of the four sets of equations.

3.6 MUTUAL INDUCTANCE

So far we have considered RLC networks without transformers or coupled coils. Figure 3.15 shows a pair of coupled coils. The voltage across coil 2 (coil 1) has two components, v_{22} and v_{21} (v_{11} and v_{12}). v_{22} (v_{11}) is the voltage induced in the coil 2 (coil 1) due to the current i_2 (i_1) flowing through coil 2 (coil 1). v_{21} (v_{12}) is the voltage induced in coil 2 (coil 1) due to current i_1 (i_2) flowing through coil 1 (coil 2). The latter is the voltage induced in a coil due to the flux produced by a current in another coil. This voltage is called the *mutually induced* voltage. We know that

$$v_{11} = L_1 \frac{di_1}{dt}$$

and

$$v_{22} = L_2 \frac{di_2}{dt}$$

where L_1 and L_2 are self inductances (or inductances) of coil 1 and coil 2, respectively. The mutually induced voltages v_{21} and v_{12} can be, similarly, written as

$$v_{21} = \frac{d\lambda_{21}}{dt}, \; v_{12} = \frac{d\lambda_{12}}{dt}$$

where λ_{21} (λ_{12}) is the flux linking coil 2 (coil 1) due to current i_1 (i_2) in coil 1 (coil 2). As in the case of self inductances we can write

$$\lambda_{21} = Mi_1 \text{ and } \lambda_{12} = Mi_2$$

where M is the *mutual inductance* between coil 1 and coil 2. M is always positive.

Table 3.1

Network Equation $MX = N$, Network graph has e edges and v vertices			
Node	*Cut-set*	*Mesh*	*Loop*
$X = \mathbf{V}_n$ – a set of node voltages	$X = \mathbf{V}_b$ – a set of branch voltages	$X = \mathbf{I}_m$ – a set of mesh currents	$X = \mathbf{I}_c$ – a set of chord currents
$M = Y_n = AY_eA^t$	$M = Y_c = Q_f Y_e Q_f^t$	$M = Z_m = B_m Z_e B_m^t$	$M = Z_1 = B_f Z_e B_f^t$
$N = -A\mathbf{J}_q$	$N = -Q_f \mathbf{J}_q$	$N = -B_m \mathbf{E}_q$	$N = -B_f \mathbf{E}_q$
$A \rightarrow$ Incidence matrix	$Q_f \rightarrow$ f-cut-set matrix	$B_m \rightarrow$ mesh matrix	$B_f \rightarrow$ f-circuit matrix
$Y_n = [y_{ij}]_{(v-1)\times(v-1)}$	$Y_n = [y_{ij}]_{(v-1)\times(v-1)}$	$Z_m = [y_{ij}]_{(e-v+1)\times(e-v+1)}$	$Z_i = [z_{ij}]_{(e-v+1)\times(e-v+1)}$
y_{ii} = sum of admittances at node i	y_{ii} = sum of admittances in f-cutset of branch i	z_{ii} = sum of impedances in mesh i	z_{ii} = sum of impedances in f-circuit of chord i
y_{ij} = –(admittance common to node i and node j)	y_{ij} = algebraic sum of admittance common to f-cutsets of branches i and j	z_{ij} = –(impedances common to mesh i and mesh j)	z_{ij} = algebraic sum of impedances common to f-circuits of chord t and j
Applicable to all networks	Applicable to all networks	Applicable to networks with planar graphs	Applicable to all networks

Y_e = Edge admittance matrix $\qquad Z_e$ = Edge impedance matrix

$\mathbf{J}_q$ = Equivalent current source vector $\qquad \mathbf{E}_q$ = Equivalent voltage source vector

Hence the overall terminal voltages of the two coils are

$$v_1 = v_{11} + v_{12} = L_1 \frac{di_1}{dt} + M \frac{di_2}{dt}$$

$$v_2 = v_{22} + v_{21} = L_2 \frac{di_2}{dt} + M \frac{di_1}{dt} \qquad \ldots(3.32)$$

The mutually induced voltage may aid or oppose the self induced voltage (voltage due to inductance) depending on the way the two coupled coils are wound. The polarity of the mutual voltage is identified by marking "dots", as shown in Fig. 3.15. If both the currents enter or leave the coil at the dots, as in Fig. 3.15, the mutual voltages aid the self voltages and the overall terminal voltage is given by the sum of the two voltages (eqn. 3.32). If one of the currents enters the coil at the dot and the other leaves the coil at the dot (Fig. 3.16), the mutual induced voltage opposes the self voltage and the overall voltage is given as follows:

$$v_1 = L_1 \frac{di_1}{dt} - M \frac{di_2}{dt}$$

$$v_1 = L_2 \frac{di_2}{dt} - M \frac{di_1}{dt} \qquad \ldots(3.33)$$

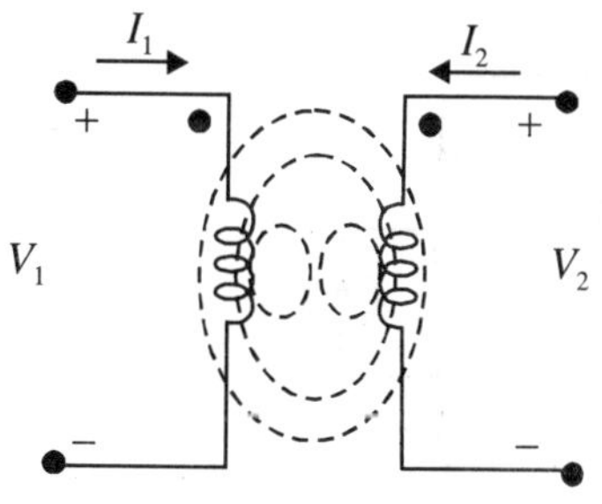

Fig. 3.15

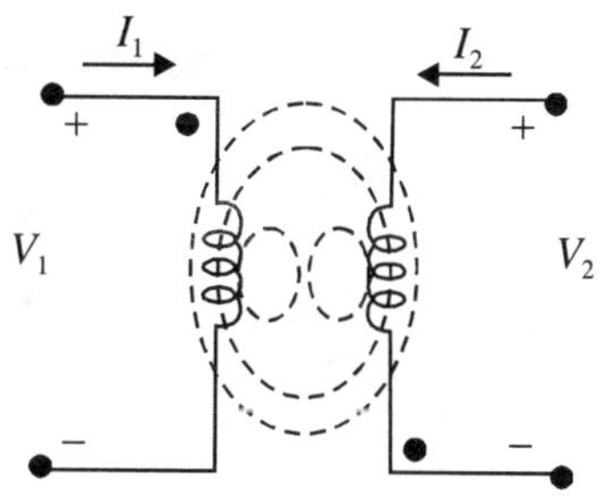

Fig. 3.16

Two coupled coils can also be characterized by the *coefficient of coupling* defined by

$$k = \frac{M}{\sqrt{L_1 L_2}} \qquad \ldots(3.34)$$

From physical considerations it can be shown that

$$0 \le k \le 1$$

$k = 0$ implies that $M = 0$ (no magnetic coupling between the coils), and $k = 1$ implies perfect coupling. In the latter case all the flux generated in one coil also links the other coil and vice versa. In a sense the coefficient of coupling is a measure of the amount of magnetic coupling between the coils.

The presence of mutual inductances makes the inductance matrix nondiagonal

i.e. $$L_e = \begin{bmatrix} L_1 & \pm M \\ \pm M & L_2 \end{bmatrix}$$

The inverse inductance matrix is then given by

$$\Gamma_e = \begin{bmatrix} \Gamma_1 & \mp\Gamma_{12} \\ \mp\Gamma_{12} & \Gamma_2 \end{bmatrix} \qquad \ldots(3.35)$$

where $$\Gamma_1 = \frac{L_2}{L_1 L_2 - M^2}, \quad \Gamma_2 = \frac{L_2}{L_1 L_2 - M^2}$$

and $$\Gamma_{12} = \frac{-M}{L_1 L_2 - M^2}$$

It can be noticed from eqn. 3.35 that the inverse inductance matrix does not exist for a pair of perfectly coupled coils, *i.e.* when $k = 1$ or $M = \sqrt{L_1 L_2}$. From this we can conclude that node and cutset equations do not exist for networks with perfectly coupled coils.

Example 3.6. Write the mesh equations for the network shown in Fig. 3.17.

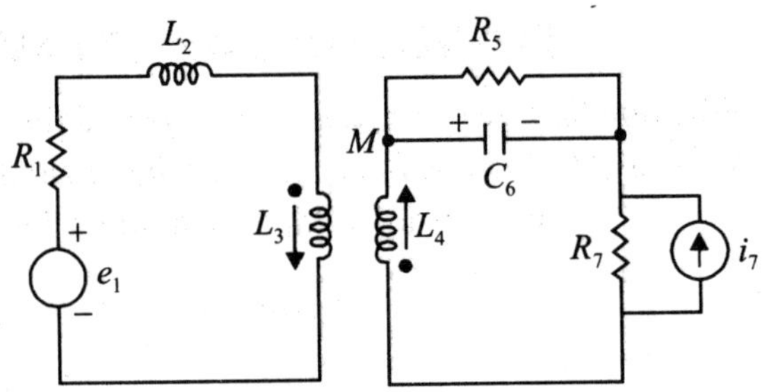

Fig. 3.17

Figure 3.18 shows the graph of the network. The three mesh currents are indicated on the graph. The mesh current vector is

$$\mathbf{I}_m(s) = \begin{bmatrix} I_1(s) \\ I_2(s) \\ I_3(s) \end{bmatrix}$$

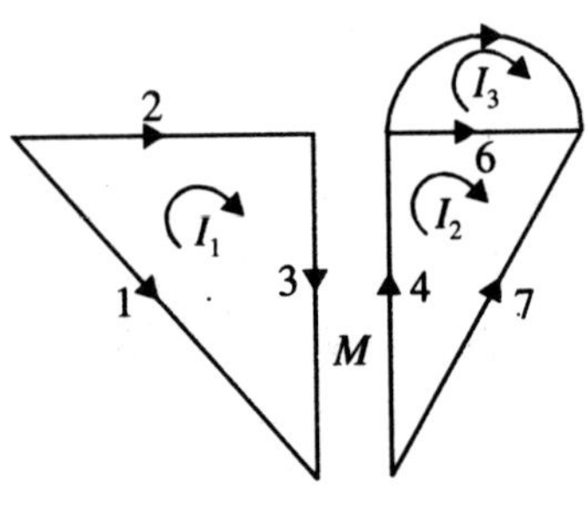

Fig. 3.18

The edge impedance matrix is

$$Z_r = \begin{bmatrix} R_1 & 0 & 0 & 0 & 0 & 0 & 0 \\ 0 & sL_2 & 0 & 0 & 0 & 0 & 0 \\ 0 & 0 & sL_3 & sM & 0 & 0 & 0 \\ 0 & 0 & sM & sL_4 & 0 & 0 & 0 \\ 0 & 0 & 0 & 0 & R_5 & 0 & 0 \\ 0 & 0 & 0 & 0 & 0 & \frac{S_6}{s} & 0 \\ 0 & 0 & 0 & 0 & 0 & 0 & R_7 \end{bmatrix}$$

The edges 3 and 4 are coupled by M. The mesh matrix is

$$R_m = \begin{array}{c} \\ 1 \\ 2 \\ 3 \end{array} \begin{array}{c} \begin{array}{ccccccc} 1 & 2 & 3 & 4 & 5 & 6 & 7 \end{array} \\ \begin{bmatrix} -1 & 1 & 1 & 0 & 0 & 0 & 0 \\ 0 & 0 & 0 & 1 & 0 & 1 & -1 \\ 0 & 0 & 0 & 0 & 1 & -1 & 0 \end{bmatrix} \end{array}$$

The mesh impedance matrix is

$$Z_m = B_m Z_e B_m^t = \begin{bmatrix} R_1 + s(L_2 + L_3) & sM & 0 \\ sM & R_7 + sL_4 + \frac{S_6}{s} & -\frac{S_6}{s} \\ 0 & -\frac{S_6}{s} & R_5 + \frac{S_6}{s} \end{bmatrix}$$

(The off-diagonal entries corresponding to the mutual inductance term are not always negative. The "dots" on the coils determine the sign of these entries.)

The equivalent voltage-source vector is

$$\mathbf{E}_q(s) = \begin{bmatrix} E_1 \\ 0 \\ -L_3 i_{L_3}(0) - M i_{L_4}(0) \\ -L_4 i_{L_4}(0) - M i_{L_2}(0) \\ 0 \\ \frac{1}{s} V_{c6}(0) \\ -R_7 i_7 \end{bmatrix}$$

(Notice that the rows 3 and 4 have the mutually induced voltages due to the initial currents in the two coils.)

Hence,

$$-B_m \mathbf{E}_q(s) = \begin{bmatrix} E_1(s) + L_3 i_{L_2}(0) + M i_{L_4}(0) \\ L_4 i_{L_4}(0) + M i_{L_4}(0) - \frac{1}{s} V_{c6}(0) - R_7 i_7 \\ \frac{1}{s} V_{c6}(0) \end{bmatrix}$$

The mesh equation for the network is

$$B_m Z_e B_m^t \mathbf{I}_m = -B_m \mathbf{E}_q$$

3.7 NETWORKS WITH CONTROLLED SOURCES

An active network is obtained by embedding one or more controlled sources in a passive RLC network. The network equations for such networks can be written based on the methods discussed here. The first step in writing the equations is to convert the network with active component to a network with controlled sources. Then the equations are formulated in the usual way considering the controlled or dependent sources as independent sources. The final step is to rearrange the equation suitably. This is illustrated by an example.

Consider the network shown in Fig. 3.19. In addition to the independent current source i_1, the network contains two active components—one a gyrator and the

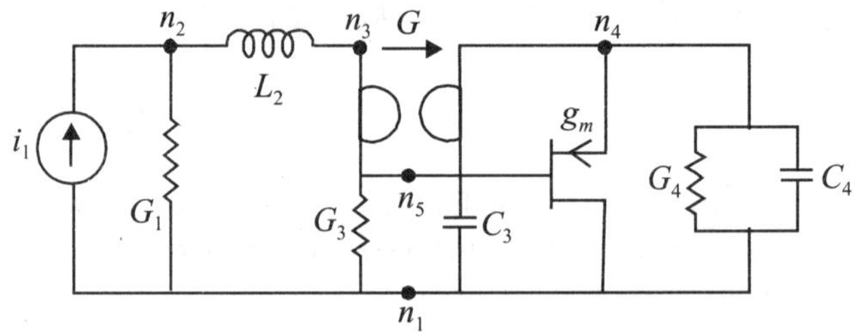

Fig. 3.19

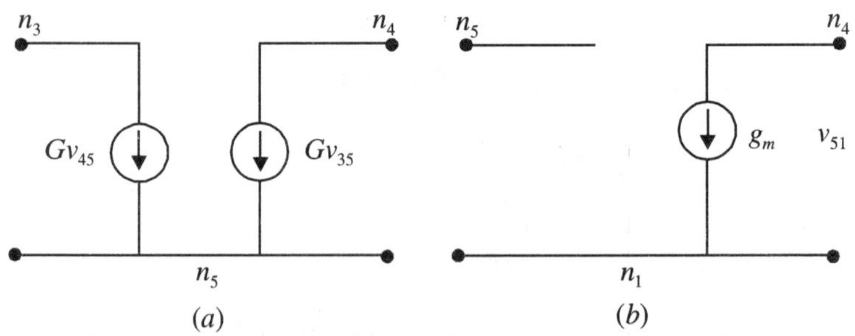

Fig. 3.20

other a field-effect-transistor (FET). It is required to write the node equations for this network. A controlled source model of gyrator is shown in Fig. 3.20(*a*) (Modelling and realisation of gyrators will be discussed in Chap. 12). A controlled source model of FET is shown in Fig. 3.20(*b*). Figure 3.21 shows the controlled source model of the network of Fig. 3.19 where the gyrator and the FET are replaced by their equivalents.

Figure 3.22 shows the oriented linear graph of Fig. 3.21 where the orientation of edges 2, 3 and 4 are arbitrary while those of others are dictated by the direction of the sources. The incidence matrix with node n_1 as datum node is

$$A = \begin{array}{c} \\ n_2 \\ n_3 \\ n_4 \\ n_5 \end{array} \begin{array}{c} \begin{array}{ccccccc} 1 & 2 & 3 & 4 & 5 & 6 & 7 \end{array} \\ \begin{bmatrix} -1 & 1 & 0 & 0 & 0 & 0 & 0 \\ 0 & -1 & 0 & 0 & 1 & 0 & 0 \\ 0 & 0 & 0 & 1 & 0 & 1 & 1 \\ 0 & 0 & 1 & 0 & -1 & -1 & 0 \end{bmatrix} \end{array}$$

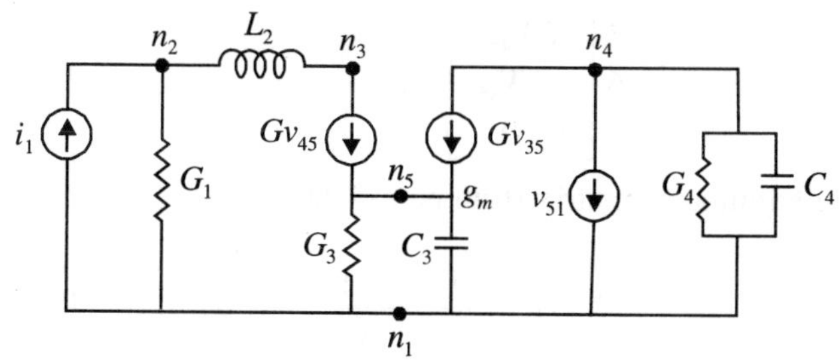

Fig. 3.21

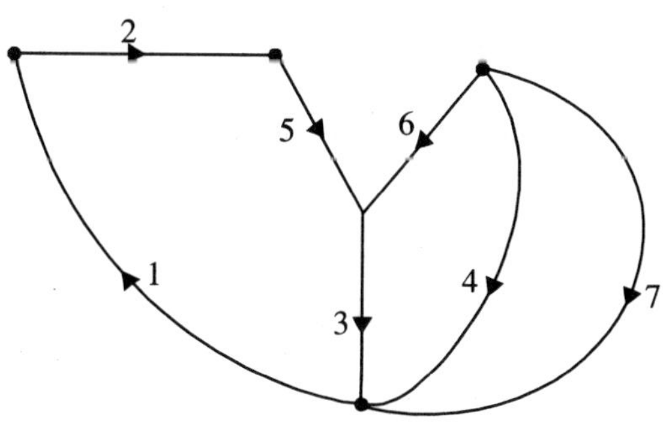

Fig. 3.22

The node voltage vector is

$$V_n = [v_2 \quad v_3 \quad v_4 \quad v_5]^t$$

The edge admittance matrix is

$$Y_e = \begin{matrix} 1 \\ 2 \\ 3 \\ 4 \\ 5 \\ 6 \\ 7 \end{matrix} \begin{bmatrix} G_1 & 0 & 0 & 0 & 0 & 0 & 0 \\ 0 & 1/sL_2 & 0 & 0 & 0 & 0 & 0 \\ 0 & 0 & G_3 + sC_3 & 0 & 0 & 0 & 0 \\ 0 & 0 & 0 & G_4 + sC_4 & 0 & 0 & 0 \\ 0 & 0 & 0 & 0 & 0 & 0 & 0 \\ 0 & 0 & 0 & 0 & 0 & 0 & 0 \\ 0 & 0 & 0 & 0 & 0 & 0 & 0 \end{bmatrix}$$

The edge admittance of edges 5, 6 and 7 are zero as they correspond to ideal current sources. The equivalent current source vector is

$$J_q(s) = \begin{matrix} 1 \\ 2 \\ 3 \\ 4 \\ 5 \\ 6 \\ 7 \end{matrix} \begin{bmatrix} I_1(s) \\ 0 \\ 0 \\ 0 \\ GV_{45} \\ -GV_{35} \\ g_m GV_{51} \end{bmatrix}$$

The node admittance matrix from eqn. 3.28 is

$$Y_n = \begin{matrix} 2 \\ 3 \\ 4 \\ 5 \end{matrix} \begin{bmatrix} G_1 + \dfrac{1}{sL_2} & -\dfrac{1}{sL_3} & 0 & 0 \\ -\dfrac{1}{sL_2} & \dfrac{1}{sL_2} & 0 & 0 \\ 0 & 0 & G_4 + sC_4 & 0 \\ 0 & 0 & 0 & G_3 + sC_3 \end{bmatrix}$$

Observe that this corresponds to the passive network only and can be written by inspection

$$AJ_q = \begin{matrix} n_2 \\ n_3 \\ n_4 \\ n_5 \end{matrix} \begin{bmatrix} -I_1 \\ GV_{45} \\ -GV_{35} + g_m V_{51} \\ -GV_{45} + GV_{35} \end{bmatrix}$$

On observing that $V_{35} = V_3 - V_5$, $V_{46} = V_4 - V_5$ and $V_{51} = V_5$, we have

$$AJ_q = \begin{bmatrix} -I_1 \\ GV_4 - GV_5 \\ -GV_3 + (G + g_m)V_5 \\ GV_3 - GV_4 \end{bmatrix}$$

The node equations from eqn. 3.27 is

$$\begin{bmatrix} G_1+\dfrac{1}{sL_2} & -\dfrac{1}{sL_2} & 0 & 0 \\ -\dfrac{1}{sL_2} & \dfrac{1}{sL_2} & 0 & 0 \\ 0 & 0 & G_4+sC_4 & 0 \\ 0 & 0 & 0 & G_3+sL_3 \end{bmatrix} \begin{bmatrix} V_2 \\ V_3 \\ V_4 \\ V_5 \end{bmatrix}$$

$$= -\begin{bmatrix} -I_1 \\ GV_4 - GV_5 \\ -GV_3 + (G+g_m)V_5 \\ GV_3 - CV_4 \end{bmatrix}$$

Transferring elements with V_3, V_4 and V_5 to LHS we have

$$\begin{bmatrix} G_1+\dfrac{1}{sL_2} & -\dfrac{1}{sL_2} & 0 & 0 \\ -\dfrac{1}{sL_2} & \dfrac{1}{sL_2} & G & -G \\ 0 & -G & G_4+sC_3 & G+g_m \\ 0 & G & -G & G_3+sC_3 \end{bmatrix} \begin{bmatrix} V_2 \\ V_3 \\ V_4 \\ V_5 \end{bmatrix} = \begin{bmatrix} I_1 \\ 0 \\ 0 \\ 0 \end{bmatrix}$$

which are the required node-equations.

From the example we can summarize the steps in writing the equations.

1. Replace the active components by their controlled sources.
2. Write the node-equations by assuming the controlled sources to be independent sources.
3. Rearrange the equations.

3.8 TRANSFORM NETWORKS

The network equations for a linear time-invariant network are generally written in the Laplace domain, as demonstrated in the previous sections. The procedure we adopted was to write the terminal relation of the elements in the time domain and then Laplace transform these equations. It is possible to obtain the network equations from the *transform network,* which is the s-domain equivalent of the time-domain network. To obtain the transform network of a given (time domain) network each element is replaced by its *transform impedance* or *admittance.* The initial conditions of the elements are replaced by independent sources in the transformed network. We first obtain the transform immittance representations of the individual elements before forming the transform network.

3.8.1 Resistors

The terminal relation of a resistor in the time domain (t-domain) is

$$V(t) = Ri(t) \text{ or } i(t) = GV(t)$$

In the s-domain (Laplace domain) this equation is

$$V(s) = RI(s) \text{ or } I(s) = GV(s)$$

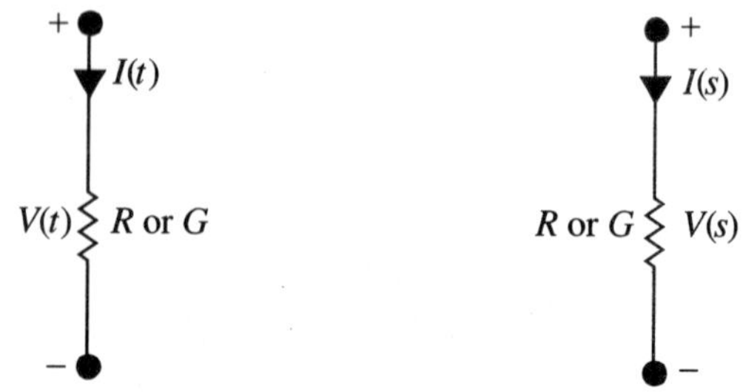

Fig. 3.23

Figure 3.23 shows the t-domain and s–domain representation of a resistor. It can be noticed that the transform network for a resistor is identical to its t-domain counterpart except that the current and voltage are expressed in s-domain.

3.8.2 Inductors

The terminal relation of an inductor in the t-domain is

$$V(t) = L\frac{di(t)}{dt}$$

In the s-domain this equation becomes

$$V(s) = L[sI(s) - i_L(0)] \qquad \text{...(3.36)}$$

or

$$I(s) = \frac{1}{sL}V(s) + \frac{i_L(0)}{s} \qquad \text{...(3.37)}$$

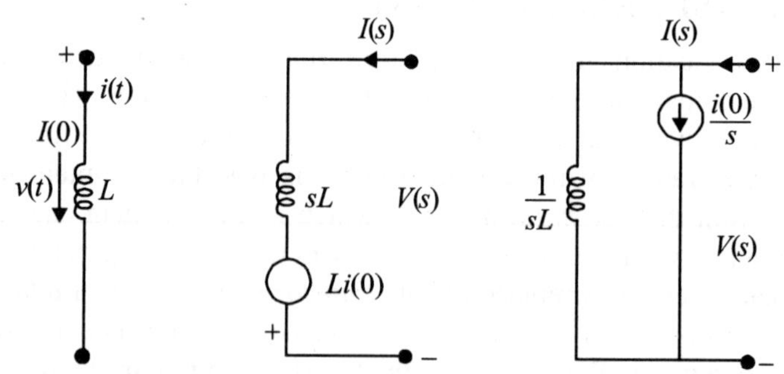

Fig. 3.24

Here $i_L(0)$ is the initial inductor current. The t-domain and the s-domain representations of an inductor are shown in Fig. 3.24. Equation 3.36 leads to a voltage source (Thevenin equivalent) model, where sL is the transform impedance of the inductor and eqn. 3.37 leads to a current source (Norton equivalent) model, the $1/sL$ is the transform admittance. In the transform network the inductor does not have any initial current through it. The initial current of the inductor in the t-domain is replaced by a voltage or a current source in the transform network.

3.8.3 Capacitors

The terminal relation of a capacitor in the t-domain is

$$i(t) = C\frac{dv(t)}{dt}$$

In the s-domain the relation is

$$I(s) = C\,[sV(s) - v_c(0)] \qquad \ldots(3.38)$$

or

$$V(s) = \frac{1}{sC}I(s) + \frac{v_c(0)}{s} \qquad \ldots(3.39)$$

Here $v_c(0)$ is the initial capacitor voltage. The t-domain and the s-domain representations of a capacitor are shown in Fig. 3.25. Equation 3.38 leads to a current source (Norton equivalent) model, where sC is the transform admittance

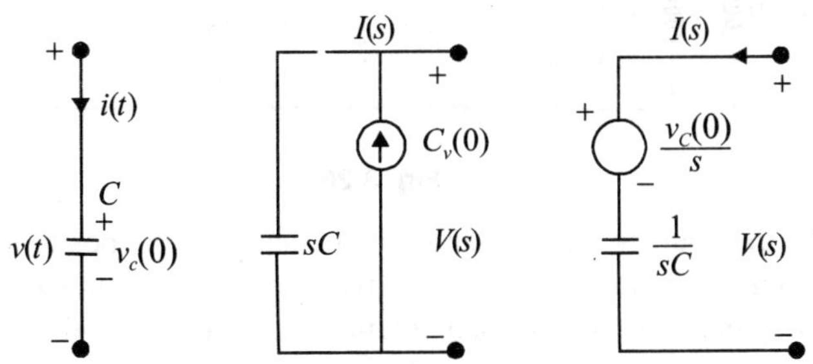

Fig. 3.25

and eqn. 3.39 leads to a voltage source (Thevenin equivalent) model, where $1/sC$ is the transform impedance. In the transform network the capacitor does not have any initial voltage across it. The initial voltage of the capacitor in the t-domain is replaced by a voltage or a current source in the transform network.

3.8.4 Coupled Coils

The terminal relations of a pair of coupled coils in the t-domain are

$$v_1(t) = L_1\frac{di_1(t)}{dt} \pm M\frac{di_2(t)}{dt}$$

$$v_2(t) = \pm M\frac{di_1(t)}{dt} + L_2\frac{di_2(t)}{dt} \qquad \ldots(3.40)$$

The corresponding s-domain equations are

$$V_1(s) = L_1\left[sI_1(s) - iL_1(0)\right] \pm M\left[sI_2(s) - i_{L_2}(0)\right]$$

$$V_2(s) = \pm M\left[sI_1(s) - i_{L_1}(0)\right] + L_2\left[sI_2(s) - i_{L_2}(0)\right] \qquad \text{...(3.41)}$$

where $i_{L_1}(0)$ and $i_{L_2}(0)$ are the initial currents through coil 1 and 2, respectively. The t-domain and s-domain representations of the coupled coils are shown in Fig. 3.26. As before, the inductors in the transform network do not have any initial conditions associated with them. The initial conditions of t-domain are represented by equivalent voltage sources in the s-domain. In Fig. 3.26 we have not marked the polarities of the coils. If the polarities of the coils are such that the two voltage components are additives (+ sign in equations 3.40 and 3.41), then the sign of the voltage sources $Mi_{12}(0)$ and $Mi_{11}(0)$ due to mutual inductances will coincide with those of $L_1 i_{L_1}(0)$, $L_2 i_{L_2}(0)$, respectively. If the polarities of the coils are such that the two voltages are subtractive (–sign in equations 3.40 and 3.41), then the

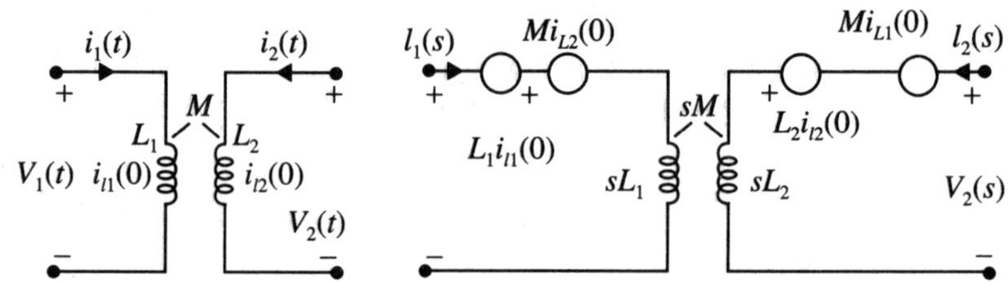

Fig. 3.26

polarities of the sources due to mutual inductances are opposite to those due to self inductances L_1 and L_2. Figure 3.26 gives the voltage source equivalent of the t-domain-coupled coils. We can also obtain a corresponding s-domain model with current source equivalents. To this end we have to write the terminal relations of the coil in the form

$$\frac{di_1(t)}{dt} = \Gamma_1 v_1 \pm \Gamma_m v_2$$

$$\frac{di_2(t)}{dt} = \pm\Gamma_m v_1 + \Gamma_2 v_2$$

where Γ_1, Γ_2 and Γ_m are inverse inductance parameters. The inverse inductance parameters do not exist if the coils are perfectly coupled ($k = 1$).

The transform network for a given time domain network can be obtained by replacing each element of the t-domain network by its equivalent s-domain representation. If we are interested in writing the mesh or loop equations, then it is more convenient to write the equivalent voltage source model. On the other hand, if we want to write node or cut-set equations, it is advantageous to write the equivalent current source model. In the final transform network the independent voltage and current sources are replaced by their corresponding Laplace transformed functions.

Example 3.7. Find the transform network shown in Fig. 3.27 and obtain its mesh equations.

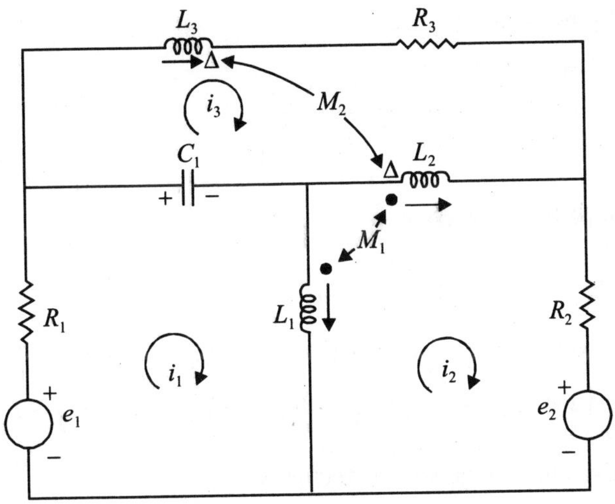

Fig. 3.27

The transform network is shown in Fig. 3.28. The mutual inductances term between the two coils 1 and 2, considering only the initial currents through these coils, is positive. Hence the two equivalent voltage sources, one due to the self

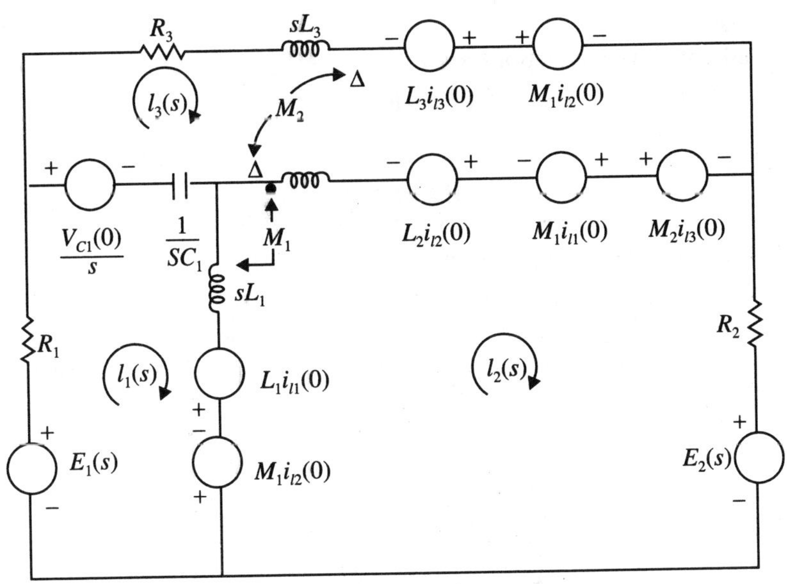

Fig. 3.28

inductances, and the other due to the mutual inductance, due to the initial currents $i_{L_1}(0)$ and $i_{L_2}(0)$ will have the same orientation. On the other hand, the mutual inductance term between coils 2 and 3 considering only the initial currents, is negative. Hence the two equivalent voltage sources due to the initial currents $i_{L_3}(0)$ and $i_{L_2}(0)$ are oppositely oriented.

Mesh current vector is

$$\mathbf{I}_m = \begin{bmatrix} I_1(s) \\ I_2(s) \\ I_3(s) \end{bmatrix}$$

and mesh impedance matrix is

$$Z_m = \begin{bmatrix} R_1 + \dfrac{1}{sC_1} + sL_1 & s(M_1 - L_1) & -\left(SM_1 + \dfrac{1}{SC_1}\right) \\ s(M_1 - L_1) & R_2 + s(L_1 + L_2 - 2M_1) & s(M_1 - M_2 - L_2) \\ -\left(sM_1 + \dfrac{1}{sC_1}\right) & s(M_1 - M_2 - L_2) & R_3 + s(L_3 + L_2 + 2M_2) \end{bmatrix}$$

and the equivalent voltage source vector is

$$-B_m E_q = \begin{bmatrix} E - \dfrac{V_{c_1}(0)}{s} + L_1 i_{L_1}(0) + M_1 i_{L_2}(0) \\ -E_2 + i_{L_1}(0)(M_1 - L_1) + i_{L_2}(0)(L_2 - M_1) - M_2 i_{L_2}(0) \\ \dfrac{V_{c_1}(0)}{s} - M_1 i_{L_1}(0) - i_{L_2}(0)(M_2 + L_2) + iL_3(0)(M_2 + L_3) \end{bmatrix}$$

Finally, the mesh equations are

$$Z_m \mathbf{I}_m = -B_m \mathbf{E}_q$$

PROBLEMS

3.1 Let N be a network with a planar graph. Prove that

$$\mathbf{I}_e = B_m^t \mathbf{I}_m$$

where $\mathbf{I}_e$ is the edge current vector, $\mathbf{I}_m$ the mesh current vector and B the mesh matrix.

3.2 For the network shown in Fig. P. 3.2 verify equations 3.3, 3.7, 3.8 and 3.10.

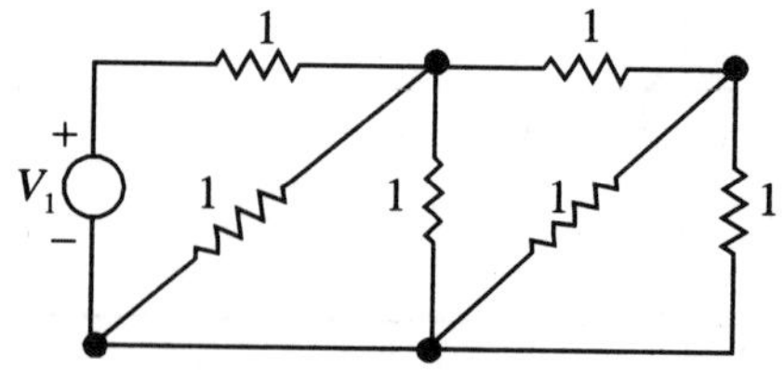

Fig. P. 3.2

3.3 Write the mesh and node equations of the network shown in Fig. P. 3.3.

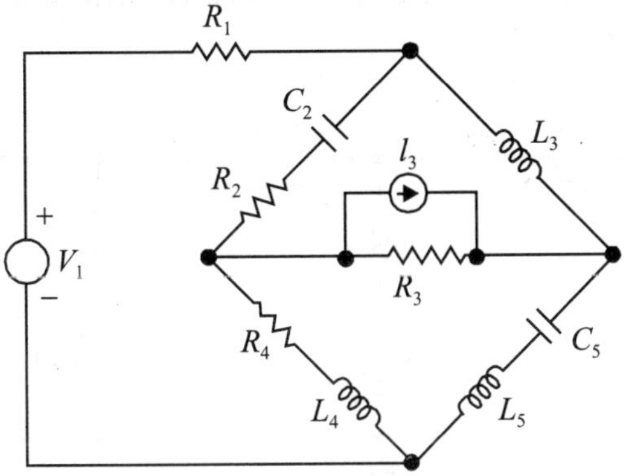

Fig. P. 3.3

3.4 Write the cut-set and loop equations for the network of Fig. P. 3.4.

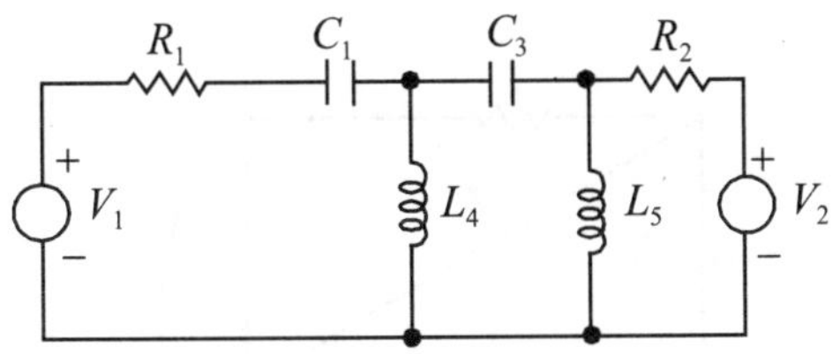

Fig. P. 3.4

3.5 Write the transform network of the network shown in Fig. P. 3.5. Write the mesh equation for the network.

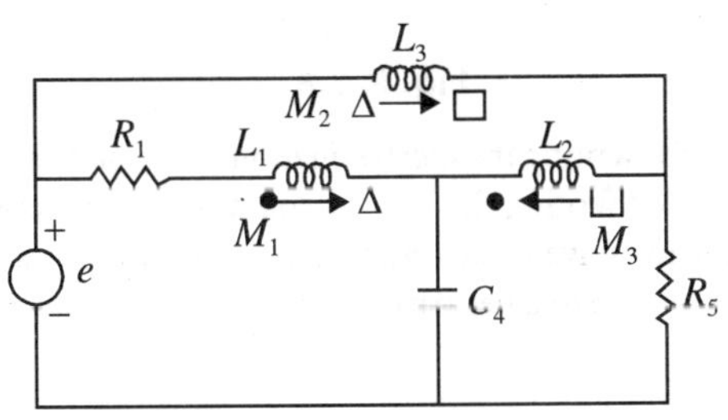

Fig. P. 3.5

3.6 Write the mesh equations the network shown in Fig. P. 3.6.

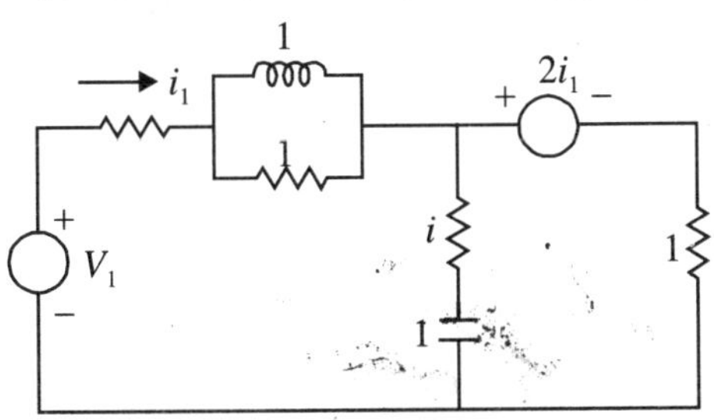

Fig. P. 3.6

3.7 Find the equivalent of the network in Fig. P. 3.7 when all the voltage sources are in the shunt arms. with one of their terminals grounded. Verify that V_0 obtained in both cases is the same.

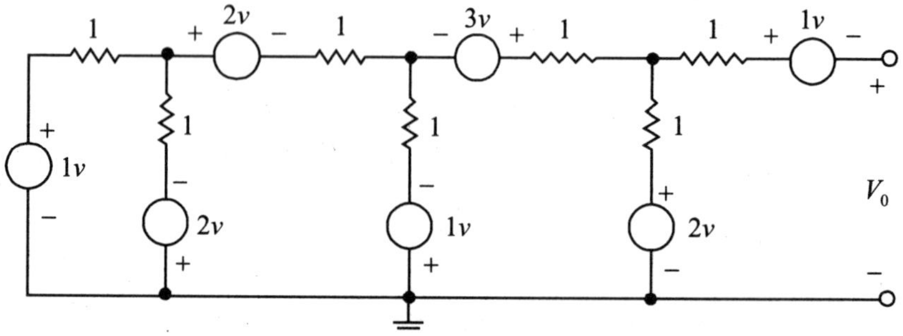

Fig. P. 3.7

3.8 Describe the current in the 2 ohm resistor of Fig. P. 3.8.

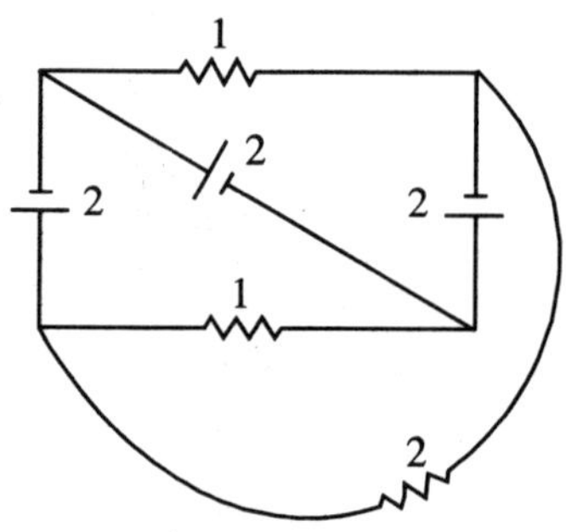

Fig. P. 3.8

3.9 Formulate the rules for writing the mesh equations for a planar network by inspection.

3.10 Formulate the rules for writing the cut-set equations for a network by inspection.

3.11 A pair of coupled coils have a self-inductance of L_1 and L_2, and mutual inductance M. If the energy delivered to a network from t_1 to t_2 is $w(t_1) - w(t_2)$, show that for the coupled coil

$$w(t) = \frac{1}{2}L_1 i_1^2 + \frac{1}{2}L_2 i_2^2 + M i_1 i_2$$

where i_1 and i_2 are the two coil currents. Also determine the condition under which $w(t) \geq 0$ for all t.

NOTES AND REFERENCES

Formulation of network equations through network graphs and their matrices is central to most, if not all, computer oriented techniques. Hence this chapter provides a natural setting for introducing computer formulation of network equations. The methods developed are general and can be applied to passive RLCM/active networks.

The classic book by Van Valkenberg[1] (Chap. 3) provides many additional examples and problems. Problems may also be supplemented from Seshu and Balbanian[2] (Chap. 4) and Desoer & Kuh[3] (Chaps. 9–11).

1. M.E. Van Valkenberg, *Network Analysis*, Pentice-Hall, 1974.
2. S. Seshu and N. Balbanian, *Linear Network Analysis*, John Wiley, 1959.
3. C.A. Desoer and E.S. Kuh, *Basic Circuit Theory*, McGraw-Hill, 1969.

CHAPTER

4

Solution of Network Equations and Sinusoidal Response

In Chapter 3, we discussed methods of formulating network equations. Time-domain network equations are a set of integro-differential equations and can be converted to a set of higher order differential equations. To solve these equations we need two sets of initial conditions—the initial currents through the inductors and the initial voltages across the capacitors. In order to avoid solving differential equations we can resort to Laplace transforms which convert the set of integro-differential equations to a set of algebraic equations which can then be solved to obtain the solution in the Laplace domain (frequency domain). The time domain solution is then obtained by inverse Laplace transforms.

The complete solution of any linear ordinary differential equation consists of two parts—the *complementary function* and the *particular integral.* The former is the solution of the homogeneous equation, *i.e.* the equation with the forcing function set to zero and the network driven only by its initial conditions. For this reason the complementary function is also called the *zero input response* (ZIR). The particular integral is a solution for the equation with the forcing function. For this case all the initial conditions are set to zero and the response is called the *zero state response* (ZSR).

The complete response of a network can also be separated into *transient and steady state* responses. In general, the initial part of the response is termed as the transient response. In contrast, steady state response is the response as $t \to \infty$. The steady state response of a network to a sinusoidal input is of particular interest in filter analysis and design.

4.1 PROPERTIES OF NETWORK MATRICES

In Chapter 3, we demonstrated that the network equations can, in general, be written as

$$M\mathbf{X} = \mathbf{N} \qquad \ldots(4.1)$$

where $\mathbf{X}$ and $\mathbf{N}$ are vectors and M is a square matrix. For example, if we are writing the node equations then

$$M = Y_n,\ \mathbf{X} = V_n,\ \mathbf{N} = -A\mathbf{J}_q.$$

In general, the matrix M is formed by a *matrix triple product* of the form

$$M = WFW^t$$

where W is some matrix of the network graph and F the *edge impedance* or *admittance matrix.* If we consider a passive network (*i.e.* network with R, L, C and coupled coils), the matrix F is a symmetric matrix. (If the network comprises R, L and C elements only then F is a diagonal matrix). For such a network M is a symmetric matrix. To show this consider M^t:

$$M^t = (WFW^t)^t$$

$$= WF^t W^t = WFW^t = M$$

indicating that M is symmetric. For a network with active elements (*i.e.* controlled sources) the matrix M may not be symmetric.

In order to solve for the vector X in equation 4.1, matrix M should be invertible. If M is invertible

$$\mathbf{X} = M^{-1}\mathbf{N} \qquad \text{...(4.2)}$$

For a matrix to be invertible it should be non-singular, *i.e.* it must have non-zero determinant. For a passive RLC network the matrix M is a *dominant matrix* (a dominant matrix is a matrix in which the diagonal element is greater than or equal to the sum of the absolute values of the off-diagonal entries in the same row or column) and hence is non-singular. To show that M is a dominant matrix consider the *nodal admittance matrix* Y_n. From equations 3.25 and 3.28 we have

$$Y_n = G_n + sC_n + \frac{1}{s}\Gamma_n$$

where G_n, and C_n and Γ_n are the corresponding node conductance, capacitance and *inverse inductance matrices,* respectively, and are given by

$$G_n = AG_e A^t,\ C_n = AC_e A^t,\ \Gamma_n = A\ \Gamma_e A^t$$

Consider the node conductance matrix G_n.

$$G_n = AG_e A^t = [g_{ij}]$$

Here

g_{ii} = sum of all conductances connected to node i.

g_{ij} = negative of the sum of the conductances connected between nodes i and j.

Since the first set of conductances must necessarily include all those in the second set, we conclude that

$$g_{ii} \geq \sum_{j=1}^{v-1} |g_{ij}|$$

i.e. the matrix G_n is dominant.

Similarly, we can show that C_n is a dominant matrix. Γ_n is a dominant matrix if we do not have any mutual inductances. Hence, we can conclude that the node admittance matrix of an RLC network is dominant for non-negative, its real value of s.

4.2 SOLUTION OF EQUATIONS

The solution for the equation 4.1 is given by $\mathbf{X} = M^{-1}\,\mathbf{N}$.

The inverse of a matrix is unique and, as such, if M is non-singular then the solution is also unique. We can state the existence and the uniqueness of the solution for a network in the following theorem.

Theorem 4.1: Let N be a linear time-invariant RLC network with mutual inductances. If all the elements of the network are positive and the inductance matrix is positive definite, then for a given set of initial conditions and inputs, the network N has a unique solution.

Of course, a passive time-invariant RLC network has a unique solution once the initial conditions and inputs are specified.

Again, consider the node equations. The solution to the node equations, *i.e.*, a set of node voltages, is given by

$$\mathbf{V}_n(s) = Y_n^{-1}\left\{-AJ_q(s)\right\} \qquad \text{...(4.3)}$$

From equation 3.26

$$\mathbf{J}_q(s) = J_e(s) - Y_e(s)\mathbf{E}_e(s) - C_e\mathbf{V}_{ce}(0) + \frac{1}{s}\mathbf{i}_{le}(0)$$

where $\mathbf{J}_e(s)$ is current source vector, $\mathbf{E}_e(s)$ the voltage source vector, $\mathbf{V}_{ce}(0)$ the initial capacitor voltage vector, $\mathbf{i}_{le}(0)$ the initial inductor current vector and $Y_e(s)$ the edge admittance matrix, and C_e the *edge capacitance matrix.*

We can regroup the terms in eqn. 4.3 as

$$\mathbf{V}_n(s) = \mathbf{V}_{n_1}(s) + \mathbf{V}_{n_2}(s) \qquad \text{...(4.4)}$$

where

$$\mathbf{V}_{n_1} = -Y_n^{-1}A\left[\mathbf{J}_e(s) - Y_e(s)\mathbf{E}_e(s)\right]$$

and

$$\mathbf{V}_{n_2} = -Y_n^{-1}A\left\{\frac{1}{s}\mathbf{i}_{le}(0) - C_e\mathbf{V}_{ce}(0)\right\}.$$

$\mathbf{V}_{n_1}$ corresponds only to the sources and is the zero state response. Similarly, $\mathbf{V}_{n_2}$ corresponds to the initial conditions only, and is the zero input response.

Hence,

$$\text{Laplace Transform of the total response} = \text{Laplace Transform of ZIR} + \text{Laplace Transform of ZSR}$$

In order to find the total response we have to find the inverse of Y_n. By definition

$$Y_n^{-1} = \frac{adj(Y_n)}{|Y_n|}$$

where $adj(Y_n)$ is the adjoint of $Y_n(s)$ and $|Y_n|$ is the determinant of $Y_n(s)$. We can further write

$$Y_n^{-1} = \frac{\Delta(s)}{\alpha(s)}$$

where $\Delta(s)$ is a polynomial matrix and $\alpha(s)$ a polynomial in s. Substitution of this equation in eqn. 4.4 yields

$$\mathbf{V}_n(s) = -\frac{\Delta(s)}{\alpha(s)} A\{\mathbf{J}_e(s) - Y_e(s)\mathbf{E}_e(s)\}$$

$$-\frac{\Delta(s)}{\alpha(s)} A\left\{\frac{1}{s}\mathbf{i}_{le}(0) - C_e\mathbf{V}_{ce}(0)\right\}$$

In general, we have

$$\mathbf{V}_{n_1}(s) = \frac{\mathbf{P}_1(s)}{q_1(s)} \text{ and } \mathbf{V}_{n_2}(s) = \frac{\mathbf{P}_2(s)}{q_2(s)}$$

where $\mathbf{P}_1$ and $\mathbf{P}_2$ are vectors and q_1 and q_2 polynomials in s. If we consider the ith node voltage, then

$$\mathbf{V}_i(s) = \frac{p_{i_1}(s)}{q_1(s)} + \frac{p_{i_2}(s)}{q_2(s)} \quad \text{...(4.5)}$$

and

$$v_i(t) = \mathcal{L}^{-1}\left\{\frac{p_{i_1}(s)}{q_1(s)}\right\} + \mathcal{L}^{-1}\left\{\frac{p_{i_2}(s)}{q_2(s)}\right\}$$

Hence, we can determine $\mathbf{V}_n(t)$, the node voltage vector in time domain. The node transformation

$$\mathbf{V}_e = A^t\,\mathbf{V}_n$$

yields the edge voltage vector $\mathbf{V}_e(t)$. The edge currents can now be determined by the v-i relationships of the corresponding elements

i.e.

$$\mathbf{I}_e(s) = Y_e(s)\,\mathbf{V}_e(s)$$

or

$$\mathbf{I}_e(t) = \mathcal{L}^{-1}\left[Y_e(s)\mathbf{V}_e(s)\right]$$

and the solution of the network is complete, as we have determined all the edge voltages and currents.

From eqn. 4.5 we see that a typical element for which we have to find the inverse Laplace transform is of the form

$$F(s) = \frac{p(s)}{q(s)} \quad \text{...(4.6)}$$

where p and q are polynomials in s. In the next section we described a method of finding the inverse Laplace transform of such *rational functions.*

4.3 PARTIAL-FRACTION EXPANSION

Inverse Laplace transform of simple functions can be obtained from a table of Laplace transforms. But the inverse transform of a complicated function will not directly appear in the tables. If the given function is a rational function, the inverse transform can be obtained by breaking the rational function into simple functions whose Laplace inverses are found in the tables. This method is called *partial-fraction expansion* (see Appendix A).

Let the given rational function be

$$F(s)=\frac{p(s)}{q(s)}=\frac{b_m s^m+b_{m-1}s^{m-1}+\ldots+b_1 s+b_0}{a_n s^n+a_{n-1}s^{n-1}+\ldots+a_1 s+a_0}$$

where p and q are polynomials in s and the coefficients a_0, $a_1 \ldots a_n$, b_0, b_1, $\ldots b_m$ are real numbers. Polynomials can also be expressed in factored form as

$$F(s)=\frac{K\prod_{i=1}^{m}(s-z_i)}{\prod_{i=1}^{n}(s-p_i)}$$

where z_i ($i = 1, 2, \ldots m$) are the roots of $p(s)$ and p_i ($i = 1, 2, \ldots n$) are the roots of $q(s)$. The $z_i's$ are called the *zeros* of $F(s)$ because $F(z_i) = 0$. The $p'_i s$ are called the *poles* of $F(s)$ because $F(p_i) = \infty$ (see Chapter 7). Here we assume that $p(s)$ and $q(s)$ do not have any common factors. If p_i is a simple root of $q(s)$, then it is a *simple pole of* $F(s)$. If p_i is a root of order r of $q(s)$, then is said to be a *multiple pole of order* r of $F(s)$.

A rational function is called a *proper* rational function, if the degree of the numerator polynomial is less than the degree of the denominator polynomial. If $F(s)$ is not proper, we can write

$$F(s)=F_1(s)+\frac{\hat{p}(s)}{q(s)} \qquad \ldots(4.7)$$

where $\hat{p}(s)/q(s)$ is a proper rational function and $F_1(s)$ is a polynomial in s. Inverse transform of $F_1(s)$ can be easily obtained. It will contain delta functions and its derivatives. Hence, in what follows we only consider proper rational functions.

In obtaining the partial-fraction expansion we need not factorize the numerator. We have to factorize the denominator polynomial and find the poles of $F(s)$. We consider three separate cases.

Case 1: $F(s)$ has simple poles

$$F(s)=\frac{p(s)}{q(s)}=p(s)\Big/\prod_{i=1}^{n}(s-p_i) \qquad \ldots(4.8)$$

(Here $q(s)$ is assumed to be a monic polynomial.)

Then the partial-fraction expansion is

$$F(s)=\sum_{i=1}^{n}\frac{k_i}{s-p_i} \qquad \ldots(4.9)$$

where k_i is the *residue* of $F(s)$ at the pole p_i. To obtain k_i, multiply both sides of 4.9 by $(s - p_i)$.

i.e.
$$(s-p_i)F(s) = k_i + \sum_{\substack{j=1 \\ j\neq 1}}^{n} \frac{k_i}{s-p_j}(s-p_i)$$

from which it follows that

$$k_i = (s-p_i)F(s)\Big|_{s=p_i} \qquad \ldots(4.10)$$

The inverse transform $F(s)$ is

$$\mathcal{L}^{-1}[F(s)] = \sum_{i=1}^{n} \mathcal{L}^{-1}\left[\frac{k_i}{s-p_i}\right]$$

i.e.
$$f(t) = \sum_{i=1}^{n} k_i e^{p_i t} \qquad \ldots(4.11)$$

(This follows from the linearity property of the Laplace transforms.)

Example 4.1.

Let
$$F(s) = \frac{s^2+6s+8}{(s+1)(s+3)(s+5)}$$

From eqn. 4.9

$$F(s) = \frac{k_1}{s+1} + \frac{k_2}{s+3} + \frac{k_3}{s+5}$$

From eqn. 4.10

$$k_1 = (s+1)F(s)\Big|_{s=-1} = \frac{s^2+6s+8}{(s+3)(s+5)}\bigg|_{s=-1} = \frac{3}{8}$$

$$k_2 = (s+3)F(s)\Big|_{s=-3} = \frac{s^2+6s+8}{(z+1)(s+5)}\bigg|_{s=-3} = \frac{-1}{-4} = \frac{1}{4}$$

$$k_3 = (s+5)F(s)\Big|_{s=-5} = \frac{s^2+6s+8}{(s+1)(s+3)}\bigg|_{s=-5} = \frac{3}{8}$$

Hence,
$$F(s) = \frac{3/8}{s+1} + \frac{1/4}{s+3} + \frac{3/8}{s+5}$$

and from eqn. 4.11

$$f(t) = \frac{3}{8}e^{-t} + \frac{1}{4}e^{-3t} + \frac{3}{8}e^{-5t}$$

Case 2: $F(s)$ has multiple poles.

Let
$$q(s) = \prod_{t=1}^{r} (s-p_i)^{n_i} \qquad \ldots(4.12)$$

where $$\sum_{i=1}^{r} n_i = n.$$

The partial-fraction expansion of $F(s)$ is of the formation

$$F(s) = \frac{k_{11}}{s-p_1} + \frac{k_{12}}{(s-p_1)^2} + \ldots + \frac{k_{1n_1}}{(s-p_1)^n} + \frac{k_{21}}{s-p_2} + \frac{k_{22}}{(s-p_2)^2}$$

$$+ \ldots + \frac{k_{2n_2}}{(s-p_2)^{n_2}} + \ldots + \frac{k_{rl}}{s-p_r} + \frac{k_{r_2}}{(s-p_r)^2} + \ldots + \frac{k_{rn_r}}{(s-p_r)^{n_r}}$$

i.e. $$F(s) = \sum_{i=1}^{r}\sum_{j=1}^{n_i} \frac{k_{ij}}{(s-p_i)^j} \qquad \ldots(4.13)$$

It can be shown that

$$k_{ij} = \frac{1}{(n_i - j)!}\frac{d^{(n_i-j)}}{ds^{(n_i-j)}}\left\{(s-p_i)^{n_i} F(s)\right\}\Big|_{s=p_i} \qquad \ldots(4.14)$$

The inverse transform of $F(s)$ is

$$f(t) = \sum_{i=1}^{r}\sum_{j=1}^{n_r} k_{ij}\frac{t^{j-1}}{(j-1)!} e^{p_i t} \qquad \ldots(4.15)$$

Example 4.2.

$$F(s) = \frac{s^2+1}{(s+1)^2(s+2)}$$

From eqn. 4.13

$$F(s) = \frac{k_{11}}{s+1} + \frac{k_{12}}{(s+1)^2} + \frac{k_{13}}{(s+1)^3} + \frac{k_2}{s+2}$$

From eqn. 4.10

$$k_2 = (s+2)F(s)\big|_{s=-2} = \frac{s^2+1}{(s+1)^3}\bigg|_{s=-3} = -5$$

From eqn. 4.14

$$k_{13} = \frac{1}{(3-3)!}\frac{d^{(3-3)}}{ds^{(3-3)}}(s+1)^3 F(s)\bigg|_{s=-1}$$

$$= (s+1)^3 F(s)\big|_{s=-1} = \frac{s^2+1}{s+2}\bigg|_{s=-1} = 2$$

$$k_{12} = \frac{1}{1}\frac{d}{ds}(s+1)^3 F(s)\Big|_{s=-1}$$

$$= \frac{d}{ds}\frac{s^2+1}{s+2}\Bigg|_{s=-1} = \frac{2s(s+2)-(s^2+1)}{(s+2)^2}\Bigg|_{s=-1}$$

$$= \frac{s^2+4s-1}{(s+2)^2}\Bigg|_{s=-1} = -4$$

$$k_{11} = \frac{1}{2}\frac{d^2}{ds^2}(s+1)^3 F(s)\Bigg|_{s=-1} = \frac{1}{2}\frac{d}{ds}\frac{s^2+4s-1}{(s+2)^2}\Bigg|_{s=-1}$$

$$= \frac{1}{2}\frac{(2s+4)(s+2)^2-2(s^2+4s-1)(s+2)}{(s+2)^2}\Bigg|_{s=-1} = 5$$

Hence,

$$F(s) = \frac{5}{s+1} - \frac{4}{(s+1)^2} + \frac{2}{(s+1)^3} - \frac{5}{s+2}$$

and from eqn. 4.15

$$f(t) = 5e^{-t} - 4te^{-t} + 2t^2e^{-t} - 5e^{-2t}$$

The evaluation of the coefficients in the partial-fraction expansion involves the differentiation of a rational function. The coefficients can be obtained, without resorting to differentiation but by expansion of simple poles only, as demonstrated by the following example.

Example 4.3.

$$F(s) = \frac{s^2+1}{(s+2)(s+1)^3}. \text{ Let } F_1(s) = \frac{s^2+1}{(s+2)(s+1)}$$

Observe that F_1 is not a proper rational function. We can rewrite $F_1(s)$ as

$$F_1(s) = 1 - \frac{3s+1}{(s+2)(s+1)}$$

By partial-fraction expansion

$$F_1(s) = 1 - \frac{5}{s+2} + \frac{2}{s+1}$$

Let
$$F_2(s) = \frac{1}{s+1}F_1(s) = \frac{1}{s+1}\left\{1 + \frac{2}{s+1} - \frac{5}{s+2}\right\}$$

$$= \frac{1}{s+1} + \frac{2}{(s+1)^2} - \frac{5}{(s+1)(s+2)}$$

Again, by partial-fraction expansion we have

$$F_2(s) = \frac{1}{s+1} + \frac{2}{(s+1)^2} - \frac{5}{s+1} + \frac{5}{s+2}$$

$$= -\frac{4}{s+1} + \frac{2}{(s+1)^2} + \frac{5}{s+2}$$

$$F_3(s) = \frac{1}{s+1} F_2(s) = \frac{1}{s+1}\left\{-\frac{4}{s+1} + \frac{2}{(s+1)^2} + \frac{5}{s+2}\right\}$$

$$= -\frac{4}{(s+1)^2} + \frac{2}{(s+1)^3} + \frac{5}{(s+1)(s+2)}$$

Further $F_3(s)$ can be expanded as

$$F_3(s) = -\frac{4}{(s+1)^2} + \frac{2}{(s+1)^3} + \frac{5}{(s+1)} - \frac{5}{(s+2)}$$

but $F(s) = F_3(s)$

Hence,

$$F(s) = \frac{5}{s+1} - \frac{4}{(s+1)^2} + \frac{2}{(s+1)^3} - \frac{5}{s+2}$$

Case 3: $F(s)$ has complex poles.

The cases 1 and 2 are applicable whether the poles are real or complex. However, if there are complex poles then, in general, the residues are complex. If $F(s)$ is a rational function with real coefficients, then the complex poles must appear in conjugate pair, *i.e.*, if $p_i = \sigma_1 + j\omega_1$ is a pole of $F(s)$ then $p_1^* = \sigma_1 - j\omega_1$ is also pole of $F(s)$. (A polynomial with real coefficients satisfies the property that $q(s^*) = [q(s^*)]$.)

Let $F(s)$ have a simple pole at $s = p_1 = \alpha + j\beta$. Then, it has another simple pole at $s = s_1^* = \alpha - j\beta$. The partial-fraction expansion of $F(s)$ has the factors

$$F_1(s) = \frac{K_1}{s-\alpha-j\beta} + \frac{K_2}{s-\alpha+j\beta} \qquad \ldots(4.16)$$

From eqn. 4.10

$$K_1 = (s-\alpha-j\beta)\,F(s)\big|_{s=\alpha+j\beta}$$

$$K_2 = (s-\alpha+j\beta)\,F(s)\big|_{s=\alpha-j\beta}$$

Hence, $K_2 = K_1^*$

Let $$K_1 = k_1 + jk_2 = |K_1| e^{j\phi_1},$$

where $$|K_1| = \sqrt{k_1^2 + k_1^2} \text{ and } \phi_1 = \tan^{-1}[k_2/k_1].$$

Then $$K_2 = |K_1| e^{-j\phi_1}$$

Now $$f_1(t) = K_1 e^{(\alpha + j\beta)t} + K_2 e^{(\alpha - j\beta)t}$$

$$= |K_1| e^{\alpha t} \left\{ e^{j(\beta t + \phi)} + e^{-j(\beta t + \phi)} \right\}$$

i.e. $$f_1(t) = 2|K_1| e^{\alpha t} \cos(\beta t + \phi) \qquad \text{...(4.17)}$$

Example 4.4.

$$F(s) = \frac{s^2 + 3s + 7}{(s+1)(s+2+j2)(s+2-j2)}$$

The partial-fraction expansion is

$$= \frac{k_1}{s+1} + \frac{k_2}{s+2+j2} + \frac{k_2^*}{s+2-j2}$$

$$k_1 = (s+1) F(s)\big|_{s=-1} = 1$$

$$k_2 = (s+2+j2) F(s)\big|_{s=-2-j2}$$

$$= \left.\frac{s^2 + 3s + 7}{(s+1)(s+2-j2)}\right|_{s=-2-j2} = -j/4$$

$$= \frac{1}{4} e^{-j\pi/2}$$

$$k_2^* = \frac{1}{4} e^{+j\pi/2}$$

Hence $$f(t) = e^{-t} - \frac{1}{2} e^{-2t} \sin 2t.$$

4.4 IMPULSE RESPONSE AND CONVOLUTION

By solving the mesh and the node equations we can determine the currents and voltages in all the elements of a network. However, in many cases we are interested in treating the network as a single input single output system. The zero state response $R(s)$ of a network can be related to the input $E(s)$ by

$$R(s) = H(s)\, E(s)$$

where $H(s)$ is a rational function of s. $H(s)$ is generally called the *transfer function* of the system. The transfer function of the system can be defined as the ratio of the Laplace transform of the output to the Laplace transform of the input when all initial conditions are zero.

$$H(s) = \frac{R(s)}{E(s)} \qquad \ldots(4.18)$$

A physical interpretation of the transfer function can be given by selecting the input as a delta function.

i.e. $$e(t) = \delta(t)$$

Then $$E(s) = \mathcal{L}[\delta(t)] = 1$$

Hence, $$H(s) = R(s)\big|_{e(t)=\delta(t)}$$

i.e. the transfer function is the Laplace transform of the response when the input to the network is a delta function. The inverse transform of $H(s)$,

i.e. $$h(t) = \mathcal{L}^{-1}[R(s)]_{|E(s)=1} \qquad \ldots(4.19)$$

is called the *impulse response.* The impulse response is defined as the zero state response to a unit impulse input. Though we have resorted to Laplace transform technique, implying that the network is linear time-invariant, we can define impulse response of nonlinear and time-variant networks.

Example 4.5. Find the impulse response for the network shown in Fig. 4.1.

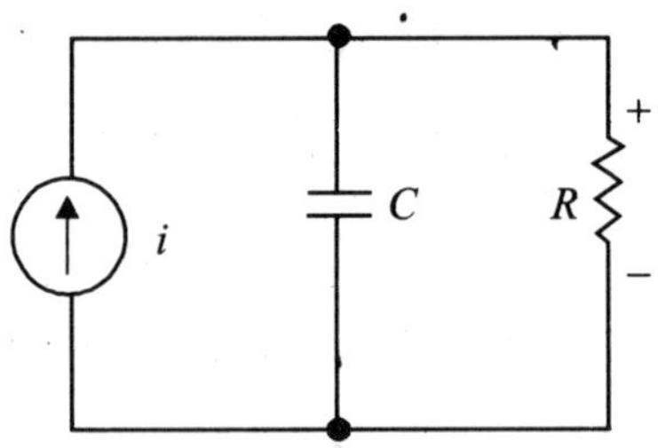

Fig. 4.1

The network equation for this can be written as (node equations)

$$C\frac{dv}{dt} + \frac{1}{R}v = i(t)$$

But, $$i(t) = \delta(t)$$

i.e. $$C\frac{dv}{dt} + \frac{1}{R}v = \delta(t)$$

Laplace transforming (initial capacitor voltage set to zero)

$$\left(sC + \frac{1}{R}\right)V(s) = 1$$

i.e. $$V(s) = \frac{R}{1 + sCR} = \frac{1/C}{s + 1/RC}$$

Inverse transforming, we have

$$v(t) = \frac{1}{C} e^{-t/RC} = h(t)$$

The impulse response is sketched in Fig. 4.2.

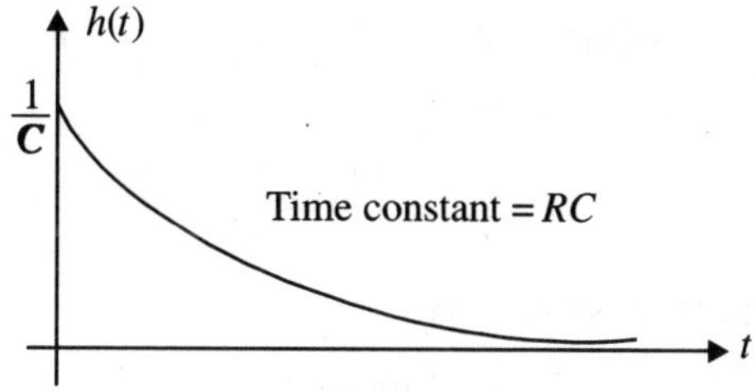

Fig. 4.2

The relation between the impulse response and the *step response* (*i.e.*, the response when the input is a unit step function) for a linear time-invariant network is contained in the following theorem.

Theorem 4.2: The impulse response of a linear time-invariant network is the time derivative of the step response, *i.e.*

$$h(t) = \frac{dS(t)}{dt} \qquad \ldots(4.20)$$

where $S(t)$ is the step response.

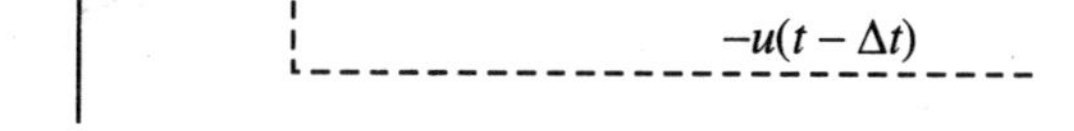

Fig. 4.3

We prove this by approximating the impulse by a pulse function, as shown in Fig. 4.3. The pulse function $p(t)$ can be represented in terms of two step functions.

$$p(t) = \frac{1}{\Delta t}\{u(t) - u(t - \Delta t\}$$

and $$\delta(t) = \operatorname*{Lim}_{\Delta t \to 0} p(t)$$

Let the response of a network to $p(t)$ be $r(t)$, *i.e.*

$$r(t)=\frac{1}{\Delta t}\{r[u(t)]-r[u(t-\Delta t)]\} \qquad \ldots(4.21)$$

where $r\,[u(t)]$ is the response of the network to an input $u(t)$. In writing eqn. 4.21 we have used the *principle of superposition* which is valid as the network is linear. $r\,[u(t)]$ is the step response of the network, *i.e.*

$$r\,[u(t)] = S(t)$$

As the network is time-invariant,

$$r\,[u(t-\Delta t)] = S(t-\Delta t)$$

Hence, from eqn. 4.21 we have

$$r(t)=\frac{S(t)-S(t-\Delta t)}{\Delta t}$$

But $h(t)$, the impulse response, is

$$h(t)=\lim_{\Delta t\to 0} r(t)$$

i.e.

$$h(t)=\lim_{\Delta t\to 0}\frac{S(t)-S(t-\Delta t)}{\Delta t}=\frac{dS(t)}{dt}$$

Hence, the theorem.

From eqn. 4.20, we have

$$S(t)=\int_{-\infty}^{t} h(\tau)\,d\tau \qquad \ldots(4.22)$$

Example 4.6. Step-response of the network of Fig. 4.1 can be obtained by solving the following equation

$$C\frac{dv(t)}{dt}+\frac{1}{R}v(t)=u(t)$$

Carrying out the Laplace transformation and rearranging we have $(\mathcal{L}u(t)=1/s)$

$$V(s)=\frac{1/C}{s(s+1/RC)}$$

By partial-fraction expansion

$$V(s)=\frac{R}{s}-\frac{R}{s+1/RC}$$

Hence,

$$S(t)=R(1-e^{-t/RC})u(t)$$

and

$$h(t)=\frac{dS(t)}{dt}=R\delta(t)(1-e^{-t/RC})+\frac{1}{C}e^{-t/RC}u(t)$$

Since $\delta(t)$ exists only for $t = 0$ the first term vanishes and, thus

$$h(t) = \frac{1}{C} e^{-t/RC} u(t) \text{ (see example 4.5)}$$

If the impulse response of a network is known, then we can find its response to any arbitrary input through the convolution integral. Linearity and time-invariance are the two crucial properties which facilitate us to employ such a procedure. To this end we obtain a representation of an arbitrary input in terms of *delta functions*. Figure 4.4 shows an arbitrary function of time. We can get an approximate representation of this function $e(t)$ by considering a number of pulse functions as shown in the figure. This is achieved by dividing the time axis into segments of width Δt.

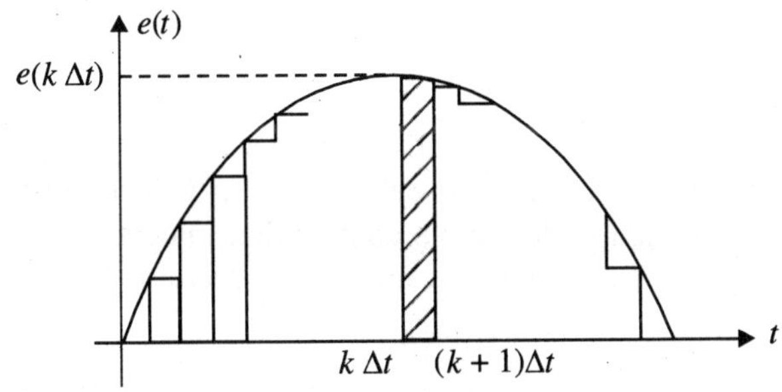

Fig. 4.4

These pulse functions, in turn, can be represented by two unit step functions as shown in Figure 4.5.

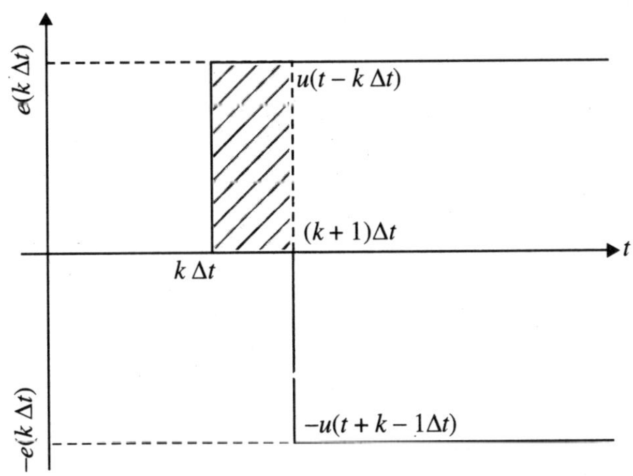

Fig. 4.5

Hence, we have

$$e(t) \approx \sum_{k=0}^{n} e(k\Delta t)\left\{u(t - k\Delta t) - u(t - \overline{k+1}\,\Delta t)\right\}$$

where n is the number of segments. The smaller the segment width the better is the approximation. In the limit $\Delta t \to 0$ ($n \to \infty$) we have an exact representation of $e(t)$. The preceding equation can be written as

$$e(t) \approx \sum_{k=0}^{n} e(k\Delta t) \frac{u(t-k\Delta t) - u\{t-(k+1)\Delta t\}}{\Delta t} \Delta t \qquad \ldots(4.23)$$

As $\Delta t \to 0$ the summation becomes an integral (infinite sum). Also noting that

$$\lim_{\Delta t \to 0} \frac{u(t) - u(t-\Delta t)}{\Delta t} = \delta(t)$$

reduces the eqn. 4.23, in the limit, to

$$e(t) = \int_0^t e(\tau)\delta(t-\tau)\, d\tau \qquad \ldots(4.24)$$

Equation 4.24 gives the representation of any function by the infinite number of weighted delta function. This is also called the '*sifting property*' of the delta function. We now proceed to derive the convolution integral.

Let a network N be described by its impulse response $h(t)$, *i.e.*

$$\delta(t) \to h(t)$$

Here $\to$ implies that $h(t)$ is the response to $\delta(t)$. If the system is time-invariant, then

$$\delta(t-\tau) \to h(t-\tau)$$

If in addition N is a linear network

$$e(\tau)\ \delta(t-\tau) \to e(\tau)\ h(t-\tau)$$

and consequently

$$\int_0^t e(\tau)\delta(t-\tau)\, d\tau \to \int_0^t e(\tau) h(t-\tau)\, d\tau$$

Then from eqn. 4.24 we have

$$e(t) \to \int_0^t e(\tau) h(t-\tau)\, d\tau$$

i.e. the response $r(t)$ to an arbitrary input $e(t)$ is given by

$$r(t) = \int_0^t e(\tau) h(t-\tau)\, d\tau \qquad \ldots(4.25)$$

This is called the *convolution integral*. The steps involved in this argument are shown in Fig. 4.6.

The right hand side of eqn. 4.25 is generally written as

$$e(t) * h(t)$$

where * indicates the convolution. Hence

$$r(t) = e(t) * h(t)$$

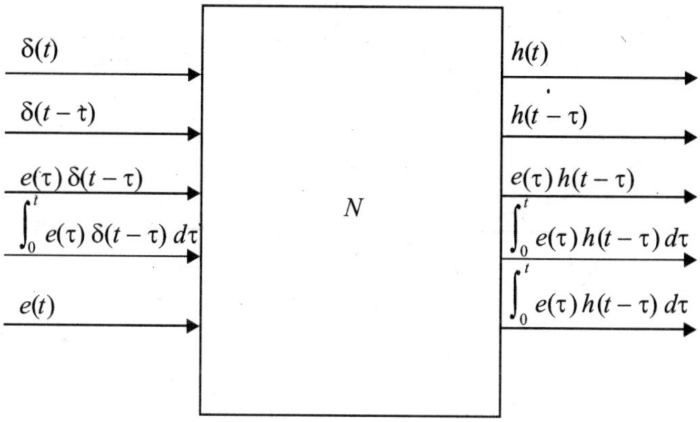

Fig. 4.6

Equation 4.25 implies that *"the response of a linear time-invariant network to an input e(t) is the convolution of the input with h(t), the impulse response of the network."* This very important property of linear time-invariant network facilitates the determination of the network response to an arbitrary input once the impulse response is known. It is hardly possible to overemphasize the importance of convolution integral in linear system analyses.

The convolution integral can also be graphically interpreted. Figure 4.7 shows the graphical development of convolution of two functions $f_1(t)$ and $f_2(t)$. Let $f_1(t) = e^{-t}\, u(t)$ and $f_2(t) = u(t)$. The convolution $f_1(t) * f_2(t)$ is given by

$$f(t) = \int_0^t f_1(t-\tau) f_2(\tau)\, d\tau = \int_0^t e^{-(t-\tau)}\, u(\tau)\, d\tau$$

Figures 4.7(*a*) and (*b*) show $f_1(\tau)$ and $f_2(\tau)$, while (*c*) shows $f_1(-\tau)$, (*d*) $f_1(t_1 - \tau)$ for a prescribed value of $t = t_1$ and (*e*) shows $f_1(f_1 - \tau)\, f_2(\tau)$. The area under the curve $f_1(t - \tau)\, f_2(\tau)$ gives the value of convolution integral for $t = t_1$. Figure 4.7(*e*) shows the convolution of $f_1(t)$ and $f_2(t)$. Thus, thc convolution of two functions can be graphically evaluated.

It can be easily shown that the convolution is symmetric (Problem 4.8), *i.e.*

$$e(t) * h(t) = h(t) * e(t)$$

It can also be shown that the Laplace transform of equation 4.25 yields $R(s) = H(s)\, E(s)$ (Problem 4.9).

Example 4.7. Consider the network shown in Fig. 4.1 with $R = 2$ and $C = 1$.

The impulse response $h(t)$ from example 4.5 is $e^{-1/2}u(t)$. (Here we have multiplied the response by $u(t)$ to indicate that the response exists for $t \geq 0$. It may be recalled that $u(t) = 0$ for $t < 0$ and $u(t) = 1$ for $t \geq 0$.)

Let $i_1(t) = u(t)$. Then the response V_2 is

$$v_2(t) = \int_0^t h(\tau) u(t-\tau)\, d\tau$$

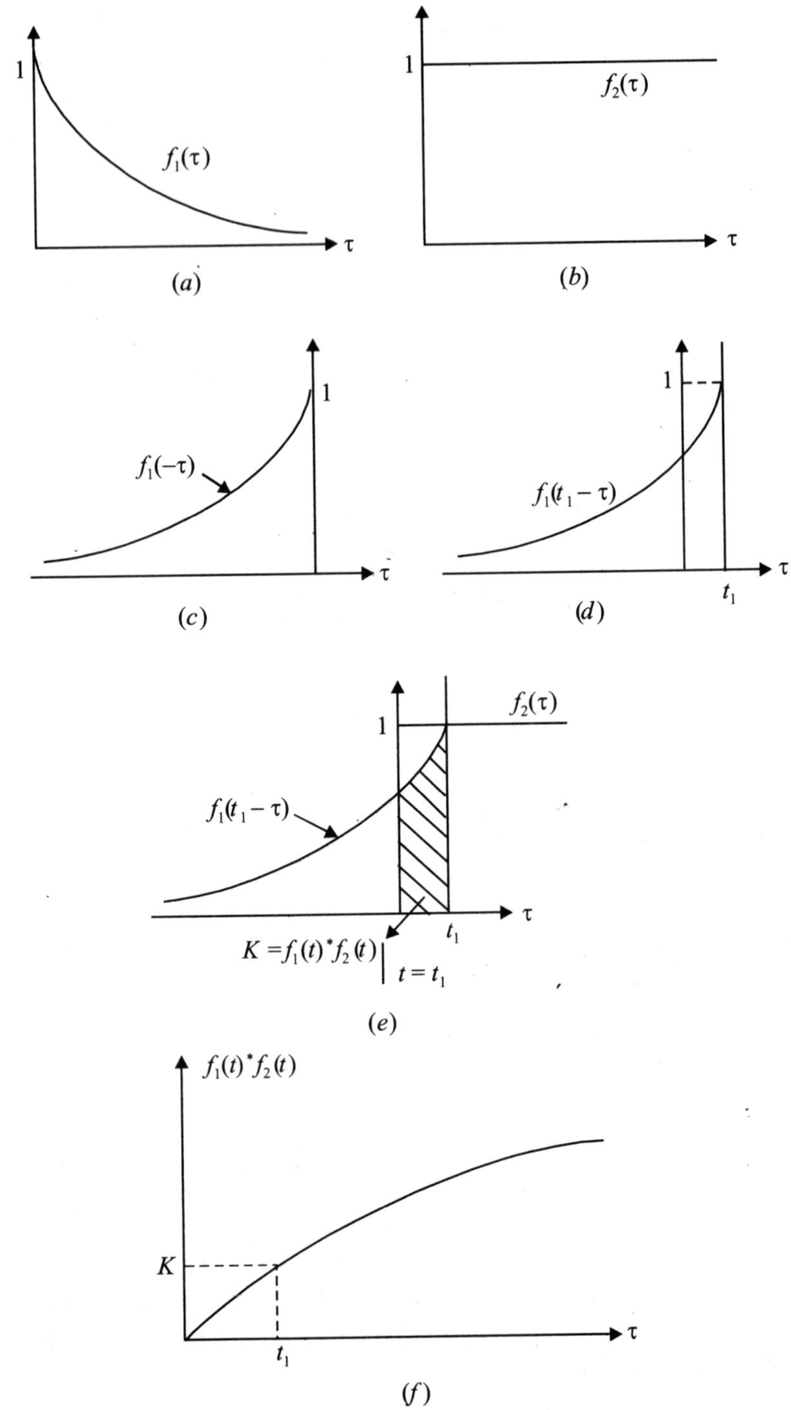
1
$f_1(\tau)$
τ
(a)
1
$f_2(\tau)$
τ
(b)
1
$f_1(-\tau)$
τ
(c)
1
$f_1(t_1-\tau)$
τ
t_1
(d)
$f_2(\tau)$
1
$f_1(t_1-\tau)$
τ
t_1
$K = f_1(t) * f_2(t)\big|_{t=t_1}$
(e)
$f_1(t) * f_2(t)$
K
τ
t_1
(f)

Fig. 4.7

$$= \int_0^t e^{-\tau/2} u(\tau) u(t-\tau)\, d\tau$$

$$= \left. \frac{e^{-\tau/2}}{-1/2} \right|_0^t$$

i.e.

$$v_2(t) = 2(1 - e^{-t/2}) u(t). \qquad \text{(see example 4.6)}$$

4.5 LINEAR TIME-INVARIANT NETWORKS

We have so far discussed some of the properties of linear time-invariant systems. In this section, we restate the properties with pertinent comments.

(*i*) For any linear time-invariant network the Laplace transform of the complete response is the sum of the Laplace transforms of the zero input response and the zero state response, *i.e.*

Complete Response = ZIR + ZSR

Because of the linearity of the Laplace transforms, the complete time response is also the sum of the ZI and ZS time responses.

(*ii*) The network function or transfer function is the ratio of the Laplace transform of the ZSR and the Laplace transform of the input, *i.e.*

$$H(s) = \left. \frac{R(s)}{E(s)} \right|_{\text{initial conditions}=0}$$

For a lumped network the network function is a rational function of s with real coefficients.

(*iii*) The initial conditions for the network are completely specified by the initial voltages of the capacitors and the initial currents of the inductors.

(*iv*) For any linear time-invariant network the network function is the Laplace transform of the inpulse response; *i.e.*

$$H(s) = \mathcal{L}[h(t)]$$

where $h(t)$ is the impulse response.

(*v*) For any linear time-invariant network, the response to any arbitrary input is given by the convolution of the input and the impulse response, *i.e.*

$$r(t) = \int_0^t h(t-\tau) e(\tau)\, d\tau$$

(*vi*) For any linear time-invariant network the impulse response is the time derivative of its step response, *i.e.*

$$h(t) = \frac{dS(t)}{dt}$$

(*vii*) A linear time–invariant network with positive resistances, capacitances and inductances has a unique solution for a given set of initial conditions and any set of inputs.

4.6 EVALUATION OF INITIAL CONDITIONS

The time domain mesh and node equations of a general RLC network are *integro-differential equations.* To obtain the general solution for such a set of equations we need to evaluate one or more arbitrary constants. To evaluate these constants some additional information must either be given with the equations or determined from the physical network itself. In the previous sections we have tacitly assumed that this information, the initial conditions, are 'somehow' specified. In this section we evaluate these initial conditions.

It a network does not contain any energy storage elements, the question of evaluating any initial conditions does not arise at all. The initial condition that need to be evaluated, for an RLC network, are the initial capacitor voltages and initial inductor currents. The initial conditions indeed depend on the history of the network. Let us suppose that the network is connected to an energy source through a switch (or switches) and let the switch (or switches) be closed at the instant of time $t = 0$. We designate $t = 0-$ and $t = 0+$ as the instant of time just before and after the operation of the switch. The values of the capacitor voltages and inductor currents at $t = 0+$, designated as $v_c(0+)$ and $i_l(0+)$, respectively are called the *initial conditions* of the network. (In the previous sections and Chapter 3 wherever we have $v_c(0)$ and $i_l(0)$ we should interpret them as $v_c(0+)$ and $i_l(0+)$). Of course, the values of $v_c(0-)$ and $i_l(0-)$ at $t = 0-$ are closely related to the corresponding values at $t = 0+$. The values at $t = 0+$ (*i.e.* $t > 0$) depend on both the past history of the network and configuration for $t > 0$.

The terminal relationship for a capacitor can be written as

$$i_c(t) = C\frac{dv_c(t)}{dt}$$

or

$$v_c(t) = \frac{1}{C}\int_{-\infty}^{t} i_c(x)\,dx$$

Hence, we have

$$v_c(0-) = \frac{1}{C}\int_{-\infty}^{0-} i_c(x)\,dx$$

and

$$v_c(0+) = \frac{1}{C}\int_{-\infty}^{0+} i_c(x)\,dx$$

$$= v_c(0-) + \frac{1}{C}\int_{0-}^{0+} i_c(x)dx \qquad \ldots(4.26)$$

For an inductor we can write the terminal relation as

$$v_l(t) = L\frac{di_1(0)}{dt}$$

or

$$i_l(t) = \frac{1}{L}\int_{-\infty}^{t} v_l(x)\,dx\cdot$$

Hence, we have

$$i_l(0) = \frac{1}{L}\int_{-\infty}^{0-} v_l(x)\,dx$$

and

$$i_l(0+) = i_l(0-) + \frac{1}{L}\int_{0-}^{0+} v_l(x)\,dx \qquad \text{...(4.27)}$$

In eqns. 4.26 and 4.27 the integrals are evaluated over an infinitesimally small interval and is always zero unless $i_c(x)$ and $v_l(x)$ are impulses at $x = 0$, respectively. This will only happen if the given network is *degenerate.* A network is said to be degenerate if there exists all capacitor circuits when voltage sources are shorted or all inductor cut-sets when current sources are opened, or both.

Consider an inductor-current source node (Fig. 4.8) in a network. The Kirchhoff's current law at $t = 0+$ is

$$i_s(0+) = i_{L_1}(0+) + i_{L_2}(0+)$$

$i_{L_1}(0-)$ and $i_{L_2}(0-)$ are, in general, independent of each other. Hence, in order to satisfy the constraint imposed by KCL the currents through the inductors must change instantaneously. An instantaneous change of current through an inductor implies impulse voltage across it.

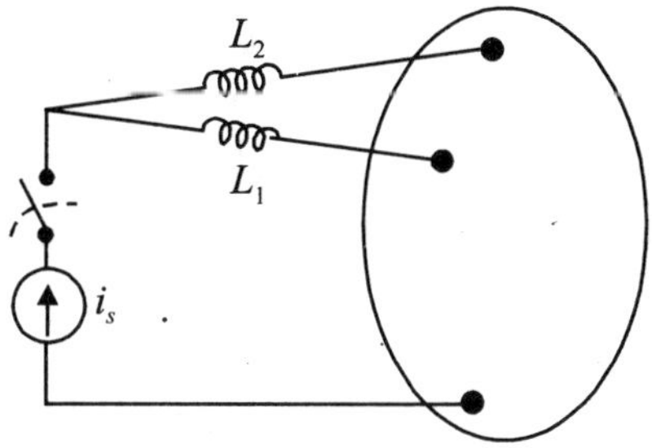

Fig. 4.8

In the case of a degenerate circuit (circuit of capacitors) the voltage across the capacitor must change instantaneously in order to satisfy the constraint imposed by the KVL at $t = 0+$. This requires an impulse current through the capacitors.

If the network is non-degenerate then, from equations 4.26 and 4.27,

$$v_c(0+) = v_c(0-) \qquad \text{...(4.28)}$$

$$i_l(0+) = i_l(0-)$$

i.e. the initial conditions are continuous at $t = 0$. If the network is degenerate, then the continuity condition no longer holds good.

In a physical network the inductances and capacitances do not appear as the idealized elements as we have represented them. The exact inductor (or capacitor) model needs a series and/or shunt resistor in addition to the inductor (capacitor). As such, the inductor currents and capacitor voltages are always continuous. Then why evaluate $i_l(0+)$ and $v_c(0+)$? We have already explained that 'modelling' involves certain idealization. Though ideal elements do not exist in physical world, a great many practical networks are designed on the idealized basis. If not, a simple network with only a few elements will become very complicated. Hence, we are stuck with ideal elements and degenerate networks. To evaluate the discontinuous initial condition we have to use two physical principles—conservation of charge and conservation of flux linkages, which cannot be derived from Kirchhoff's laws.

4.6.1 Network with Degenerate Circuits

We have seen that the voltage across a capacitor is discontinuous if the network has a circuit of capacitors only or capacitors and voltage sources. As we need to compute the discontinuities only, we can remove (open) all the resistors, inductors, current sources and capacitors not in the degenerate circuit and obtain a reduced network. The reduced network has only capacitors and voltage sources. To this network we apply KVL and equation of conservation of charge. To obtain the latter we apply the principle of conservation of charge to the reduced network:

The principle of conservation of charge states that the total charge transferred to a node of a network at any time is zero, *i.e.*

$$\sum_k q_k = 0 \text{ at a node} \qquad \ldots(4.29)$$

The charge transferred by the kth element from $t = 0-$ to $t = 0+$ is

$$\Delta q_k = \int_{0-}^{0+} i_k(t)\,dt \qquad \ldots(4.30)$$

This charge transfer is zero except for a capacitor in a degenerate circuit (*i.e.*, a capacitor in the reduced network). From eqns. 4.26 and 4.30

$$v_c(0+) = v_c(0-) + \frac{\Delta q}{C} \qquad \ldots(4.31)$$

The value of $v_c(0+)$ can be determined from equations 4.29, 4.31 and KVL.

Example 4.8. We are required to find the initial condition in the network shown in Fig. 4.9.

The network has degenerate circuits and the reduced network is obtained by opening all elements which do not have charge transfer capability at $t = 0$, *i.e.* elements R_1, R_2, R_3. The reduced network is shown in Fig. 4.10.

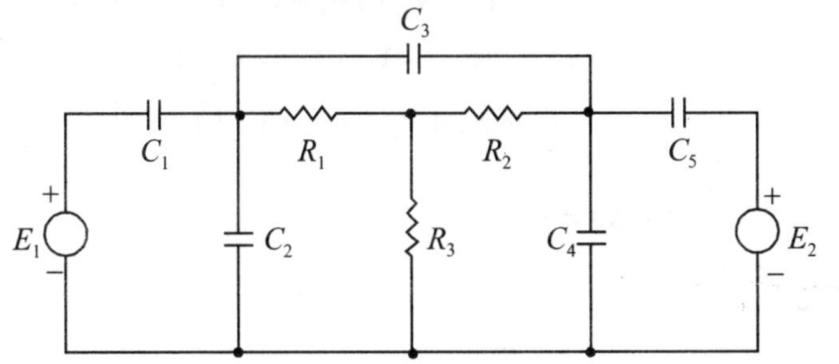

Fig. 4.9

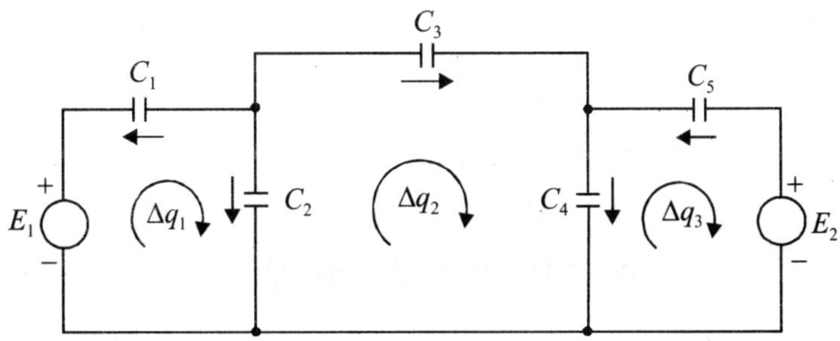

Fig. 4.10

(*i*) For all the capacitors the initial conditions are discontinuous at $t = 0$.

(*ii*) Let Δq_1, Δq_2 and Δq_3 be the three loop charges. Then from eqn. 4.29

$$\Delta q_{c_1} = \Delta q_1, \quad \Delta q_{c_2} = \Delta q_1 - \Delta q_2$$

$$\Delta q_{c_3} = \Delta q_2, \quad \Delta q_{c_4} = \Delta q_2 - \Delta q_3$$

$$\Delta q_{c_5} = -\Delta q_3$$

(*iii*) The KVL applied to the three circuits at $t = 0+$ yields

$$E_1 = V_{c_1}(0+) + V_{c_2}(0+)$$

$$0 = V_{c_3}(0+) + V_{c_4}(0+) - V_{c_2}(0+)$$

$$E_2 = V_{c_5}(0+) + V_{c_4}(0+)$$

(*iv*) From eqn. 4.31 and (*ii*) and (*iii*) we have

$$\begin{bmatrix} \frac{1}{C_1} + \frac{1}{C_2} & -\frac{1}{C_2} & 0 \\ -\frac{1}{C_2} & \frac{1}{C_2} + \frac{1}{C_3} + \frac{1}{C_4} & \frac{1}{C_4} \\ 0 & -\frac{1}{C_4} & \frac{1}{C_4} + \frac{1}{C_5} \end{bmatrix} \begin{bmatrix} \Delta q_1 \\ \Delta q_2 \\ \Delta q_3 \end{bmatrix}$$

$$= \begin{bmatrix} E_1 - V_{c_1}(0-) - V_{c_2}(0-) \\ V_{c_2}(0-) - V_{c_3}(0-) - V_{c_4}(0-) \\ -E_2 + V_{c_4}(0-) + V_{c_5}(0-) \end{bmatrix}$$

i.e. $C\Delta q = Eq$

Hence, we have

$$\Delta q = \begin{bmatrix} \Delta q_1 \\ \Delta q_2 \\ \Delta q_3 \end{bmatrix} = C^{-1}Eq$$

Let

$$C^{-1}\mathbf{E}q = \begin{bmatrix} Q_1 \\ Q_2 \\ Q_3 \end{bmatrix}, \text{ then}$$

$$\Delta q_1 = Q_1, \Delta q_2 = Q_2, \Delta q_3 = Q_3$$

(*v*) Hence, from (*ii*)

$$\Delta q_{c_1} = Q_1, \Delta q_{c_2} = Q_1 - Q_2, \Delta q_{c_3} = Q_2$$

$$\Delta q_{c_4} = Q_2 - Q_3, \Delta q_{c_5} = -Q_3$$

(*vi*) Finally, from (*v*) and eqn. 4.31

$$V_{c_1}(0+) = V_{c_1}(0-) + \frac{Q_1}{C_1}, V_{c_2}(0+) = V_{c_2}(0-) + \frac{Q_1 - Q_2}{C_2}$$

$$V_{c_3}(0+) = V_{c_3}(0-) + \frac{Q_2}{C_3}, V_{c_4}(0+) = V_{c_4}(0-) + \frac{Q_2 - Q_3}{C_4}$$

$$V_{c_5}(0+) = V_{c_5}(0-) - \frac{Q_3}{C_5}$$

4.6.2 Network with Degenerate Cut-Set

We have seen that the current through an inductor is discontinuous if the network has a cut-set of inductors only or inductors and current sources. As we need to compute the discontinuities, we can remove (short) all the resistors, capacitors, voltage sources and inductors not in the degenerate cut-set and obtain a reduced network. The reduced network has only inductors and current sources. To this network we apply KCL and equation of continuity of flux linkages. To obtain the latter we apply the principle of continuity of flux linkages to the reduced network.

The principle of continuity of flux linkages states that *the summation of all the flux linkages at any time over any circuit (closed loop) of a network is continuous, i.e.*

$$\sum_j L_j i_j(t) = 0 \text{ in a loop} \qquad \text{...(4.32)}$$

$L_j\, i_j$ is the flux linkage associated with an inductance L_j carrying a current $i_j(t)$. Let $\Delta\lambda_j$ be the flux linkage transfer at $t = 0$, *i.e.*

$$\Delta\lambda_j = L_j\,[i_j\,(0+) - i_j\,(0-)]$$

which yields

$$i_j(0+) = i_j(0-) + \frac{\Delta\lambda_j}{L_j} \qquad \text{...(4.33)}$$

Now, the sum of $\Delta\lambda_j$ around a loop, from eqn. 4.32 must be zero.

The value of $i_l\,(0+)$ can be determined from eqns. 4.32, 4.33 and KCL.

Example 4.9. The network shown in Fig. 4.11(*a*) has degenerate cut-sets and hence the initial inductor currents are not continuous.

(*i*) For elements with no flux transfer the initial conditions are continuous at $t = 0$, *i.e.*

$$V_{c_2}(0+) = V_{c_2}(0-)$$

(*ii*) The reduced network is shown in Fig. 4.11(*b*). Let $\Delta\lambda_4$ and $\Delta\lambda_6$ be the flux transferred by the current sources and $\Delta\lambda_3$ and $\Delta\lambda_5$ those of inductors. Then

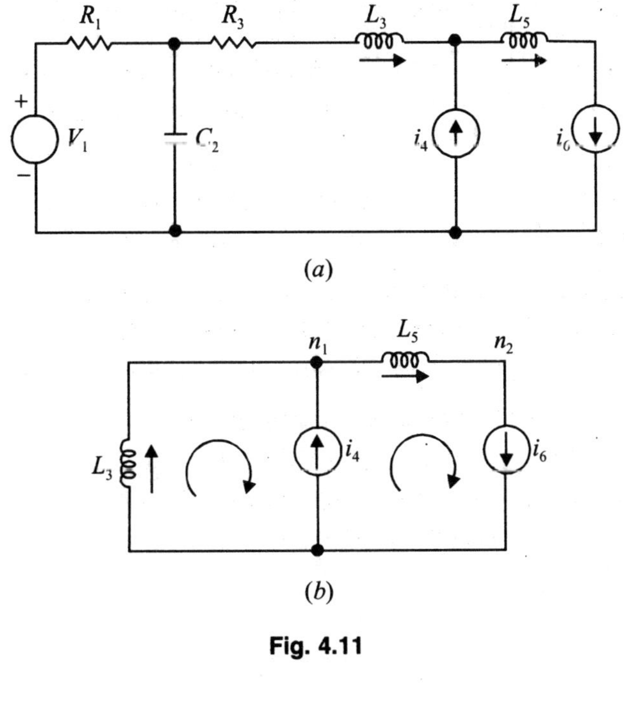

Fig. 4.11

from eqn. 4.32 with the loops as marked
$\Delta\lambda_3 = \Delta\lambda_4$, $\Delta\lambda_5 = -(\Delta\lambda_1 + \Delta\lambda_6)$

(*iii*) The KVL applied at nodes n_1 and n_2 at $t = 0+$ yields

$$i_4 = -iL_3(0+) + i_{L_5}(0+)$$

$$i_6 = i_{L_5}(0+)$$

(*iv*) From eqn. 4.33 and (*ii*) and (*iii*), we have

$$\begin{bmatrix} -\frac{1}{L_3} - \frac{1}{L_5} & -\frac{1}{L_5} \\ -\frac{1}{L_5} & -\frac{1}{L_5} \end{bmatrix} \begin{bmatrix} \Delta\lambda_4 \\ \Delta\lambda_6 \end{bmatrix} = \begin{bmatrix} i_4 + i_{L_3}(0-) - i_{L_5}(0-) \\ i_6 - i_{L_5}(0-) \end{bmatrix}$$

i.e. $L\Delta\lambda = j_q.$

Hence, we have $\Delta\lambda = \begin{bmatrix} \Delta\lambda_4 \\ \Delta\lambda_6 \end{bmatrix} = L^{-1} j_q$

Let $L^{-1} j_q = \begin{bmatrix} P_1 \\ P_2 \end{bmatrix}$

then $\Delta\lambda_4 = P_1,\ \Delta\lambda_6 = P_2$

(*v*) Hence, from (*ii*)

$$\Delta\lambda_3 = P_1, \Delta\lambda_5 = -(P_1 + P_2)$$

(*vi*) Finally, from (*v*) and eqn. 4.33

$$i_{L_3}(0+) = i_{L_3}(0-) + \frac{P_1}{L_3}$$

$$i_{L_5}(0+) = i_{L_5}(0-) - \frac{P_1 + P_2}{L_5}$$

If a given network has both degenerate circuits and cut-sets, there will be two separate reduced networks. The initial condition of the elements of these reduced networks is evaluated as in Secs. 4.5.1 and 4.5.2. For those elements not in either of these two reduced networks the initial conditions are continuous.

4.7 SINUSOIDAL STEADY STATE

The natural response of any network is, in general, an exponential or sum of exponential functions. If the network function of a passive network has complex poles, the response is a damped sinusoid. For example, consider the response of a network with respect to a pair of complex poles, *i.e.* let

$$F(s) = \frac{P(s)}{(s - p_1)(s - p_1^*)}$$

If we expand $F(s)$ into partial-fractions, we have

$$F(s) = \frac{A}{s - p_1} + \frac{A^*}{s - p_1^*}$$

If we let $p_1 = -\sigma + j\omega$ and $A = a + jb$, it can be shown that

$$f(t) = 2a\, e^{-\sigma t} \cos \omega t + 2b\, e^{-\sigma t} \sin \omega t$$

Thus, a sinusoid appears naturally. Indeed the motion of a pendulum, a bouncing ball and the vibration of a string all display a reasonably sinusoidal characteristic. Also, we know that any periodic function can be expressed as a sum of sinusoids, the well known Fourier series. Hence, the response of a linear system to a periodic input can be expressed as a sum of the responses to a series of sinusoidal inputs. Thus, sinusoidal waveforms play an important role in science and engineering. A mathematical advantage in using sinusoidal function is that their derivatives and integrals are also sinusoids.

The complete response of a linear time-invariant network for a sinusoidal input can be written as

$$y(t) = y_h(t) + y_p(t)$$

where $y_h(t)$ is the solution of the homogeneous differential equation (ZIR) and $y_p(t)$ the particular solution (ZSR). If the poles of the network function are all distinct, we can write

$$y_h(t) = \Sigma\, k_i \exp\, (p_i\, t)$$

where p_i is a pole of the network function and k_i is the arbitrary constant determined by the initial conditions. The response of a linear network to a sinusoidal input being a sinusoid of the same frequency as that of the input y_p will also be a sinusoidal function. If all the poles of the network function are restricted to the open left-half of the complex plane, the ZIR tends to zero as $t \to \infty$, *i.e.*

$$\lim_{t\to\infty} y(t) = \lim_{t\to\infty} y_p(t)$$

In other words, we can state that if all the poles of a network function are restricted to the open-left half of the complex plane, then irrespective of the initial conditions, the response of a linear network to a sinusoidal input will become sinusoidal as $t \to \infty$. This sinusoidal response is called the *sinusoidal steady-state response.*

If the network function has all its poles in the open left-half plane, we say that the network is *asymptotically stable.* If any of the poles are in the right-half plane, the network is *unstable.* An asymptotically stable network implies that the ZIR of this network tends to be zero as $t \to \infty$. Obviously, for an unstable network ZIR tends to ∞ as $t \to \infty$. For an asymptotically stable network driven by a sinusoidal source, the response of the network tends to its sinusoidal steady state independent of the initial conditions.

Sinusoidal steady state response can be easily determined by resorting to phasor method (see references). This method is illustrated by the following example.

Example 4.10. The network shown in Fig. 4.12 is in the sinusoidal steady state and we wish to find the current through the 2-ohm resistor.

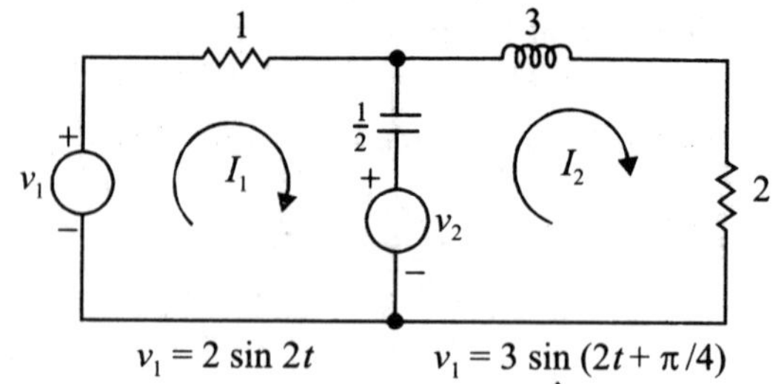

Fig. 4.12

(*a*) The phasors of v_1 and v_2 are $V_1 = 2e^{jo}$ and $V_2 = 3e^{j\pi/4}$ and $\omega = 2$.

(*b*) Let the phasors of the two mesh currents be I_1 and I_2.

(*c*) The mesh impedance of mesh 1 is $(1 - j1)$ and that of mesh 2 is $2 + j(6 - 1)$. Hence, the mesh equations are

$$\begin{bmatrix} 1-j1 & +j1 \\ +j1 & 2+j5 \end{bmatrix} \begin{bmatrix} I_1 \\ I_2 \end{bmatrix} = \begin{bmatrix} V_1 - V_2 \\ V_2 \end{bmatrix}$$

(*d*) Solving for I_2 by Cramers rule

$$I_2 = \frac{\begin{vmatrix} 1-j1 & V_1 - V_2 \\ j1 & V_2 \end{vmatrix}}{\Delta}$$

where $$\Delta = (1-j1)(2+j5) - (j1)(j1)$$
$$= 8 + 3j$$

i.e. $$\Delta = \sqrt{73} \tan^{-1}\left(\frac{3}{8}\right).$$

Hence, $$I_2 = \frac{(1-j1)V_2 - j1(V_1 - V_2)}{\Delta}$$

$$= \frac{(1-j1)3e^{j\pi/4} - j1(2 - 3e^{j\pi/4})}{\sqrt{73}\,\underline{|\tan^{-1}\frac{3}{8}}}$$

$$= \frac{\sqrt{2e^{-j\pi/4}}\, 3e^{j\pi/4} - j1[2 - (2.121 + j2.121)]}{\sqrt{73}\,\underline{|20.56}}$$

$$= \frac{(3\sqrt{2} - j(-0.121 - j2.121)}{\sqrt{73}\,\underline{|20.56}}$$

$$= \frac{(3\sqrt{2} - 2.121) + j0.121}{\sqrt{73}\,\underline{|20.56}}$$

$$= \frac{2.121 + j0.121}{\sqrt{73}\ \underline{|20.56}} = \frac{2.125\underline{|3.275}}{8.544\ \underline{|20.56}}$$

i.e. $$I_2 = 0.249\ \underline{|-17.29}$$

(*e*) Finally

$$i_2 = 0.249 \sin(2t - 0.096\pi).$$

4.8 RESONANCE

Resonance is a phenomenon encountered in circuits with both capacitors and inductors. Resonant frequency can be defined as "the frequency at which an RLC network behaves like a purely resistive network" or "the frequency at which the current and the voltage at the terminals of the circuit are in phase." At resonance the network has unity power factor. Besides discussing parallel and series resonances, we introduce the very important aspect of '*Q*' of a circuit.

4.8.1 Parallel Resonance

Consider the parallel RLC circuit shown in Figure 4.13. The circuit is driven by a sinusoidal current source of frequency ω. The terminal relationship of the circuit can be written as

$$I = Y(j\omega)\ V$$

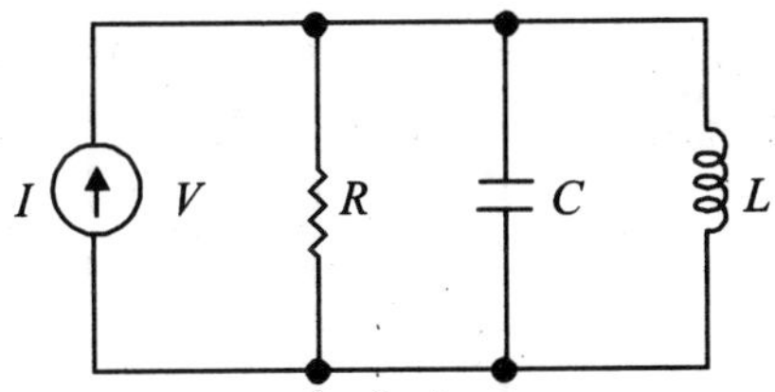

$I = I_m \sin \omega t$

Fig. 4.13

where $$Y(j\omega) = \frac{1}{R} + j\left(\omega C - \frac{1}{\omega L}\right) \quad \text{...(4.34)}$$

The current and voltage will be in phase when

$$\omega C = \frac{1}{\omega L}$$

Designating the *resonant frequency* as ω_0 we have

$$\omega_0 = \frac{1}{\sqrt{LC}} \quad \text{or} \quad f_0 = \frac{1}{2\pi\sqrt{LC}} \quad \text{...(4.35)}$$

The admittance and susceptance variations are plotted in Figs. 4.14(*a*) and (*b*), respectively. It can be observed from the figure and equation 4.34 that for $\omega < \omega_0$

the circuit is inductive and for $\omega > \omega_0$ it is capacitive. At low frequencies most of the current flows through the inductor and at high frequencies through the capacitor. At resonance, all the current goes through the resistor and the loop formed by the inductor and capacitor has a local circulating current (and hence the name tank circuit). Fig. 4.15 shows the locus of the impedance of the parallel circuit. The

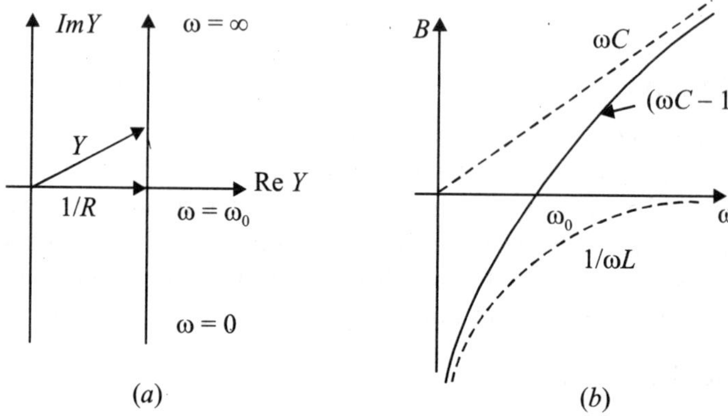

Fig. 4.14

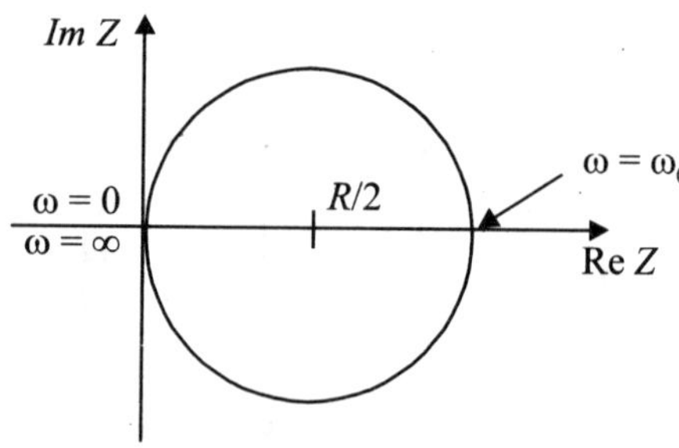

Fig. 4.15

locus is a circle with radius $R/2$ and at the centre at $(R/2, 0)$. If the parallel circuit consists of only capacitor and inductor, then at the resonance it will be an open circuit and thus a parallel circuit has maximum impedance at resonance.

The resonant frequency can also be found from the pole zero plot of the admittance function. The admittance of the parallel circuit is

$$Y(s) = \frac{1}{R} + sC + \frac{1}{sL}$$

$$= C\frac{s^2 + s/RC + 1/LC}{s}$$

The zeros of the function are at

$$s_1 = -\alpha + j\omega_d$$

$$s_2 = -\alpha - j\omega_d$$

where $$\alpha = \frac{1}{2RC} \text{ and } \omega_d = \sqrt{\omega_0^2 - \alpha^2} \qquad ...(4.36)$$

Figure 4.16 shows the *pole zero plot* of the admittance. Here α is called the *damping factor* and ω_d the *damped natural frequency.* Once s_1 and s_2 are known α, ω_α and ω_0 can be easily determined. Observe that ω_0 is the radial distance of the zero from the origin.

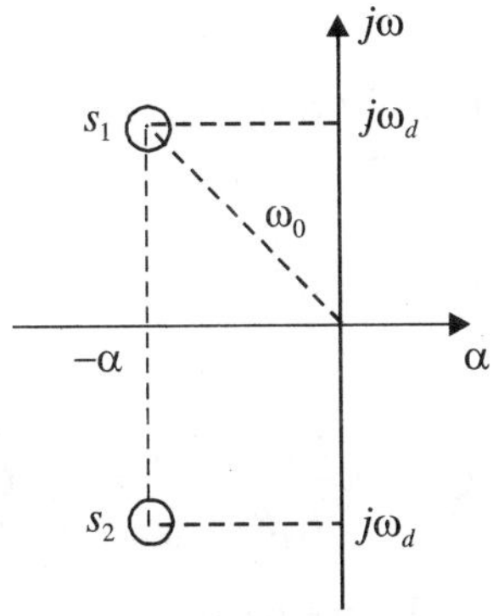

Fig. 4.16

Q of the circuit can be introduced through the voltage response of the parallel circuit as shown in Fig. 4.17. The maximum value of the response depends only on R, but the width and the steepness of the response depends on other elements

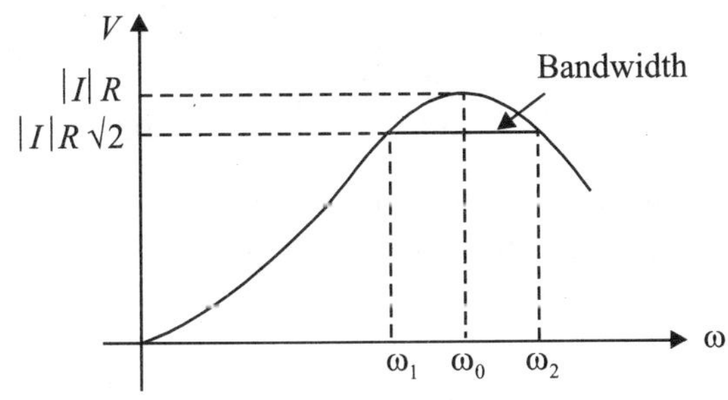

Fig. 4.17

also. The width of the response is usually measured at the *half power points* (or 3 dB points) and this width is called the *Bandwidth* of the circuit. The bandwidth of the circuit depends on the *quality factor* of the circuit. The quality factor Q is defined as

$$Q = 2\pi \frac{\text{Max. energy stored at resonance}}{\text{Total energy lost/cycle at resonance}}$$

$$= 2\pi \frac{[W_L + W_C]_{\max}}{P_R}$$

where W_L and W_C are the energy stored in the inductor and the capacitors, and P_R the energy dissipated per cycle by the resistor, respectively.

Let $i(t) = I_m \cos \omega t$; then the voltage response at resonance is

$$V(t) = RI_m \cos \omega t$$

Hence,

$$W_C = \frac{1}{2} CR^2 I_m^2 \cos^2 \omega t$$

$$W_L = \frac{1}{2} CR^2 I_m^2 \sin^2 \omega t$$

and

$$P_R = \frac{1}{2} \frac{I_m^2 R}{f_0}$$

Thus, finally,

$$Q = 2\pi \frac{I_m^2 R^2 C/2}{I_m^2 R/2f_0} = \omega_0 RC \qquad \text{...(4.37)}$$

From equations 4.35 and 4.37, we have

$$Q = \omega_0 RC = R\sqrt{\frac{C}{L}} = \frac{R}{\omega_0 L} \qquad \text{...(4.38)}$$

Equation 4.38 gives the Q of the parallel resonant circuit, and together with 4.36, it yields

$$\alpha = \frac{\omega_0}{2Q}, \quad \omega_d = \omega_0 \sqrt{1 - \left(\frac{1}{2Q}\right)^2} \qquad \text{...(4.39)}$$

indicating that the Q-factor is inversely proportional to the damping α. From eqns. 4.34 and 4.38 it can be shown that the two *half-power frequencies* ω_1 and ω_2 (Fig. 4.17) are given by (Problem 4.17)

$$\omega_1 = \omega_0 \left[\sqrt{1 + \frac{1}{4Q^2}} - \frac{1}{2Q} \right]$$

and

$$\omega_2 = \omega_0 \left[\sqrt{1 + \frac{1}{4Q^2}} + \frac{1}{2Q} \right] \qquad \text{...(4.40)}$$

For high Q circuits we can approximate the expressions further

$$\omega_1 \approx \omega_0 \left(1 - \frac{1}{2Q}\right) \qquad \text{...(4.41)}$$

$$\omega_2 \approx \omega_0 \left(1 + \frac{1}{2Q}\right)$$

and we see that $\omega_0 \approx \omega_d$. Equation 4.40 (or 4.41) yields the 3 dB bandwidth as

$$BW = \omega_2 - \omega_1 = \frac{\omega_0}{Q} \qquad \text{...(4.42)}$$

showing the relationship between BW, W_0 and Q.

Now, we are in a position to draw some conclusions from the response curve of 4.17. If the circuit Q is small, the BW is large, the response curve is flat and the *frequency selectivity* of the circuit is poor. On the other hand, a high Q circuit has a low bandwidth, the response curve is high peaked and hence the circuit is highly selective. The parallel resonant circuit fundamentally behaves like a *band-pass filter* (*i.e.* a network which transmits signals with frequencies between ω_1 and ω_2 with very little attenuation and signals at other frequencies with large attenuation).

As stated earlier, at resonance all the current flows through the resistor. The capacitor and inductor currents do not affect the overall current input to the circuit. However, these currents can be much larger than the input current, depending upon the Q of the circuit. At resonance, the voltage across the parallel combination is given by

$$V_0 = IR.$$

The currents through the capacitor and inductor are

$$I_C = j\omega_0 C V_0 = jQI, \qquad \text{...(4.43)}$$

$$I_L = \frac{V_0}{j\omega_0 L} = -jQI.$$

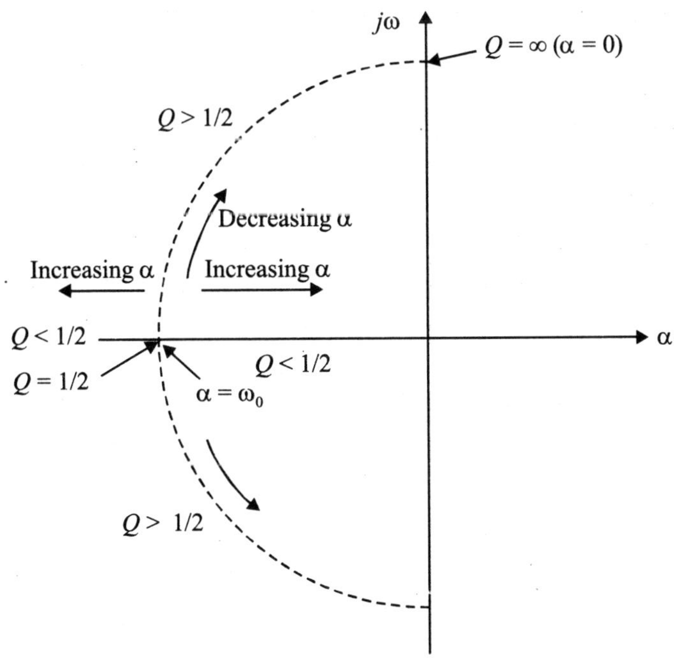

Fig. 4.18

Thus, we see that I_C and I_L can be much larger than I, the overall circuit current.

The relationship between Q and the damping factor α is an interesting one to study and is shown in Fig. 4.18. Q is inversely proportional to damping. When the damping is zero, the Q is infinity indicating a *loss-less circuit.* At the *critical damping, i.e.* $\alpha = \omega_0$, Q is $\frac{1}{2}$. (The two roots are coincident at $Q = \frac{1}{2}$).

4.8.2 Series Resonance

Consider the series RLC circuit shown in Fig. 4.19. This circuit is the dual of the one shown in Fig. 4.13. The impedance of the circuit is

$$Z(j\omega) = R + j\left(\omega L - \frac{1}{\omega C}\right)$$

The resonance occurs at the frequency (the resonance frequencies of parallel and series circuits are the same)

$$\omega_0 = \frac{1}{\sqrt{LC}} \qquad \ldots(4.44)$$

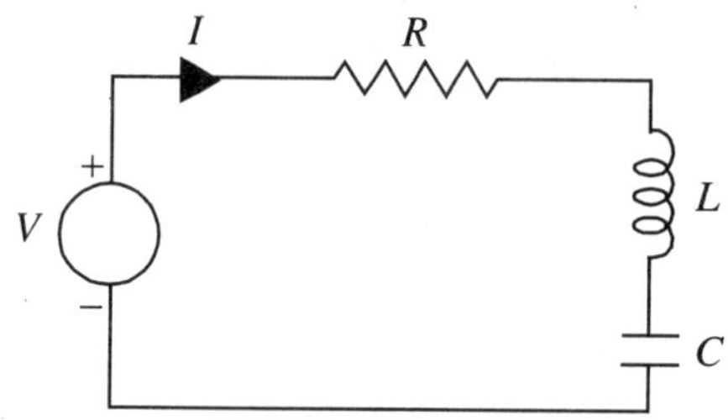

Fig. 4.19

We can draw figures similar to the one shown in Figs. 4.14 to 4.17. From a knowledge of dual networks we can write (Problem 4.18)

$$Q = \frac{\omega_0 L}{R} = \frac{1}{R}\sqrt{\frac{L}{C}} \qquad \ldots(4.45)$$

$$BW = \frac{\omega_0}{Q}$$

At resonance the voltage drop across the resistor is the supply voltage. The voltage drops across the capacitor and inductor are equal and opposite, and are given by

$$V_C = -jQV$$

$$V_L = jQV \qquad \ldots(4.46)$$

The voltage across L or C can be considerably larger than the supply voltage, depending upon the Q of the circuit. If the series circuit consists of only capacitor and inductor, at resonance it will be a short circuit. The series resonant circuit has maximum admittance at resonance.

In the design of frequency selective electric filters (low-pass, high-pass, band-pass and band elimination) the specifications are generally given in the frequency domain, *i.e.* the frequency response (the response of the network to a unit sinusoid of varying frequency) of the filter is specified. Highly selective filters have a large Q-factor. The Q-factor of an RLC network to a great extent depends on the quality of coils used in the design. The Q-factor of a coil, with an inductance of L henries and a resistance of r Ω at a frequency ω is given by $\omega L/r$.

4.9 FREQUENCY AND IMPEDANCE SCALING

In most of the examples we have considered the values of the elements R, L and C, which have been of the order of unity. It is very difficult, if not impossible, to build a capacitor of 1 farad. Beside, the practical value of capacitors available in electronic circuit is of the order of microfarads. The circuits considered so far have normalized the element values. There are mainly two reasons for resorting to normalized designs. The first reason is simplicity in numerical computation. It is easier to manipulate numbers of the order of unity. The round-off errors that occur in normalized designs is less severe. The second reason is that if we have a normalized design of, say a band-pass filter, it is easy to generate band-pass filters of similar characteristics of varying centre frequencies and impedance levels without re-designing the whole circuit. To obtain the element values of the required band-pass filter we have to resort amplitude (impedance), and frequency scaling the normalized design.

Consider the series resonant circuit of Sec. 4.8.2

$$\text{Impedance level} = Z_0 = R$$

$$\text{Resonance frequency} = \omega_0 = \sqrt{\frac{1}{LC}}$$

$$\text{Bandwidth} = \Delta\omega = \frac{\omega_0}{Q} = \frac{R}{L}$$

Hence, we have

$$L = \frac{R}{\Delta\omega} = \frac{R}{\omega_0}Q$$

$$C = \frac{I}{L\omega_0^2} = \frac{\Delta\omega}{R\omega_0^2} = \frac{1}{\omega_0 QR} \qquad \ldots(4.47)$$

Let the normalized impedance be 1 Ω and the normalized resonant frequency be 1 rad/sec. Then the normalized elemental values of the resonant circuit are

$$R_n = 1$$

$$L_n = \frac{R}{\omega_0}Q = Q \qquad \ldots(4.48)$$

$$C_n = \frac{1}{Q}$$

The values of *R, L* and *C* are different, but the value of *Q* is unaffected by normalization. To obtain the actual or de-normalized values we have to change the impedance level to Z_0 and the frequency to ω_0. The actual values of the elements are obtained as follows:

Let γ_n and Ω_n be the *impedance* and *frequency normalizing factors.*

$$\gamma_n = \frac{\text{desired impedance level}}{\text{normalized impedance level}}$$

$$\Omega_n = \frac{\text{desired frequency}}{\text{normalized frequency}}$$

(It is not always necessary to work with a normalized impedance and frequency level of 1 Ω and 1 Hz, respectively.) Then

$$R = \gamma_n R_n$$

$$L = \frac{\gamma_n}{\Omega_n} L_n \qquad \text{...(4.49)}$$

$$C = \frac{C_n}{\gamma_n \Omega_n}$$

It is easy to verify these results. Insofar as changes in impedance level are concerned, the capacitor gets divided by γ_n as the impedance of the capacitor is $1/(j\omega)C$. Let L_n and L_1 be the normalized and denormalized inductances, respectively. Let $\omega_n(X_n)$ and $\omega_1(X_1)$ be the normalized and denormalized frequencies (reactances), respectively. In order not to change the property of the circuit we must have

$$X_n = X_1$$

i.e. $$\omega_n L_n = \omega_1 L_1$$

i.e. $$L_1 = \frac{\omega_n L_n}{\omega_1} = \frac{L_n}{\Omega_n}$$

Similarly, we can show that the transformation for the capacitance is as given by equation 4.49. Of course, the resistance is not affected by frequency normalization.

Example 4.11. The circuit shown in Fig. 4.20 is to be scaled to an impedance level of 5 kΩ and a resonant frequency of 5 M rad/sec.

For the given circuit

$$\omega_0 = \frac{1}{\sqrt{LC}} = 1 \text{ rad/sec}$$

$$Q = R\sqrt{\frac{C}{L}} = 5$$

$$BW = \frac{1}{5} = 0.5 \text{ rad/sec}$$

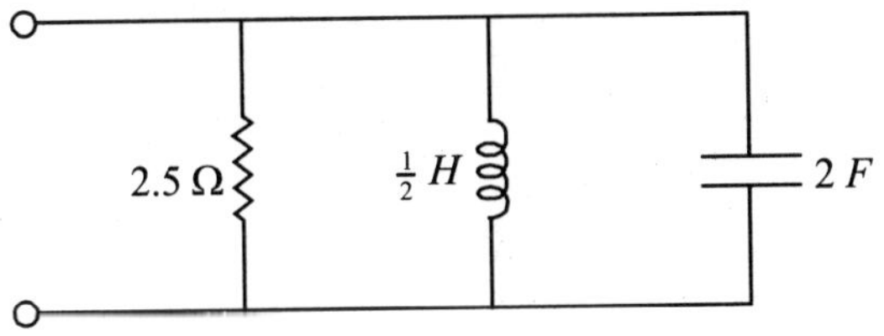

Fig. 4.20

The impedance and frequency normalizing factors are

$$\gamma_n = \frac{5\times 10^3}{2.5} = 2\times 10^3;$$

$$\Omega_n = \frac{5\times 10^6}{1}$$

$$= 5\times 10^6$$

Hence, the value of the elements of the denormalized network use

$$R = \gamma_n R_n = 2\times 10^3 \times 2.5 = 5\,\text{k}\Omega$$

$$L = \frac{\gamma_n}{\Omega_n} L_n = 0.2 \text{ mH}$$

$$C = \frac{C_n}{\gamma_n \Omega_n} = 200 \text{ pF}$$

It can be observed that Q is the same as before. The bandwidth of the denormalized circuit is

$$BW = \frac{5\times 10^6}{5} = 10^6 \text{ rad/sec.}$$

PROBLEMS

4.1 The following matrices are supposed to be node admittance matrices of passive linear time-invariant networks. Which ones do you accept as correct, and why?

$$\begin{bmatrix} 4 & -3 \\ -3 & 6 \end{bmatrix} \begin{bmatrix} 4 & -3 \\ -2 & 5 \end{bmatrix}$$

$$\begin{bmatrix} -1+s & -s \\ s & 4+s \end{bmatrix} \begin{bmatrix} 6s+3/s & -4s \\ -4s & 5s+3 \end{bmatrix}$$

$$\begin{bmatrix} 1+1/s+2s & -(s+2/s) \\ -(s+2/s) & 1+s+3/s \end{bmatrix} \begin{bmatrix} 2+s+2/s & -2/s \\ 3s & 2/s+4 \end{bmatrix}$$

4.2 Find the inverse transforms of the following rational functions:

(*i*) $\dfrac{s^2+10s+16}{s^2+8s+7}$ (*ii*) $\dfrac{(s+1)(s+5)}{(s+2)(s+4)(s+6)}$

(*iii*) $\dfrac{s^2+3s+2}{s(s+3)^2(s+4)}$ (*iv*) $\dfrac{s}{(s+9)(s+1)}$

(*v*) $\dfrac{s+1}{(s+2)(s^2+s+1)}$ (*vi*) $\dfrac{s^2+2s+1}{(s+2)^2(s^2+4)(s^2+s+1)}$

4.3 Find the elemental currents and voltages in the networks shown in Fig. P. 4.3.

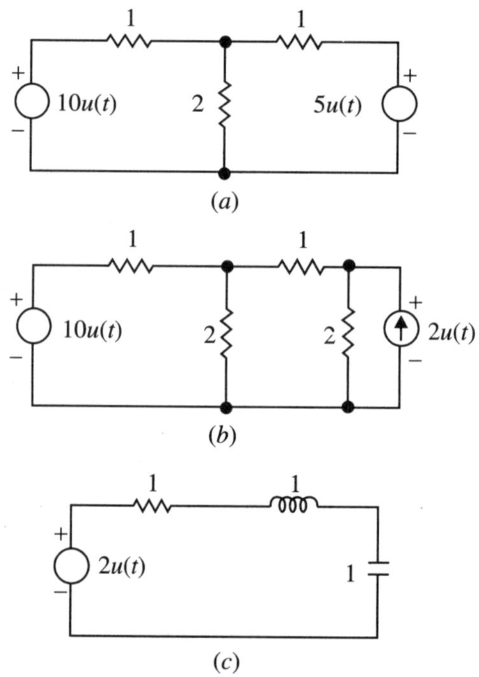

Fig. P. 4.3

4.4 Sketch the step and the impulse response of the circuits shown in Fig. P. 4.4.

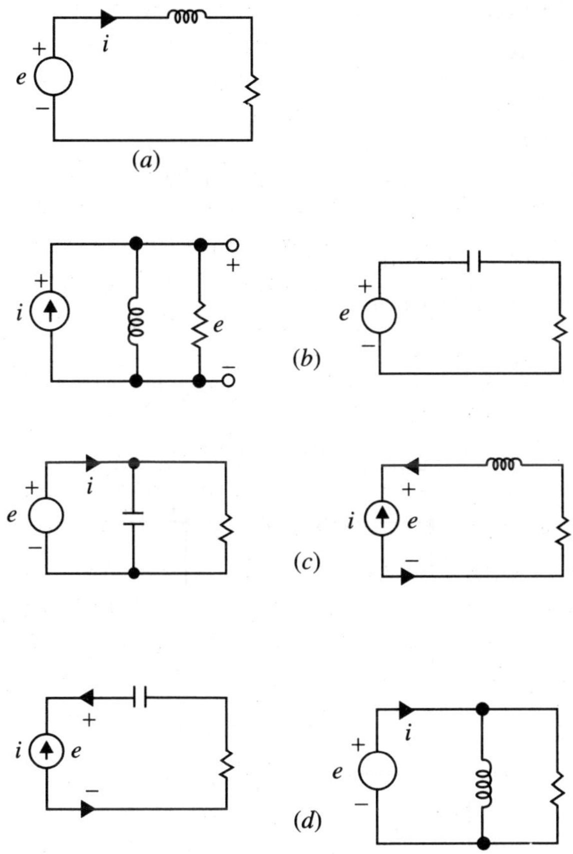

Fig. P. 4.4

4.5 Find the impulse for the network shown in Fig. P. 4.5.

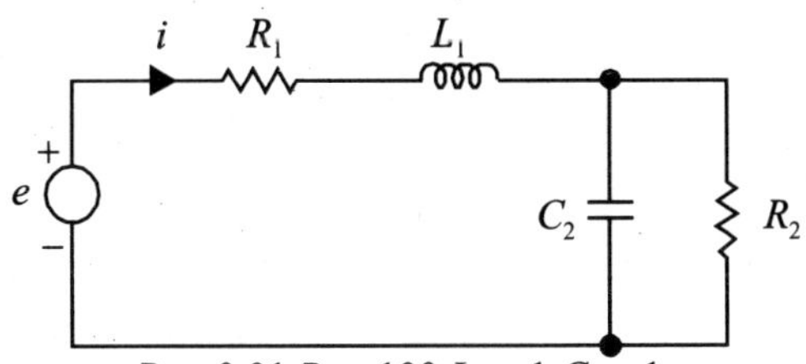

$R_1 = 0.01$ $R_2 = 100$ $L_2 = 1$ $C_2 = 1$

Fig. P. 4.5

4.6 The impulse response of a network is

$$h(t) = e^{-1},\ t \geq 0$$
$$= 0,\ t < 0$$

Find the response to an input $e(t) = u(t) - u(t - t_1)$.

4.7 Evaluate the convolution integrals

(*a*) $f(t) = e^{-t} * \delta(t - t_1)$. (What do you conclude?)

(*b*) $f(t) = f_1(t) * f_2(t)$

$f_1(t) = e^{-2t}$

$f_2(t) = u(t) - u(t-3)$

(*c*) $f(t) = f_1(t) * f_2(t)$; $f_1(t) = f_2(t) = u(t) - u(t - T)$

4.8 Show that

$$f_1(t) * f_2(t) = f_2(t) * f_1(t)$$

4.9 If $r(t) = h(t) * e(t)$, show that

$$R(s) = H(s)\,E(s)$$

4.10 Find the initial conditions in the networks shown in Fig. P. 4.10.

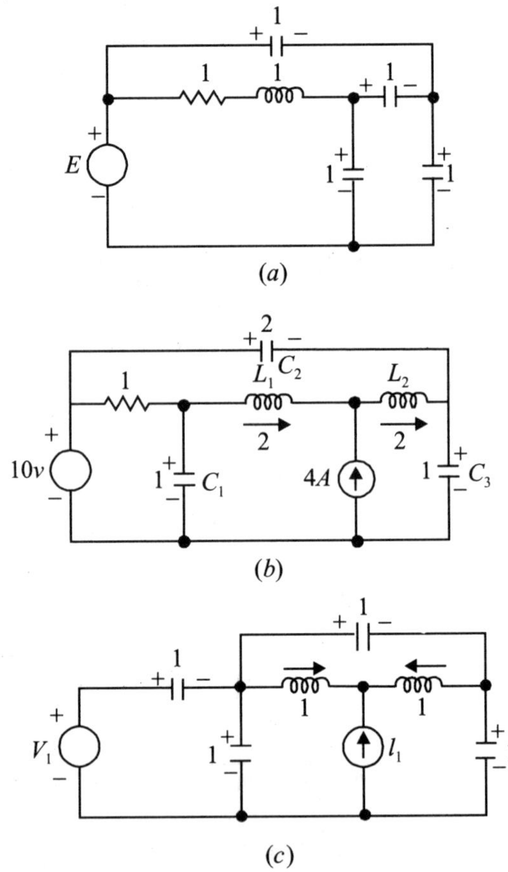

Fig. P. 4.10

4.11 Figure P. 4.11 shows an equivalent circuit of a common base transistor amplifier. Find V_e/V_i for the amplifier.

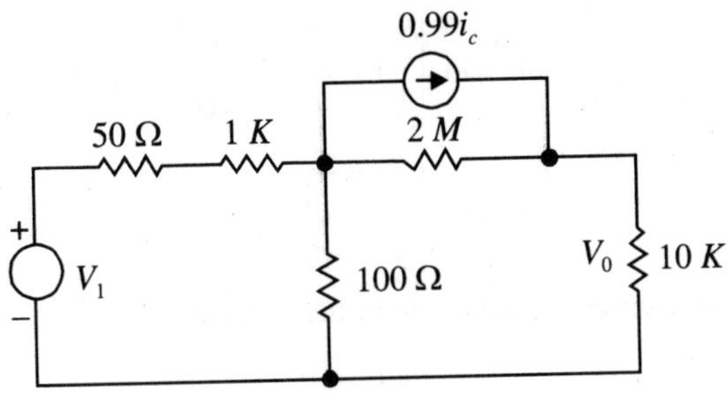

Fig. P. 4.11

4.12 The response of a linear time-invariant network for an input $x(t)$ is $y(t)$. Show that the response of the network to the input $\frac{dx(t)}{dt}$ is $\frac{dy(t)}{dt}$.

4.13 Determine the current through R, L and C and the source in the network of Fig. P. 4.13

$R = 10$ kΩ, $L = 1$ mH, $C = 1$ μF, $V = 10 \sin \omega t$,

$\omega = 4$ k rad/sec,

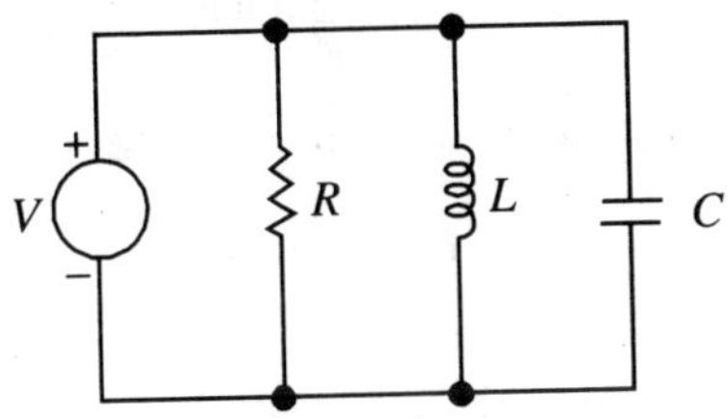

Fig. P. 4.13

4.14 Determine the loop currents I_1 and I_2 (Fig. P. 4.14). $V = 2 \sin t$, $V_2 = 3 \sin (t + \pi/4)$.

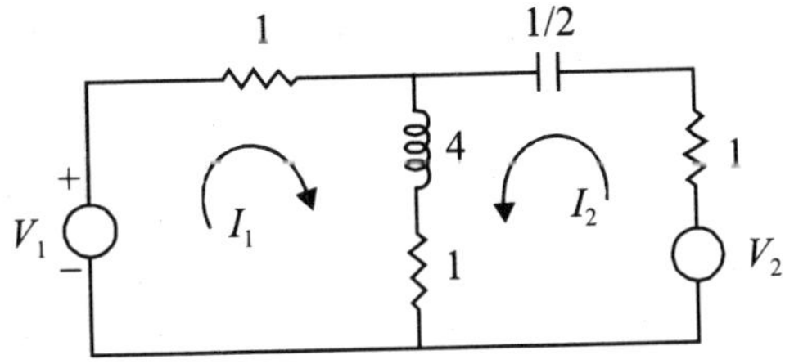

Fig. P. 4.14

4.15 A parallel resonant circuit has $R = 10$ kΩ, $L = 150$ mH and $C = 0.47$ μF. Find its resonant frequency, Q and bandwidth.

4.16 A parallel resonant circuit driven by a current source of $0.12 \sin \omega t$ shows a maximum voltage response amplitude of 100 V at $\omega = 5$ k rad/sec and the response at $\omega = 4$ k rad/sec is 60 volts. Find the parameters of the resonant circuit. What is its Q?

4.17 Show that for a parallel resonant circuit the upper and lower cut-off frequencies are given by

$$\omega_1 = \omega_0 \left[\sqrt{1 + \frac{1}{4Q^2}} - \frac{1}{2Q} \right]$$

$$\omega_2 = \omega_0 \left[\sqrt{1 + \frac{1}{4Q^2}} + \frac{1}{2Q} \right]$$

4.18 Show that for a series resonant circuit the Q-factors is

$$Q = \frac{\omega_0 L}{R} = \frac{1}{R}\sqrt{\frac{L}{C}}$$

4.19 Design a parallel resonant circuit to resonate at 1 rad/sec with a Q of 5 and a resistance of 1 Ω.

Frequency and impedance scale have an impedance of 10 kΩ and a bandwidth of 500 Hz.

4.20 Determine the Q of the circuit shown in Fig. P. 4.20 when

(*a*) $g = 0$, (*b*) $g = 3.5 \times 10^{-4}$, and (*c*) $g = -4 \times 10^{-3}$

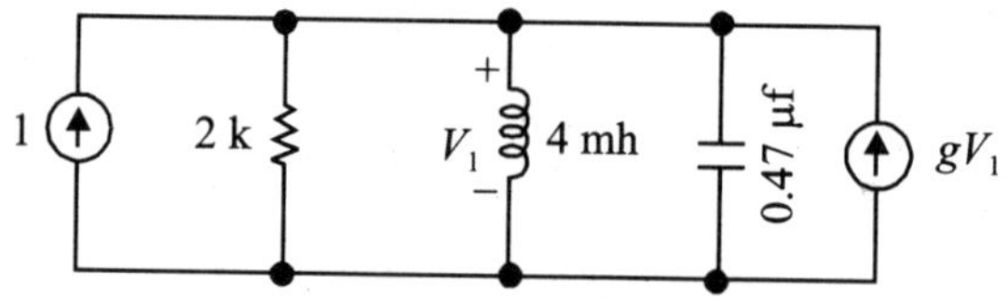

Fig. P. 4.20

4.21 Determine ω_0 and Q of the circuit shown in Fig. P. 4.21 when

(*a*) $r = 0$, (*b*) $r = 6 \times 10^3$ and (*c*) $r = -99$.

What can you conclude?

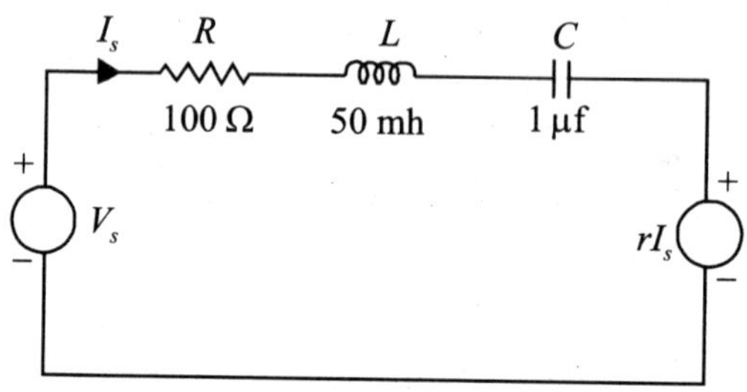

Fig. P. 4.21

4.22 A parallel resonant circuit (R_p, L and C) and a series resonant circuit (R_s, L and C) have the same Q. Find the relation between R_p and R_s.

4.23 Figure P. 4.23 shows a parallel resonant circuit using a physical inductor. What is its resonant frequency?

If L = mH, R_1 = 0.1Ω, C = 0.11 μF and R_2 = 1 MΩ find the resonant frequency, impedance at resonance, the maximum impedance and the frequency at which it occurs and the Q of the circuit.

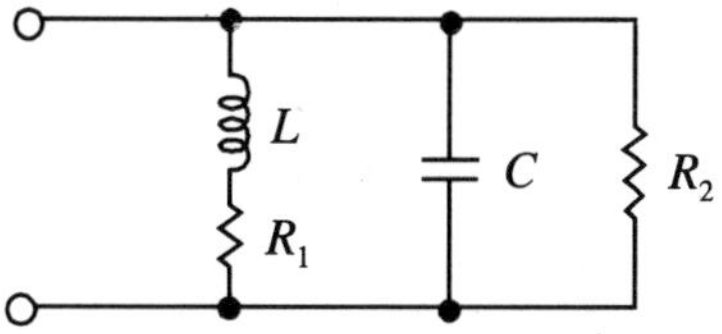

Fig. P. 4.23

NOTES AND REFERENCES

Solution of network equations and subsequent determination of network response involves the inversion of a matrix. Except for very simple networks, limited to 3 or 4 independent nodes or meshes, the inversion of these polynomial matrices by long hand is prohibitive, if not impossible. Computer has become an indispensable tool in obtaining the response of a network. Gaussian elimination, numerical integration, *LU* factorization and other numerical techniques have become standard tools. Directors book[1] (Chaps. 3, 13, 15 and 16) lists a number of simple computer programs that can be implemented easily. Packaged program like ECAP, SPICE are available for determining network response.

Sinusoidal steady state response is generally discussed in the first course on circuit theory. Here it is introduced for completeness and as a prelude to resonance phenomena. The concept and the measurement of the Q-factor and the bandwidth are fundamental to the design of filters. A set of suitably designed experiments are essential to crystallise these concepts. A suitable set is:

1. Parallel Resonance; variable capacitor (to control ω_0) and variable resistance (to control Q). Measurement of frequency response, Q and bandwidth.
2. Measurement of Q of an iron cored coil by series resonance.
3. Step response of a variable Q resonant circuit to bring out the correlation between the damping factor and Q-factor.

The books cited in Chapter 3 may be referenced for further information and problems.

1. S.W. Director: *Circuit Theory—A Computational Approach*, John Wiley, 1975.

CHAPTER

5

Network Theorems

The loop and node equations completely characterize a network. On solving these equations, we can determine the current and voltage of any element of the network. Many times we are interested in the current through and voltage across a particular element of the network. In this case, we can replace the rest of the network by an equivalent source and an impedance. This leads to the familiar Thevenin or Norton equivalent networks. In this chapter, we study Thevenin-Norton's, Tellegen's and other network theorems which are applicable to a large class of networks. The study of these theorems places in evidence many general properties of the network response and helps us gain insight into many of the network problems.

5.1 SUPERPOSITION THEOREM

A linear system can be defined as a system which satisfies *the principle of superposition.* The only requirement for the application of the superposition theorem is that the system or the network be linear. The network can be either time-invariant or time-varying.

Theorem 5.1: At any point in a linear network the response to a number of excitations acting simultaneously is equal to the algebraic sum of the responses at that point due to each excitation acting alone in the network.

We prove the theorem for a time-invariant network. A linear time-invariant network can be characterized by a set of equations in the Laplace domain as

$$M\mathbf{X} = \mathbf{N}$$

where M is a matrix $\mathbf{X}$ and $\mathbf{N}$ are vectors (Ch. 3). Without loss of generality, let us consider the node equations. For this case

$$M = Y_n, \ \mathbf{X} = \mathbf{V}_n, \ \mathbf{N} = -A\mathbf{J}_q$$

where Y_n is the *node-admittance matrix,* $\mathbf{V}_n$ a set of node voltages ($(v - 1)$ in number), A the *incidence matrix* of the network graph and $\mathbf{J}_q$ the equivalent current source, *i.e.*

$$Y_n \mathbf{V}_n = -A\mathbf{J}_q$$

from which
$$\mathbf{V}_n = -Y_n^{-1} A\mathbf{J}_q \qquad \ldots(5.1)$$

If we further consider the *ZSR* only, then

$$\mathbf{J}_q = [j_{qk} + j_{eqk}] = \mathbf{J}_e - Y_e\mathbf{E}_e \qquad \ldots(5.2)$$

where $\mathbf{J}_e$ is the current source vector, $\mathbf{E}_e$ the voltage source vector, Y_e the *edge admittance matrix* and j_{qk} and j_{eqk} are the equivalent current sources at the *k*th node due to independent current sources and independent voltage sources of the network. If we let $Y_n^{-1} = [\beta_{ij}]$, then from eqns. 5.1 and 5.2, we have

$$v_{ni} = \sum_{k=1}^{v=1} \sum_{l=1}^{e} \beta_{ik} a_{kl} \left[j_{ql} + j_{eql} \right]$$

where V_{ni} is the *i*th node voltage and a_{kj} is the (*k*, *j*)th entry of *A* (*e* is the number of elements in the network). The right hand side of this equation is obviously the sum of response due to each source acting alone in the network. Hence, the node voltage responses satisfy the superposition theorem. We already know that

$$\mathbf{V}_e = A^t \mathbf{V}_n$$

$$\mathbf{I}_e = Y_e \mathbf{V}_e$$

where $\mathbf{V}_e$ and $\mathbf{I}_e$ are the edge voltage and current vectors, respectively. Hence, the theorem.

Notice that the theorem is true even if we include the equivalent sources due to initial conditions. We have omitted them here due to notational simplicity.

Theorem 5.1 is true for any general input. Let *f* (*x*) be a complex periodic input to a linear network. *f*(*x*) can be expanded into Fourier series as

$$f(x) = \frac{a_0}{2} + \sum_{n=1}^{\infty} \{a_n \cos nx + b_n \sin nx\}$$

Let the response of the network to dc input $\frac{a_0}{2}$ be ψ_0 and to sinusoidal input cos *nx* and sin *nx* be $\psi(\cos nx)$ and $\psi(\sin nx)$ respectively. Then the response $\psi(x)$ to the input *f*(*x*) from superposition theorem is

$$\psi(x) = \psi_0 + \sum_{n=1}^{\infty} \{\psi(\cos nx) + \psi(\sin nx)\}$$

This demonstrates the usefulness of superposition theorem in the analysis of linear networks when the input is a complex function.

5.2 SUBSTITUTION THEOREM

The *substitution theorem* allows us to replace any particular edge or element of a network by a suitably chosen independent source without changing any voltage or current in the network. This method is sometimes useful in solving complex networks.

Consider an arbitrary edge of a network, say the *k*th edge. Let this edge be connected between the nodes *a* and *b* and the edge voltage be e_k [Fig. 5.1(*a*)]. Let

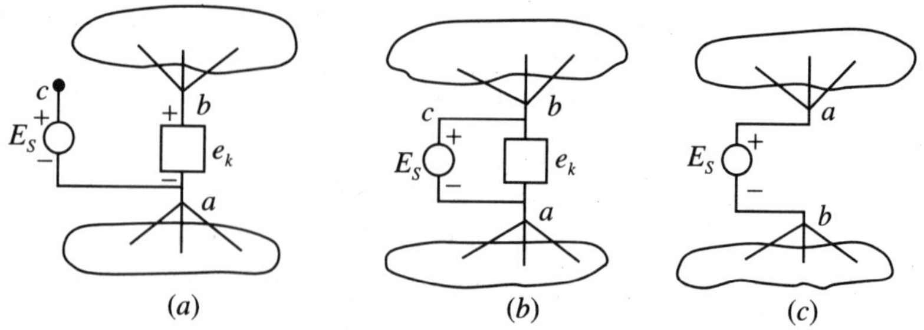

Fig. 5.1

us connect a voltage source E_s of value e_k to node a [Fig. 5.1(a)]. Now the nodes b and c can be shorted as they are at the same potential [Fig. 5.1(b)], The kth element is in parallel with a voltage source and hence can be removed without affecting the rest of the network [Fig. 5.1(c)]. Thus we have replaced an element by an independent voltage source without affecting the reminder of the network. The element can also be replaced by a current source. Let the edge current be i_k [Fig. 5.2(a)]. Let us connect a current source i_s between b and c [Fig. 5.2(b)], the value of i_s being equal to the edge current i_k. The current through the short-circuit is zero and hence the short-circuit can be removed. Now the kth element is in series with a current source and hence this element can be removed sofaras the rest of the network is concerned [Fig. 5.2(c)]. Thus we have replaced the edge by a current source.

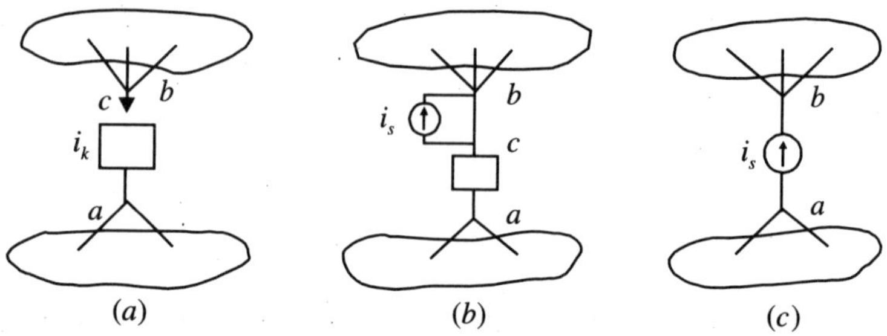

Fig. 5.2

In the foregoing discussion we have made two assumptions. First, the kth element is not coupled to any other element of the network, *i.e.* kth element is neither an arm of a coupled inductor nor an element of a dependent source. Secondly, the network prior to and after the replacement of the kth element has unique solution. We can summarize these discussions in the form of a theorem.

Theorem 5.2: Let an arbitrary network N have a unique solution. Let i_k and e_k be the current and voltage of a particular element k. Let a new network N' be formed by replacing the kth element of N by a current source i_k or a voltage source e_k. If N' has a unique solution, then the element currents and voltages of N' are identical with those of N.

Substitution theorem is particularly useful in the analysis of networks which contain a single non-linear or a time-varying element.

Example 5.1. The network of Fig. 5.3 is initially at rest. At $t = 0$ the switch is suddenly closed. We determine the voltage across *ab* by using substitution theorem. The network has a degenerate circuit and hence the initial voltage of the capacitors not continuous. The conditions at $t = 0-$ are

$$V_{c_1}(0-) = V_{c_2}(0-) = 0$$

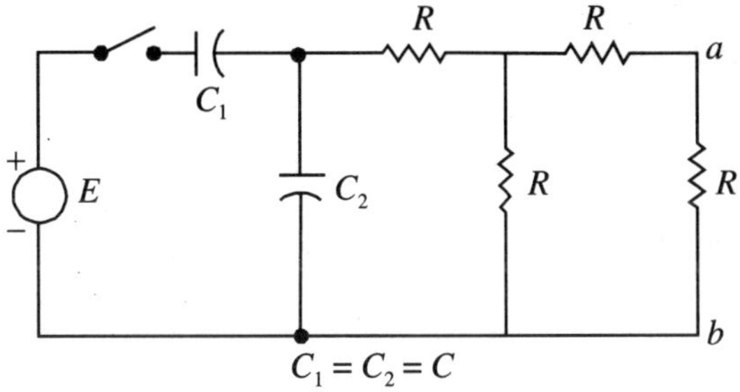

Fig. 5.3

It is easy to see that for this network

$$V_{c_1}(0+) = V_{c_2}(0+) = \frac{E}{2}$$

The transform network of the network under consideration is shown in Fig. 5.4.

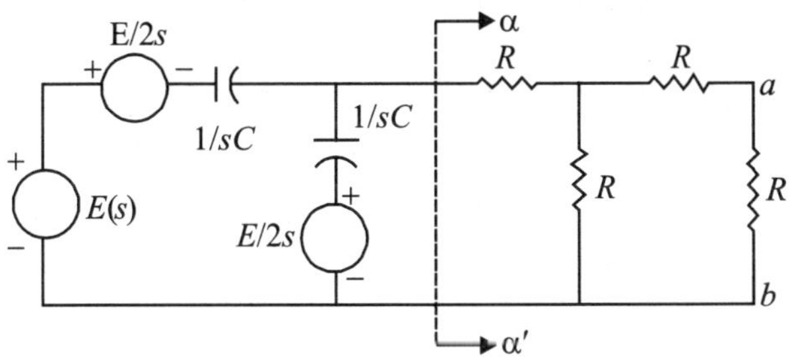

Fig. 5.4

The impedance of the network to the right of $\alpha\alpha'$ is

$$Z_1 = R + \frac{R \cdot 2R}{R + 2R} = \frac{5}{3}R$$

Hence the network can be rewritten as shown in Fig. 5.5.

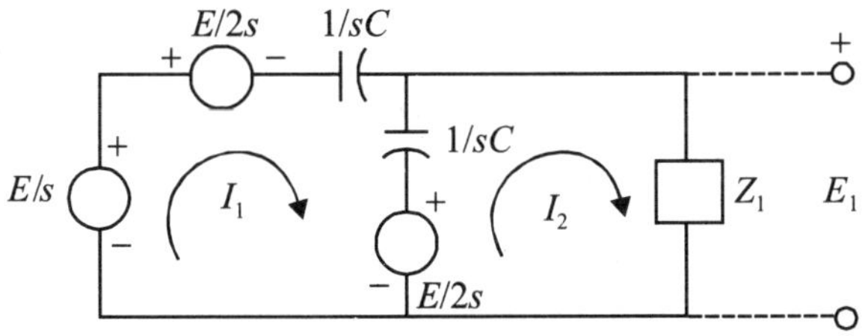

Fig. 5.5

The mesh equations are

$$\begin{bmatrix} \dfrac{2}{sC} & -\dfrac{1}{sC} \\ -\dfrac{1}{sC} & \dfrac{1}{sC}+\dfrac{5}{3}R \end{bmatrix} \begin{bmatrix} I_1 \\ I_2 \end{bmatrix} = \begin{bmatrix} 0 \\ E/2s \end{bmatrix}$$

Solving this for I_2 yields

$$I_2 = \frac{E}{s(1/sC + 10R/3)}$$

Hence E_1, the voltage across Z_1, is

$$E_1 = \frac{E}{s(1/sC + 10R/3)} \frac{5}{3}R$$

The network to the left of $\alpha\alpha'$ can now be substituted by the equivalent voltage source E_1 as shown in Fig. 5.6. (Here we have used a more general form of the substitution theorem in that we have replaced the entire network to the left of $\alpha\alpha'$ by an equivalent voltage source. The current drawn by that part of the network to the right of $\alpha\alpha'$ is independent of the current to the left of $\alpha\alpha'$ and only depends on the voltage across C_2. Hence we are justified in replacing the network to the left of $\alpha\alpha'$ by a voltage source as we are only interested in the current through *ab*.)

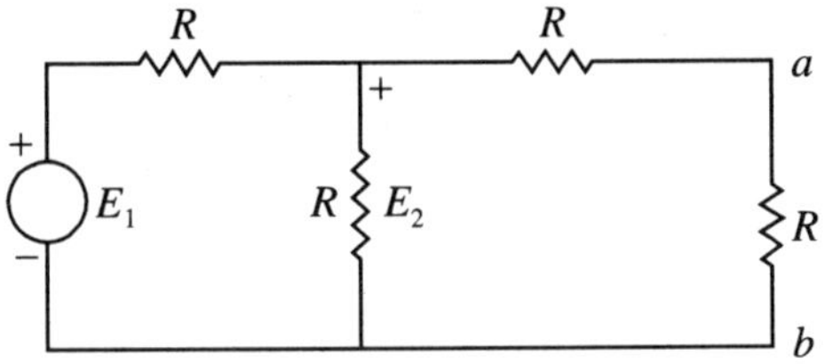

Fig. 5.6

Hence $$E_2 = \frac{E_1}{5/3R} \frac{R2R}{R+2/R} = \frac{2E_1}{5}$$

and $$E_{ab} = E_2/2 = E_1/5$$

On substituting for E_1, we have

$$E_{ab} = \frac{5ER}{3s(1/sC + R/3)} \cdot \frac{1}{5} = \frac{ERC}{3 + 10RCs} = \frac{E}{10\left(s + \dfrac{3}{10RC}\right)}$$

i.e. $$e_{ab}(t) = \left(\frac{E}{10}\right) e^{-3t/10RC}.$$

5.3 THEVENIN-NORTON THEOREM

Consider a linear network N and let Z_L be an arbitrary element connected between the terminals 1 and 1′ (Fig. 5.7). Let us assume that Z_L is not coupled to any element of the network *i.e.* the only interaction between N and Z_L comes from the current through the terminals 1 and 1′. * *Thevenin-Norton theorem* asserts that the current and voltage, $i(t)$ and $v(t)$, will not be affected if N is replaced by either a *"Thevenin equivalent network"* or a *"Norton equivalent network"*.

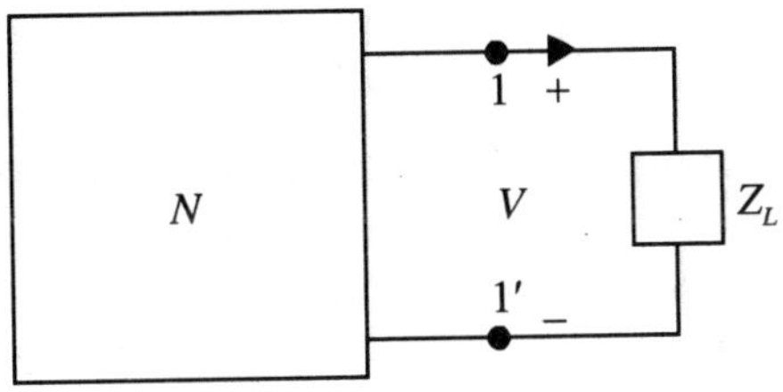

Fig. 5.7

The *Thevenin equivalent* network is shown in Fig. 5.8. It consists of a network N_0 in series with a voltage source V_{oc}. V_{oc} is the *open circuit voltage across 11′, i.e.,* the voltage across the terminals 11′ when Z_L is disconnected. N_0 is obtained from N by *removing all independent sources, i.e.* shorting all independent voltage sources and opening all independent current sources, and setting all initial condition to zero.

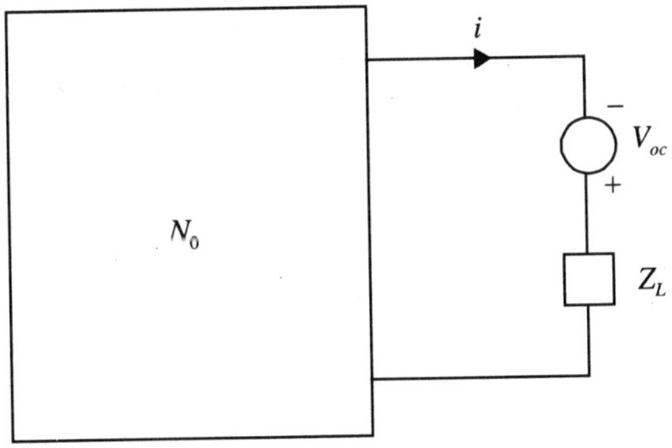

Fig. 5.8

The *Norton equivalent network,* shown in Fig. 5.9, consists of the network N_0 in parallel with a current source I_{sc}. I_{sc} is the current through a short circuit placed across thc terminals 11′.

Theorem 5.3: Let a linear network N be connected by two of its terminals 1 and 1′ to an arbitrary load and let N have a unique solution. Let N_0 be the network obtained from N by setting all independent sources and initial conditions to zero. Let v_{oc} and i_{sc} be the open circuit voltage and the short-circuit current at 11′. Then

* This implies that Z_L is neither coupled magnetically to N nor a part of a controlled source.

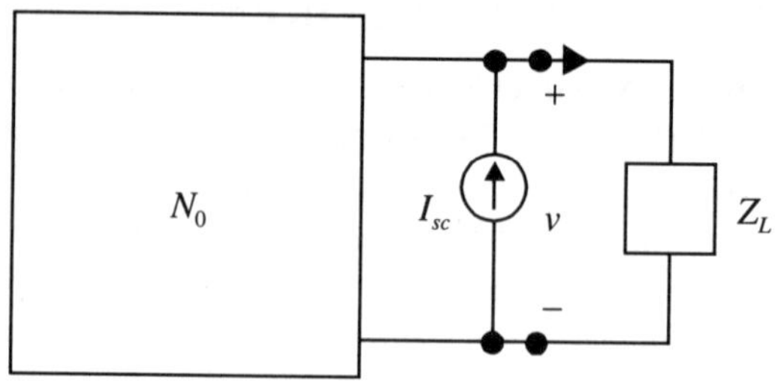

Fig. 5.9

the voltage and current at the terminals remain unchanged when N is replaced by its Thevenin or Norton equivalent networks.

In the network N, let us replace all initial conditions by independent sources. The initial capacitor voltage is replaced by a voltage source in series with the capacitor and the initial inductor current is replaced by a current source in parallel with the inductor.

Let us replace the load by a current source $i(t)$ (Fig. 5.10). From substitution theorem we are assured (the assumption that the network has unique solution is essential for this) that current and voltage distribution in N is unchanged. In particular, the voltage $v(t)$ at $11'$ is unchanged.

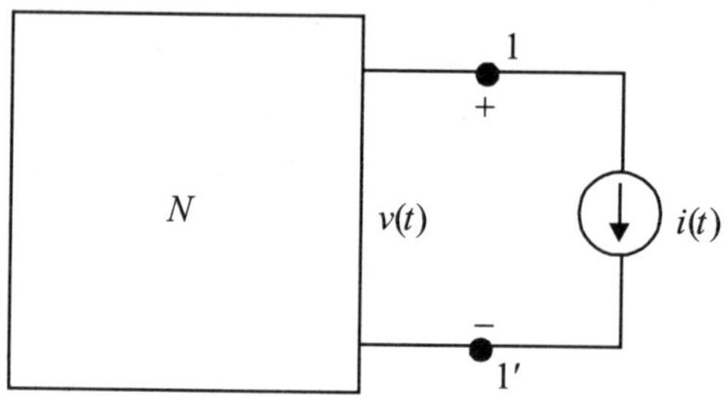

Fig. 5.10

Now the network N is driven by the load current source $i(t)$ and all the independent sources of N. Thus $v(t)$ is the zero state response.

Let $v_1(t)$ be the *ZSR* of N due to $i(t)$ acting alone, *i.e.*, we set all the independent sources of N to zero. This results in network N_0. Hence we can write

$$v_1(t) = \int_0^t h(t, \tau) i(\tau) d\tau$$

where $h(t, \tau)$ represents the impulse response of N_0. (If we assume the network to be time-invariant the equation reduces to the familiar convolution integral. This assumption is not necessary for Thevenin-Norton theorem).

Let v_{oc} be the *ZSR* due to the independent sources of N only, *i.e.* the response obtained by setting $i(t)$ to zero. From the superposition theorem (the network is assumed to be linear) we have

$$v(t) = v_{oc}(t) + v_1(t) = v_{oc}(t) + \int_0^t h(t, \tau)i(\tau)d\tau$$

Now consider the Thevenin equivalent network shown in Fig. 5.11. Writing the *KVL* for the mesh we have

$$v(t) = v_1(t) + v_{oc}(t) = v_{oc}(t) + \int_0^t h(t, \tau)i(\tau)d\tau$$

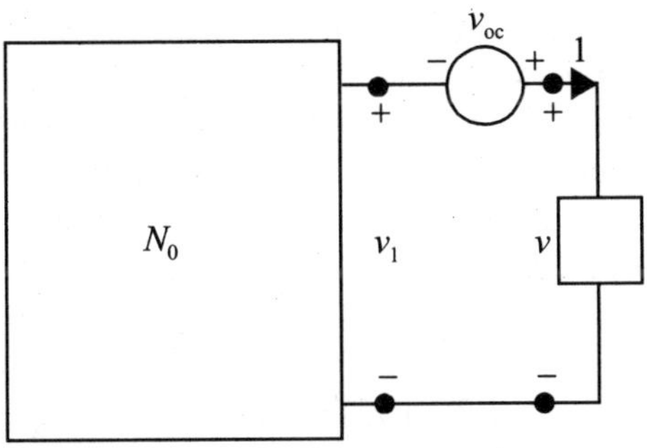

Fig. 5.11

Hence the theorem.

The proof for the Norton-equivalent theorem can be easily developed, on dual lines (prob. 5.8).

Thevenin-Norton theorem is extremely general. The only restriction on the network is that it should be linear and the only restriction on the load is that it should not be coupled to the network. From this theorem we conclude so far as one of its terminal pairs is concerned the network can be replaced by a voltage source in series with a relaxed network or a current source in parallel with a relaxed network.

If the network N is time-invariant, then the equivalent network N_0 can be replaced by an impedance or admittance. The Thevenin-equivalent when N is time-invariant is shown in Fig. 5.12(a), where z_{eq} is the impedance looking into N_0 at the

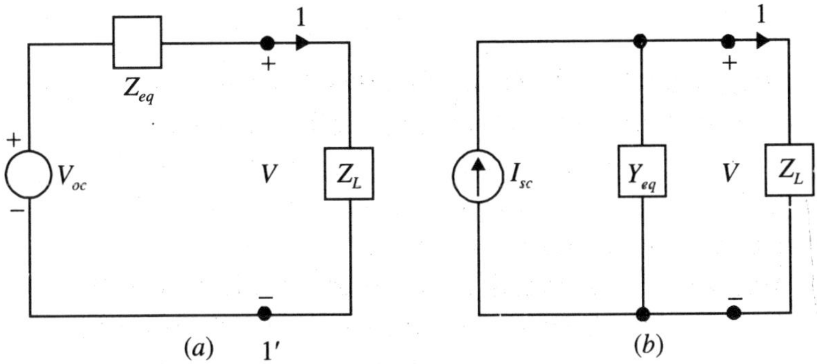

(a) (b)

Fig. 5.12

terminals 11′. The Norton equivalent is shown in Fig. 5.12(*b*), where Y_{eq} is the admittance of N_0 at the terminals 11′. Obviously $Z_{eq} = 1/Y_{eq}$. From Fig. 5.12 it can be easily seen that

$$V_{oc} = Z_{eq}\, I_{sc}$$

and $$I_{sc} = Y_{eq}\, V_{oc}$$

Thevenin equivalent is sometimes referred to as the *voltage source equivalent* and the Norton equivalent as the *current source equivalent.* We have already used this concept in Chapter 3 in formulating the network equations. (In all the preceeding discussions, we have assumed that we are working in the Laplace domain. There is no loss of generality in this.)

We now demonstrate an application of Thevenin theorem. Figure 5.13(*a*) shows a linear time-invariant network, and we wish to determine the change of current in the *k*th element due to a small change in the *k*th element impedance. Such problems are designated as 'sensitivity studies' and play an important role in practical circuit design. Let the network *N* to the left of the terminal *kk*′ be replaced by its Thevenin equivalent, as shown in Fig. 5.13(*b*). Let us also assume that the network is in sinusoidal steady state. The phasor of the current in the element Z_k is given by

$$I(Z_{eq} + Z_k) = V_{oc} \qquad \ldots(5.3)$$

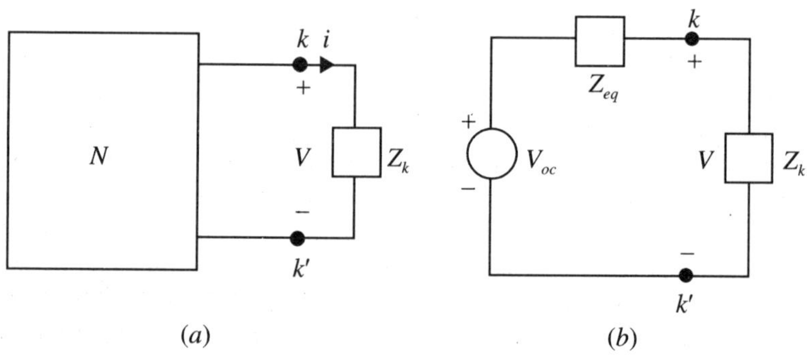

Fig. 5.13

Let the *k*th element impedance change by a small amount to $Z_k + \Delta Z_k$ and the current change to $I' = I + \Delta I$. Then, we have

$$(I + \Delta I)\,(Z_{eq} + Z_k + \Delta Z_k) = V_{oc} \qquad \ldots(5.4)$$

From eqns. 5.3 and 5.4 neglecting the second order term $\Delta I \cdot \Delta Z_k$, we get

$$I\Delta Z_k + \Delta I(Z_{eq} + Z) \simeq 0$$

which reduce to $$\Delta I = -\frac{I\Delta Z_k}{Z_{eq} + Z}$$

This leads to a simple interpretation of ΔI. ΔI is the current that would flow in a circuit with impedance $(Z_{eq} + Z_k)$ driven by a voltage source of $I\Delta Z_k$ (Fig. 5.14).

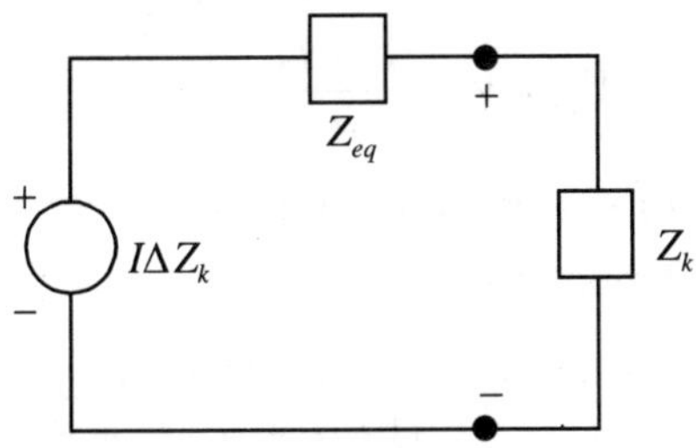

Fig. 5.14

Example 5.2. Determine the current through the resistor R_3 of the network shown in Fig. 5.15 by replacing the rest of the network by its Thevenin equivalent.

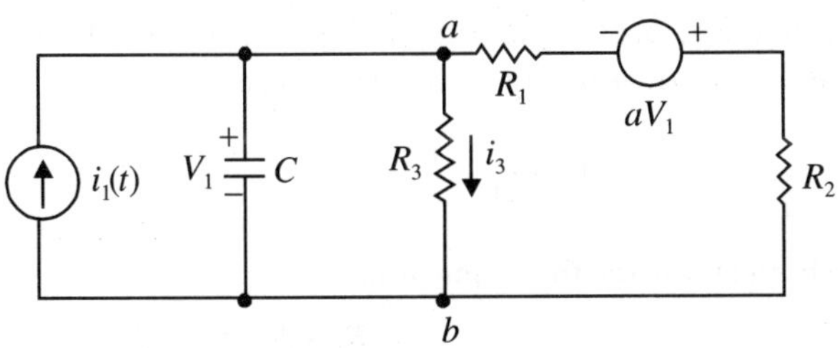

Fig. 5.15

The circuit to determine i_3 is shown in Fig. 5.16 where V_{ab} is the open circuit voltage at ab and Z_{eq} is the Thevenin equivalent impedance.

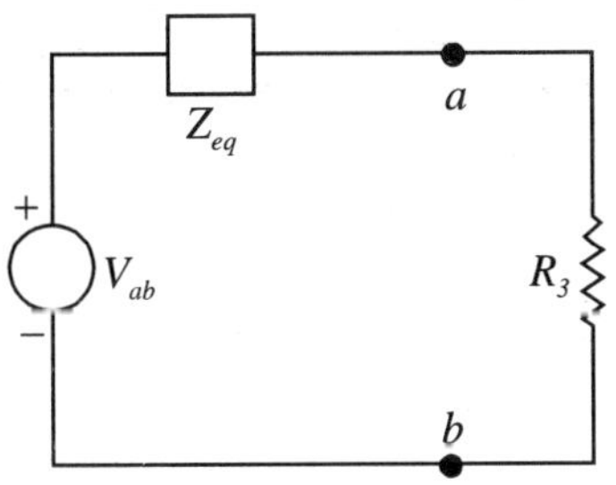

Fig. 5.16

(*a*) To find V_{ab} remove R_3 and solve the network. The network with R_3 removed is rewritten in Fig. 5.17.

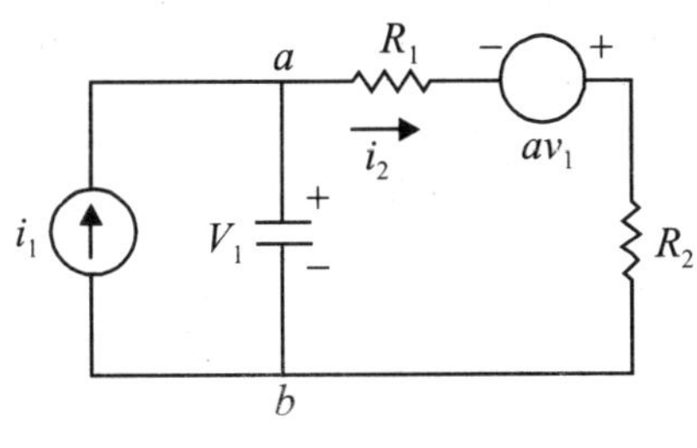

Fig. 5.17

$$V_{ab} = V_1 = \frac{I_1 - I_2}{sC}$$

and $\quad I_2(R_1 + R_2) = V_1\,(1 + a)]$

Solving for I_2 in terms of I_1 we have

$$I_2 = \frac{1+a}{(1+a)+sC(R_1+R_2)} I_1$$

Hence $$V_1 = \frac{R_1+R_2}{(1+a)+sC(R_1+R_2)}$$

which gives V_{ab}.

(*b*) To determine Z_{eq}, open R_3 and the independent current source i_1 and apply a voltage *V*, as shown in Fig. 5.18. As before

$$V = V_1 = \frac{I - I_2}{sC}$$

The mesh equation for the I_2 mesh is

$$V = (R_1+R_2)I_2 - aV$$

Solving for *V/I* we have

$$Z_{eq} = \frac{V}{I} = \frac{R_1+R_2}{(1+a)+sC(R_1+R_2)}$$

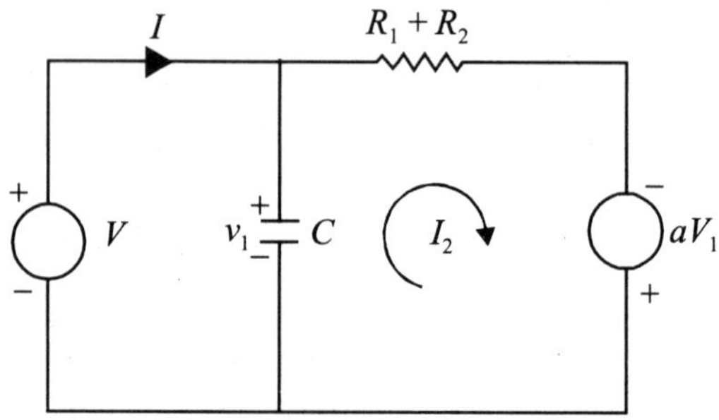

Fig. 5.18

(*c*) Thevenin equivalent voltage and impedance are given by V_{ab} and Z_{eq}. Hence

$$i_3 = \frac{V_{ab}}{Z_{eq} + R_3}$$

i.e. $$i_3 = \frac{(R_1+R_2)i_1}{(R_1+R_2)(1+sCR_3)+(1+a)R_3}$$

5.4 TELLEGEN'S THEOREM

This theorem, introduced by B.D.H. Tellegen, is extremely general and is applicable to any lumped network. The only restriction on the application of the theorem is that the voltages and currents of the elements of the network satisfy the constraints imposed by the Kirchhoff's laws. As such, the nature of the elements is irrelevant.

Theorem 5.4: Consider an arbitrary network N whose graph G has v nodes and e edges. Let i_k and v_k represent the voltage and current of kth edge. If the edge voltages v_k, $k = 1, 2, \ldots e$ and edge currents i_k, $k = 1, 2, \ldots e$ satisfy KVL and KCL respectively, then

$$\sum_{k=1}^{e} v_k i_k = 0 \qquad \ldots(5.5)$$

The theorem can be proved in two ways:

(*i*) Without loss of generally we assume that G is connected and has no parallel edges. Let the nodes be numbered 1, 2, … v and node numbered 1 be selected as the reference node. Let e_α and e_β be the voltages of nodes α and β w.r.t. node 1 ($e_1 = 0$). Let kth edge connect node α and β (Fig. 5.19). Let the current flow from the node α to the node β. Then

$$v_k\, i_k = (e_\alpha - e_\beta)\, i_{\alpha\beta}$$

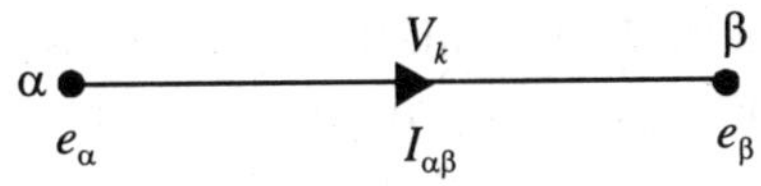

Fig. 5.19

In fact, we can write this relation as

$$v_k\, i_k = (e_\beta - e_\alpha)\, i_{\beta\alpha}$$

since $i_{\alpha\beta} = -\, i_{\alpha\beta}$. Adding the two equations yields

$$v_k i_k = \frac{1}{2}\left[(e_\alpha - e_\beta) i_{\alpha\beta} + (e_\beta - e_\alpha) i_{\beta\alpha}\right]$$

We can write one such equation for every edge. Summing over all the edges of G we have

$$\sum_{k=1}^{e} v_k i_k = \frac{1}{2} \sum_{\alpha=1}^{v-1} \sum_{\beta=1}^{v-1} (e_\alpha - e_\beta) i_{\alpha\beta}$$

i.e.

$$\sum_{k=1}^{e} v_k i_k = \frac{1}{2}\left[\sum_{\alpha=1}^{v-1} e_\alpha \left(\sum_{\beta=1}^{v-1} i_{\alpha\beta}\right) - \sum_{\beta=1}^{v-1} e_\beta \left(\sum_{\alpha=1}^{v-1} i_{\alpha\beta}\right)\right]$$

In this equation $\sum_{\alpha} i_{\alpha\beta}$ and $\sum_{\beta} i_{\alpha\beta}$ are identically zero as these represent KCL at nodes α and β, respectively. Hence the theorem.

(*ii*) The second proof is based on using *node* and *chord transformations* (Chap. 3). We can write the left hand side of eqn. 5.5 as the product of two vectors,

$$\sum_{k=1}^{e} v_k i_k = \mathbf{V}_e^t \mathbf{I}_e$$

where $\mathbf{V}_e$ is the column vector of the edge voltages and $\mathbf{I}_e$ the column vector of the edge currents. From eqns. 3.7 and 3.9 we have

$$\mathbf{V}_e = A^t\,\mathbf{V}_n,\ \mathbf{I}_e = B_f^t\,\mathbf{I}_c$$

Hence we have

$$\sum_{k=1}^{e} v_k i_k = \left(A^t \mathbf{V}_n\right)^t \left(B_f^t \mathbf{I}_c\right)$$

$$= \mathbf{V}_n^t \left(A B_f^t\right) \mathbf{I}_c$$

$$= 0$$

Hence the theorem.

We now discuss some of the implications of *Tellegen's theorem.*

(*i*) In proving the Tellegen's Theorem only two laws were used, namely, *KCL* and *KVL* which depend only on the topology of the network. The characteristics of individual elements were not used and as such are not important. Hence, Tellegen's Theorem is applicable to any network—linear or non-linear, active or passive, time-varying or invariant, as long as Kirchhoff's laws are not violated.

(*ii*) Since $v_k\, i_k$ is the power delivered to an edge of the network at any time, the theorem may be interpreted as: the sum of the power delivered to the network is zero which is a re-statement of the *law of conservation of energy.* Hence we conclude that Kirchhoff's laws imply law of conservation of energy.

(*iii*) Since the element characteristics were not used in deriving Tellegen's Theorem, it is applicable even if the set of voltages and currents were evaluated at two different instants of time, *i.e.*

$$\sum_{k=1}^{e} v_k(t_1) i_k(t_2) = 0 \qquad \ldots(5.6)$$

where $v_k(t_1)$, $k = 1, 2, \ldots e$ and $i_k(t_2)$, $k = 1, 2, \ldots$ are the edge voltages and currents of the network at time t_1 and t_2, respectively. This result is obvious as all that we need in proving the theorem is that $\Sigma I_{\alpha\beta}$ be zero.

(*iv*) We can further generalize the results of (*iii*). Let N_1 and N_2 be two different networks having the same oriented graph *G*. Let v_k and i_k, $k = 1, 2, \ldots e$ be the edge voltages and currents of N_1 and $\hat{v}_k$ and $\hat{i}_{\hat{k}}$ be those of N_2. Then it is easy to show that

$$\sum_{k=1}^{e} v_k(t_1)\hat{i}_{\hat{k}}(t_2) = \sum_{k=1}^{e} \hat{v}_k(t_2) i_k(t_1) = 0 \qquad \ldots(5.7)$$

Though Tellegen's theorem holds good, the relationship expressed in eqn. 5.7 does not directly imply conservation of energy as $v_k\hat{i}_{\hat{k}}$ or $i_k\hat{v}_k$ is not the energy of the edge k in either N_1 or N_2.

Example 5.3. Consider a linear time-invariant *RLC* three-port network *N* shown in Fig. 5.20. Let the sinusoidal steady state of the terminal variables, V_j, $j = 1, 2, 3$ and

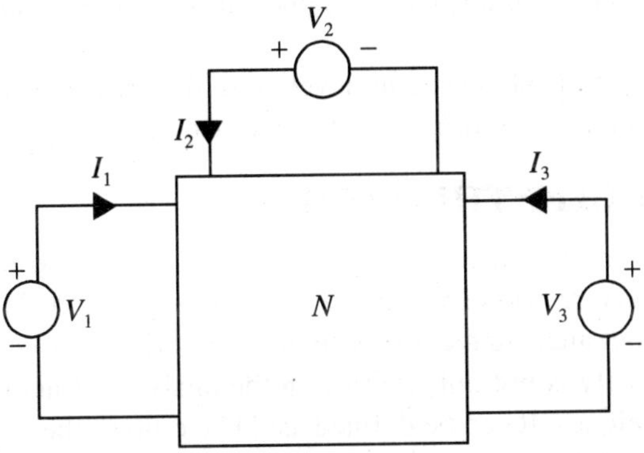

Fig. 5.20

I_j, $j = 1, 2, 3$ be measured for two sets of sinusoidal voltages, V_1, V_2 and V_3, but at the same frequency ω. Let the first set of measurements be represented by *V*, *I* and the second set by $\hat{V}$, $\hat{I}$, respectively. Let the network *N* have *n* elements. Then from Tellegen's theorem we have

$$\sum_{k=1}^{n+3} V_k(\omega)\hat{I}_k(\omega) = \sum_{k=1}^{n+3} \hat{V}_k(\omega)I_k(\omega)$$

This equation can be written in the expanded form as

$$\sum_{j=1}^{3} V_j(\omega)\hat{I}_j(\omega) + \sum_{k=4}^{n+3} V_k(\omega)\hat{I}_k(\omega) = \sum_{j=1}^{3} \hat{V}_j(\omega)I_j(\omega) + \sum_{k=4}^{n+3} \hat{V}_k(\omega)I_k(\omega) \quad \ldots(5.8)$$

In the preceeding equation the first summation on both sides of the equation is with respect to the terminal variables and second summation is with respect to the other elements of *N*. For the *r*th element of *N* we have

$$V_r(\omega)\hat{I}_r(\omega) = Z_r(\omega)I_r(\omega)\hat{I}_r(\omega)$$

and
$$\hat{V}_r(\omega)I_r(\omega) = Z_r(\omega)\hat{I}_r(\omega)I_r(\omega)$$

where $Z_r(\omega)$ is the impedance of the *r*th element at the frequency ω. Hence, we have

$$V_r(\omega)\hat{I}_r(\omega) = \hat{V}_r(\omega)I_r(\omega)$$

Combining this with eqn. 5.8, we have

$$\sum_{j=1}^{3} V_j(\omega)\hat{I}_j(\omega) = \sum_{j=1}^{3} \hat{V}_j(\omega) I_j(\omega)$$

which can be written as

$$V_1\hat{I}_1 + V_2\hat{I}_2 + V_3\hat{I}_3 = \hat{V}_1 I_1 + \hat{V}_2 I_2 + \hat{V}_3 I_3$$

where the index ω is dropped for convenience. The last equation contains only the terminal variables. Through this example one can see the generality of Tellegen's theorem.

Tellegen's Theorem is also used in developing the *sensitivity coefficients* for a network from the concept of *adjoint networks* (see 5.6).

5.5 RECIPROCITY THEOREM

Reciprocity is a useful property found in many linear time-invariant passive networks. Intuitively reciprocity implies that the input and the output can be interchanged without altering the response of a network to a prescribed input. The property of reciprocity is not only helpful in the analysis of networks but also in measurement techniques. It can be defined as: 'The ratio of the response observed at a point to the excitation at another point is invariant to a change of position of observation and excitation as long as the topology of the network is unaltered'. Any network which satisfies this property is called a *reciprocal network.*

The property of reciprocity is satisfied by a wide class of linear time-invariant passive networks. If a network consists of RLCM (resistors, inductors, capacitors and mutual inductors) and excludes either dependent or independent sources, then the network satisfies the property of reciprocity. This can be stated in the form of a theorem.

Theorem 5.5: A linear time-invariant passive network containing only RLCM elements is reciprocal.

Before proving the theorem, we explore the implications of reciprocity a little further. Consider a linear time-invariant RLCM network with terminals $11'$ and $22'$ (Fig. 5.21) which are accessible for applying input and for measuring the response.

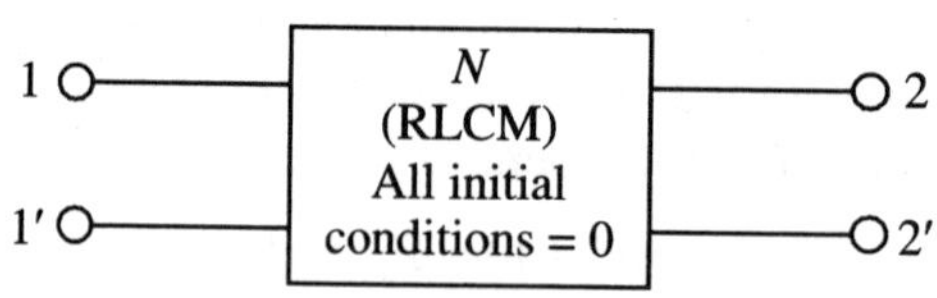

Fig. 5.21

For reciprocity three conditions of input and response are possible.

1. Let a voltage source v_1 be connected at $11'$ and the short circuit i_2 be measured at $22'$. For the second set of measurements, let a voltage source $\hat{v}_2$ be applied at $22'$ and the short-circuit current $\hat{i}_1$ at $11'$ be measured. Then the reciprocity implies that

$$\frac{i_2}{v_1} = \frac{\hat{i}_1}{\hat{v}_2}$$

Here a combination of voltage source with zero internal impedance and short circuit current measurements assures that the same terminal conditions, *i.e.* short circuit, is maintained.

2. Let a current source i_1 be applied at $11'$ and open-circuit voltage v_2 at $22'$ be measured. Then let a current source $\hat{i}_2$ be applied at $22'$ and the open-circuit voltage $\hat{v}_1$ at $11'$ be measured. The assertion of reciprocity theorem implies that

$$\frac{v_2}{i_1} = \frac{\hat{v}_1}{\hat{i}_2}$$

Here with current source of infinite impedance and open circuit the same terminal condition of open-circuit is maintained throughout the measurement.

3. Let a voltage source v_1 be applied at $11'$ and open-circuit voltage v_2 at $22'$ be measured. Then a current source of $\hat{i}_2$ be applied at $22'$ and short circuit current $\hat{i}_1$ at $11'$ be measured. Then reciprocity theorem asserts

$$\frac{v_2}{v_1} = \frac{\hat{i}_1}{\hat{i}_2}$$

Here in both sets of measurements, terminal $11'$ is short circuited and terminal $22'$ is open circuited.

The reciprocity relationship as defined here is applicable to any two pair of nodes of the network.

To prove the reciprocity theorem consider a linear time-invariant passive RLCM network *N* shown in Fig. 5.22(*a*). Let the point of observation and excitation be *j* and *k*, respectively. Let a voltage source be connected at the *k*th element as shown

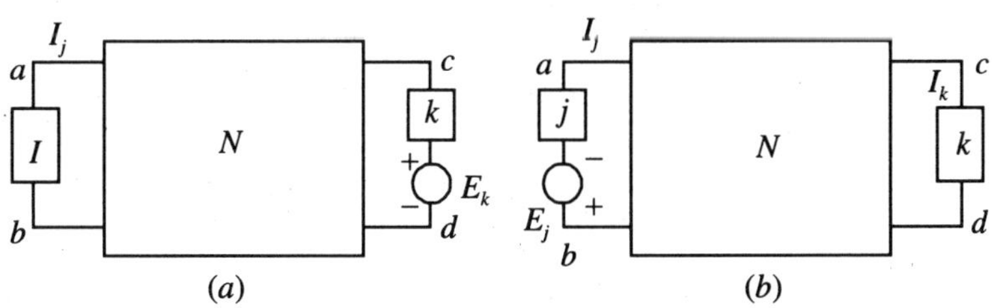

Fig. 5.22

in the figure. Let the response be the current I_j through the element *j*. The mesh equation for the network can be written as

$$\mathbf{Z}_m(s)\, \mathbf{I}_m(s) = \mathbf{E}_s \qquad \ldots(5.9)$$

where

$$\mathbf{E}(s) = \begin{bmatrix} 0 \\ 0 \\ \vdots \\ 0 \\ E_k \\ 0 \\ 0 \\ \cdot \\ 0 \end{bmatrix} \begin{matrix} 1 \\ 2 \\ \\ k-1 \\ k \\ k+1 \\ \\ \\ \\ \end{matrix}$$

Let $|Z_m(s)| = \Delta(s)$ and the cofactor of the (k, j)th entry of $Z_m(s)$ be $\Delta_{kj}(s)$. Then solving eqn. 5.9 by Cramer's rule we have

$$I_j(s) = \frac{\Delta_{kj}(s)}{\Delta(s)} E_k(s)$$

Hence the ratio of the response to excitation is

$$\frac{I_j(s)}{E_k(s)} = \frac{\Delta_{kj}(s)}{\Delta(s)} \qquad \ldots(5.10)$$

Now let us place a source E_j in the jth element and observe the response I_k, the current through the kth element [see Fig. 5.22(b)]. Then in eqn. 5.10 we have

$$\mathbf{E}(s) = \begin{bmatrix} 0 \\ 0 \\ \vdots \\ 0 \\ E_j \\ 0 \\ \vdots \\ 0 \end{bmatrix} \begin{matrix} 1 \\ 2 \\ \\ j-1 \\ j \\ j+1 \\ \\ \\ \end{matrix}$$

The response I_k is, by Cramer's rule, given by

$$I_k(s) = \frac{\Delta_{jk}(s)}{\Delta(s)} E_j(s)$$

Hence

$$\frac{I_k(s)}{E_j(s)} = \frac{\Delta_{jk}(s)}{\Delta(s)} \qquad \ldots(5.11)$$

The network N being linear time-invariant passive RLCM network, from Chap. 4, $Z_m(s)$ is a symmetric matrix and

$$\Delta_{jk}(s) = \Delta_{kj}(s)$$

The eqns. 5.10 and 5.11 imply the theorem.

The theorem can also be proved for other combination of excitation and response. It is possible that certain active networks may satisfy reciprocity relationship at their terminals. But in general active networks are not reciprocal. The reciprocity theorem can also be proved using Tellegen's theorem (Prob. 5.9).

Example 5.4. Two sets of measurements are made on a linear passive resistive two-port network, as shown in Fig. 5.23(*a*) and (*b*). Our object is to find current through the 2Ω resistor.

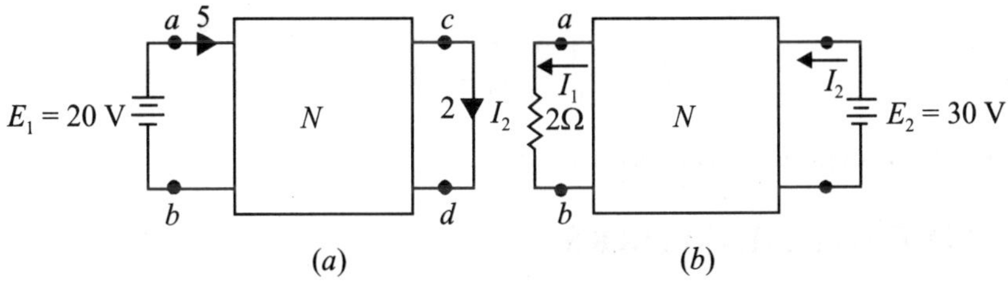

Fig. 5.23

We solve this problem by finding the Norton equivalent at *ab*. To find the Norton equivalent at *ab* of Fig. 5.23(*b*) we place a short circuit at *ab* (Fig. 5.24). Now from reciprocity theorem (*N* is a linear passive resistive network and topology of the network is the same for Fig. 5.23(*a*) and Fig. 5.24).

$$\frac{E_2'}{I_2'} = \frac{E_1}{I_2}$$

i.e.

$$I_1' = \frac{I_2}{E_1} E_2' = \frac{2}{20} 30 = 3\text{A}$$

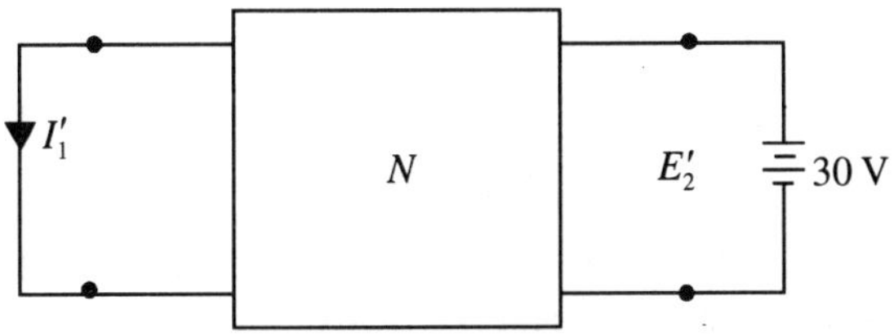

Fig. 5.24

To find Z_{eq} for the Norton equivalent of Fig. 5.23(*b*) we short E'_2 and measure the impedance at *ab*.

Hence

$$Z_{eq} = \frac{E_1}{I_1} = \frac{20}{5} = 4$$

Finally the Norton equivalent to determine the current through 2 Ω resistor is shown in Fig. 5.25.

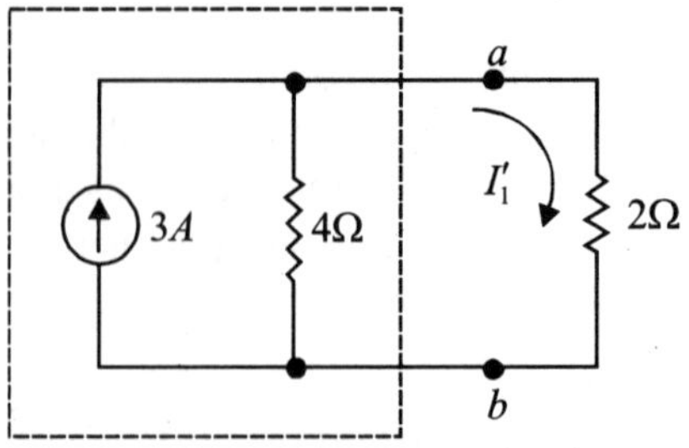

Fig. 5.25

Hence $$I_1' = 3\frac{4 \cdot 2}{4+2} \cdot \frac{1}{2} = 2A$$

i.e. current through the 2Ω resistor is 2*A*.

5.6 ADJOINT NETWORKS

Sensitivity of a network variable, current or voltage, with respect to network parameters like *R*, *L* and *C* is an important consideration in the design of networks and filters. The concept of sensitivity introduced in Section 5.3 will be elaborated in Chap. 12. Sensitivity can be computed by analysing the *adjoint network* N_A of a given network *N*. Here adjoint network is derived as an application of Tellegen's theorem.

Let *N* and N_A be two linear time-time-invariant networks having identical linear graphs (*i.e.* topology). Let **V** and **I** be the voltage and the current vectors of *N* (Fig. 5.26). In the figure source variables are shown separately and are considered as a

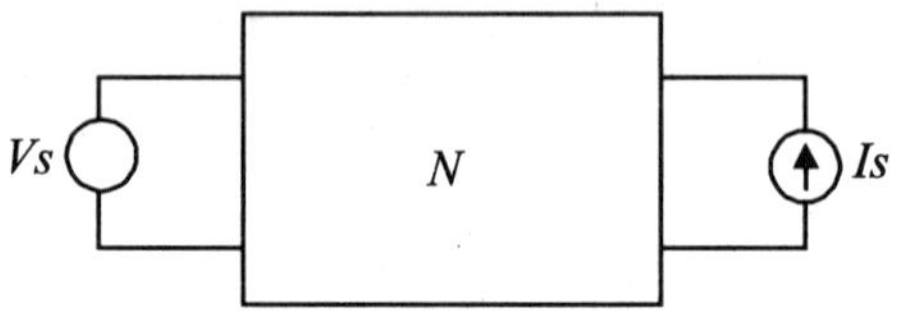

Fig. 5.26

part of the network. Let **E** and **J** be the voltage and the current vectors of N_A. Applying Tellegen's theorem to *N* and N_A yields

$$\mathbf{V}^t\,\mathbf{J} = 0 \text{ and } \mathbf{E}^t\,\mathbf{I} = 0$$

$$\mathbf{V}^t\,\mathbf{J} - \mathbf{E}^t\,\mathbf{I} = 0 \qquad \text{...(5.12)}$$

Due to the variation in the elements of *N* let **V** and **I** change by **ΔV** and **ΔI**, respectively. Now application of the Tellegen's theorem for *N* and N_A yields

$$(\mathbf{V} + \Delta\mathbf{V})^t\;\mathbf{J} - \mathbf{E}^t\,(I + \Delta I) = 0 \qquad \text{...(5.13)}$$

From equations 5.12 and 5.13

$$\Delta\mathbf{V}^t\,\mathbf{J} - \mathbf{E}^t\,\Delta\mathbf{I} = 0$$

This is the differential form of Tellegen's theorem. This can be written as

$$\sum_{k=\text{sources}} (J_k \Delta V_k - E_k \Delta I_k) + \sum_{k=\text{elements}} (J_k \Delta V_k - E_k \Delta I_k) = 0 \quad \text{...(5.14)}$$

The elements of the adjoint network N_A are so selected that the second summation in eqn. 5.14 becomes independent of variations in element currents and voltages. This is illustrated by considering a resistive network N. For a resistor

$$V = RI$$

or

$$\Delta V = R\,\Delta I + I\,\Delta R$$

Then from eqn. (5.14) we have

$$\sum_{\text{sources}} (J_k \Delta V_k - E_k \Delta I_k) + \sum_{\text{elements}} \{(J_k (R_k \Delta I_k + I_k \Delta R_k) - E_k \Delta I_k\} = 0$$

or

$$\sum_{\text{elements}} \{(R_k J_k - E_k)\Delta I_k + J_k I_k \Delta R_k\} + \sum_{\text{sources}} (E_k \Delta I_k - J_k \Delta V_k) \quad \text{...(5.15)}$$

The elements of N_A are now chosen such that the LHS is independent of ΔI_k *i.e.* N_A is defined by

$$R_k\, J_k \; - E_k = 0$$

or

$$E_k = R_k\, J_k,\; k = 1, 2, \ldots, n$$

With this eqn. 5.15 reduces to

$$\sum_{\text{elements}} J_k I_k \Delta R_k = \sum_{\text{sources}} (E_k \Delta I_k - J_k \Delta V_k) \quad \text{...(5.16)}$$

The sensitivity component with respect to R_k is $J_k\, I_k$. By proper selection of the sources in N and N_A we can determine the sensitivity of any network functions. As N_A is the adjoint network, we can conclude that the adjoint network of a resistive network is another resistive network with identical element values.

Here we have derived the adjoint network for a resistive network. Details of adjoint network and of sensitivity computation are discussed elsewhere (see notes).

5.7 MAXIMUM POWER TRANSFER THEOREM

A problem that arises in the design of circuits is to match the impedance of the load and the source so that maximum power transfer takes place. In the network

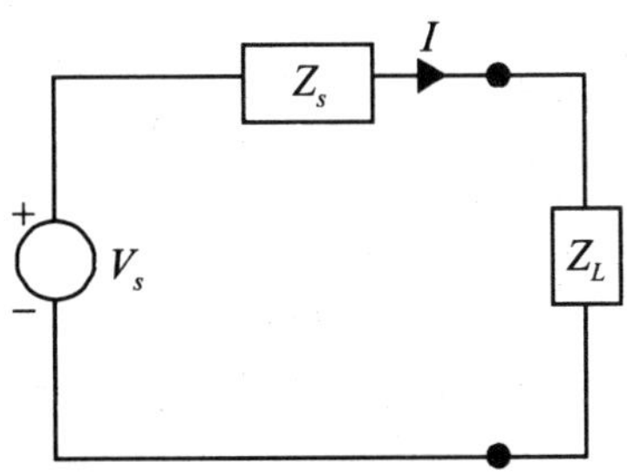

Fig. 5.27

shown in Fig. 5.27, V_s is the *rms* value of the sinusoidal voltage source at an angular frequency ω, Z_s is the given source impedance and Z_L is the passive load impedance. The problem is to select Z_L such that the power entering load is maximum. The condition for such a power transfer is contained in the following theorem.

Theorem 5.6: The optimum load impedance for maximum power transfer is equal to the complex conjugate of Z_s, *i.e.*

$$Z_L = Z_s^* \qquad \text{...(5.17)}$$

Let $Z_s = R_s + jx_s$ and $Z_L = R_L + jx_L$ be the source and load impedance at the frequency ω. Let I be the *rms* current through the load. Then the average power delivered to the load is

$$P_L = |I|^2 R_L$$

since

$$I = \frac{V_s}{Z_s + Z_L}$$

$$P_L = \frac{|V_s|^2 R_L}{(R_s + R_L)^2 + (X_s + X_L)^2}$$

In this equation V_s, R_s and X_s are known. R_L and X_L are chosen to maximize P. Let R_L be constant and X_L varying. Then to maximize P

$$\frac{\partial P_L}{\partial X_L} = 0$$

i.e.

$$\frac{-2|V_s|^2 R_L (X_s + X_L)}{\left\{(R_s + R_L)^2 + (X_s + X_L)^2\right\}^2} = 0$$

which yields $X_L = -X_s$.

Under this condition the average power is

$$P_a = \frac{|V_s|^2 R_L}{(R_s + R_L)^2}$$

Now if R_L is varying, to maximize P_a we have

$$\frac{\partial P_L}{\partial R_L} = 0$$

i.e.

$$|V_s|^2 = \frac{R_S^2 - R_L^2}{(R_s + R_L)^4} = 0$$

which yields

$$R_s = R_L$$

Hence the theorem.

The power absorbed by the load is maximum when the load impedance is *conjugately matched* to the source impedance. Under matched condition the power absorbed is

$$P_{\max} = \frac{|V_s|^2}{4R_s} \qquad \text{...(5.18)}$$

The average power delivered by the source is

$$P_s = |I|^2 (R_s + R_L)$$

Under matched conditions

$$|I| = \frac{|V_s|}{2R_s}$$

Hence $$P_s = \frac{|V_s|^2}{2R_3} \qquad \text{...(5.19)}$$

We may define the *efficiency of the circuit* as the ratio of the average power delivered to the load and the average power delivered by the source. From eqns. 5.18 and 5.19, the efficiency of the conjugately matched circuit is 50 per cent.

PROBLEMS

5.1 Let N be a linear passive network. Suppose the network is in the sinusoidal steady state. Show that

$$\sum_{k=1}^{e} \frac{1}{2} V_k I_k^* = 0$$

where V_k is phasor of the edge voltage and I_k is the phasor of the edge current and* indicates complex conjugate.

5.2 Two sets of measurements are taken on a resistive network (see Fig. P. 5.2). Find V_2. (*a*) $R_2 = 1\Omega$, $V_1 = 5V$, $i_1 = 2A$, $V_2 = 1V$; (*b*) $\mathbf{R}_2 = 10\ \Omega$, $\mathbf{V}_1 = 6V$, $i_1 = 6A$.

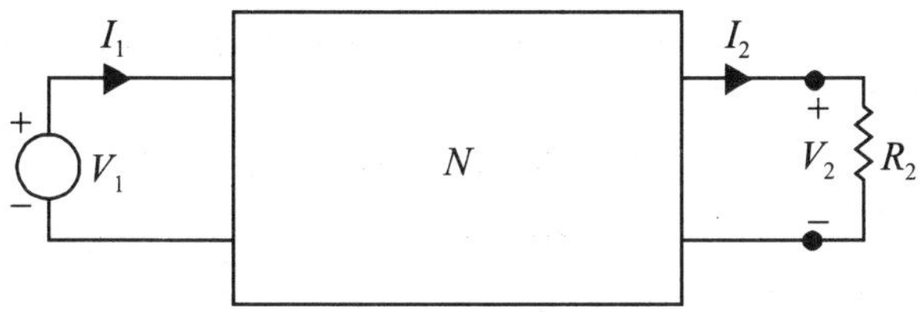

Fig. P. 5.2

5.3 Determine the current through R_3 and the voltage across R_1 in Fig. P. 5.3 by superposition theorem.

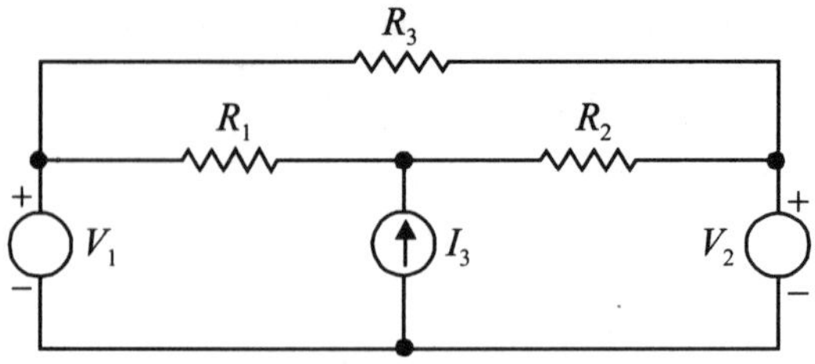

Fig. P. 5.3

5.4 A set of measurements is made on a linear time-invariant passive network, as shown in Fig. P. 5.4(a). The network is then reconnected as shown in Fig. P. 5.4(b). Find the current through the 5 Ω resistor.

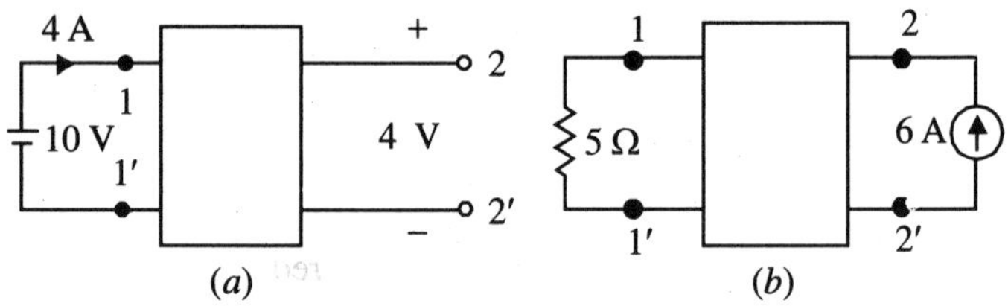

Fig. P. 5.4

5.5 The terminal currents and voltages of a resistor network are shown in Fig. P. 5.5(a). The network is reconnected, as shown in Fig. P. 5.5(b), if $I_2 = 0$, find I_1.

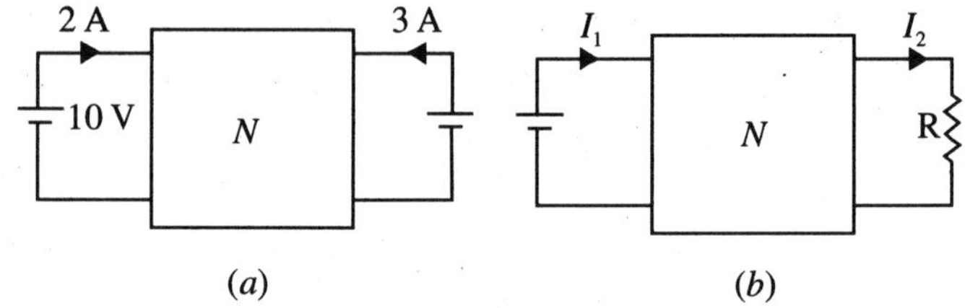

Fig. P. 5.5

5.6 Figure P. 5.6 shows the equivalent circuit of a common emitter transistor amplifier driving a resistive load. Find the Thevenin equivalent network at the load terminals and determine the load voltage.

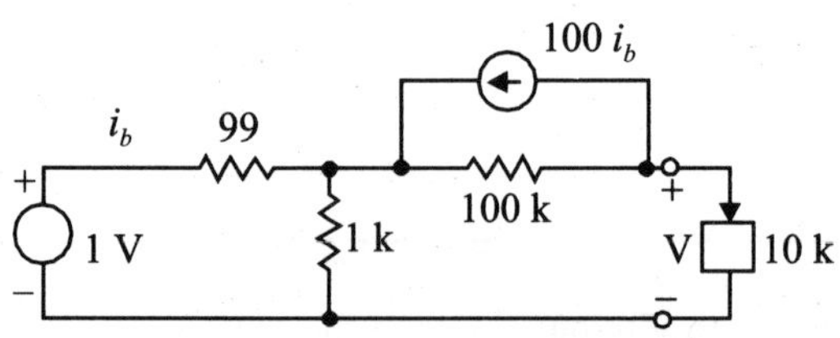

Fig. P. 5.6

5.7 Two sets of measurements are made on a linear time-invariant resistive network (Fig. P. 5.7). Find the value of R_1.

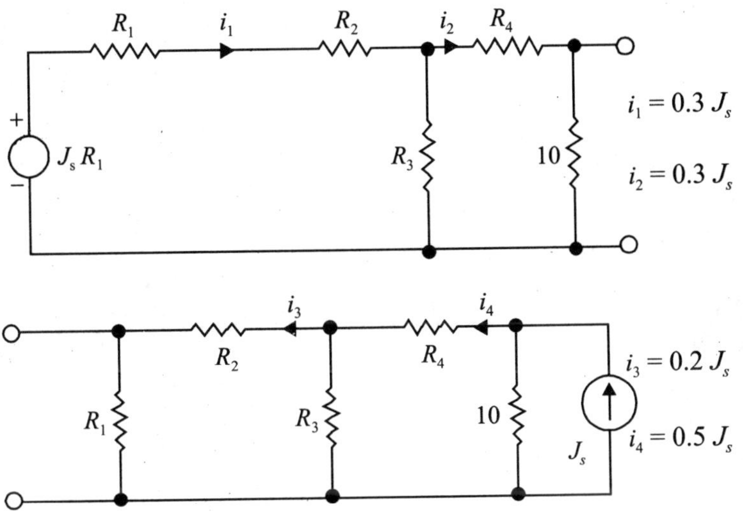

Fig. P. 5.7

5.8 Prove Norton equivalent theorem.

5.9 Show that a linear time-invariant passive *RLC* two-port network is reciprocal by using Tellegen's theorem. (Show this for all combinations of sources at the ports.)

5.10 A linear-invariant passive one-port network is driven by a current source whose phasor is J_1. Show that

$$Z_{in}(j\omega) = \frac{2P_{av} + 4j\omega[E_M - E_E]}{|J_1|^2}$$

where J_1 is the peak amplitude of the sinusoidal input current, P_{av}, E_M and E_E are the average power dissipated, the average magnetic energy stored and the average electric energy stored over one period, respectively.

NOTES AND REFERENCES

Network theorems 'except Tellegen's theorem' is generally covered in the first course on circuit theory. Application of Thevenin-Norten's theorem can be demonstrated by experiments on simple electronic circuits; for instance a common emitter transistor amplifier. Most circuit theory books-cover these topics and additional problems may be supplemented from reference cited in Chap. 3 (Ref. 1-Chap. 9, Ref. 2-Chap. 5, Ref. 3–Chap. 16).

Tellegen's theorem has many elegent applications and they are well discussed in Penfield et, all's book[1]. The concept of adjoint network is an important outcome of this theorem and further details can be had from references[2, 3]. Temes and Lapatra's book[4] discusses several example of deriving adjoint networks besides delineating applications of the same for sensitivity calculations (Chaps. 9–11).

1. P. Penfield, R. Spence and S. Duinker, *Tellegen's Theorem and Electric Networks*, M.I.T. Press, 1970.
2. S.W. Director and R.A. Rohrer, *The Generalized Adjoint Network and Network Sensitivities*, IEEE Trans. on CT, Vol. CT-1969, pp. 318–323.
3. S.W. Director and R.A. Rohrer, *Automated Network Design–The Frequency Domain Case*, IEEE Trans. on CT, Vol. CT-1969, pp. 330–337.
4. G.C. Temes and J.W. Lapatra, *Circuit Synthesis and Design*, McGraw-Hill, 1977.

CHAPTER

6 Natural Frequencies and Network Functions

The response of a linear network is made up of two components—the zero input response (*ZIR*) and the zero state response (Chap. 4). The *ZIR* of the network gives its natural response. If a series or parallel *LC* network is driven by a sinusoidal source of frequency $\omega_0 = 1/\sqrt{LC}$, we observe the phenomenon of resonance. This frequency ω_0 (the resonant frequency) is called the *natural frequency* of the *LC* network. The natural frequencies of a network are closely related to the poles and zeros of certain network functions. In this chapter, we discuss this concept and the method determining the natural frequencies. We also discuss network functions and frequency response.

6.1 NATURAL FREQUENCY

Let a linear network be characterized by the set of equations,

$$M\mathbf{X} = \mathbf{N}$$

where M is a network matrix, $\mathbf{X}$ a vector of certain network variables and $\mathbf{N}$ an equivalent source vector (see Chaps. 3 and 4). We set all the independent sources to zero (*i.e.* open current sources and short voltage sources) and determine the *j*th network variable (for example *j*th mesh current or *j*th node voltage). The *j*th network variable is given by

$$X_j = \sum_k \frac{\Delta_{jk}}{\Delta} N_k \qquad \ldots(6.1)$$

where X_j and N_k are the *j*th and *k*th elements of the vectors X and N, respectively, Δ is the determinant and Δ_{jk} the (k, j)th cofactor of M. N_k consist of initial capacitor voltages and inductor currents. The determinant of M, being a polynomial in the Laplace variable s, can be written as

$$\Delta(s) = \prod_{t=1}^{n} (s - p_i)$$

Inverse transforming eqn. 6.1 by partial fractions we have

$$x_j(t) = \sum_{l=1}^{n} k_i e^{p_i t}$$

where k_i are constants determined by the initial conditions. We can, in general, write the *ZIR* of a network variable $x(t)$, either a current or a voltage, as

$$x(t) = \sum_{i=1}^{n} k_i e^{p_i t} \qquad \text{...(6.2)}$$

The constant p_i depends on the network elements and the topology, and the constants k_i depend, in addition, on the initial conditions. The constants p_i, which may be complex, determine the natural response of the network and are called the *natural frequencies* of the network variable $x(t)$.

Example 6.1. Consider the network shown in Fig. 6.1. The differential equation for i is

$$L\frac{di}{dt} + Ri = 0$$

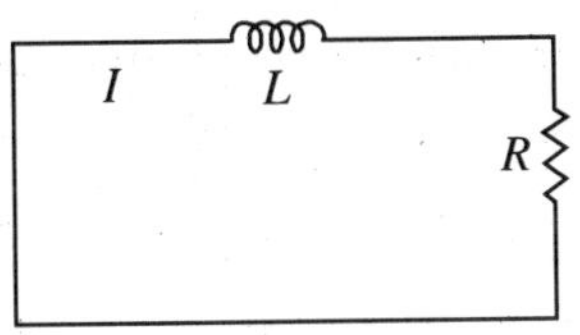

Fig. 6.1

If $i(0)$ is the initial current through the inductor, then

$$i(t) = i(0)e^{-tR/L}$$

The natural frequency is $-R/L$.

In eqn. 6.2 we have assumed that all the roots are simple. This need not be the case. If the natural frequency p_i is repeated n_i times and there are r distinct roots, we can rewrite eqn. 6.2 as

$$x(t) = \sum_{i=1}^{r}\sum_{j=1}^{n_i} k_{ij} t^{j-1} e^{p_i t} \qquad \text{...(6.3)}$$

and we say that p_i is a *natural frequency of order* n_i of the network variable x. (There is one further assumption we have made in he foregoing discussion that Δ_{jk} and Δ have no common factors. If such a situation occurs, the network has some degenerate circuits and for cut-sets. In most cases, we have to find the minimum order differential equation satisfied by the network variable before we determine the natural frequencies).

We can extend the concept of natural frequency to the entire network. We say that p_t is a *natural frequency of the network* if p_i is a natural frequency of some voltage or current in the network. If $p_i(p_i \neq 0)$ is a natural frequency of the current of a network element, it is also the natural frequency of the voltage of that element and vice versa. As a consequence, in order to find the nonzero natural frequencies we may resort to any method of network analysis. For example, consider the mesh equations. The mesh currents are given by

$$\mathbf{I}_m = Z_m^{-1}\mathbf{E}_q$$

i.e.

$$\mathbf{I}_m = \frac{Adj\, Z_m}{|\mathbf{Z}_m|}\mathbf{E}_q$$

Hence, the nonzero natural frequencies of the network are given by the zeros, (roots) of $|Z_m(s)|$. It can be shown that, but for the zeros at the origin, the zeros of the determinant $\Delta(s)$ are the same for other methods of analysis. Hence, we conclude that '*the nonzero natural frequencies of a linear time invariant network are given by the nonzero roots of* $\Delta(s)$, *where* $\Delta(s)$ *is the determinant of the matrix of any of the network equations* (eqn. 6.1).

Natural frequencies of simple networks can be written by inspection. For example, the natural frequency of networks shown in Fig. 6.1 is $-R/L$. It can be easily seen that the natural frequency of the networks in Figs. 6.2(*a*), (*b*), (*c*) and (*d*) are $-1/RC$, $\pm j1/\sqrt{LC}$, $-R/2L \pm \sqrt{R^2/4L^2 - 1/LC}$ and $-\frac{1}{2}RC \pm \sqrt{\left(\frac{1}{4}R^2C^2 - 1/LC\right)}$, respectively. It can be shown that the natural frequencies of *RC and RL networks are all negative real and those of LC networks are imaginary.*

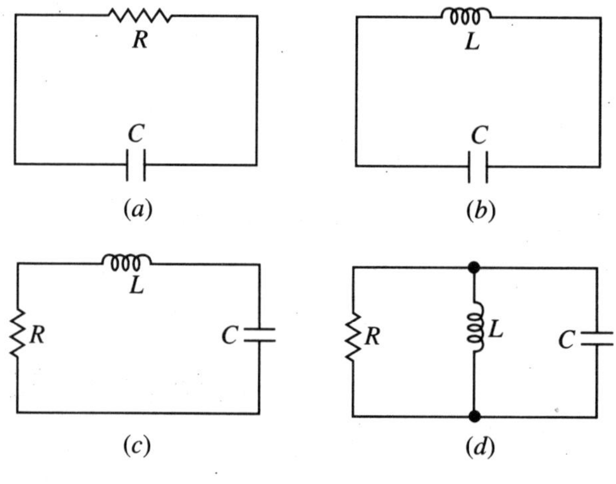

Fig. 6.2

A natural frequency $p_i = 0$ of a network variable indicates that *ZIR* may contain a constant term. This situation may arise in two cases.

Consider the network shown in Fig. 6.3. Let the network have a cut-set of capacitors as shown. In a network, a particular distribution of current and voltage can exist as long as the elemental voltages and currents satisfy KVL and KCL, respectively, and the voltage across and current through an element satisfy the terminal relation of the particular element. We claim that the network of Fig. 6.3

has the following current and voltage distribution:

$$V_1 = V_2 = V_3 = Ve^{0t} = V$$

$$V_k = 0, \quad k = 4, 5, \ldots n$$

$$i_k = 0, \quad k = 1, 2, \ldots n$$

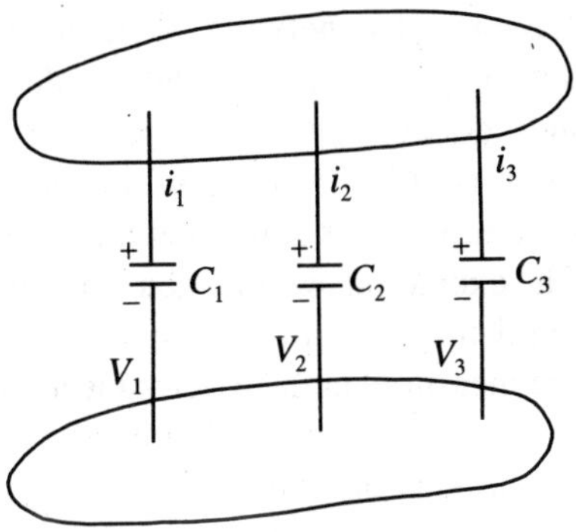

Fig. 6.3

where we have assumed that the network has n elements including the three capacitors in the cut-set. To justify our assertion we have to check the terminal relations, KCL and KVL. All the currents in the network being zero, KCL is obviously satisfied. Since $V_k = 0$, $k = 4, 5, \ldots n$, we have only to check elements 1, 2, 3 for terminal relations. The current through the capacitor C_k is given by

$$C_k \frac{dV_k}{dt} = C_k \frac{dV}{dt} = 0$$

in accordance with the assertion. To check KVL we are only to see that any circuit of the network includes two of the capacitors C_1, C_2 and C_3 or none at all. In the latter case, KVL is obviously satisfied, as all voltages are zero. In the former case, the two capacitor voltages are opposing each other in a circuit and, hence, KVL is satisfied. Thus, we conclude that in a source-free network containing a cut-set of capacitors, there is a natural frequency $p_i = 0$. Similarly, it can be shown that in a source-free network containing a circuit of inductors (Fig. 6.4), there is a natural frequency $p_i = 0$ (prob. 6.2).

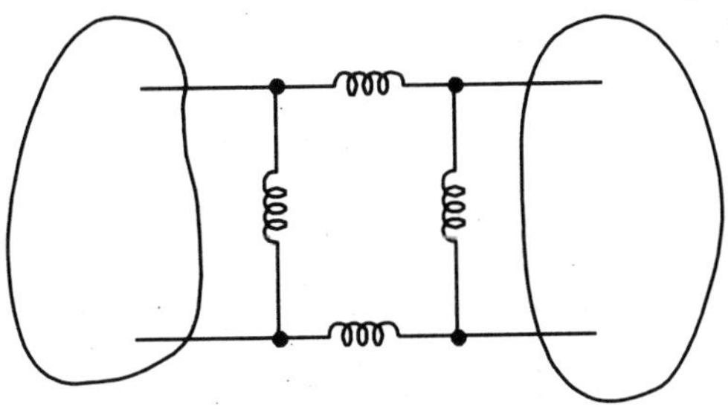

Fig. 6.4

The number of natural frequencies of a network is called the *"Order of Complexity"* of the network. The order of complexity of a network cannot exceed the number of energy storage elements in the network. The order of complexity is also equal to the number of independent initial conditions needed to determine uniquely all the voltages and the currents in the network. The natural frequencies specify the general form of the *ZIR* of the network variables, *i.e.* the general form of the time-domain behaviour of a network variable when excited by initial conditions. Thus, the addition or deletion of independent sources to a network will not change its natural frequencies, if the sources are so added as not to change the topology of the resultant network. (How can this be done?)

The concept of natural frequency is fundamental to the understanding of the behaviour of a network. The excitation and the measurement of a natural frequency depend on its topology and the elemental values. If it has an internal natural frequency which cannot be measured at the external terminals of the network then we can conclude that there must be a cancellation of a factor, corresponding to this natural frequency, in the network function (*i.e.* Δ_{jk} and Δ have a common factor). This will be treated in detail in Appendix C.

6.2 NETWORK FUNCTIONS

A *network function* is defined as the ratio of the zero-state response to the input, both the response and the input expressed in Laplace domain, *i.e.*

$$H(s) = \frac{R(s)}{E(s)}$$

where *s*, the Laplace variable (see Appendix A), is the *complex frequency variable.*

$$s = \sigma + j\omega$$

and ω is the *real frequency* in radians per second. The network function is the same as the transfer function defined in Sec. 4.4. Here, we reserve the same transfer function for a particular network function (see Sec. 4.5). We define two different sets of network functions.

(i) Driving point function

If the excitation and the response are measured at the same set of terminals, the network function is called the *driving point* (DP) function. There are two DP functions; impedance and admittance.

$$z_{dp}(s) = \frac{V_1(s)}{I_1(s)} \quad \text{(DP impedance)}$$

$$y_{dp}(s) = \frac{I_1(s)}{V_1(s)} \quad \text{(DP admittance)}$$

From the definition of DP impedance and admittance it can be seen that

$$y_{dp}(s) = \frac{1}{z_{dq}(s)} \qquad \ldots(6.4)$$

i.e. the reciprocal of a DP function is another DP function.

(ii) Transfer functions

If the excitation and response are measured at different sets of terminals (see Fig. 6.5), the corresponding network function is called a *transfer function.*

$$z_{21}(s) = \frac{V_2(s)}{I_2(s)} \rightarrow \text{Transfer impedance}$$

$$y_{21}(s) = \frac{I_2(s)}{V_1(s)} \rightarrow \text{Transfer admittance}$$

$$G_{21}(s) = \frac{V_2(s)}{V_1(s)} \rightarrow \text{Transfer voltage ratio}$$

$$\alpha_{21}(s) = \frac{I_2(s)}{I_1(s)} \rightarrow \text{Transfer current ratio}$$

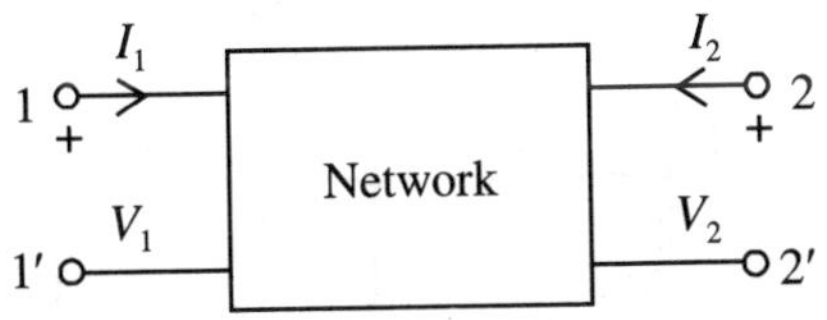

Fig. 6.5

In the preceding relations, we have used the second set of terminals, terminals 22′, as the terminals at which the response is measured and the first set, terminals 11′, as the excitation point. It can be observed from the definition of DP and transfer functions that for an impedance function excitation is a current source and response a voltage, and for an admittance function excitation is a voltage source and response a current.

For simple networks, DP functions can be obtained by inspection. For example, consider the ladder network shown in Fig. 6.6. The DP impedance of the network is

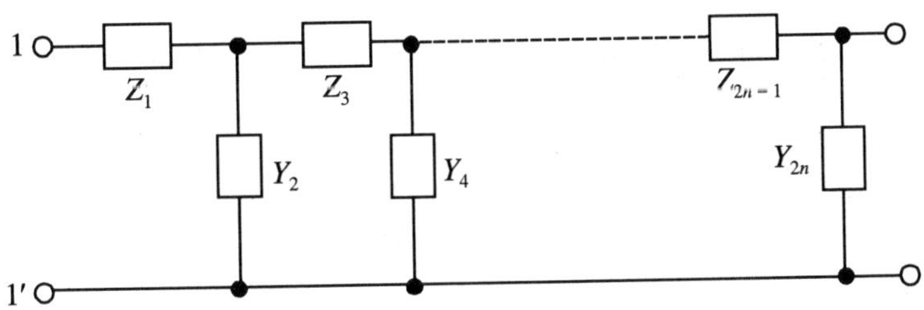

Fig. 6.6

$$z_{11}(s) = z_1(s) + \cfrac{1}{y_2(s) + \cfrac{1}{z_3(s) + \cfrac{1}{y_4(s) + \cfrac{1}{\cdots}}}}$$

i.e. the DP impedance of a ladder can be written in the form of a *continued fraction expansion.* Such an expansion is very useful in network synthesis (see Chap. 10).

For more complicated networks we have to resort to the standard method of network analysis in order to find the network functions. (We elaborate on this when we discuss two-port parameters and indefinite admittance matrices.)

Example 6.2. Find the voltage transfer ratio of the active network shown in Fig. 6.7.

We solve this problem by writing node equations at nodes n_1 and n_2.

At n_1 :
$$\frac{V_3 - V_1}{R} + \frac{V_3 - V_4}{R} + \frac{V_3 - V_2}{1} sC = 0$$

i.e.
$$V_3\,(2 + sCR) - V_1 - V_4 - sCRV_2 = 0$$

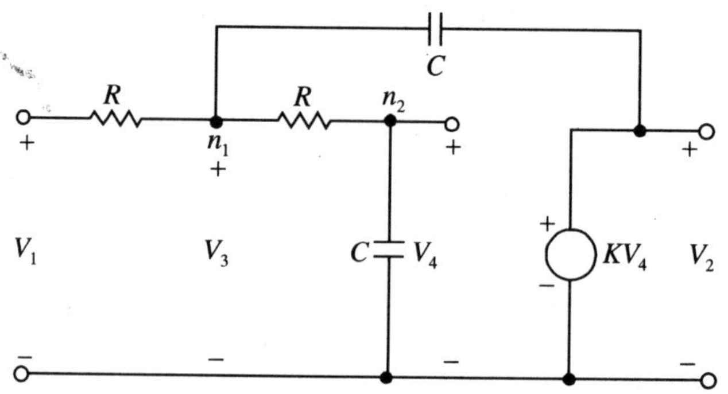

Fig. 6.7

At n_2 :
$$\frac{V_4 - V_3}{R} + sCV_4 = 0$$

i.e.
$$V_4\,(1 + sCR) - V_3 = 0$$

The constraint imposed by the active element (*voltage-controlled voltage source*) is

$$V_2 = KV_4$$

Combining this with the two node equations we have

$$(V_2/K)\,(1 + sCR)\,(2 + sCR) - V_1 - (V_2/K) - sCR\,V_2 = 0$$

which yield

$$\frac{V_2}{V_1} = \frac{K}{s^2C^2R^2 + sCR(3 - K) + 1} \qquad \ldots(6.5)$$

6.3 POLES AND ZEROS

Consider a linear time-invariant network. Let $x(t)$ be the excitation and $y(t)$ the response. Then, in general, we can express the relationship between the variables

$x(t)$ and $y(t)$ in the form of a differential equation,

$$(a_n p^n + a_{n-1} p^{n-1} + ... + a_1 p + a_0) y(t)$$
$$= (b_m p^m + b_{m-1} p^{m-1} + ... + b_1 p + b_0) x(t) \qquad ...(6.6)$$

where $p^i f(t) = \frac{d^i}{dt^i} f(t)$

The coefficient in this differential equation depend on the elements and the topology of the network. Laplace transforming and rearranging (with all initial conditions assumed to be zero) we have

$$H(s) = \frac{Y(s)}{X(s)} = \frac{b_m s^m + b_{m-1} s^{m-1} + ... + b_1 s + b_0}{a_n s^n + a_{n-1} s^{n-1} + ... + a_1 s + a_0}$$

i.e.

$$H(s) = \frac{p(s)}{q(s)} \qquad ...(6.7)$$

where p and q are polynomials in the complex frequency variable s. Thus, we conclude that a network function is a rational function of s with real coefficients (see Sec. 4.5). Alternatively, we can write eqn. 6.7 as

$$H(s) = k \frac{\prod_{i=1}^{m} (s - z_i)}{\prod_{i=1}^{m} (s - p_i)} \qquad ...(6.8)$$

where $k = b_m / a_n$ is the scale factor z_i the complex frequency at which $H(s)$ becomes zero, is called a *zero of* $H(s)$ and p_i the complex frequency at which $H(s)$ becomes infinity is called a *pole of* $H(s)$. The poles and zeros of a network function need not be distinct. The coefficients of the polynomials $p(s)$ and $q(s)$ being real any complex zeros and poles must appear in conjugate pairs. From eqn. 6.8 it is clear that any network function is completely specified by its poles and zeros and the scale factor.

Example 6.3. Consider the network shown in Fig. 6.8. Let $x(t)$ be the input and $y(t)$ (mesh current of the mesh II) be the output. The mesh equations (the currents are assumed to be in clockwise direction) in integro-differential equation form are:

Mesh I
$$x(t) = R_1 i_1 + L_1 \frac{di_1}{dt} + \frac{1}{C_1} \int (i_1 - y) dt$$

Mesh II
$$0 = R_2 y + \frac{1}{C_2} \int y\, dt + \frac{1}{C_1} \int (y - i_1) dt$$

where i_1 is the current of mesh I. The second equation, after differentiation, can be written as

$$i_1 = R_2C_1\frac{dy}{dt} + \left(1 + \frac{C_1}{C_2}\right)y$$

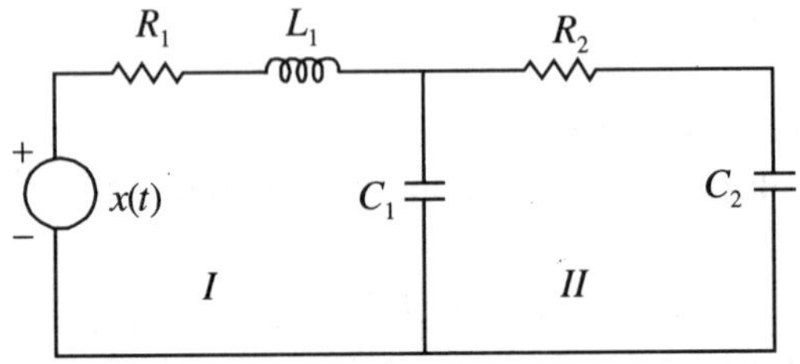

Fig. 6.8

Hence, we have

$$\frac{di_1}{dt} = R_2C_1\frac{d^2y}{dt^2} + \left(1 + \frac{C_1}{C_2}\right)\frac{dy}{dt}$$

and
$$\frac{d^2i_1}{dt^2} = R_3C_1\frac{d^3y}{dt^3} + \left(1 + \frac{C_1}{C_2}\right)\frac{d^2y}{dt^2}$$

Differentiating the equation for mesh I, we have

$$\frac{dx}{dt} = R_1\frac{di_1}{dt} + L_1\frac{d^2i_1}{dt^2} + \frac{1}{C_1}i_1 - \frac{1}{C_1}y$$

Substituting for i_1, $\frac{di_1}{dt}$ and $\frac{d^2i_1}{dt}$ and rearranging we have

$$\frac{d^3y}{dt^3} + \left[\frac{C_2 + C_1}{R_2C_1C_2} + \frac{R_1}{L_1}\right]\frac{d^2y}{dt^2} + \left[\frac{R_2(C_2 + C_1)}{R_2L_1C_1C_2} + \frac{1}{L_1C_1}\right]\frac{dy}{dt}$$

$$+\frac{1}{R_2L_1C_1C_2}y = \frac{1}{R_2L_1C_1}\frac{dx}{dt}$$

which is of the form of eqn. 6.6. The transfer function (assuming $L_1 = R_2 = C_1 = C_2 = 1$, and $R_1 = 2$) is

$$H(s) = \frac{s}{s^3 + 4s^2 + 5s + 1}$$

$$= \frac{s}{(s + 0.2451)(s^2 + 3.7549s + 4.0797)}$$

The poles and zeros of a network function can be represented by a *pole-zero plot.* The pole-zero plot of the network function of example 6.3 is shown in Fig. 6.9. In the figure, x represents a pole and 0 (a zero).

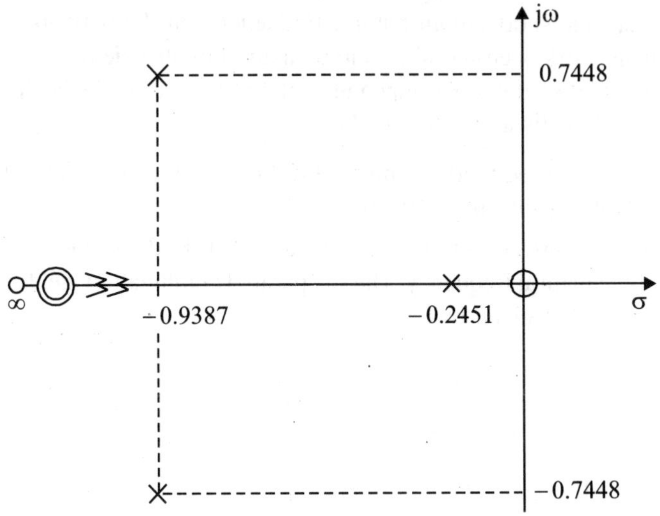

Fig. 6.9

If $x(t)$ and $y(t)$ are measured at the same set of terminals in eqn. 6.6, then $H(s)$ in eqn. 6.7 would be a driving point function. Without loss of generality we assume that $x(t)$ is current, and the response $y(t)$ is voltage. Then

$$H(s) = Z(s) = \frac{V(s)}{I(s)} = \frac{p(s)}{q(s)}$$

i.e. $$q(s)\,V(s) = p(s)\,I(s)$$

If we make the current $x(t)$ go to zero, *i.e.* the network is being tested under open circuit conditions, the zero input voltage response is governed by the equation

$$q(s)\,V(s) = 0$$

Hence, the natural frequencies of the voltage response are determined by the roots of the equation,

$$q\,(s) = 0$$

Thus, the roots of $q(s)$ are the natural frequencies of the network under open circuit conditions and are called the *open-circuit natural frequencies* (OCNF). On the other hand, if we make the voltage $y(t)$ go to zero, *i.e.* the network is being tested under short circuit conditions, the zero input current response is governed by the equation

$$p(s)\,I(s) = 0$$

and the natural frequencies of the current response are determined by the roots of the equation

$$p(s) = 0$$

Hence, the roots of $p(s)$ are called the *short-circuit natural frequencies* (SCNF). We could have carried out the discussion starting from an admittance function. Thus, we conclude: 'The poles (zeros) of an impedance (admittances) function are the open-circuit natural frequencies and the zeros (poles) of the impedance (admittance) are the short-circuit natural frequencies of the network.'

For simple network open-and short-circuit natural frequencies can be determined by inspection and then the input impedance of admittance can be determined. The following examples illustrates the method.

Example 6.4. We are required to find the OCNF, SCNF and the input impedance of the network shown in Fig. 6.10(*a*).

Under open circuit condition the inductor of 1 H does not come into the calculation of the natural frequency. The LC parallel combination yields two natural frequencies: s_1 and s_2 given by

$$s_1, s_2 = \pm\frac{j}{\sqrt{2\times 4}} = \pm j\frac{1}{\sqrt{8}}$$

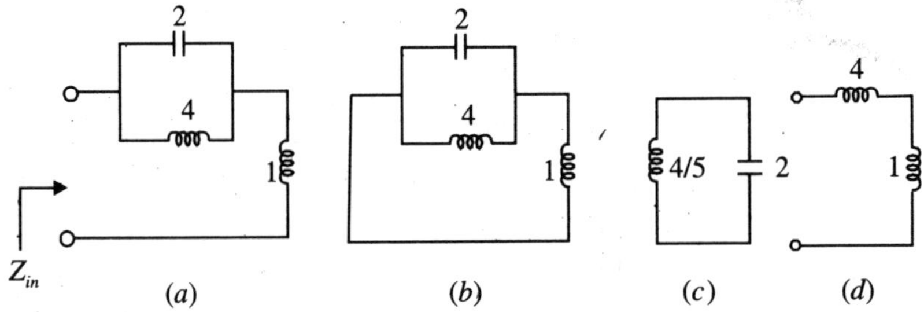

Fig. 6.10

Under short-circuit condition, the network reduces to the one shown in Fig. 6.10(*b*). There is a circuit of inductors yielding a natural frequency $p_1 = 0$. The short-circuited network reduces to the one shown in Fig. 6.10(*c*) yielding two natural frequencies p_2, p_3;

$$p_2, p_3 = \pm\frac{j}{\sqrt{\frac{4}{5}\cdot 2}} = \pm j\sqrt{5/8}$$

$$OCNF: s_1, s_2 = \pm\frac{j}{\sqrt{8}}$$

$$SCNF: \ p_1 = 0, p_3, p_s = \pm j\sqrt{5/8}$$

Hence,

$$Z_{in} = k\frac{(s-p_1)(s-p_2)(s-p_3)}{(s-s_1)(s-s_2)}$$

$$= k\frac{s(s^2+5/8)}{(s^2+1/8)}$$

where k is a constant we have to determine. Z_{in} tends to 5 ks as $s \to 0$. As $s \to 0$ the network of Fig. 6.10(a) tends to the one shown in Fig. 6.10(d) as the capacitor acts like an open circuit and the impedance of network of $5s$. Hence $k = 1$. The input impedance is,

$$Z_{in} = \frac{s(s^2 + 5/8)}{(s^2 + 1/8)}$$

Example 6.5. We are required to find the input admittance of the network shown in Fig. 6.11.

Under open-circuit condition, there exists a cut-set of capacitors yielding a natural frequency $s_1 = 0$. On combining the two series capacitors we get the

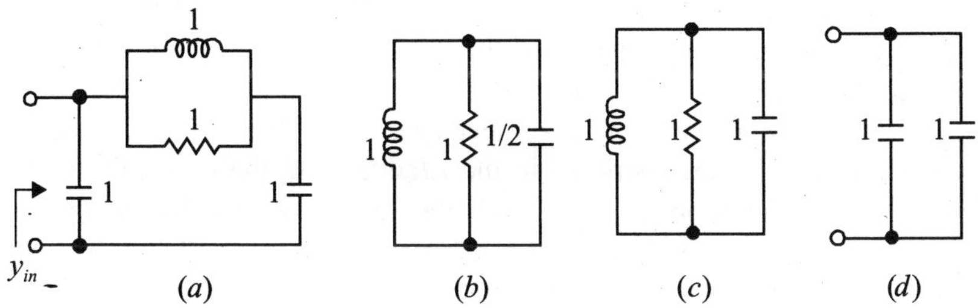

Fig. 6.11

network in Fig. 6.11(b). The natural frequencies are

$$s_2, s_3 = -\frac{1}{2RC} \pm \sqrt{\frac{1}{4R^2C^2} - \frac{1}{LC}} = -1 \pm j1.$$

Under short-circuit conditions, the first capacitor does not contribute to the natural frequency of the network. The reduced network is shown in Fig. 6.11(c). The natural frequencies are

$$p_1, p_2 = -\frac{1}{2} \pm j\sqrt{3/2}$$

$$OCNF: s_1 = 0,\ s_2, s_3 = -1 \pm j1$$

$$SCNF: p_1, p_2 = -\frac{1}{2} \pm j\sqrt{3/2}$$

Hence,

$$y_{in} = k\frac{(s - s_1)(s - s_2)(s - s_3)}{(s - p_1)(s - p_2)}$$

$$= k\frac{s(s^2 + 2s + 2)}{(s^2 + s + 1)}$$

y_{in} tends to $2ks$ as $s \to 0$.

As $s \to 0$ the network of Fig. 6.11(a) tends to the one shown in Fig. 6.10(d) whose admittance is $2s$. Hence $k = 1$

$$y_{in} = \frac{s(s^2 + 2s + 2)}{(s^2 + s + 1)}$$

Since the natural response of a passive network must decrease to zero with time as the energy initially stored in the network dissipates gradually through the resistors in the form of heat, the poles and zeros of a *DP* function of such a network must have negative real parts. However, in the case of lossless (*i.e. LC*) networks the stored energy is conserved and the poles and zeros are on the imaginary axis. Hence, we conclude that 'the poles and zeros of a passive *DP* function are restricted to the left half of the complex frequency plane (*LHP*).'

If the poles of the impedance function (the open-circuit natural frequencies) are restricted to the *LHP* but the zeros (the short circuit natural frequencies) are not, the natural response (current response) of the network will not die out when the network is under short-circuit condition. We say that such a network is *open-circuit stable* but *short-circuit unstable.* Such networks should not be driven by a voltage source as it has low internal impedance. On the other hand, if the zeros of impedance functions are restricted to the *LHP* but not the poles, the network is *short-circuit stable* but *open-circuit unstable.* Such networks should not be driven by a current source as it has a large internal impedance. Evidently, a passive immittance is both open-circuit and short-circuit stable.

If in eqn. 6.6 $x(t)$ and $y(t)$ are measured at different sets of terminals, then $H(s)$ in eqn. 6.7 will be a transfer function. Without loss of generality if we assume that both $x(t)$ and $y(t)$ are voltages, then

$$H(s) = G(s) = p(s)/q(s)$$

If the excitation $x(t)$ is set to zero, then the natural response of the network is given by

$$y(t) = \sum_{i=1}^{n} c_i e^{p_i t}$$

where p_i are the poles of $H(s)$ and c_i are constants determined by the initial conditions. For a stable network the poles must be restricted to the left half plane. Hence, we conclude that 'the poles of the transfer function of a stable network are restricted to the left half of the s-plane'. In particular, the poles of the transfer function of passive networks are restricted to the left half plane, as passive networks are always stable. On the other hand, active networks are not always stable, which can be demonstrated by studying eqn. 6.5. In eqn. 6.5, if $K > 3$, then the network is unstable. Also, the natural frequencies and the poles of the network function, depend on the active parameter K.

Now, that the reciprocal of a transfer function is not, in general, a network function, there is no restriction on the zeros of the transfer function. The zeros can be anywhere in the complex plane. The zeros of a transfer function are called the *transmission zeros.*

Consider the ladder network shown in Fig. 6.5. Let the impedance z_{2i-1} have a pole at the frequencies s_i. At this frequency, the series arm corresponding to z_{2i-1}

is open and thus creates a transmission zero at $s = s_i$. Let the admittance y_{2i} have a pole at the frequency p_i. At this frequency, the shunt arm corresponding to y_{2i} is a short and thus creates a transmission zero at $s = p_i$. Hence, we conclude that the transmission zeros of a ladder network can only exist at the pole frequencies of the series impedances and shunt admittances.' If we have a passive ladder, all the transmission zeros are restricted to the left half plane. There is one restriction on the creation of transmission zeros by the poles of z_{2i-1} and y_{2i}. If the impedance looking to the right of z_{2i-1}, *i.e.* to the output end, has a pole at $s = s_i$, the pole s_i, of z_{2i-1} is not a transmission zero. Similarly, if the admittance looking to the right of y_{2i-1}, *i.e.* to the output end, has a pole at $s = p_i$, the pole p_i of y_{2i} is not a transmission zero (prob. 6.12).

Filter networks can be classified by the location of their transmission zeros. If all the transmission zeros are at infinity, the network is a *low-pass network* and if the transmission zeros are at the origin, the network is a *high-pass network.* A *band-pass network* has some of its transmission zeros at the origin and others at infinity. Thus, ladder network with series inductors' and shunt capacitors is a low-pass network and a ladder network with series capacitors and shunt inductors is a high-pass network.

6.4 TIME RESPONSE

Consider the network shown in Fig. 6.12. The current response in Laplace domain is given by

$$I(s) = \frac{V(s)}{L(s + R/L)}$$

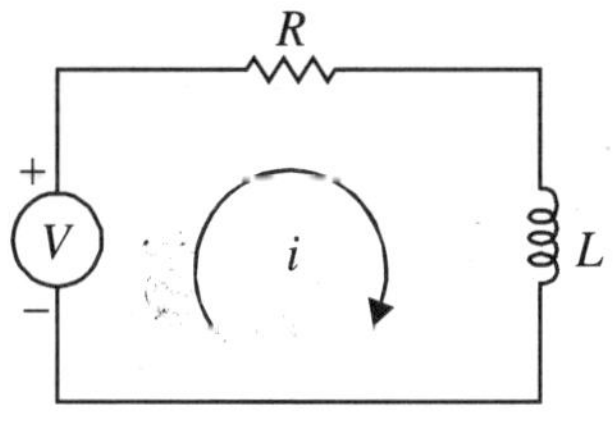

Fig. 6.12

The pole of the admittance function is at

$$s = -R/L$$

and the response to a unit impulse voltage input is

$$i(t) = \frac{1}{L} e^{-Rt/L}$$

The response to a unit step input is

$$i(t) = \frac{1}{R}[1 - e^{-Rt/L}]\, u(t)$$

But for a constant multiplier the impulse and step responses of the network are

$$e^{-Rt/L} \text{ and } 1 - e^{-Rt/L}$$

respectively. This illustrates that the impulse and step responses can be correlated with the location of the poles of a network function.

Figure 6.13 and Table 6.1 give the impulse and step responses for several pole locations. Response curves of Fig. 6.13(*a*), (*b*), (*c*) and (*d*) correspond to stable

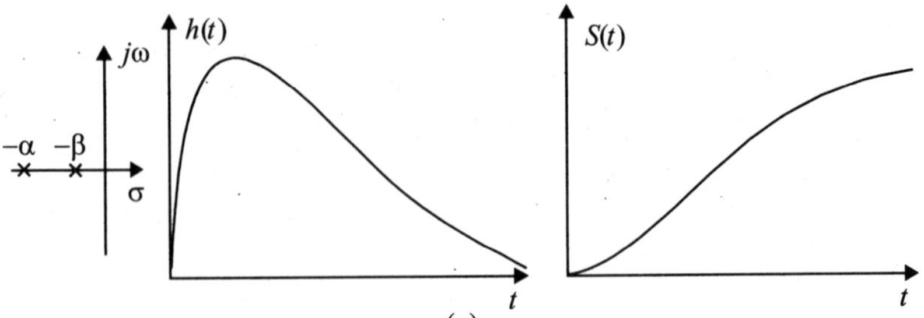

(*a*)

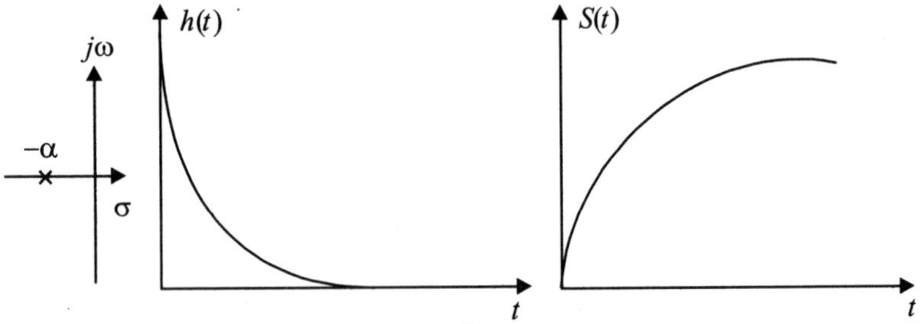

(*b*)

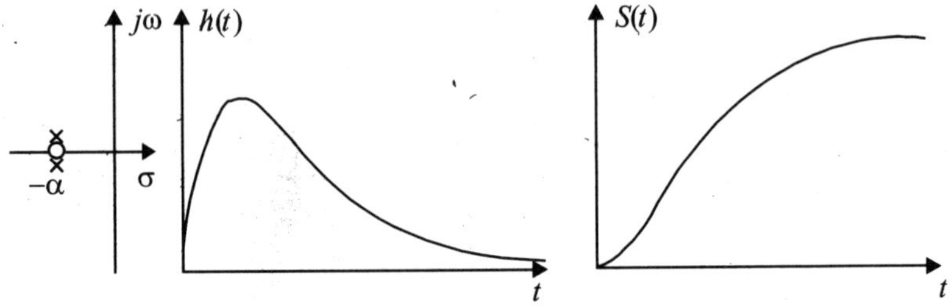

(*c*)

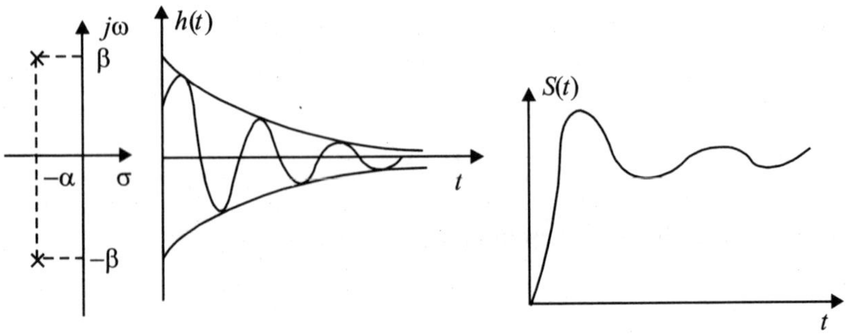

(*d*)

Fig. 6.13 (continued)

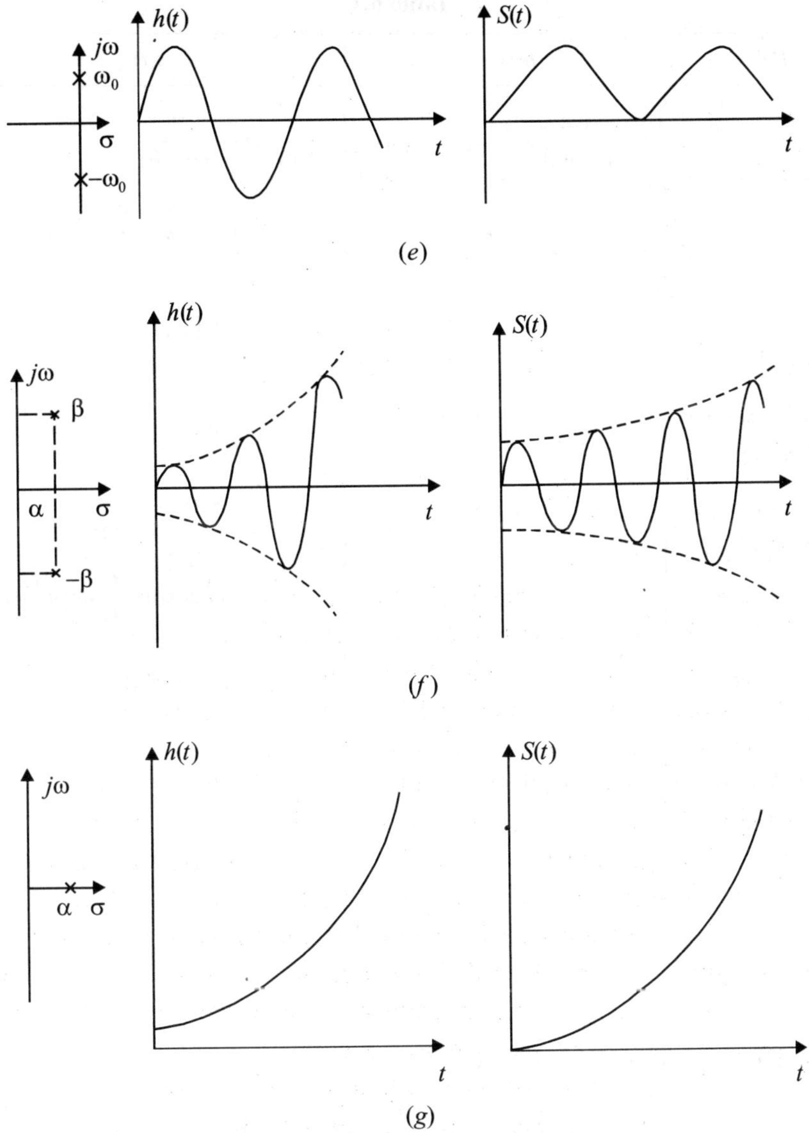

Fig. 6.13

networks and those of (*f*) and (*g*) unstable networks. The response of Fig. (*e*) is that of an oscillatory (*LC*) network and is marginally stable. The pole locations of Fig. 6.13(*a*) is for an overdamped network) ($Q < 1/2$) and that of Fig. (*c*) corresponds to a critically damped system ($Q = 1/2$). The network corresponding to Fig. (*d*) is an underdamped ($Q > 1/2$, ringing circuit) network. The Q for the network corresponding to Fig. (*e*) is infinity (zero damping).

The presence of finite zeros will change the time response. (The zeros for the networks corresponding to Fig. 6.13 are all assumed to be at 'infinity'.)

Table 6.1

Poles	$h(t)$	$s(t)$
$s=-\alpha,-\beta$ Fig. 6.13 (*a*)	$\frac{1}{(\alpha-\beta)}\left[e^{-\beta t}-e^{-\alpha t}\right]u(t)$	$\frac{1}{\alpha\beta}u(t)+\frac{1}{(\alpha-\beta)} \times\left[\frac{e^{-\alpha t}}{\alpha}-\frac{e^{-\beta t}}{\beta}\right]u(t)$
$s=-\alpha$ Fig. 6.13(*b*)	$e^{-\alpha t}u(t)$	$\frac{1}{\alpha}\left[1-e^{-\alpha t}\right]u(t)$
$s=-\alpha,-\alpha$ Fig. 6.13(*c*)	$te^{-\alpha t}u(t)$	$\left[\frac{1}{\alpha^2}-\frac{t}{\alpha}e^{-\alpha t}-\frac{1}{\alpha^2}e^{-\alpha t}\right]u(t)$
$s=-\alpha,+j\beta$ Fig. 6.13(*d*)	$\frac{1}{\beta}e^{-\alpha t}\sin\beta t\,u(t)$	$\frac{1}{\alpha^2+\beta^2}u(t)-\frac{e^{-\alpha t}}{\alpha^2+\beta^2} \times\left[\cos\beta t+\frac{\alpha}{\beta}\sin\beta t\right]u(t)$
$s=\pm j\omega_0$ Fig. 6.13(*e*)	$\frac{1}{\omega_0}\sin\omega_0 t\,u(t)$	$\frac{1}{\omega_0^2}\left[1-\cos\omega_0 t\right]u(t)$
$s=\alpha\pm j\beta$ Fig. 6.13(*f*)	$\frac{1}{\alpha}e^{\alpha t}\sin\beta t\,u(t)$	$\frac{1}{\alpha^2+\beta^2}u(t)-\frac{e^{\alpha}}{\alpha^2+\beta^2} \times\left[\cos\beta t-\frac{\alpha}{\beta}\sin\beta t\right]u(t)$
$s=\alpha$ Fig. 6.13(*g*)	$e^{\alpha t}u(t)$	$\alpha\left[e^{\alpha t}-1\right]u(t)$

6.5 FREQUENCY RESPONSE

Consider the network shown in Fig. 6.14. The transfer function of the network is

$$H(s)=1/(1+s)$$

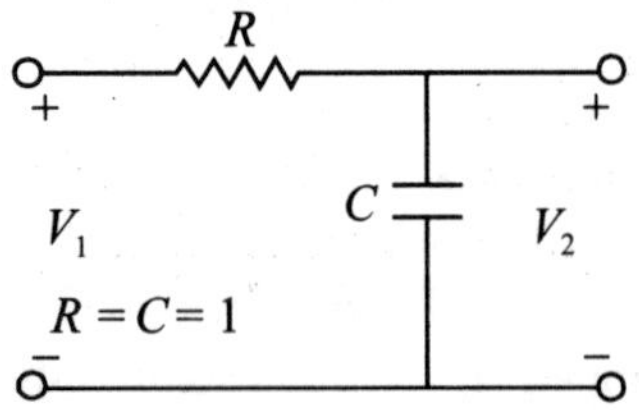

Fig. 6.14

Let $s = j\omega$, where $\omega = 2\pi f$ and f the frequency in Hz. Then

$$H(j\omega) = \frac{1}{1 + j\omega} = |H(\omega)| e^{j\phi(\omega)}$$

The magnitude $|H(\omega)|$ is

$$|H(j\omega)| = \sqrt{\frac{1}{1 + \omega^2}}$$

and the phase is

$$\phi(\omega) = -\tan^{-1}\omega$$

Figure 6.15 shows the plot of the magnitude and phase as function of the radian frequency ω. These curves give the frequency response of the network. As can be observed from the figure, the network passes lower frequency signals with little attenuation and high frequency signals with a large attenuation (low-pass filter) and the phase becomes progressively more lagging.

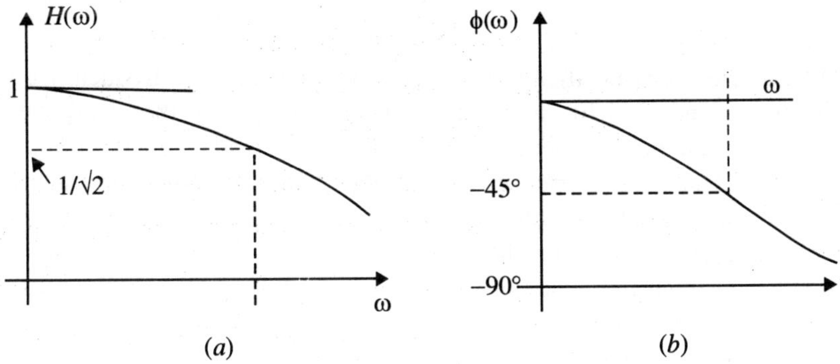

Fig. 6.15

Let
$$H(s) = \frac{p(s)}{q(s)}$$

be any network function. Then

$$H(j\omega) = \frac{p(j\omega)}{q(j\omega)} \qquad \ldots(6.9)$$

is the frequency response of the network. The frequency response of the network is the value of the network function evaluated on the imaginary axis (real frequency axis) of the complex s-plane. The frequency response is, in general, complex and hence can be written as

$$H(j\omega) = |H(j\omega)| \quad \phi(\omega) \qquad \ldots(6.10)$$

Most often the frequency response curves are plotted with frequency (ω or f) represented on logarithmic scale. To this end we define

$$H(\omega) = 20 \log |H(j\omega)|$$

and

$$\phi(\omega) = \angle p(j\omega) - \angle q(j\omega) \qquad \ldots(6.11)$$

The units for $H(\omega)$ are the decibels (dB). The plots of $(H\omega)$ and $\phi(\omega)$ or equivalently $H(f)$ and $\phi(f)$ vs. log ω or log f are called *Bode plots* (Appendix D). There are other methods of representing the frequency response (polar plots Nyquist plots).

Figure 6.16 shows typical Bode magnitude plots for LP and HP networks.

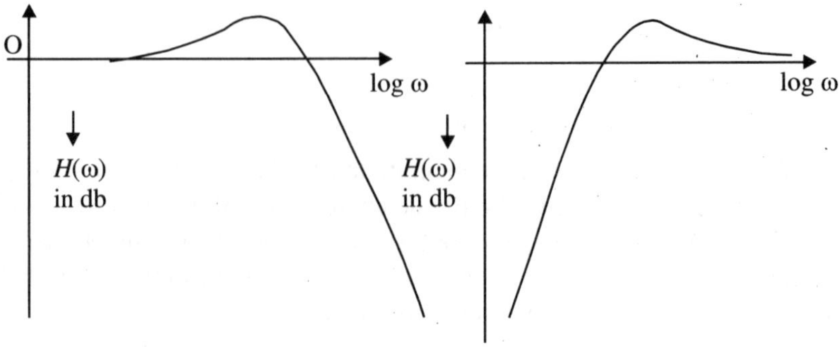

Fig. 6.16

Most of the filter specifications are generally given in the frequency domain, *i.e.* a filter is specified by its frequency response. Indeed, the classification of filters is based on their frequency response. From sec. 6.4 and 6.5 it is obvious that the time and frequency responses are correlated with the pole-zero pattern. The main concern of the approximation problem, which is a prelude to network synthesis and filter design, is to obtain the poles and zeros of a network function which satisfy the specified frequency or time response. This problem will be discussed in Chap. 11.

PROBLEMS

6.1 Find the natural frequencies of the network shown in Fig. P. 6.1.

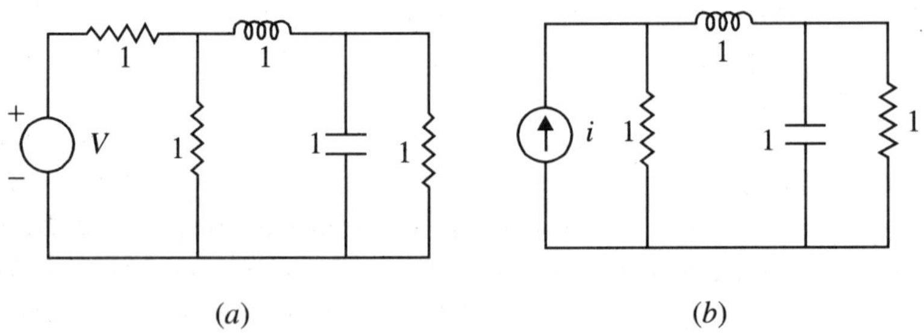

Fig. P. 6.1

6.2 Show that a network containing a circuit of inductors can support a natural frequency at $p_i = 0$.

6.3 The natural frequencies of a linear time-invariant network are given by:

(*a*) $s_1 = -2, s_2 = -3$

(*b*) $s_1 = s_2 = -2$

(c) $s_1 = 2j$, $s_2 = -2j$

(d) $s_1 = -2 + j3$, $s_2 = -2 - j3$

Give expressions for the zero input response in terms of real time functions.

6.4 Calculate the driving point impedance of the networks shown in Fig. P. 6.4 and plot their pole zero diagrams.

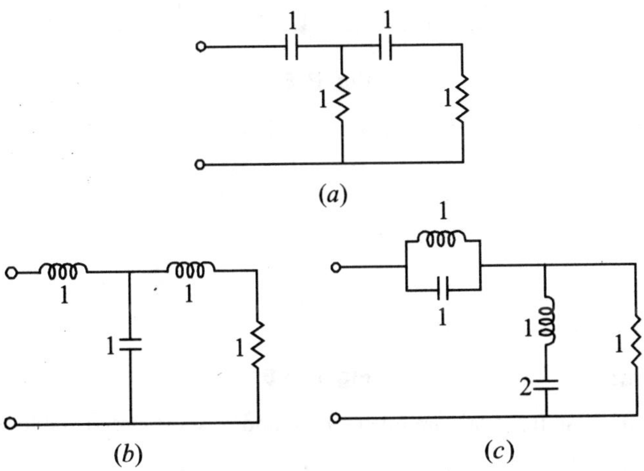

Fig. P. 6.4

6.5 Find the voltage transfer function of the network shown in Fig. P. 6.5.

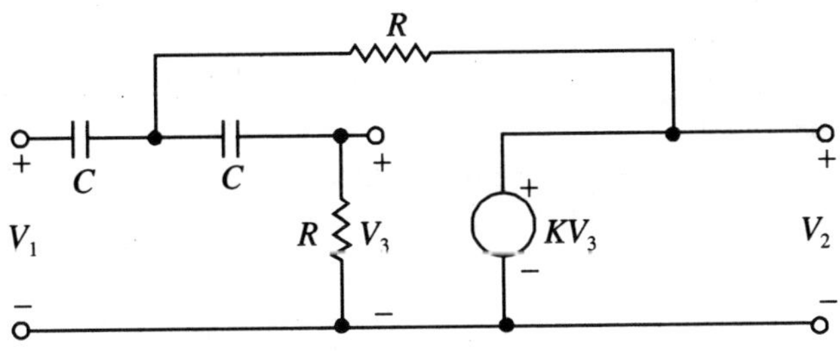

Fig. P. 6.5

6.6 The pole-zero plot of a voltage transfer function is shown in Fig. P. 6.6. The dc gain is to be 10. Find the transfer function.

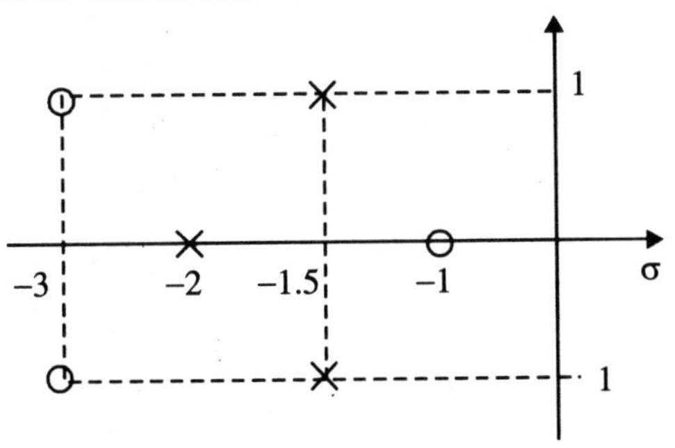

Fig. P. 6.6

6.7 Fig. P. 6.7 shows as infinite resistive ladder. Find the input resistance of the ladder.

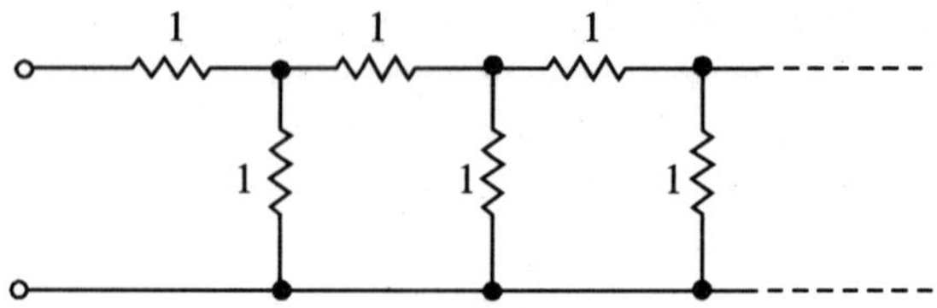

Fig. P. 6.7

6.8 Find the input impedance of the network shown in Fig. P. 6.8.

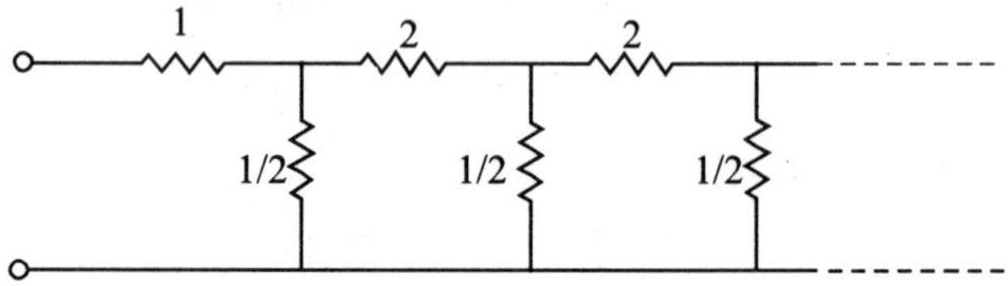

Fig. P. 6.8

6.9 Which of these functions are open circuit and/or short circuit stable?

$$Z(s)=\frac{(s+1)(s+3)}{(s+2)(s+4)} \qquad Z(s)=\frac{(s^2+1)(s+3)}{(s-2)(s+4)}$$

$$Z(s)=\frac{(s-1)(s+3)}{(s+2)(s+4)} \qquad Z(s)=\frac{(s-1)(s+3)}{(s+2)(s-4)}$$

$$Y(s)=\frac{s^2-s+2}{s^2+s+2} \qquad Y(s)=\frac{s^3+8s^2+11s-20}{s^3+6s^2+11s+6}$$

6.10 Find the open circuit and short circuit natural frequencies of the networks shown in Fig. P. 6.10 and thus find their input impedances.

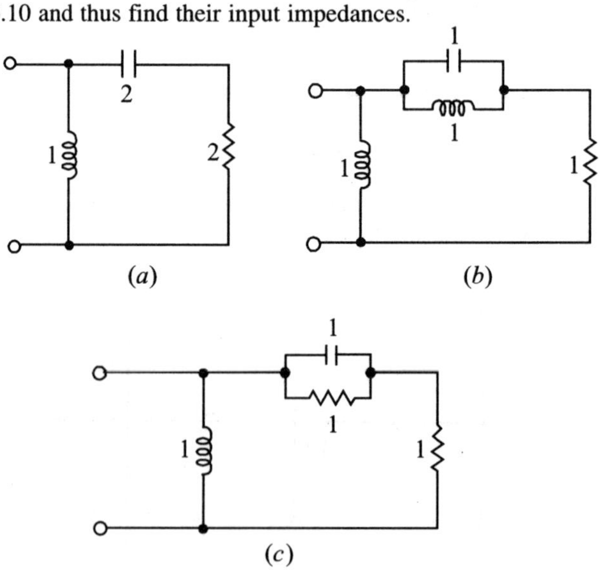

Fig. P. 6.10

6.11 Find the open circuit and short circuit natural frequencies of the networks shown in Fig. P. 6.11 and thus find the input admittances.

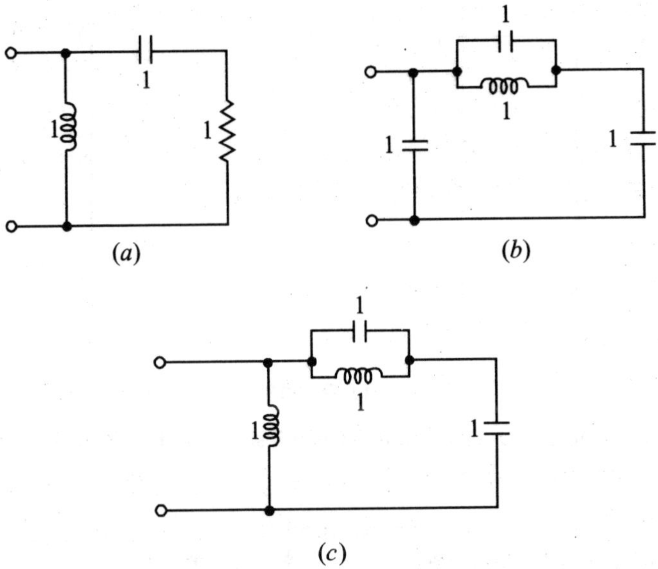

Fig. P. 6.11

6.12 Consider the network shown in Fig. P. 6.12. Show that a transmission zero exists at a pole frequency of z_1 if z_2 does not have the same pole frequency.

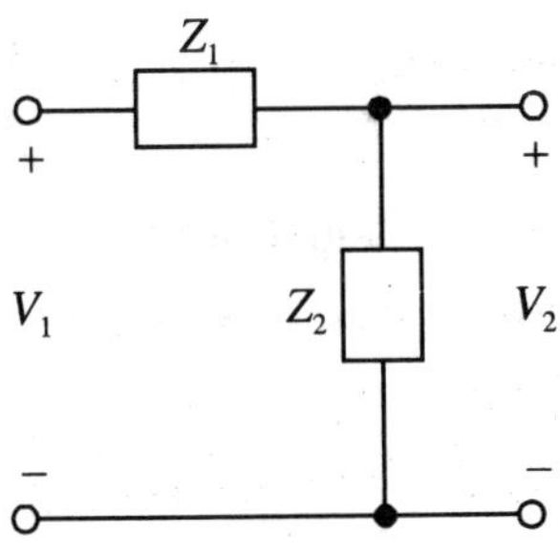

Fig. P. 6.12

6.13 Plot the impulse end step responses of networks with pole-zero pattern shown in Fig. P. 6.13.

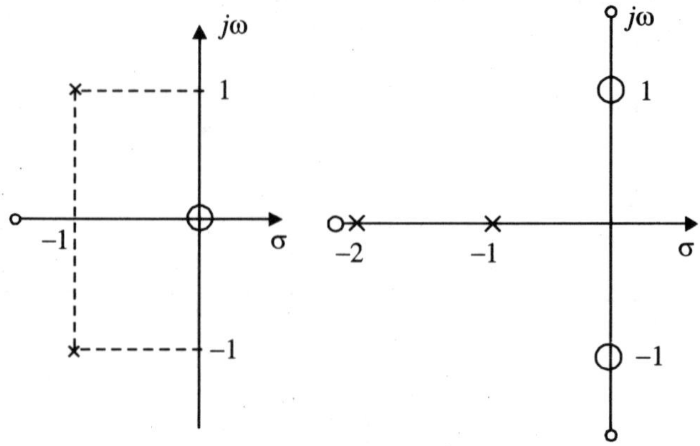

Fig. P. 6.13

6.14 Find the frequency response of the network shown in Fig. P. 6.14.

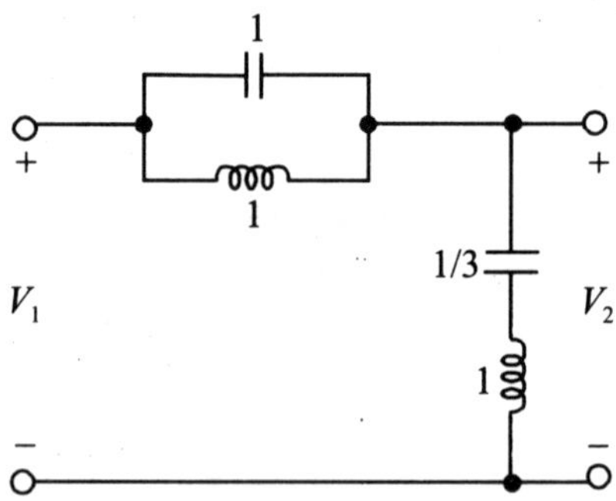

Fig. P. 6.14

6.15 Find the magnitude and phase plots of the transfer function.

$$H(a) = \frac{s^2 + 2}{s^2 + 0.1s + 1}$$

NOTES AND REFERENCES

Many concepts like natural frequency and Pole-zero pattern introduced here are important in network and in filter design. The concepts of OCNF and SCNF are particularly useful in active network design.

The correlation between time and frequency response is a topic generally covered in detail in control system course and the book by Kuo[1] provides the details. This book also provides analytical characterisation of "Controlability" and "Observability" which are closely related to excitation and decoupling of natural frequencies (Chap. 4).

A thorough discussion of natural frequencies and modes in a network is found in Huang and Parkers book[2] (Chaps. 5–7). Incidentally, this book presents a unique approach to the teaching of circuit and network theory. Desoer and Kuh's book[3] discusses the determination of minimal order differential equation and the order of complexity of a network (Chap. 14).

1. B.C. Kuo, *Automatic Control System*, Prentice-Hall, 1975.
2. T.S. Huang and S.R. Parker, *Network Theory—An Introductory Course,* Addison Wesley. 1971.
3. C.A. Desoer and E.S. Kud, *Basic Circuit Theory*, McGraw-Hill, 1969.

CHAPTER

7 Multiterminal Networks

In many applications of electric networks a pair of terminals are designated as the input terminals and another pair of terminals as the output terminals. If we consider an audio amplifier, we can designate the microphone end and the speaker end as the input and output terminals, respectively. The nodes internal to the amplifier itself are of secondary importance. In this case, the amplifier is considered as a block. Of course, we can think of systems which are made up of a number of such blocks. Each block is characterized by relationships between its terminal currents and voltages.

A pair of terminals such that the current entering one of the terminals is the same as the current leaving the other is designated as a "port" (see Chap. 1). Fig. 7.1 shows a *one-port* (or two terminal) network, which is characterized by its driving point impedance or admittance function (see Sec. 6.2)

$$Z_{11} = \frac{V_1}{I_1} = \frac{1}{Y_{11}}$$

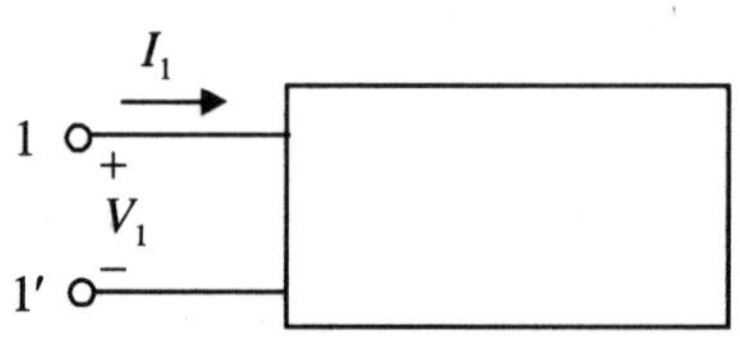

Fig. 7.1

A *two-port* network has two accessible terminal pairs, *i.e.* there are two pairs of terminals which can be used as input and output ports. Figure 7.2 shows a two-port network which acts as a low-pass filter. The study of two-port networks is very important, as they form the building blocks of most of the electric systems. Figure 7.3 shows a general two-port and the standard convention adopted in designating the terminal voltages and currents. In this chapter, we consider several methods of characterizing a two-port network.

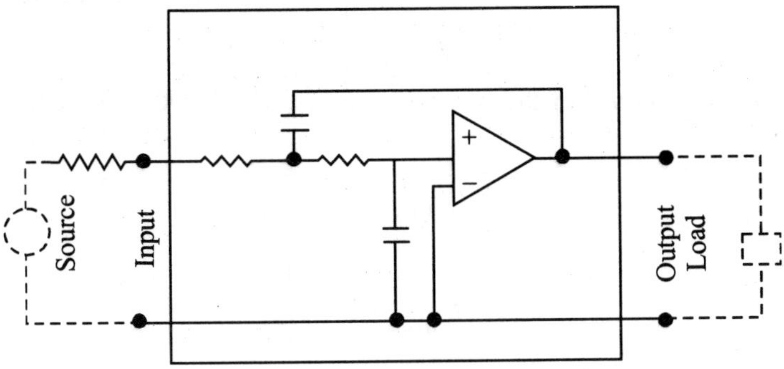

Fig. 7.2

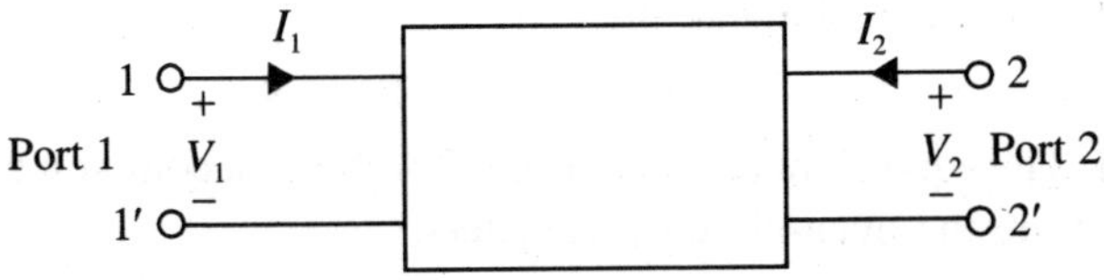

Fig. 7.3

A general multiterminal network can be represented as a "n-port" network by identifying n terminal-pairs as the n-ports. An alternative way to describe a general network is to treat it as a multiterminal network ratner than a multiport network. In this case, it can be assumed that an external terminal is selected as the reference node and that ports are formed by the terminals of the network and the external reference terminal. This leads to "Indefinite" matrix and is discussed in sections 7.4 to 7.7.

7.1 TWO-PORT PARAMETERS

There are a number of methods of characterizing a two-port network. Of the four variables—V_1, I_1, V_2 and I_2—any two can be considered as independent variables and the other two as dependent variables. Hence, in all we have 6 different methods of choosing the independent variable. There is seventh set—the scattering parameters. These parameters are defined analogous to a transmission line (see Appendix E). In what follows six sets of two-port parameters are introduced.

7.1.1 Impedance Parameters

If we choose I_1 and I_2 as the independent variables, we can characterize the network by the following set of equations.

$$\begin{bmatrix} V_1 \\ V_2 \end{bmatrix} = \begin{bmatrix} z_{11} & z_{12} \\ z_{21} & z_{22} \end{bmatrix} \begin{bmatrix} I_1 \\ I_2 \end{bmatrix} \qquad \ldots(7.1)$$

where $$Z_{oc} = \begin{bmatrix} z_{11} & z_{12} \\ z_{21} & z_{22} \end{bmatrix} \qquad \text{...(7.2)}$$

is the *impedance matrix* of the two-port. From eqn. 7.1 we observe that

$$z_{11} = \left.\frac{V_1}{I_2}\right|_{I_2=0}, \quad z_{12} = \left.\frac{V_1}{I_2}\right|_{I_1=0}$$

$$z_{21} = \left.\frac{V_2}{I_1}\right|_{I_2=0}, \quad z_{22} = \left.\frac{V_1}{I_2}\right|_{I_2=0} \qquad \text{...(7.3)}$$

All these parameters are measured under open-circuit conditions. Hence, this set of parameters is called *open-circuit impedance parameters.* z_{11} and z_{22} are driving point impedances, and z_{12} and z_{21} are transfer impedances. If the two-port under consideration is reciprocal, then

$$z_{12} = z_{21}$$

Example 7.1. The equivalent circuits of Fig. 7.4 shows a common base transistor. The following are the open-circuit parameters:

$$z_{11} = \left.\frac{V_e}{I_e}\right|_{I_c=0} = r_e + r_b$$

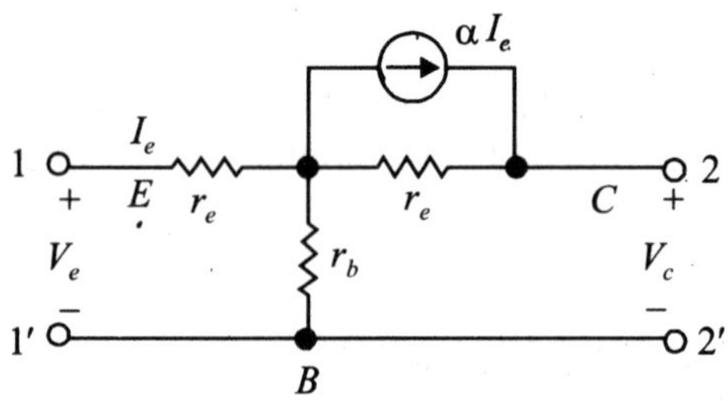

Fig. 7.4

$$z_{12} = \left.\frac{V_e}{I_c}\right|_{I_e=0} = r_b$$

$$z_{21} = \left.\frac{V_c}{I_c}\right|_{I_c=0} = r_b + \alpha r_c$$

$$z_{22} = \left.\frac{V_c}{I_c}\right|_{I_e=0} = r_c + r_b$$

This clearly is not a reciprocal network.

7.1.2 Admittance Parameters

If we choose V_1 and V_2 as independent variables, we can characterize the network by

$$\begin{bmatrix} I_1 \\ I_2 \end{bmatrix} = \begin{bmatrix} y_{11} & y_{12} \\ y_{21} & y_{22} \end{bmatrix} \begin{bmatrix} V_1 \\ V_2 \end{bmatrix} \quad \text{...(7.4)}$$

where
$$Y_{sc} = \begin{bmatrix} y_{11} & y_{12} \\ y_{21} & y_{22} \end{bmatrix} \quad \text{...(7.5)}$$

is the admittance matrix of the two-port. From eqn. 7.4 we have

$$y_{11} = \left.\frac{I_1}{V_1}\right|_{V_2=0}, \quad y_{12} = \left.\frac{I_1}{V_2}\right|_{V_1=0}$$

$$y_{21} = \left.\frac{I_2}{V_1}\right|_{V_2=0}, \quad y_{22} = \left.\frac{I_2}{V_2}\right|_{V_1=0} \quad \text{...(7.6)}$$

All the y-parameters are measured under short-circuit conditions, and hence these parameters are called *short-circuit admittance parameters.* y_{11} and y_{22} are driving point admittances, and y_{12} and y_{21} are transfer admittances. For a reciprocal network

$$y_{12} = y_{21}$$

Example 7.2. Consider the common base transistor shown in Fig. 7.4. The networks corresponding to eqn. (7.6) and the expressions for the y-parameters are shown here.

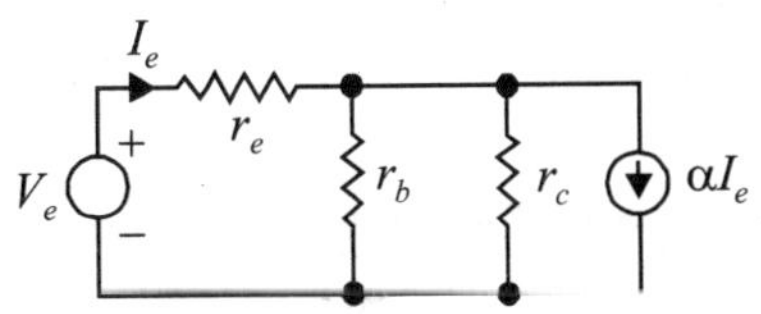

y_{11}

$$y_{11} = \left.\frac{I_e}{V_e}\right|_{V_e=0} = \frac{1}{r_e + \frac{r_b r_c}{r_b + r_c}(1-\alpha)}$$

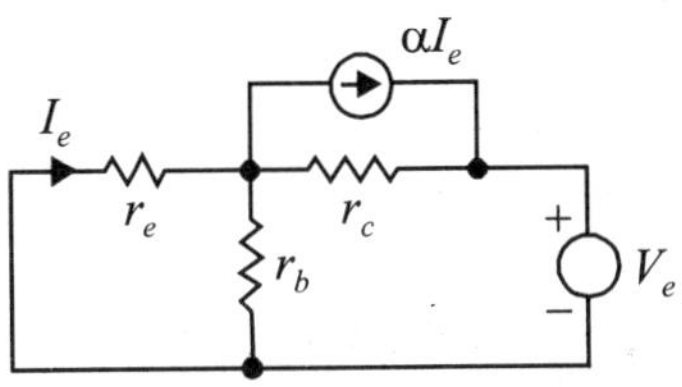

y_{12}

$$y_{12} = \left.\frac{I_e}{V_c}\right|_{V_e=0} = \frac{1}{(\alpha r_c + r_b) - \{(r_b + r_c)(r_c + r_b)/r_b\}}$$

(Contd.)

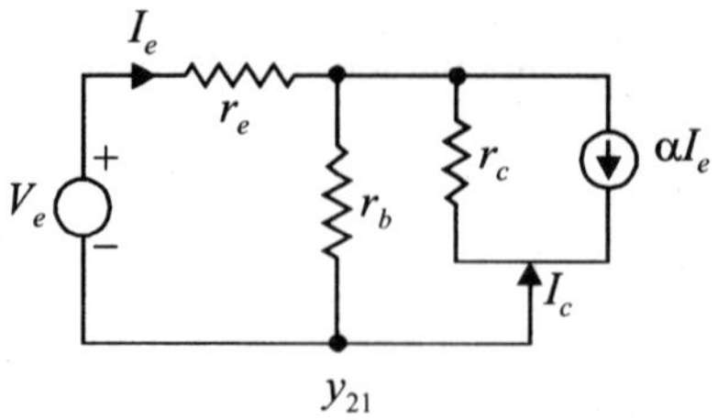

$$y_{21} = \left.\frac{I_c}{V_e}\right|_{V_e=0}$$

$$= \frac{1}{r_b - \{(r_b + r_c)(r_e + r_b)/(r_b + \alpha r_c)\}}$$

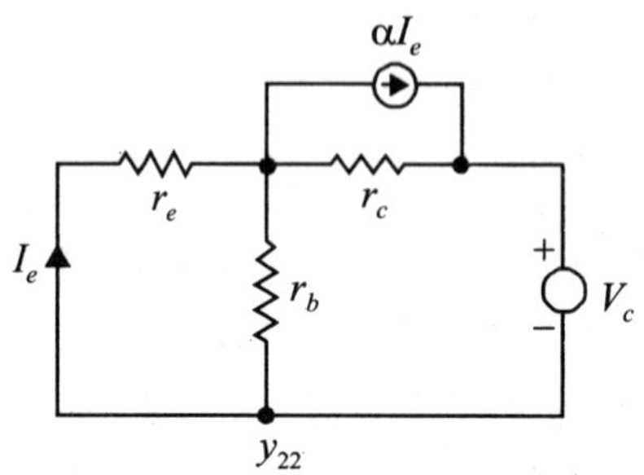

$$y_{22} = \left.\frac{I_c}{V_c}\right|_{V_e=0}$$

$$= \frac{1}{(r_c + r_b) - \dfrac{(\alpha r_c + r_b)r_b}{r_e + r_b}}$$

Example 7.2

7.1.3 Transmission Parameters

If we consider V_2 and I_2 as the independent variables, we can write

$$\begin{bmatrix} V_1 \\ I_1 \end{bmatrix} = \begin{bmatrix} A & B \\ C & D \end{bmatrix} \begin{bmatrix} V_2 \\ -I_2 \end{bmatrix} \quad \ldots(7.7)$$

where

$$T = \begin{bmatrix} A & B \\ C & D \end{bmatrix} \quad \ldots(7.8)$$

is the Transmission Parameter matrix of the two-port. Note that here $-I_2$ is used instead of I_2. From eqn. 7.7, we observe that

$$A = \left.\frac{V_1}{V_2}\right|_{I_2=0}, \quad B = \left.\frac{V_1}{-I_2}\right|_{V_2=0}$$

$$C = \left.\frac{I_1}{V_2}\right|_{I_2=0}, \quad D = \left.\frac{I_1}{-I_2}\right|_{V_2=0} \quad \ldots(7.9)$$

A and *C* are measured under open-circuit conditions, and *B* and *D* are measured under short-circuit conditions. *A* and *D* are dimensionless; *B* has the dimensions of an impedance and *C* that of an admittance. For a reciprocal network

$$AD - BC = 1$$

Example 7.3. We will find the transmission parameter of the common base transistor.

The circuits for the evaluation of parameters are shown along with the parameters.

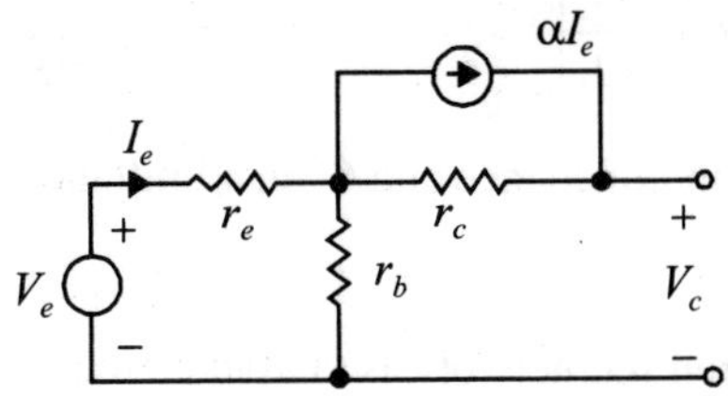

$$A = \left.\frac{V_e}{V_c}\right|_{I_c=0} = \frac{r_e + r_b}{\alpha r_c + r_b}$$

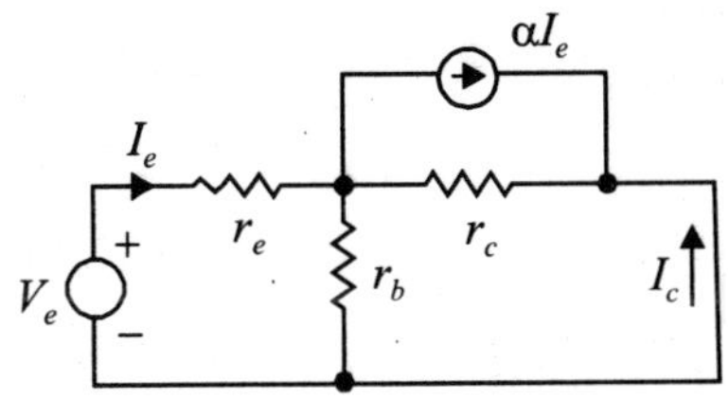

$$B = \left.\frac{V_e}{-I_c}\right|_{V_c=0}$$

$$= -r_b + \frac{(r_e + r_b)(r_c + r_b)}{(\alpha r_c + r_b)}$$

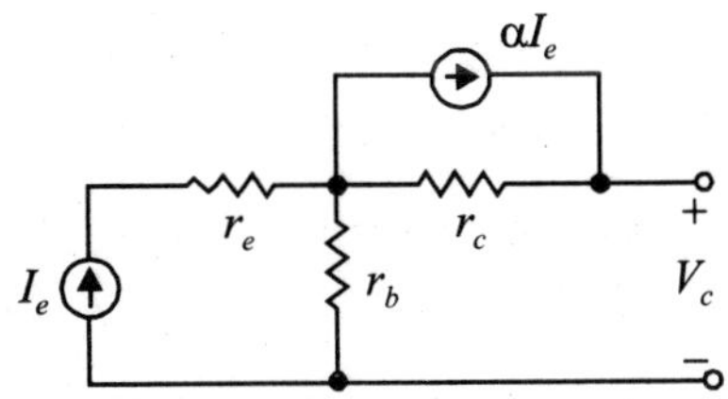

$$C = \left.\frac{I_e}{V_c}\right|_{I_c=0} = \frac{1}{\alpha r_c + r_b}$$

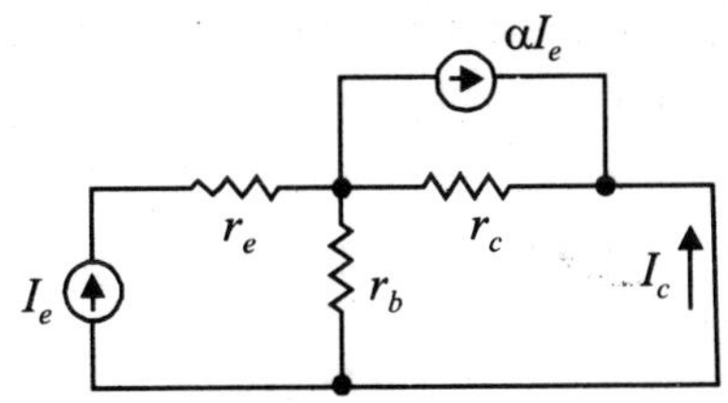

$$D = \left.\frac{I_e}{-I_c}\right|_{V_c=0} = \frac{r_c + r_b}{\alpha r_c + r_b}$$

Example 7.3

If we consider V_1 and I_1 as independent variables, we can define a new set of parameters as

$$\begin{bmatrix} V_2 \\ I_2 \end{bmatrix} = \begin{bmatrix} A' & B' \\ C' & D' \end{bmatrix} \begin{bmatrix} V_1 \\ -I_1 \end{bmatrix} \qquad \ldots(7.10)$$

where

$$T' = \begin{bmatrix} A' & B' \\ C' & D' \end{bmatrix} \qquad \ldots(7.11)$$

is the *reverse transmission matrix* of the two-port. From eqn. 7.10, we have

$$A' = \left.\frac{V_2}{V_1}\right|_{I_1=0}, \quad B' = \left.\frac{V_2}{-I_1}\right|_{V_1=0}$$

$$C' = \left.\frac{I_2}{V_1}\right|_{I_1=0}, \quad D' = \left.\frac{I_2}{-I_1}\right|_{V_1=0} \qquad \text{...(7.12)}$$

As before A' and D' are dimensionless, B' is an impedance and C' an admittance.

7.1.4 Hybrid Parameters

Let I_1 and V_2 be the independent variables. Then the network is characterised by

$$\begin{bmatrix} V_1 \\ I_2 \end{bmatrix} = \begin{bmatrix} h_{11} & h_{12} \\ h_{21} & h_{22} \end{bmatrix} \begin{bmatrix} I_1 \\ V_2 \end{bmatrix} \qquad \text{...(7.13)}$$

where

$$H = \begin{bmatrix} h_{11} & h_{12} \\ h_{21} & h_{22} \end{bmatrix} \qquad \text{...(7.14)}$$

is the *hybrid matrix* of the network. Equation 7.13 yields

$$h_{11} = \left.\frac{V_1}{I_1}\right|_{V_2=0}, \quad h_{12} = \left.\frac{V_1}{V_2}\right|_{I_1=0}$$

$$h_{21} = \left.\frac{I_2}{I_1}\right|_{V_2=0}, \quad h_{22} = \left.\frac{I_2}{V_2}\right|_{I_1=0} \qquad \text{...(7.15)}$$

h_{11} and h_{21} are measured under short-circuit conditions, h_{12} and h_{22} are measured under open-circuit conditions. h_{12} and h_{21} are dimensionless parameters, whereas h_{11} is an impedance and h_{22} an admittance. For a reciprocal network

$$h_{12} = -\,h_{21}$$

Example 7.4. Consider the same common base transistor of Fig. 7.4.

$$h_{11} = \left.\frac{V_e}{I_e}\right|_{V_c=0} = (r_e + r_b) - \frac{\alpha r_c + r_b}{r_c + r_b} r_b$$

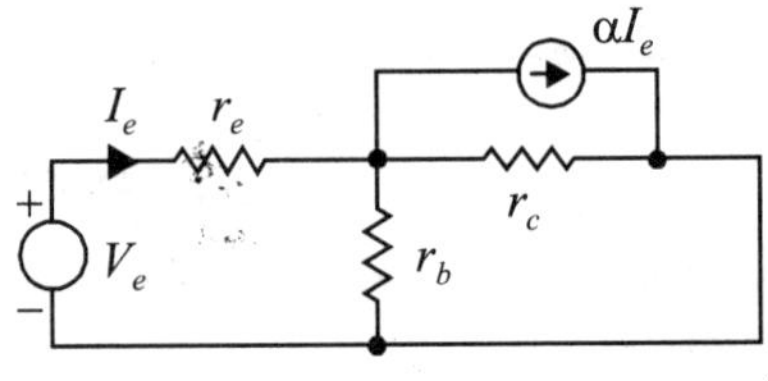

$$h_{12} = \left.\frac{V_e}{V_c}\right|_{I_c=0} = \frac{r_b}{r_c + r_b}$$

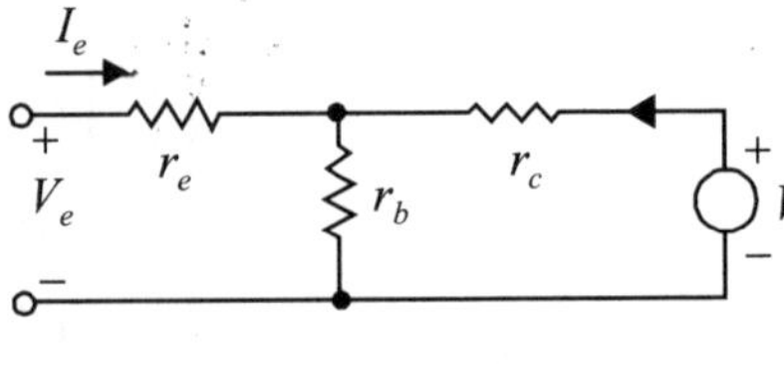

(Contd.)

$$h_{21} = \left.\frac{I_c}{I_e}\right|_{V_c=0} = -\frac{r_b + \alpha r_c}{r_c + r_b}$$

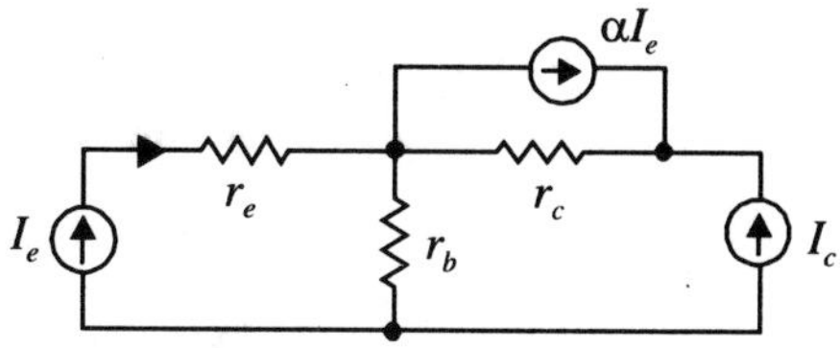

$$h_{22} = \left.\frac{I_c}{V_c}\right|_{I_e=0} = \frac{1}{r_c + r_b}$$

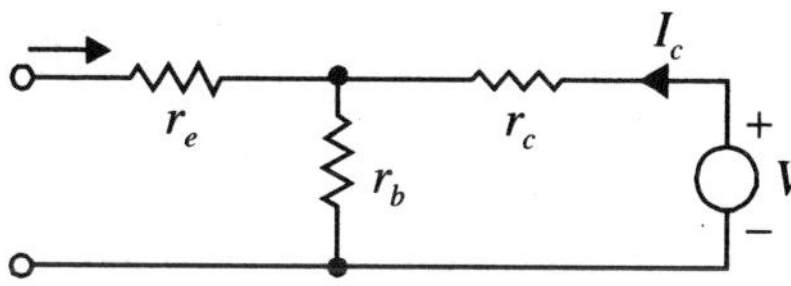

Example 7.4

The hybrid parameters are the most commonly used ones in transistor circuit theory. The notations generally employed are different from the ones used here. For the common base configuration the notations used h_{ib} (input impedance), h_{rb} (reverse transmission), h_{fb} (forward transmission) and h_{ob} (output admittance) correspond to h_{11}, h_{12}, h_{21} and h_{22}, respectively. In the former notation the second subscript corresponds to the configuration under consideration. Similarly, the parameters for common emitter configuration are h_{ie}, h_{re}, h_{fe} and h_{oe} respectively.

By selecting V_1 and I_2 as independent parameters we can define the sixth set of parameters as

$$\begin{bmatrix} I_1 \\ V_2 \end{bmatrix} = \begin{bmatrix} g_{11} & g_{12} \\ g_{21} & g_{22} \end{bmatrix} \begin{bmatrix} V_1 \\ I_2 \end{bmatrix} \qquad ...(7.16)$$

where $$G = \begin{bmatrix} g_{11} & g_{12} \\ g_{21} & g_{22} \end{bmatrix} \qquad ...(7.17)$$

is the *inverse hybrid* or *g-parameters* of the network.

7.1.5 Interrelationship among Parameters

Though we have defined six different sets of parameters, these parameters are all interrelated. From eqns. 7.1 and 7.4 it is seen that

$$Z_{oc} = Y_{sc}^{-1}$$

i.e. $$z_{11} = \frac{y_{22}}{\Delta y}, \; z_{12} = \frac{-y_{12}}{\Delta y}$$

$$z_{21} = \frac{-y_{21}}{\Delta y}, \; y_{22} = \frac{y_{11}}{\Delta y}$$

where Δy is the determinant of Y_{sc}. Observe that $y_{11} \neq 1/z_{11}$.

Finding the relationship between $Z(Y)$ and T parameters is little more involved.

Consider eqn. 7.7. This can be rewritten as

$$\begin{bmatrix} 1 & 0 \\ 0 & 1 \end{bmatrix}\begin{bmatrix} V_1 \\ I_1 \end{bmatrix} = \begin{bmatrix} A & B \\ C & D \end{bmatrix}\begin{bmatrix} V_2 \\ -I_2 \end{bmatrix}$$

or
$$\begin{bmatrix} 1 & 0 & -A & B \\ 0 & 1 & -C & D \end{bmatrix}\begin{bmatrix} V_1 \\ I_1 \\ V_2 \\ I_2 \end{bmatrix} = 0$$

On rearranging the matrix we have

$$\begin{bmatrix} 1 & -A & 0 & B \\ 0 & -C & 1 & D \end{bmatrix}\begin{bmatrix} V_1 \\ V_2 \\ I_1 \\ I_2 \end{bmatrix} = 0$$

i.e.
$$\begin{bmatrix} 1 & -A \\ 0 & -C \end{bmatrix}\begin{bmatrix} V_1 \\ V_2 \end{bmatrix} = \begin{bmatrix} 0 & -B \\ -1 & -D \end{bmatrix}\begin{bmatrix} I_1 \\ I_2 \end{bmatrix}$$

or
$$\begin{bmatrix} V_1 \\ V_2 \end{bmatrix} = \begin{bmatrix} 1 & -A \\ 0 & -C \end{bmatrix}^{-1}\begin{bmatrix} 0 & -B \\ -1 & -D \end{bmatrix}\begin{bmatrix} I_1 \\ I_2 \end{bmatrix}$$

On simplifying this, we have

$$\begin{pmatrix} V_1 \\ V_2 \end{pmatrix} = \begin{pmatrix} \dfrac{A}{C} & \dfrac{AD-BC}{C} \\ \dfrac{1}{C} & \dfrac{D}{C} \end{pmatrix}\begin{pmatrix} I_1 \\ I_2 \end{pmatrix} \qquad \text{...(7.18)}$$

Comparing eqns. 7.1 and 7.18.

$$z_{11} = \frac{A}{C},\ z_{12} = \frac{AD-BC}{C}$$

$$z_{21} = \frac{1}{C},\ z_{22} = \frac{D}{C} \qquad \text{...(7.19)}$$

which gives the relationship between impedance and transmission parameters; similarly, the relationship between the other sets of parameters can be derived. Table 7.1 gives these relationships.

If we are considering a finite lumped linear network, all the parameters are real rational functions of the complex frequency variable s.

Table 7.1

	Z	Y	T	H	*Reciprocity*
Z	$\begin{matrix} z_{11} & z_{12} \\ z_{21} & z_{22} \end{matrix}$	$\begin{matrix} \frac{y_{22}}{\Delta y} & \frac{-y_{12}}{y} \\ \frac{-y_{21}}{\Delta y} & \frac{y_{11}}{\Delta y} \end{matrix}$	$\begin{matrix} \frac{A}{C} & \frac{\Delta T}{C} \\ \frac{1}{C} & \frac{D}{C} \end{matrix}$	$\begin{matrix} \frac{\Delta H}{h_{22}} & \frac{h_{12}}{h_{22}} \\ \frac{-h_{21}}{h_{22}} & \frac{1}{h_{12}} \end{matrix}$	$Z_{12} = Z_{21}$
Y	$\begin{matrix} \frac{z_{22}}{\Delta z} & \frac{-z_{12}}{\Delta z} \\ \frac{-z_{21}}{\Delta z} & \frac{z_{11}}{\Delta z} \end{matrix}$	$\begin{matrix} y_{11} & y_{12} \\ y_{21} & y_{22} \end{matrix}$	$\begin{matrix} \frac{D}{B} & -\frac{\Delta T}{B} \\ \frac{-1}{B} & \frac{A}{B} \end{matrix}$	$\begin{matrix} \frac{1}{h_{11}} & \frac{-h_{12}}{h_{11}} \\ \frac{h_{21}}{h_{11}} & \frac{\Delta H}{h_{11}} \end{matrix}$	$y_{12} = y_{21}$
T	$\begin{matrix} \frac{z_{11}}{z_{21}} & \frac{\Delta z}{z_{21}} \\ \frac{1}{z_{21}} & \frac{z_{22}}{z_{21}} \end{matrix}$	$\begin{matrix} \frac{-y_{22}}{y_{21}} & \frac{-1}{y_{21}} \\ \frac{-\Delta y}{y_{21}} & \frac{-y_{11}}{y_{21}} \end{matrix}$	$\begin{matrix} A & B \\ C & D \end{matrix}$	$\begin{matrix} \frac{-\Delta H}{h_{21}} & \frac{-h_{11}}{h_{21}} \\ \frac{-h_{22}}{h_{21}} & \frac{-1}{h_{21}} \end{matrix}$	$AD - BC = 1$
H	$\begin{matrix} \frac{\Delta z}{z_{22}} & \frac{z_{12}}{z_{22}} \\ \frac{-z_{21}}{z_{22}} & \frac{1}{z_{22}} \end{matrix}$	$\begin{matrix} \frac{1}{y_{11}} & \frac{-y_{12}}{y_{11}} \\ \frac{y_{21}}{y_{11}} & \frac{\Delta y}{y_{11}} \end{matrix}$	$\begin{matrix} \frac{B}{D} & \frac{\Delta T}{D} \\ \frac{1}{D} & \frac{C}{D} \end{matrix}$	$\begin{matrix} h_{11} & h_{12} \\ h_{21} & h_{22} \end{matrix}$	$h_{12} = h_{21}$

Example 7.5. Figure 7.5 shows a passive *RLC* network. We are required to find *Z*, *Y*, *T* and *H* parameters.

We find *y* parameters directly and then use the table to find the other parameters.

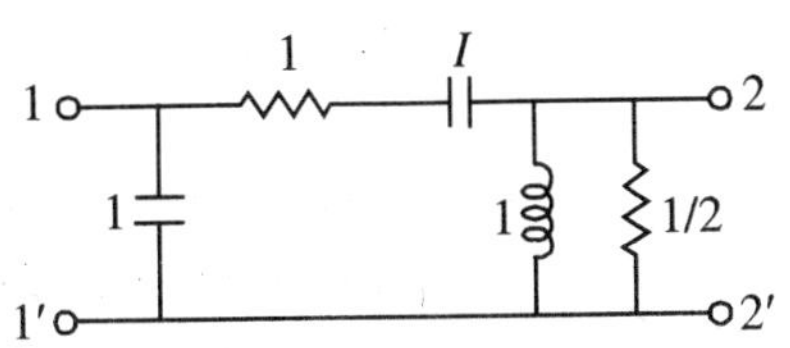

Fig. 7.5

$$y_{11} = \left.\frac{I_1}{V_1}\right|_{V_2=0} = s + \frac{1}{1+1/s} = \frac{s^2+2s}{s+1}$$

$$y_{22} = \left.\frac{I_2}{V_2}\right|_{V_1=0} = 2 + \frac{1}{s} + \frac{1}{1+1/s}$$

$$= \frac{3s^2+3s+1}{s(s+1)}$$

$$y_{12} = \left.\frac{I_1}{V_2}\right|_{V_1=0} = -\frac{1}{1+1/s} = -\frac{s}{s+1}$$

$y_{21} = y_{12}$ (the network is reciprocal)

Hence,
$$Y = \begin{bmatrix} \dfrac{s^2+2s}{s+1} & \dfrac{-s}{s+1} \\ -\dfrac{s}{s+1} & \dfrac{3s^2+3s+1}{s(s+1)} \end{bmatrix}$$

Now, we can find Z, T and H matrices from Table 7.1.

$$\Delta y = \frac{3s^2+5s+2}{s+1}$$

$$z_{11} = \frac{y_{22}}{\Delta y} = \frac{(3s^2+3s+1)}{s(s+1)} \frac{(s+1)}{3s^2+5s+2}$$

$$= \frac{3s^2+3s+1}{s(3s^2+5s+2)}$$

$$z_{22} = \frac{y_{11}}{\Delta y} = \frac{(s^2+2s)}{3s^2+5s+2}$$

$$z_{12} = z_{21} = \frac{y_{12}}{\Delta y} = \frac{s}{3s^2+5s+2}$$

Hence,
$$Z = \begin{bmatrix} \dfrac{(3s^2+3s+1)}{s(3s^2+5s+2)} & \dfrac{s}{3s^2+5s+2} \\ \dfrac{s}{3s^2+5s+2} & \dfrac{(s^2+2s)}{3s^2+5s+2} \end{bmatrix}$$

Similarly,
$$T = \begin{bmatrix} \dfrac{3s^2+3s+1}{s^2} & \dfrac{s+1}{s} \\ \dfrac{3s^2+5s+2}{s} & s+2 \end{bmatrix}$$

and
$$H = \begin{bmatrix} \dfrac{s+1}{s^2+2s} & \dfrac{1}{s+2} \\ -\dfrac{1}{s+2} & \dfrac{3s^2+5s+2}{(s^2+2s)} \end{bmatrix}$$

7.1.6 Ladder Networks

From the previous example it is clear that the evaluation of the two-port parameters is rather involved, especially when the network is complicated. Sometimes it may be advantageous to use the special property of network structure in evaluating the parameters. Consider the ladder network shown in Fig. 7.6. The driving point

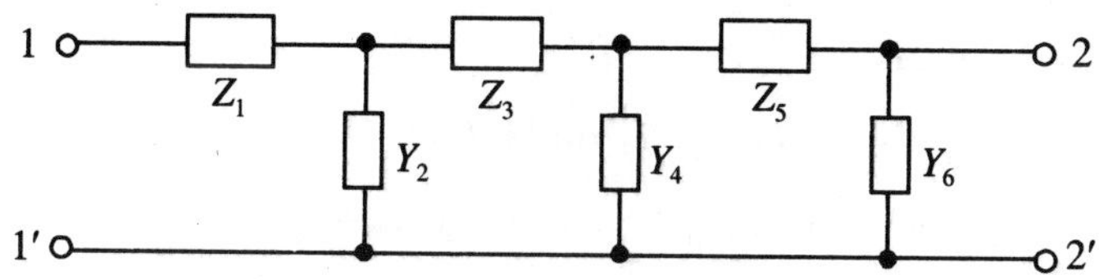

Fig. 7.6

impedance of the ladder is given by the continued fraction expansions (see sec. 6.2). The transmission parameters for this ladder are:

$$A = 1 + z_1(y_2 + y_4 + y_6) + z_3(y_4 + y_6) + z_5 y_6 + z_1 z_3 y_2 (y_4 + y_6)$$
$$+ z_1 z_5 (y_2 + y_4) y_6 + z_3 z_5 y_4 y_6 + z_1 z_3 z_5 (y_2 y_4 y_6) \qquad \ldots(7.20)$$

$$B = \frac{\partial A}{\partial Y_6} = z_1 + z_3 + z_5 + z_1 z_3 y_2 + z_1 z_5 (y_2 + y_4)$$
$$+ z_3 z_5 y_4 + z_1 z_3 z_5 y_2 y_4$$

$$C = \frac{\partial A}{\partial z_1} = y_2 + y_4 + y_6 + z_3 y_2 (y_4 + y_6) + z_5 (y_2 + y_4) y_6 + z_3 z_5 y_2 y_4 y_6$$

$$D = \frac{\partial B}{\partial z_1} = 1 + z_3 y_2 + z_5 (y_2 + y_4) + z_3 z_5 y_2 y_4$$

For an n-section ladder in general

$$B = \frac{\partial A}{\partial y_2 n}, \; C = \frac{\partial A}{\partial z_1} \text{ and } D = \frac{\partial B}{\partial z_1} \qquad \ldots(7.21)$$

From eqn. 7.21 it is clear that once we write A parameter for a ladder we can determine B, C, D and all the Z, Y and H parameters by mere inspection. Based on eqn. 7.20 certain simple rules to write A parameters can be formulated (see prob. 7.8). Another well-known method to write A parameter is the so-called *unit-voltage method* introduced by Guillemin. We illustrate the method by considering the ladder of Fig. 7.7.

Let
$$V_2 = 1 \text{ volt}$$
$$I_6 = y_6 V_2 = y_6,\ I_5 = I_6$$

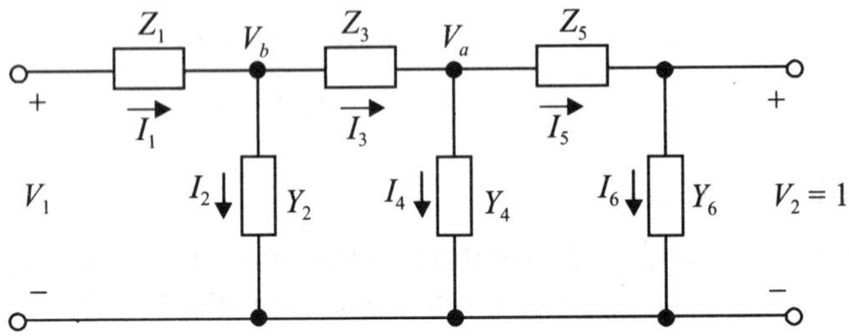

Fig. 7.7

$$V_a = V_2 + z_5 I_5 = 1 + z_5 y_6$$

$$I_4 = y_4 V_a = v_4 + z_5 y_6 y_4,\ I_3 = I_4 + I_5$$

$$V_b = V_a + I_3 z_3 = 1 + z_5 y_6 + z_3 (y_4 + z_5 y_6 y_4 + y_6)$$

$$= 1 + z_3 (y_4 + y_6) + z_5 y_6 + z_3 z_5 y_4 y_6$$

$$I_2 = y_2 V_b = y_2 + z_3 (y_4 + y_6) y_2 + z_5 y_6 y_2 + z_3 z_5 y_6 y_4 y_2$$

$$I_1 = I_2 + I_3$$

$$V_1 = V_b + I_1 z_1$$

$$= 1 + z_3 (y_4 + y_6) + z_5 y_6 + z_3 z_5 y_6 y_4 + z_1 \{ y_2 + z_3 (y_4 + y_6) y_2$$

$$+ z_5 y_6 z_2 + z_3 z_5 y_6 y_4 y_2 + y_6 + y_4 + z_5 y_6 y_4 \}$$

Hence, $$A = \frac{V_1}{V_2} = V_1 = 1 + z_5 y_6 + z_3 (y_4 + y_6) + z_1 (y_2 + y_4 + y_6) + z_5 z_3 y_4 y_6$$

$$+ z_5 z_1 (y_2 + y_4) y_6 + z_3 z_1 y_2 (y_4 + y_6) + z_1 z_3 z_5 y_2 y_4 y_6$$

Many of the active filter structures are obtained by embedding an operational amplifier in a pseudo ladder structure. The transfer function ($1/A$) of the active network can also be derived by the unit voltage method. To illustrate this we consider the structure of Fig. 7.2 which is repeated here for convenience.

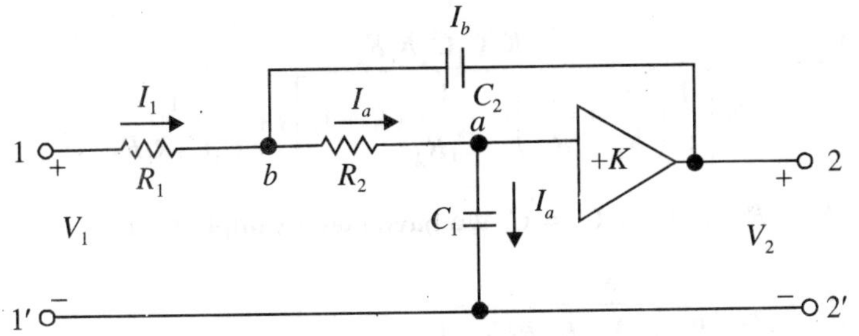

Let $V_2 = 1$ volt. The voltage at point *a*, V_a, is V_2/K, *i.e.* $1/K$. The current entering the operational amplifier is zero (ideally it has infinite input impedance).

Hence,

$$I_a = V_a sC_1 = \frac{sC_1}{K}$$

The voltage at *b* is

$$V_b = V_a + I_a R_2$$

$$= \frac{1}{K} + \frac{1}{K} sC_1 R_2 = \frac{1}{K}(1 + sC_1 R_2)$$

The current I_b is

$$I_b = (V_b - V_2)\, sC_2$$

$$= \left[\frac{1}{K}(1 + sC_1 R_2) - 1\right] sC_2$$

$$= \frac{(1 + sC_1 R_2) - K}{K} sC_2$$

From Kirchhoff's Law at node *b*,

$$I_1 = I_a + I_b$$

$$= \frac{sC_1}{K} + \frac{\left[(1 + sC_1 R_2) - K\right]}{K} sC_2$$

$$= \frac{sC_1 + \left[(1 + sC_1 R_2) - K\right] sC_2}{K}$$

Finally,

$$V_1 = V_b + I_1 R_1$$

$$= \frac{1}{K}(1 + sC_1 R_2) + \frac{sC_1 + \left[(1 + sC_1 R_2) - K\right] sC_2}{K} R_1$$

or

$$V_1 = \frac{(1 + sC_1 R_2) + sC_1 R_1 + sC_2 R_1 \left[1 + sC_1 R_2 - K\right]}{K}$$

Hence, we have

$$\frac{V_2}{V_1} = \frac{K/C_1C_2R_1R_2}{s_2 + \left[\frac{1}{C}\left(\frac{1}{R_1} + \frac{1}{R_2}\right) - \frac{1}{C_1R_2}(K-1)\right]s + \frac{1}{C_1C_2R_1R_2}}$$

If $R_1 = R_2 = R$ and $C_1 = C_2 = C$, we have (see example 6.2)

$$\frac{V_2}{V_1} = \frac{K}{s^2C^2R^2 + (3-K)RCs + 1}$$

7.2 INTERCONNECTION OF TWO-PORTS

From elementary circuit theory we know that the overall resistance (conductance) of a series (parallel) connection of resistors is the sum of the individual resistances (conductances). Indeed, we can generalize this result to one-port impedances

$$Z = \sum_{i=1}^{n} z_i$$

where Z is the equivalent impedance of a series and connection of impedances z_1, $i = 1, n$, and

$$Y = \sum_{i=1}^{n} y_i$$

where Y is the equivalent admittance of a parallel connection of admittances y_1, $i = 1, n$. In this section, we study the interconnection of two-ports.

7.2.1 Series Connection

In the series connection of two-ports the inputs and outputs of the two-ports are connected in series individually, as shown in Fig. 7.8.

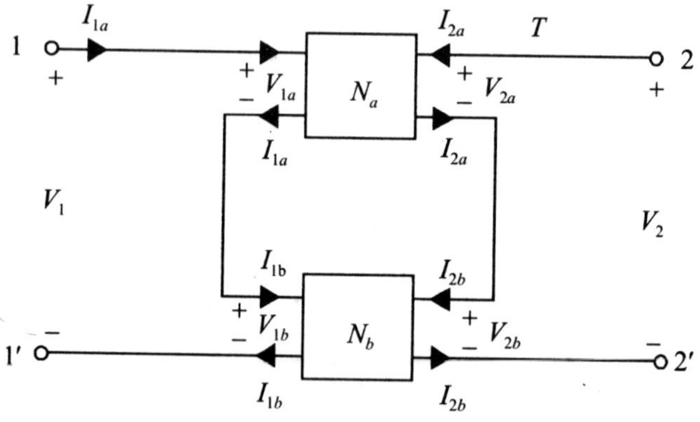

Fig. 7.8

The constraints imposed by the interconnection are:

$$I_1 = I_{1a} = I_{1b}, \quad V_1 = V_{1a} + V_{1b}$$

$$I_2 = I_{2a} = I_{2b}, \quad V_2 = V_{2a} + V_{2b} \qquad \ldots(7.22)$$

Hence, we have

$$Z = Z_a + Z_b \qquad \ldots(7.23)$$

where Z is the open-circuit impedance matrix of the interconnected two-ports, and Z_a and Z_b are the impedance matrices of the individual two-ports N_a and N_b, respectively. For eqn. 7.23 to hold good the constraints imposed by eqn. 7.22 must be satisfied. For these conditions to be satisfied, V_{ab} and V_{cd} in Fig. 7.9 must be identically zero.

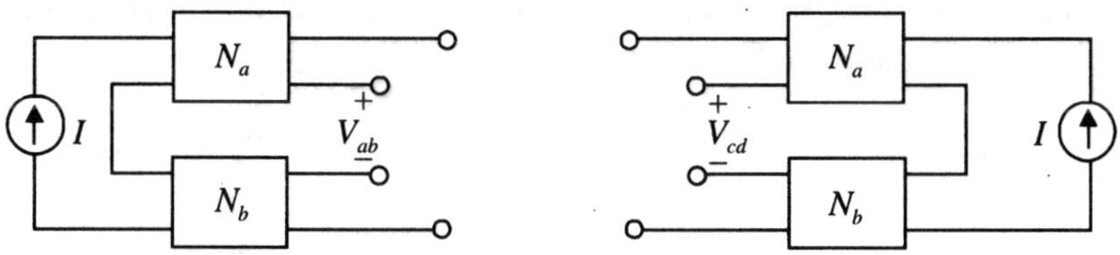

Fig. 7.9

7.2.2 Parallel Connection

If the inputs and outputs of the individual two-ports are connected in parallel, the overall connection is called the parallel connection of two-ports (Fig. 7.10).

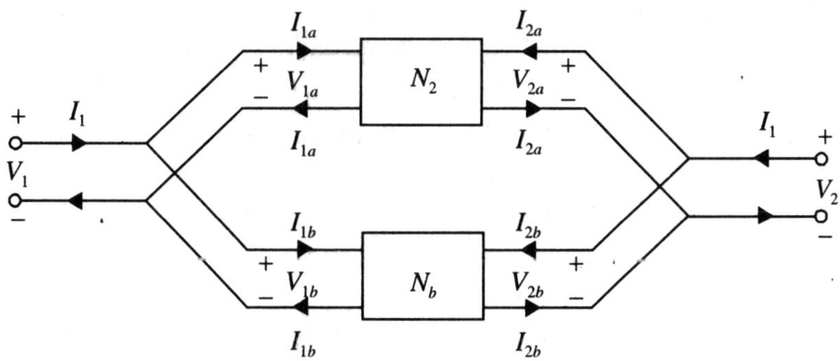

Fig. 7.10

The constraints imposed by the interconnection are:

$$V_1 = V_{1a} = V_{1b}, \quad I_1 = I_{1a} + I_{1b}$$

$$V_2 = V_{2a} = V_{2b}, \quad I_2 = I_{2a} + I_{2b} \qquad \ldots(7.24)$$

Then the overall admittance matrix of the interconnected two-ports (Y) is the sum of the two admittance matrices Y_a and Y_b of N_a and N_b, respectively, *i.e.*

$$Y = Y_a + Y_b \qquad \ldots(7.25)$$

To check the validity of the constraints in eqn. 7.25, which must be satisfied for equation to be true, consider Fig. 7.11. Here V_{ab} and V_{cd} must be zero.

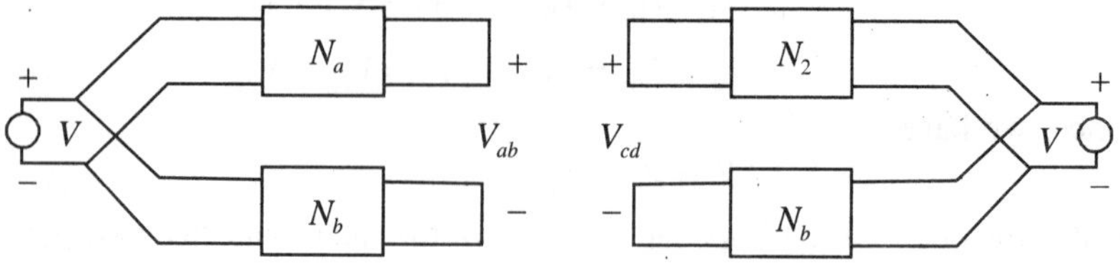

Fig. 7.11

7.2.3 Cascade Connections

If the connection of the two-ports is such that the output of one is the input to the other (Fig. 7.12) then it is known as the cascade connection.

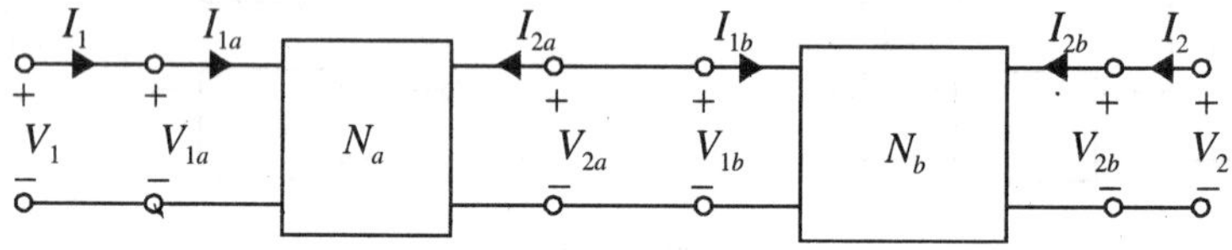

Fig. 7.12

The constraints imposed by the interconnection are:

$$V_1 = V_{1a},\ I_{1a} = I_1, \qquad I_{2a} = -I_{1b}$$
$$V_2 = V_{2b},\ I_{2b} = I_2, \qquad V_{2a} = V_{1b}$$

It can be easily shown that the transmission matrix of the overall network is the product of the transmission matrices of the individual networks, *i.e.*

$$T = T_A \cdot T_B \qquad \ldots(7.26)$$

Of course, there are other possible interconnections of two-ports, also.

(*a*) Inputs in series—outputs in parallel,

$$H = H_a + H_b$$

where H_a and H_b are the hybrid matrices of N_a and N_b, respectively, and H is the hybrid matrix of the combination.

(*b*) Inputs in parallel— outputs in series, $G = G_a + G_b$
where G_a and G_b are the inverse hybrid matrices of N_a and N_b respectively, and G that of the combination.

7.3 EQUIVALENT NETWORKS

Consider the network shown in Figs. 7.13 and 7.14. The driving

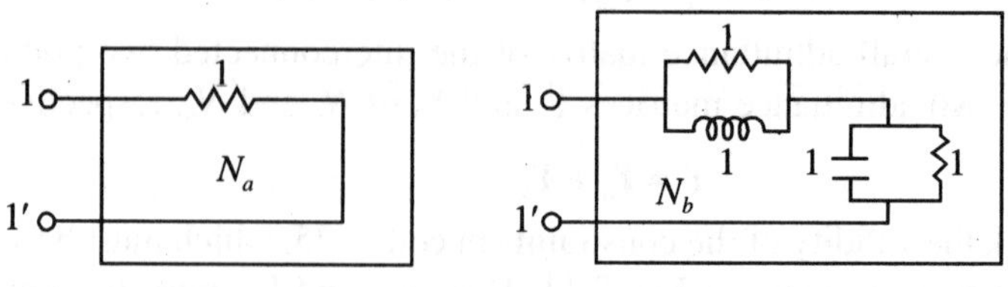

Fig. 7.13

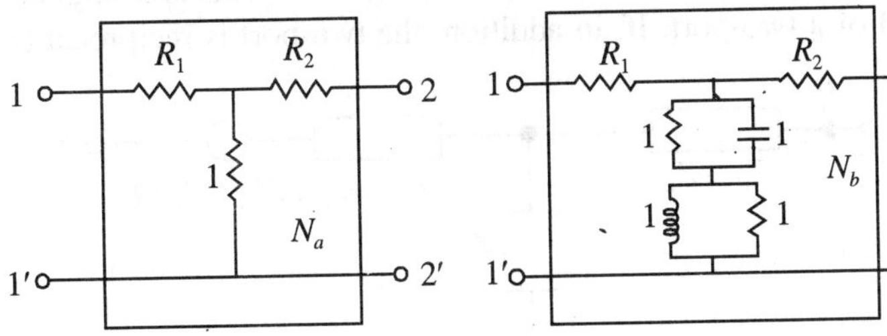

Fig. 7.14

point impedance of the two networks N_a and N_b shown in Fig. 7.13 are both given by

$$Z_{in} = 1\Omega$$

As a consequence, the impedance matrix of the two networks N_a and N_b of Fig. 7.14 is given by

$$Z = \begin{bmatrix} R_1 + 1 & 1 \\ 1 & R_2 + 1 \end{bmatrix}$$

In both cases, the network N_a is equivalent to N_b.

Two one-port networks are said to be equivalent if they have the same driving point impedance or admittance. Two two-port networks are said to be equivalent if their two-port parameters are identical. In this section, we derive a few equivalent circuits for two ports.

From eqn. 7.1, we have

$$V_1 = z_{11}\ I_1 + z_{12}\ I_2$$
$$V_2 = z_{21}\ I_1 + z_{22}\ I_2 \qquad \text{...(7.27)}$$

A two-controlled source equivalent of this two-port is shown in Fig. 7.15. If the two-port network has a common terminal (*i.e.* port 1 and port 2 have a common

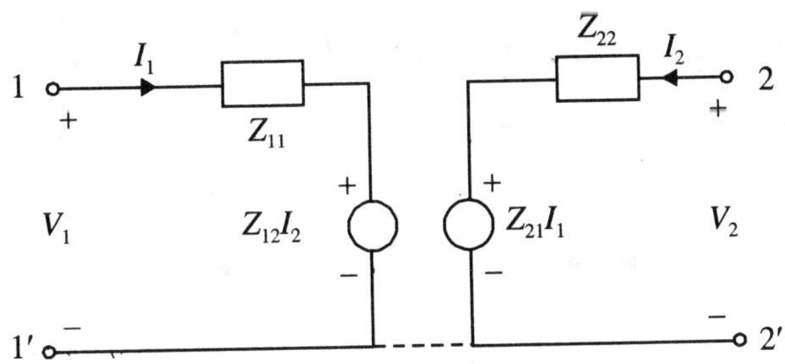

Fig. 7.15

terminal), then in the equivalent circuit also port 1 and port 2 have a common terminal (broken line in Fig. 7.15). We can rewrite eqn. 7.27 as

$$V_1 = (z_{11} - z_{12})\ I_1 + z_{12}\ (I_1 - I_2)$$
$$V_2 = (z_{22} - z_{12})\ I_2 + (z_{21} - z_{12})\ I_1 + z_{12}\ (I_1 + I_2)$$

which yields the equivalent circuit shown in Fig. 7.16. This is a single-controlled source model of a two-port. If, in addition, the two-port is reciprocal ($z_{12} = z_{21}$),

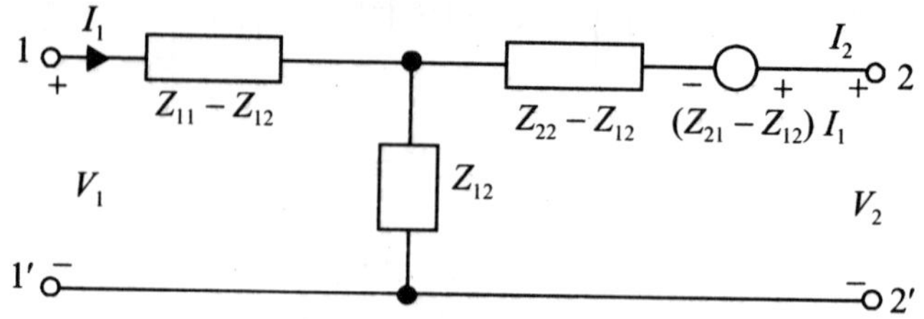

Fig. 7.16

the equivalent network reduces to the one shown in Fig. 7.17, which is the *T-equivalent* of a common ground (or terminal) two-port.

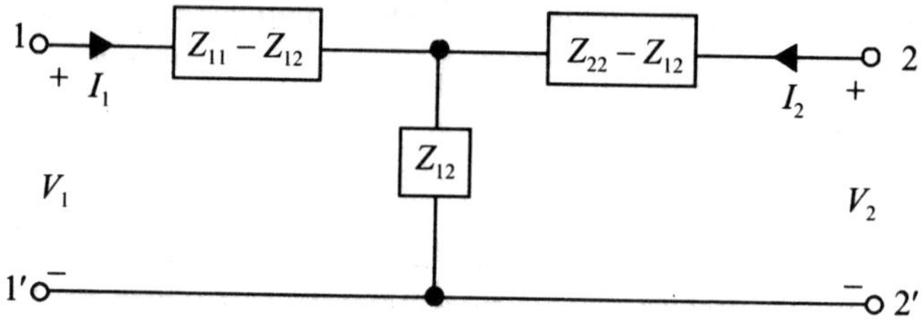

Fig. 7.17

We can obtain another set of equivalent networks starting from the *y* parameter. Fig. 7.18 shows these equivalent network—(*a*) the general equivalent, (*b*) the single controlled source model for a common ground network, and (*c*) the π-equivalent of a reciprocal two-port network.

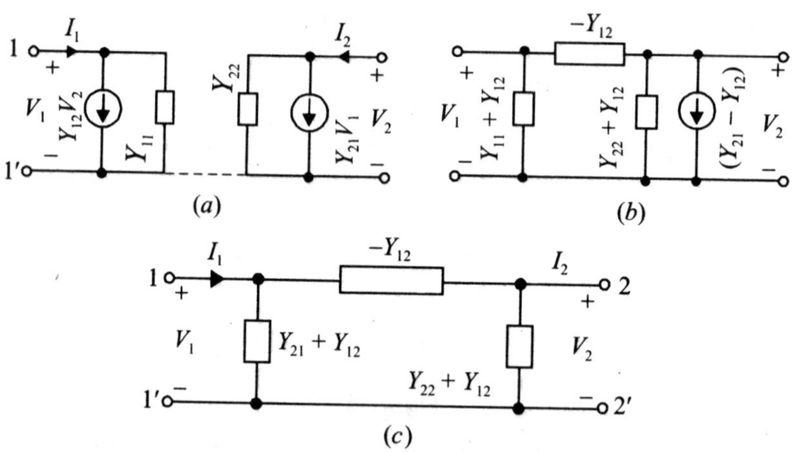

Fig. 7.18

If we consider an electronic component like operational amplifier it is generally made up of a cascade of a number of individual components and the actual analysis of the overall network is rather involved. A common method employed is to generate equivalent networks for the individual components and analyse them as two (or three) ports and then find the overall network terminal relations by applying the principles of interconnections developed in this chapter.

The concept of equivalent network is basic to the analysis and synthesis of electric networks. In the analysis of electronic circuits, the active components are replaced by their equivalent circuits and the resulting network is then analysed.

7.4 INDEFINITE ADMITTANCE MATRIX

Let a network have n terminals numbered 1 to n. Let the voltage of terminal k, with reference to the external reference node, be V_k and the current entering the terminal k be I_k. Then we can write

$$I_k = \sum_{i=1}^{n} y_{ki} V_i,\ k = 1, 2, ..., n$$

or

$$\mathbf{I} = Y_I \mathbf{V} \qquad \text{...(7.28)}$$

where

$$\mathbf{I} = [I_1\ I_2 \ldots I_n]^t,\ \mathbf{V} = [V_2\ V_2 \ldots V_n]^t$$

and

$$Y_I = \begin{bmatrix} y_{11} & y_{12} & y_{ln} \\ y_{21} & y_{22} & y_{2n} \\ \vdots & & \\ y_{nl} & y_{n2} & y_{nn} \end{bmatrix}$$

The matrix Y_I defined by eqn. 7.28 is called the *indefinite admittance matrix* (IAM) of the network. The word indefinite indicates that the reference node is outside the network. The y parameters, y_{ij}, are defined by

$$y_{ij} = \left.\frac{I_i}{V_j}\right|_{V_k=0,\ k=1,2...,n,k\neq j}$$

The y parameters here are defined in the same way as in Sec. 7.12.

Consider the three-terminal network shown in Fig. 7.19. If $V_1 = 1$ and $V_2 = V_3 = 0$, then $y_{11} = I_1$, $y_{21} = I_2$ and $y_{31} = I_3$. I_2 and I_3 leave the network and hence

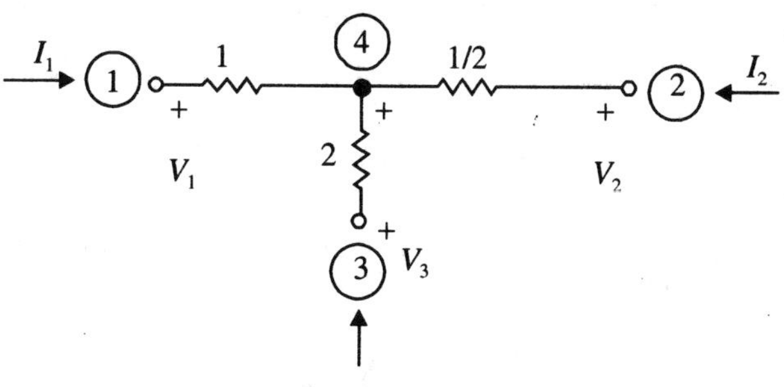

Fig. 7.19

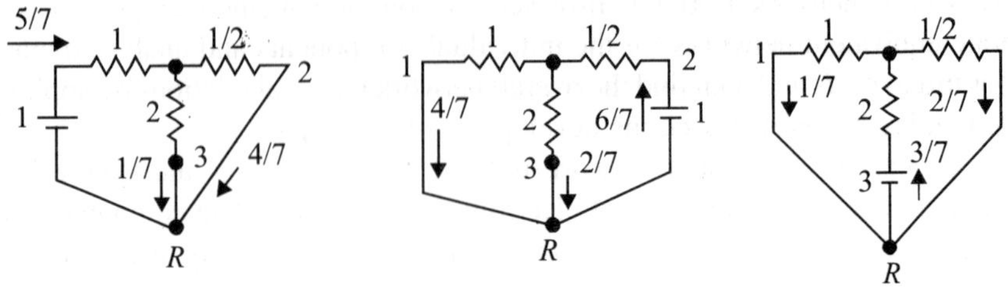

Fig. 7.20

y_{21} and y_{31} are negative. Figure 7.20 shows the circuits required to evaluate the various y_{ij}'s. The indefinite admittance matrix of Fig. 7.19 is

$$Y_1 = \begin{array}{c} \\ 1 \\ 2 \\ 3 \end{array}\begin{array}{c} \begin{array}{ccc} 1 & 2 & 3 \end{array} \\ \begin{bmatrix} 5/7 & -4/7 & -1/7 \\ -4/7 & 6/7 & -2/7 \\ -1/7 & -2/7 & 3/7 \end{bmatrix} \end{array} \quad \text{...(7.29)}$$

When all the nodes of the network are accessible IAM can be determined by mere inspection.

y_{ii} = sum of all the admittances connected to node i

y_{ij} = negative of the admittance connected between nodes i and j.

If the network is reciprocal, then obviously

$$y_{ij} = y_{ji}$$

The network of Fig. 7.19 can be considered as a four node network. In this case the IAM is

$$Y_I = \begin{array}{c} \\ 1 \\ 2 \\ 3 \\ 4 \end{array}\begin{array}{c} \begin{array}{cccc} 1 & 2 & 3 & 4 \end{array} \\ \begin{bmatrix} 1 & 0 & 0 & -1 \\ 0 & 2 & 0 & -2 \\ 0 & 0 & 1/2 & -1/2 \\ -1 & -2 & -1/2 & 7/2 \end{bmatrix} \end{array} \quad \text{...(7.30)}$$

In Sec. 7.5, we show how to obtain eqn. 7.29 from 7.30.

7.5 PROPERTIES OF IAM

IAM has some interesting properties given below, which can be easily proved.

(*i*) The sum of the elements in any column is zero.

In eqn. 7.28 assume that $V_i = 1$ and all other voltages are zero. Then summing all the equations we have

$$\sum_{k=1}^{n} I_k = \sum_{k=1}^{n} y_{ki}$$

The sum on L.H.S. is the total current input to the network or the current at the reference node and hence must be zero, *i.e.*

$$\sum_{k=1}^{n} y_{ki} = 0 \text{ for any } i$$

The sum of elements in any column being zero, the rows of Y_I are not all independent and Y_I is a singular matrix.

(*ii*) The sum of elements in any row is zero. Let a nonzero constant ΔV be added to all the terminal voltages in eqn. 7.28. This does not change the currents of the network as the potential difference between the various terminal pairs are not changed. Hence we have

$$I_k = \sum_{i=1}^{n} y_{ki}(V_i + \Delta V),\ k = 1, 2, ..., n.$$

Comparing this equation with eqn. 7.18 we see that

$$\sum_{i=1}^{n} y_{ki}\Delta V = 0 \text{ for all } k$$

As ΔV is an arbitrary increment we have

$$\sum_{i=1}^{n} y_{ki} = 0 \text{ for all } k.$$

(*iii*) If one of the terminals of the network is made common to all ports, the short-circuit admittance matrix can be obtained by removing the row and column corresponding to the common terminal. If terminal *k* is identified with the reference terminal, then $V_k = 0$, and I_k is the sum of all the currents and hence zero. Setting V_k and I_k to zero is equivalent to removing the *k*th column and row.

In Fig. 7.19 if terminal 3 is chosen as the reference, the short-circuit admittance matrix of the resulting two ports (terminal pairs 13 and 23 constitute the two ports) is obtained by deleting row 3 and column 3 of eqn. 7.29.

(*iv*) If two terminals of the network are coalesced, then the entries of the row and the column corresponding to the new terminal is obtained by summing the entries of the row and the column, corresponding to the two terminals being coalesced.

Let *k* and *l* be the two terminals being coalesced. Then from eqn. 7.28 we have

$$I_k = y_{k1}V_1 + ... + y_{kk}V_k + ... + y_{kl}V_l + ...$$

$$I_l = y_{l1}V_1 + ... + y_{lk}V_k + ... + y_{ll}V_l + ...$$

Let the new terminal be *m*. Then

$$V_m = V_k = V_l \text{ and } I_m = I_k + I_l$$

Hence

$$I_m = (y_{kl} + y_{l1})V_1 + ... + (y_{kk} + y_{lk} + y_{kl} + y_{ll})V_m + ...$$

and the property immediately follows.

In the network of Fig. 7.19 if terminals 2 and 3 are coalesced (say, the new terminal is 2′), then the IAM of the resulting network is

$$Y_I = \begin{matrix} & 1 & 2' \\ 1 & 5/7 & -5/7 \\ 2' & -5/7 & 5/7 \end{matrix}$$

(*v*) If a terminal of the network is suppressed (say, *k*th terminal) then the IAM of the resulting network can be obtained by pivotal condensation of the original IAM with y_{kk} as the pivot. (The order of the new IAM is one less than that of the original IAM.)

Let the new IAM be designated as Y_I', and its elements as y_{ij}'. Then we have

$$y_{ij}' = y_{ij} - \frac{y_{ik}y_{kj}}{y_{kk}} \qquad ...(7.31)$$

Suppression of a terminal implies that the current input at that terminal is zero. From eqn. 7.28 we have

$$I_k = \sum_{i \neq k}^{n} y_{ki}V_i + y_{kk}V_k$$

If we want to suppress the *k*th terminal then $I_k = 0$. This yields

$$V_k = -\sum_{\substack{i=1 \\ i \neq k}}^{n} \frac{y_{ki}}{y_{kk}} V_i \qquad ...(7.32)$$

The new IAM is obtained by substituting for V_k in the other current equations of eqn. 7.28. Consider current at the *i*th terminal

$$I_i = \left[\sum_{\substack{j=1 \\ j \neq k}}^{n} y_{ij}V_j \right] + y_{ik}V_k$$

On substituting for V_k from eqn. 7.32

$$I_i = \sum_{\substack{j=1 \\ j \neq k}}^{n} y_{ij}V_j - y_{ik} \sum_{\substack{j=1 \\ j \neq k}}^{n} \frac{y_{kj}}{y_{kk}} V_j$$

i.e.

$$I_i = \sum_{\substack{j=1 \\ j \neq k}}^{n} \left(y_{ij} - \frac{y_{ik}y_{kj}}{y_{kk}} \right) V_j$$

and hence the property.

In the example considered earlier the IAM shown in eqn. 7.29 should be the IAM obtained after suppressing terminal 4 or pivotal condensing the IAM of eqn. 7.30 about the (4, 4)th element. Hence, from eqn. 7.31 we have ($k = 4$)

$$y'_{ij} = y_{ij} - \frac{y_{i4}y_{4i}}{7/2}$$

$$y'_{11} = 1 - \frac{(-1)(-1)}{7/2} = 1 - 2/7 = 5/7$$

$$y'_{12} = 0 - \frac{(-1)(-2)}{7/2} = -4/7$$

$$y'_{13} = 0 - \frac{(-1)\left(-\frac{1}{2}\right)}{7/2} = -1/7$$

$$y'_{22} = 2 - \frac{(-2)(-2)}{7/2} = 2 - \frac{8}{7} = \frac{6}{7}$$

$$y'_{23} = 0 - \frac{(-2)\left(-\frac{1}{2}\right)}{7/2} = -2/7$$

$$y'_{33} = \frac{1}{2} - \frac{\left(-\frac{1}{2}\right)\left(-\frac{1}{2}\right)}{7/2} = \frac{1}{2} - \frac{1}{14} = \frac{3}{7}$$

which checks with eqn. 7.29.

Property *v* can be generalized when more than one terminal is suppressed. Let I_2 and V_2 be the current and voltage vectors of corresponding to the nodes or terminals to be suppressed. Then we can write Y_i in partitioned form as

$$\begin{bmatrix} \mathbf{I}_1 \\ \mathbf{I}_2 \end{bmatrix} = \begin{bmatrix} Y_{11} & Y_{12} \\ Y_{21} & Y_{22} \end{bmatrix} \begin{bmatrix} \mathbf{V}_1 \\ \mathbf{V}_2 \end{bmatrix}$$

With the suppression of the nodes corresponding to $\mathbf{V}_2$, we have

$$\mathbf{I}_2 = 0$$

Hence, from the preceding equation

$$\mathbf{V}_2 = -Y_{22}^{-1}\, Y_{21}\, \mathbf{V}_1$$

Hence, from the other equation

$$\mathbf{I}_1 = (Y_{11} - Y_{12}\, Y_{22}^{-1}\, Y_{21})\, \mathbf{V}_1$$

and the new IAM is

$$Y_{i2} = Y_{11} - Y_{12}\, Y_{22}^{-1}\, Y_{21}$$

(*vi*) All first cofactors of the matrix are equal. Hence, Y_I is often called the *equicofactor matrix.*

If a network is formed by interconnecting a number of sub-networks, then the overall IAM can be easily obtained from the IAMs of the sub-networks. Let N be network formed by the interconnection of the sub-networks N_i, $i = 1, 2, \ldots n$. The nodes of N are either the terminal nodes of some N_i or a node obtained by merging a terminal of some N_i with the terminal node of some N_j, N_k,... N_r ($j \neq i$, $k \neq i$,...$r \neq i$). Then the entry y_{pq} of the IAM of N is given by

$$y_{pq} = \sum_{i=1}^{n} \delta y_{\alpha_i} \beta_i (N_i) \qquad \ldots(7.33)$$

where $\delta = 1$ only if node p of N is the same as the α_i of the sub-network N_i and node q of N is the same as node β_i of N_i. Otherwise $\delta = 0$. $y_{\alpha_i}\beta_i$ (N_i) corresponds to the (α_i, β_i)th entry in the IAM of N_i. Also, observe that there is no interconnection permitted between the internal nodes of N_i and N_j $(i \neq j)$.

Consider the interconnection of 4 sub-networks as shown in Fig. 7.21. Two nodes of N are shown.

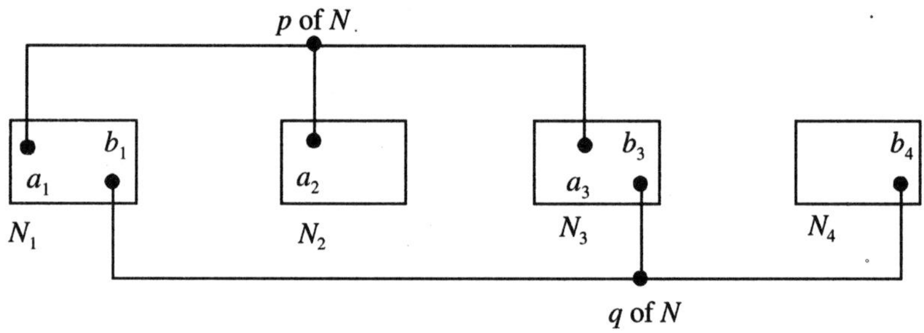

Fig. 7.21

Then eqn. 7.33 implies that

$$y_{pp}(N) = y_{a_1a_1}(N_1) = y_{a_2a_2}(N_2) + y_{a_3a_3}(N_3)$$

$$y_{pq}(N) = y_{a_1b_1}(N_1) = y_{a_3b_3}(N_3)$$

$$y_{qq}(N) = y_{b_1b_1}(N_i) + y_{b_3b_3}(N_3) + y_{b_4b_4}(N_4)$$

Example 7.6. Consider the interconnection of two two-ports shown in Fig. 7.22. The node of N_1 and N_2 are a_1, b_1, c_1 and a_2, b_2, c_2, respectively. The nodes of N are

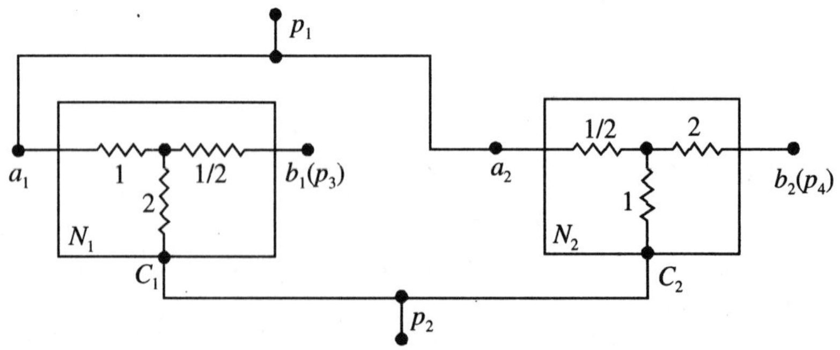

Fig. 7.22

p_1, p_2, p_3 and p_4. The IAM of N_1 and N_2 are

$$Y_{N_1} = \begin{array}{c} \\ a_1 \\ b_1 \\ c_1 \end{array} \begin{array}{c} \begin{array}{ccc} a_1 & b_1 & c_1 \end{array} \\ \begin{bmatrix} 5/7 & -4/7 & -1/7 \\ -4/7 & 6/7 & -2/7 \\ -1/7 & -2/7 & 3/7 \end{bmatrix} \end{array}$$

and

$$Y_{N_2} = \begin{array}{c} \\ a_2 \\ b_2 \\ c_2 \end{array} \begin{array}{c} \begin{array}{ccc} a_2 & b_2 & c_2 \end{array} \\ \begin{bmatrix} 6/7 & -2/7 & -4/7 \\ -2/7 & 3/7 & -1/7 \\ -4/7 & -1/7 & 5/7 \end{bmatrix} \end{array}$$

The elements of Y, the IAM of N, can now be evaluated from eqn. 7.33. For example,

$$y_{p_1p_1} = y_{a_1a_1} + y_{a_2a_2} = 5/7 + 6/7 = 11/7$$

$$y_{p_1p_2} = y_{a_1c_1} + y_{a_2c_2} = -1/7 - 4/7 = -5/7$$

$$y_{p_1p_3} = y_{a_1b_1} = -4/7$$

$$y_{p_1p_4} = y_{a_2a_2} = -2/7$$

The IAM of N is

$$Y = \begin{array}{c} \\ p_1 \\ p_2 \\ p_3 \\ p_4 \end{array} \begin{array}{c} \begin{array}{cccc} p_1 & p_2 & p_3 & p_4 \end{array} \\ \begin{bmatrix} 11/7 & -5/7 & -4/7 & -2/7 \\ -5/7 & 8/7 & -2/7 & -1/7 \\ -4/7 & -2/7 & 6/7 & 0 \\ -2/7 & -1/7 & 0 & 3/7 \end{bmatrix} \end{array}$$

7.6 NETWORK FUNCTIONS

The prime function of network analysis is to find the response of the given network to certain inputs. To this end we have defined network functions (Chap. 6) and characterised the network in terms of the measurements at its terminals. These network functions can also be determined from the IAM of the network.

The transfer impedance Z_{mn}^{ij} is defined as the voltage across the terminals i and j, i being positive, when a unit current source is applied at the terminals m and n, m being the terminal of entry of the current. It can be shown that

$$Z_{mn}^{ij} = \frac{V_{ij}}{I_{mn}} = sgn(m-n)\, sgn(i-j) \frac{Y_{ij}^{mn}}{Y_i^i} \qquad \ldots(7.34)$$

where $Y_{ij}^{mn} = (-1)^{I+j+m+n} \times$ Determinant obtained by deleting the rows corresponding to nodes m and n, and the columns corresponding to nodes i and j from the IAM of the network; Y_i^i = cofactor of the IAM with respect to (i, i)th entry; and

$$sgn(x) = 1, \text{ if } x > 0$$
$$= -1, \text{ if } x < 0$$

Consequently, the input impedance at the terminals $m - n$ is given by

$$Z_{mn}^{mn} = \frac{V_{mn}}{I_{mn}} = \frac{Y_{mn}^{mn}}{Y_n^n} \quad \ldots(7.35)$$

Similarly, it can be shown that the voltage transfer ratio t is

$$t = \frac{V_{ij}}{V_{mn}} = \frac{Z_{mn}^{ij}}{Z_{mn}^{mn}} = sgn(m-n)\,sgn(i-j)\frac{Y_{tj}^{mn}}{Y_{mn}^{mn}} \quad \ldots(7.36)$$

Example 7.7. We are required to find the input impedance at a, a' and the voltage transfer ratio V_2/V_1 for the network shown in Fig. 7.23.

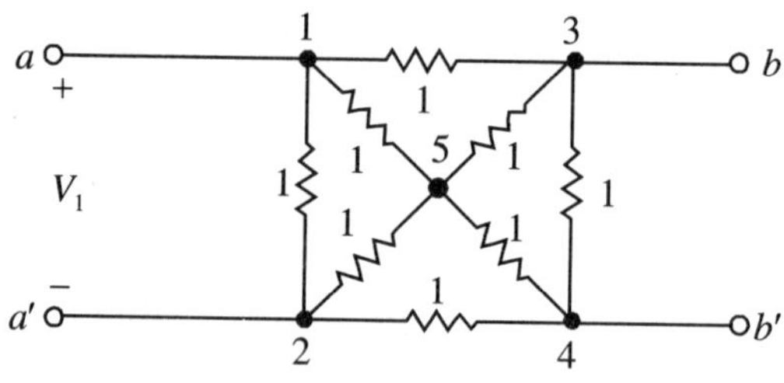

Fig. 7.23

The IAM for the network is

$$Y = \begin{array}{c} \\ 1 \\ 2 \\ 3 \\ 4 \\ 5 \end{array}\begin{array}{c} \begin{array}{ccccc} 1 & 2 & 3 & 4 & 5 \end{array} \\ \begin{bmatrix} 3 & -1 & -1 & 0 & -1 \\ -1 & 3 & 0 & -1 & -1 \\ -1 & 0 & 3 & -1 & -1 \\ 0 & -1 & -1 & 3 & -1 \\ -1 & -1 & -1 & -1 & 4 \end{bmatrix} \end{array}$$

From eqn. 7.35

$$Z_{jn} = Z_{12}^{12} = \frac{V_{12}}{I_{12}} = sgn(1-2)\,sgn(1-2)\frac{Y_{12}^{12}}{Y_2^2}$$

$$Y_{12}^{12} = (-1)^{1+2+1+2}\begin{vmatrix} 3 & -1 & -1 \\ -1 & 3 & -1 \\ -1 & -1 & 4 \end{vmatrix} = 24$$

$$Y_2^2 = (-1)^{2+2} \begin{vmatrix} 3 & -1 & 0 & -1 \\ -1 & 3 & -1 & -1 \\ 0 & -1 & 3 & -1 \\ -1 & -1 & -1 & 4 \end{vmatrix} = 45$$

Hence $$Z_{in} = \frac{24}{45} = \frac{8}{15}\,\Omega$$

From eqn. 7.36

$$\frac{V_2}{V_1} = \frac{V_{34}}{V_{12}} = \frac{Z_{12}^{34}}{Z_{12}^{12}} = sgn(1-2)\,sgn(3-4)\frac{Y_{34}^{12}}{Y_{12}^{12}}$$

$$Y_{34}^{12} = (-1)^{1+2+3+4} \begin{vmatrix} -1 & 0 & -1 \\ 0 & -1 & -1 \\ -1 & -1 & 4 \end{vmatrix} = 6$$

$$\frac{V_2}{V_1} = \frac{6}{24} = \frac{1}{4}$$

7.7 OPERATIONAL AMPLIFIER NETWORKS

IAM can also be used to analyse active networks. Here, we demonstrate its application to operational amplifier networks.

A finite gain operational amplifier (op. amp.) does not possess either impedance or admittance characterization. Ideally, we consider the input impedance of the op. amp. to be ∞ and the output impedance to be zero. Hence, an op. amp. cannot be directly described by an IAM.

In the network shown in Fig. 7.24(*a*) we have added two resistors, one of -1Ω and the other of 1Ω, in series with the output of the op. amp. As far as the terminal conditions are concerned, the network is equivalent to a regular op. amp. of Fig. 7.24(*b*). But the addition of these resistors enables us to write the IAM of the op. amp. The admittance matrix of the network, considering terminals 1 and 1′

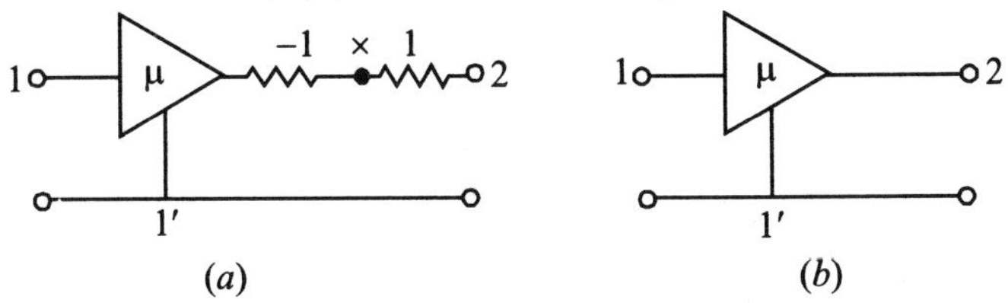

Fig. 7.24

as the input port and the terminals x and $1'$ as the output port is

$$\begin{array}{c} \\ 1 \\ x \end{array}\begin{array}{c} \begin{array}{cc} 1 & x \end{array} \\ \begin{bmatrix} 0 & 0 \\ \mu & -1 \end{bmatrix} \end{array}$$

Hence, the IAM of the same is

$$\begin{array}{c} \\ 1 \\ x \\ 1' \end{array}\begin{array}{c} \begin{array}{ccc} 1 & x & 1' \end{array} \\ \begin{bmatrix} 0 & 0 & 0 \\ \mu & -1 & 1-\mu \\ -\mu & 1 & \mu-1 \end{bmatrix} \end{array}$$

The overall IAM of the network of Fig. 7.24(a) is

$$Y_I = \begin{array}{c} \\ 1 \\ x \\ 1' \\ 2 \end{array}\begin{array}{c} \begin{array}{cccc} 1 & x & 1' & 2 \end{array} \\ \begin{bmatrix} 0 & 0 & 0 & 0 \\ \mu & 0 & 1-\mu & -1 \\ -\mu & 1 & \mu-1 & 0 \\ 0 & -1 & 0 & 1 \end{bmatrix} \end{array} \qquad \ldots(7.37)$$

From eqn. 7.37 it can be seen that the terminal x cannot be suppressed as the (x, x)th entry is 0. Of course, the terminal x *does not physically exist* and, as such, *cannot be used for external connection.* Also, the overall network being a non-reciprocal one, the IAM is not symmetric. (This method of using a series combination of resistors may be necessary to write the IAM of many of the active elements.)

Example 7.8. Figure 7.25 shows an op. amp. network, and we are required to find V_2/V_1. We adopt IAM approach to solve this problem.

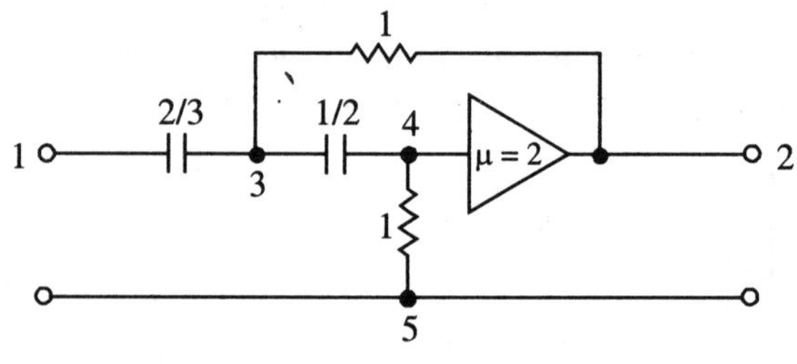

Fig. 7.25

IAM of the passive network (*i.e.* the network obtained by removing the op. amp.) is

$$Y_1 = \begin{array}{c} \\ 1 \\ 2 \\ 3 \\ 4 \\ 5 \end{array} \begin{array}{c} \begin{array}{ccccc} 1 & 2 & 3 & 4 & 5 \end{array} \\ \begin{bmatrix} \frac{2s}{3} & 0 & \frac{-2s}{3} & 0 & 0 \\ 0 & 1 & -1 & 0 & 0 \\ \frac{-2s}{3} & -1 & \frac{7s}{6}+1 & -\frac{s}{2} & 0 \\ 0 & 0 & -\frac{s}{2} & \frac{s}{2}+1 & -1 \\ 0 & 0 & 0 & -1 & 1 \end{bmatrix} \end{array}$$

The IAM of the amplifier is (eqn. 7.37)

$$Y_2 = \begin{array}{c} \\ 4 \\ x \\ 5 \\ 2 \end{array} \begin{array}{c} \begin{array}{cccc} 4 & x & 5 & 2 \end{array} \\ \begin{bmatrix} 0 & 0 & 0 & 0 \\ \mu & 0 & 1-\mu & -1 \\ -\mu & 1 & \mu-1 & 0 \\ 0 & -1 & 0 & 0 \end{bmatrix} \end{array}, \ \mu = 2$$

The overall IAM obtained by adding the corresponding elements of the matrices is

$$Y = \begin{array}{c} \\ 1 \\ 2 \\ 3 \\ 4 \\ 5 \\ x \end{array} \begin{array}{c} \begin{array}{cccccc} 1 & 2 & 3 & 4 & 5 & x \end{array} \\ \begin{bmatrix} \frac{2s}{3} & 0 & \frac{-2s}{3} & 0 & 0 & 0 \\ 0 & 2 & -1 & 0 & 0 & -1 \\ \frac{-2s}{3} & -1 & \frac{7s}{6}+1 & \frac{-s}{2} & 0 & 0 \\ 0 & 0 & \frac{-s}{2} & 1+\frac{s}{2} & 1 & 0 \\ 0 & 0 & 0 & -3 & 2 & 1 \\ 0 & -1 & 0 & 2 & -1 & 0 \end{bmatrix} \end{array}$$

From eqn. 7.36

$$\frac{V_2}{V_1} = \frac{V_{25}}{V_{15}} = \frac{Z_{15}^{25}}{Z_{15}^{15}} = sgn(2-5)\ sgn\ (1-5)\frac{Y_{25}^{15}}{Y_{15}^{15}}$$

$$Y_{25}^{15} = (-1)^{1+5+2+5} \begin{vmatrix} 0 & -1 & 0 & -1 \\ -\frac{2s}{3} & \frac{7s}{6}+1 & \frac{-s}{2} & 0 \\ 0 & \frac{-s}{2} & 1+\frac{s}{2} & 0 \\ 0 & 0 & 2 & 0 \end{vmatrix}$$

$$= -(-1)\left(\frac{-2s}{3}\right) \begin{vmatrix} 1 & 0 & -1 \\ \frac{-s}{2} & 1+\frac{s}{2} & 0 \\ 1 & 2 & 0 \end{vmatrix}$$

$$= \frac{-2s}{3}(-2) \begin{vmatrix} -1 & -1 \\ \frac{-s}{2} & 0 \end{vmatrix} = \frac{4s}{3}\left(\frac{-s}{2}\right) = \frac{-2s^2}{3}$$

$$Y_{11}^{55} = (-1)^{1+5+1+5} \begin{vmatrix} 2 & -1 & 0 & -1 \\ -1 & \frac{7s}{6}+1 & \frac{-s}{2} & 0 \\ 0 & \frac{-s}{2} & 1+\frac{s}{2} & 0 \\ -1 & 0 & 2 & 0 \end{vmatrix}$$

$$= -(-1) \begin{vmatrix} -1 & \frac{7s}{6}+1 & \frac{-s}{2} \\ 0 & \frac{-s}{2} & 1+\frac{s}{2} \\ -1 & 0 & 2 \end{vmatrix}$$

$$= (-1) \begin{vmatrix} \frac{-s}{2} & 1+\frac{s}{2} \\ 0 & 2 \end{vmatrix} + (-1) \begin{vmatrix} \frac{7s}{6}+1 & \frac{-s}{2} \\ \frac{-s}{2} & 1+\frac{s}{2} \end{vmatrix}$$

$$= s - \left(\frac{7s}{6}+1\right)\left(1+\frac{s}{2}\right) + \frac{s^2}{4}$$

$$= \frac{-s^2}{3} - \frac{2}{3}s - 1$$

Hence,
$$\frac{V_2}{V_1} = \frac{-2s^2/3}{-\left(\frac{s^2}{3} + \frac{2}{3}s + 1\right)} = \frac{2s^2}{s^2 + 2s + 3}$$

This is the transfer function of a high-pass filter. Design of various active filters will be considered in Chap. 12.

PROBLEMS

7.1 Find the impedance, admittance, hybrid and transmission matrices for the network shown is Fig. P. 7.1.

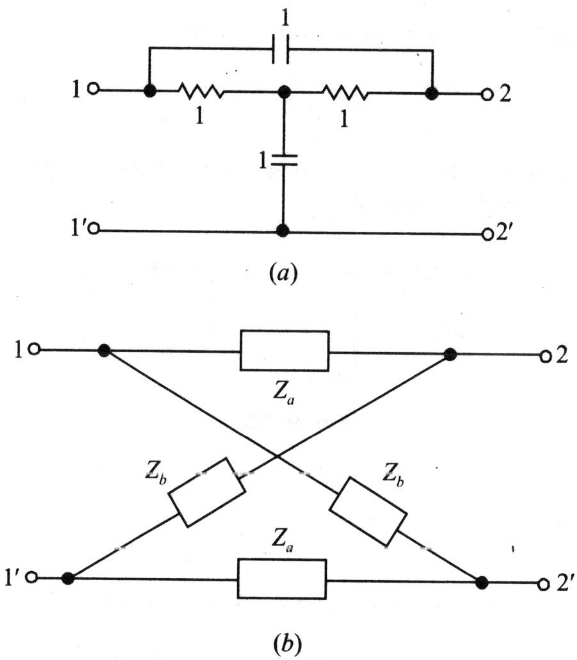

Fig. P. 7.1

7.2 Derive the relationship between the admittance and transmission parameters.

7.3 Show that the input impedance of a two-port terminated in a load Z_1 is bilinear in Z_1.

7.4 A simplified linear h-parameter equivalent circuit of a common emitter transistor is shown in Fig. P. 7.4. Two such transistors are connected in cascade and the combination is terminated in a load of 10 $k\Omega$. Find the voltage transfer ratio.

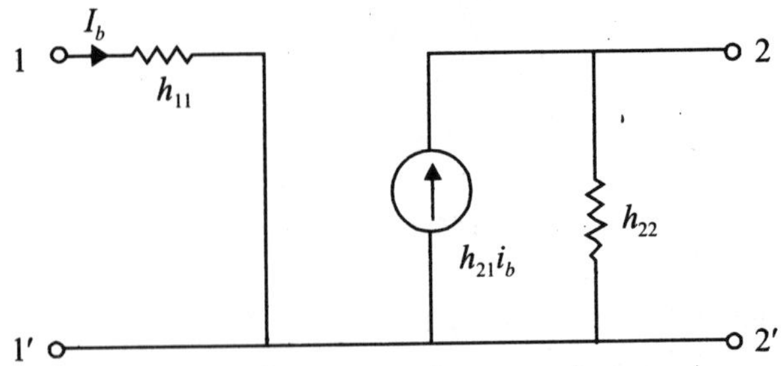

$$h_{11} = 10^3\ \Omega,\ h_{21} = 10^2,\ h_{22} = 10^{-5}\ \mho$$

Fig. P. 7.4

7.5 A two-port network is terminated in an impedance Z_l. Show that

$$\left(\frac{I_2}{I_1}\right)^2 = \frac{\partial Z_{in}}{\partial Z}$$

(this is called Vratsanov's theorem). Also show that

$$\frac{\partial^2 Z_{in}}{\partial Z_l^2} = 2C\left(\frac{I_2}{I_1}\right)^3$$

7.6 Figure P. 7.6 shows a terminated two-port network. Obtain the Thevenin equivalent as viewed from the load terminals.

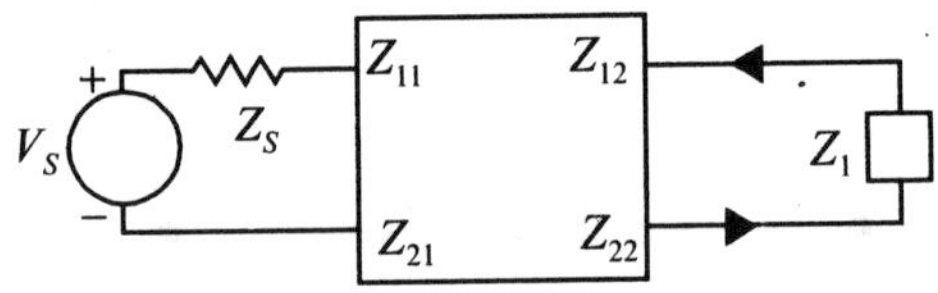

Fig. P. 7.6

7.7 A network N is formed by the cascade connected of N_a and N_b. Show that

$$z_{21} = \frac{z_{21a} z_{21b}}{z_{22a} + z_{11b}} T_{12} = \frac{V_1}{V_2} = \frac{-y_{12a} z_{12b}}{1 + y_{22a} z_{11b}}$$

7.8 Find the A, B, C and D parameters for the network of Fig. P. 7.8.

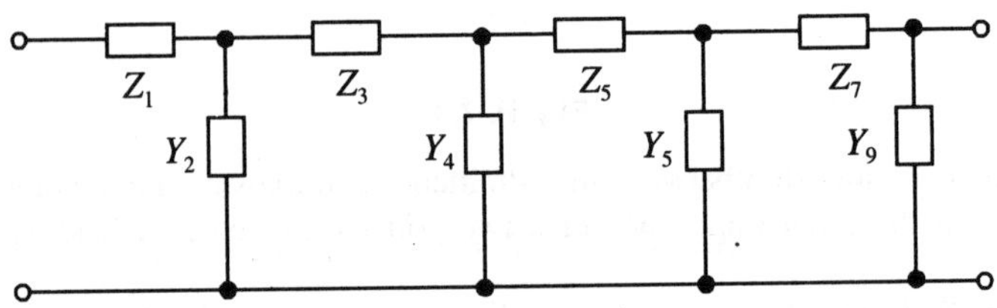

Fig. P. 7.8

7.9 Find the voltage transfer function of the network shown in Fig. P. 7.9.

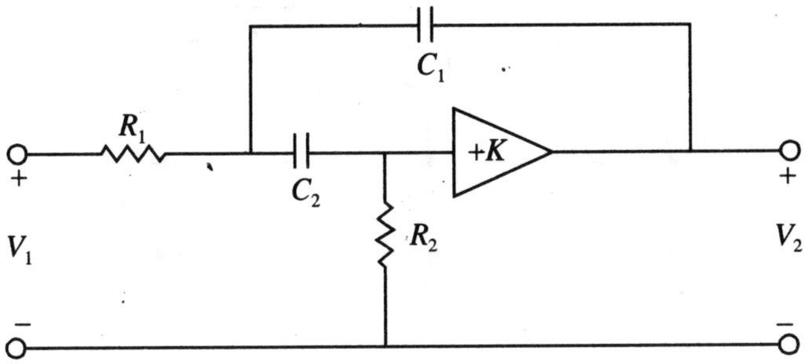

Fig. P. 7.9

7.10 Two two-ports networks N_1 and N_2 are interconnected such that their input ports are series connected and output ports are parallel connected. If H_1 and H_2 are the hybrid parameter matrices of N_1 and N_2, respectively, show that the hybrid parameter matrix of the interconnection is

$$H = H_1 + H_2$$

7.11 Two networks N_1 and N_2 are interconnected as shown in Fig. P. 7.11. Find the overall impedance matrix.

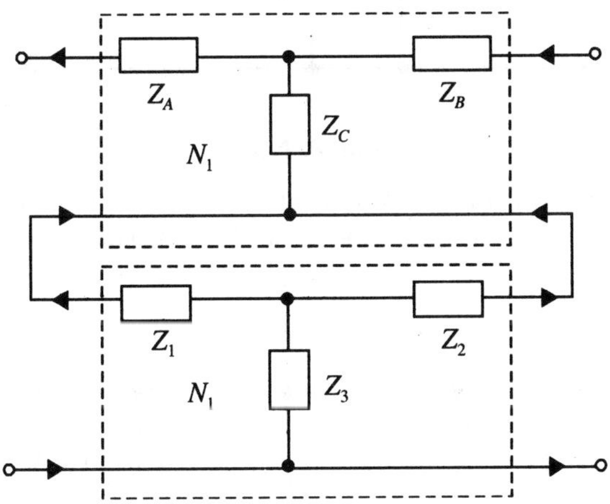

Fig. P. 7.11

7.12 Find the overall impedance matrix of two interconnected two-ports N_1 and N_2 shown in Fig. P. 7.12. Compare this with prob. P. 7.11.

7.13 Find the indefinite admittance matrix of the network shown in Fig. P. 7.13. Suppress the terminals 3 and 4 and obtain the new IAM. If terminal 5 is identified with the reference node, what is the short-circuit admittance matrix of the resulting two-port?

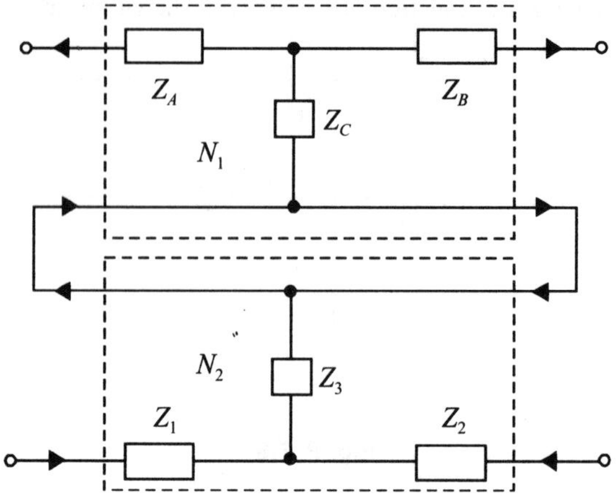

Fig. P. 7.12

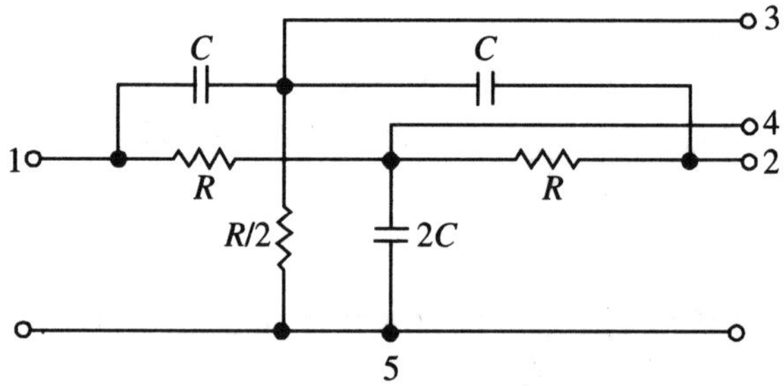

Fig. P. 7.13

7.14 Determine the IAM of the overall interconnection of the network shown in Fig. P. 7.14. p_1, p_2, p_3 and p_4 are the terminals of the interconnected network.

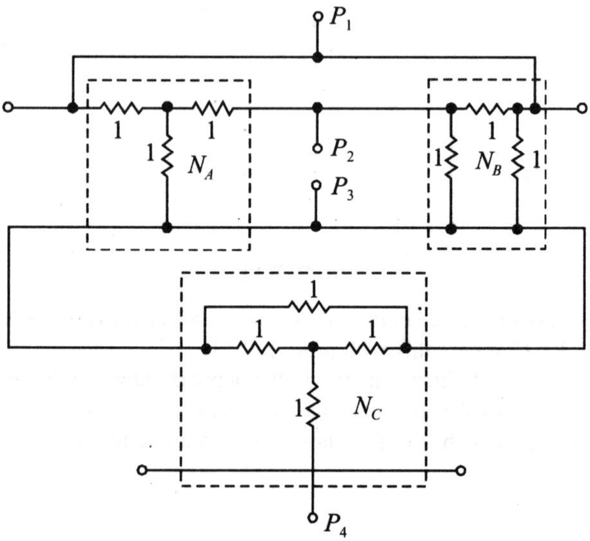

Fig. P. 7.14

7.15 Compute the IAM of the field effect transistor shown in Fig. P. 7.15.

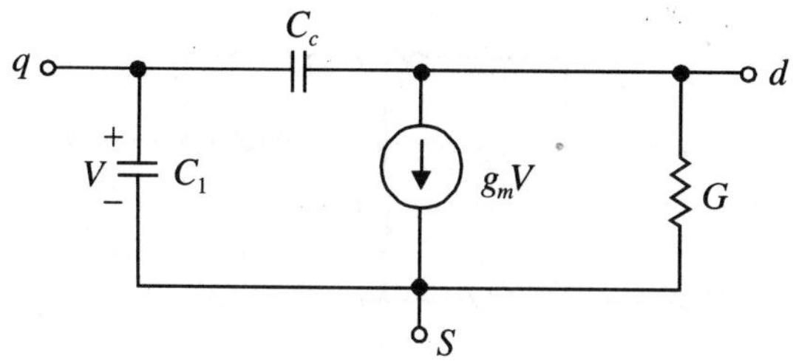

Fig. P. 7.15

7.16 Determine the IAM of the transistor shown in Fig. P. 7.16. Find the short-circuit admittance matrix for the three different modes of operation.

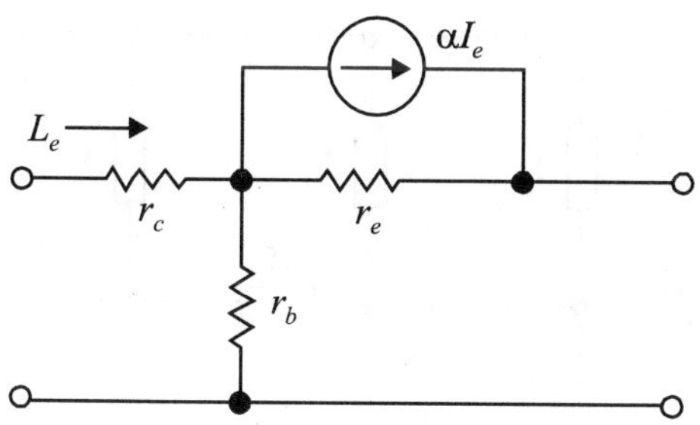

Fig. P. 7.16

7.17 The common emitter hybrid parameters of a transistor are given by

$$h_e = \begin{bmatrix} h_{ie} & h_{re} \\ h_{fe} & h_{ee} \end{bmatrix}$$

Find the common collector hybrid parameters in terms of h_e. Typical hybrid parameters are given by $h_{ie} = 2.540\ \Omega$, $h_{re} = 2.18 \times 10^{-4}$, $h_{fe} = 278$, $h_{ee} = 65.5 \times 10^{-5}$ *mho.* Find the corresponding common emitter parameters.

7.18 The two-port network shown in Fig. P. 7.18 is reciprocal. Shown that the two-port obtained by making terminal 1 or 2 as common terminal is also reciprocal.

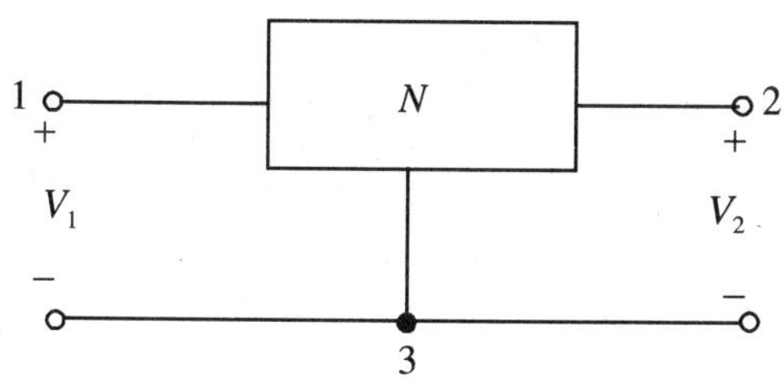

Fig. P. 7.18

7.19 Find the input impedance at 11′ and the voltage ratio V_2/V_1 for the network shown in Fig. P. 7.19.

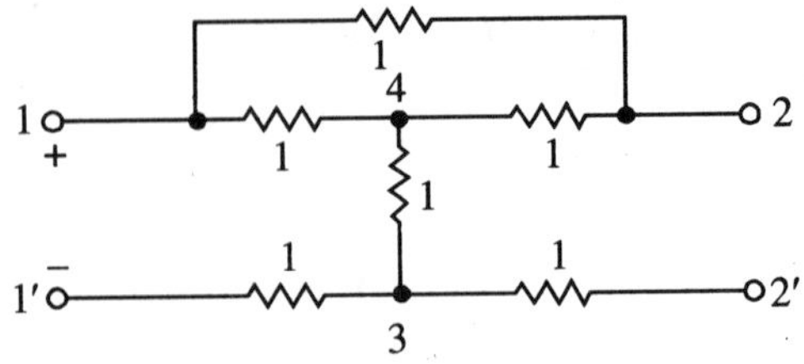

Fig. P. 7.19

7.20 Find the voltage transfer ratio V_2/V_1 of the op. amp. network shown in Fig. P. 7.20. From this obtain V_2/V_1 as $\mu \to \infty$.

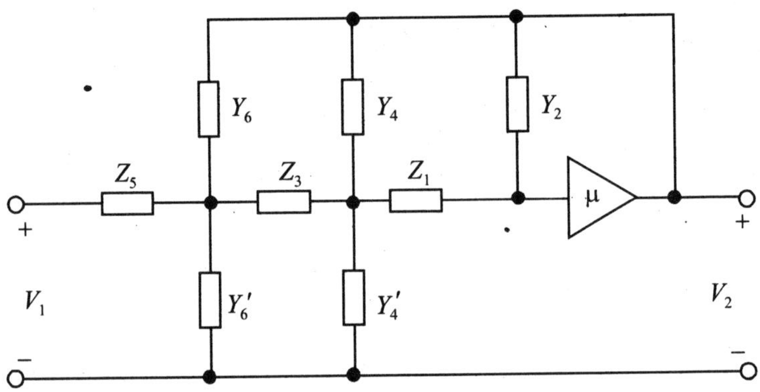

Fig. P. 7.20

NOTES AND REFERENCES

This chapter lends itself for simple laboratory experiments with electronic circuits like

(1) Measurement of input and output immittance parameters and transfer voltage ratio.
(2) Cascade connection of amplifiers and overall transfer voltage ratio.
(3) Loading effect in cascaded passive networks.

Ladder network analysis has received considerable attention from circuit theorists. Starting from the 'continuants' introduced by Prof. Gullemin there have been many methods of evaluating ladder parameters. Closed form expressions for the transmission parameters of the ubiqitous ladder network can be found elsewhere[1].

Two-port parameters are covered in detail in all network theory books. For additional information and problems the references cited in Chapter 3 can be consulted (Ref. 1–Chap. 11, Ref. 2–Chap. 8, Ref. 3–Chap. 17).

The solution of networks by equicofactor matrix was first considered by Sharp and Spain[2]. Weinberg's book[3] (Chap. 1) besides providing the proof for this property discusses the IAM in detail. Mitra's book[4] (Chap. 4) considers the application of IAM to active networks.

1. G. Prabhavathi and V. Ramachandran, *Topological Interpretation of Ladder Network Formulas Proc.*, IEEE, Vol. 54, 1966, pp. 700–701.
2. G.E. Shape and B. Spain, *On the Solution of Networks by Means of Equicofactor Matrix, IRE Trans. on Circuit Theory*, CT-7, 1960, pp. 230–239.
3. L. Weinberg, *Network Analysis and Synthesis*, McGraw-Hill, 1962.
4. S.K. Mitra, *Analysis and Synthesis of Linear Active Networks*, John Wiley, 1969.

CHAPTER

8 State Equations

The network equations, as discussed in Chap. 4, are all Laplace or frequency-domain equations. The analysis of the network through these equations involves the solutions of algebraic equations and then inverse Laplace transforming the solution to obtain the time-domain functions. As opposed to this, one can write the equations in the time-domain itself and solve for the network variable directly in the time-domain. To this end, network equations are written in the form of a set of first order differential equations. Such equations are called *state equations*. The concept of state and of state equations form an important part of the study of control systems.

8.1 CONCEPT OF STATE

As discussed in Chapter 4, response of a network consists of two parts: *zero-input* and the *zero-state*. The zero-input response in an *RLC* network is completely determined once the initial inductor currents and capacitor voltages are known. Hence, we can call the initial capacitor voltages and inductor currents (initial conditions) as the initial states of the system. Indeed, the knowledge of capacitor voltages and inductor currents, at a given time is sufficient to calculate any of the network variables (currents and voltages) at that particular time. Hence, we call the capacitor voltages and the inductor currents, as the *state variables* of the network.

We formally define the state of a network as a set of real or complex quantities that satisfy the following two conditions:

(*a*) The state at any time t_1, and the inputs from t_1 to t $(t > t_1)$ uniquely determines the state at time t.

(*b*) The state at time t and the inputs at time t determine uniquely the value of any network variable at that time t.

For example, consider the network shown in Fig. 8.1. If the capacitor voltage is known at time t, we can replace it by a known voltage source $v_c(t)$ (recall

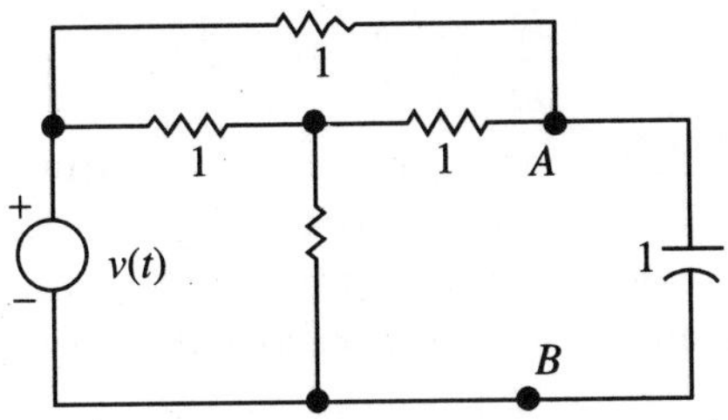

Fig. 8.1

Substitution Theorem (Sec. 5.2) as shown in Fig. 8.2. Now, the voltages across and the currents through the resistors can be found by solving a set of algebraic equations (say, mesh equations). Hence, $v_c(t)$ is the state of the network.

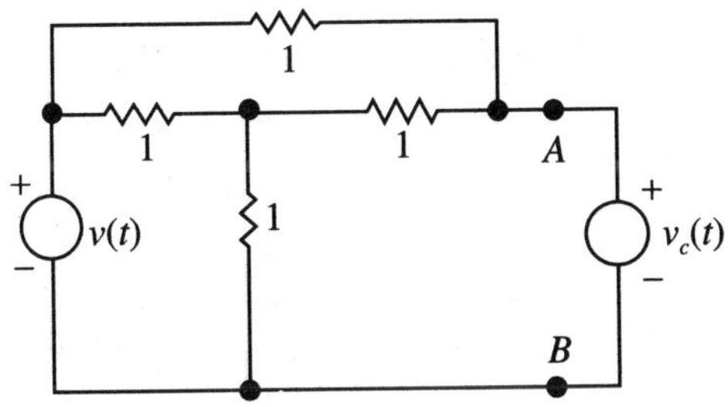

Fig. 8.2

States can also be considered as variables which carry sufficient information about the history of a system. Obviously, a purely resistive network has no states at all. Only a reactive network or a network with energy storage elements has states.

8.2 STATE EQUATIONS

For the network shown in Fig. 8.1 we can write a differential equation governing the voltage across the capacitor. To this end, we replace the network to the left of

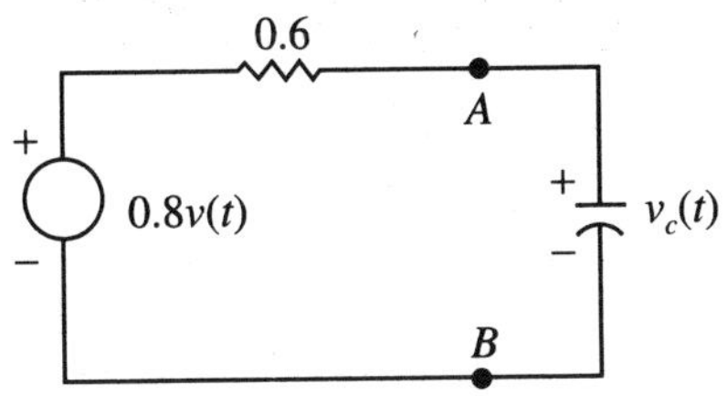

Fig. 8.3

AB by its Thevenin's equivalent (Fig. 8.3). The mesh equation of this network, in time-domain, yields

$$0.6\frac{dv_c(t)}{dt} + v_c(t) = 0.8\,v(t)$$

or $$\frac{dv_c}{dt} = -\frac{5}{3}v_c(t) + \frac{4}{3}v(t)$$

This equation is called the state equation of the network.

Consider another example, the network of Fig. 8.4. Here we have two energy storage elements. We can write two first-order differential equations, one for the current through the inductor and the other for the voltage across the capacitor, as

$$\frac{dv_c}{dt} = \frac{1}{(1/3)}(v - v_c) + i_L$$

and $$0.5\frac{di_L}{dt} = v - v_c$$

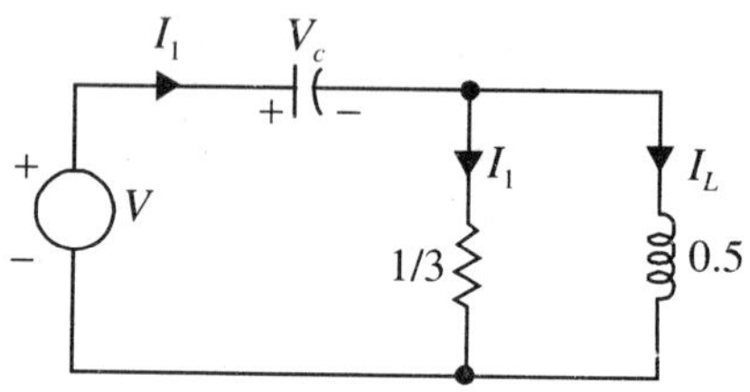

Fig. 8.4

or in matrix form

$$\begin{bmatrix} \dfrac{dv_c}{dt} \\ \dfrac{di_L}{dt} \end{bmatrix} = \begin{bmatrix} -3 & 1 \\ -2 & 0 \end{bmatrix} \begin{bmatrix} v_c \\ i_L \end{bmatrix} + \begin{bmatrix} 3 \\ 2 \end{bmatrix} v \qquad \ldots(8.1)$$

(Here we have written v for $v(t)$. This simplifies notation.)

Let i_s and i_r be the outputs. Then we have

$$\begin{bmatrix} i_s \\ i_r \end{bmatrix} = \begin{bmatrix} -3 & 1 \\ -3 & 0 \end{bmatrix} \begin{bmatrix} v_c \\ i_L \end{bmatrix} + \begin{bmatrix} 3 \\ 3 \end{bmatrix} v \qquad \ldots(8.2)$$

Equations 8.1 and 8.2 are called the state and the output equations of the network, respectively.

The *state equations* of a linear time-invariant network can be written as

$$\frac{d\mathbf{X}(t)}{dt} = A\mathbf{X}(t) + B\mathbf{U}(t) \qquad \ldots(8.3)$$

and the *output equations* as

$$\mathbf{Y}(\mathbf{t}) = C\mathbf{X}(\mathbf{t}) + D\mathbf{U}(\mathbf{t}) \qquad \ldots(8.4)$$

where **X** is the state vector, **U(t)** the input vector and **Y(t)** the output vector. *A*, *B*, *C* and *D* are matrices of appropriate dimensions. If the network is time-varying, the entries of these matrices are functions of time. An advantage of the state equations is that similar equations can be written even for a nonlinear network where the conventional network function technique is in general not applicable. In writing eqns. 8.3 and 8.4 we have made a tacit assumption in that the network does not have any circuit (cut-set) of capacitors (inductors), or capacitor (inductor)-voltage (current) sources. If it were not the case, the state and output equations take the general form (here the time t is not written to simplify notation)

$$\frac{d\mathbf{X}}{dt} = A\mathbf{X} + B\mathbf{U} + E\frac{d\mathbf{U}}{dt} \qquad \ldots(8.5)$$

$$\mathbf{Y} = C\mathbf{X} + D\mathbf{U} + F\frac{d\mathbf{U}}{dt}$$

Example 8.1. We are required to find the state equations of the network shown in Fig. 8.5.

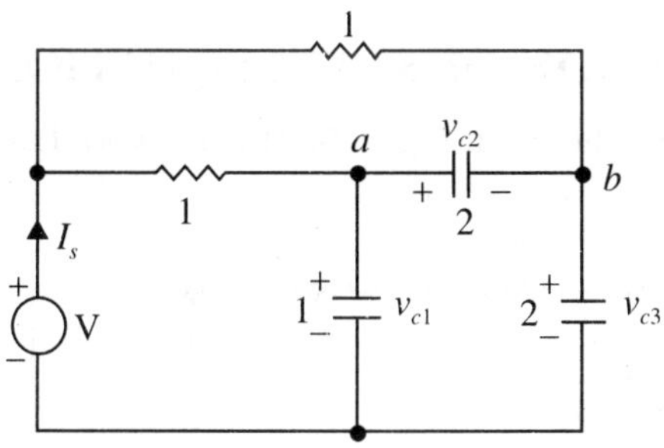

$v_{c1} = v_{c2} + v_{c3}$

Fig. 8.5

Evidently, $\quad v_{c_1} = v_{c_2} + v_{c_3}$

and hence there are only two state variables for the network (since there is a circuit of capacitors). Summing currents at nodes a and b,

at a: $$\frac{dv_{c_1}}{dt} + 2\frac{dv_{c_2}}{dt} + (v_{c_1} - v) = 0$$

at b: $$\frac{2dv_{c_2}}{dt} - \frac{2dv_{c_2}}{dt} + (v_{c_3} - v) = 0$$

But $$\frac{dv_{c_3}}{dt} = \frac{dv_{c_1}}{dt} - \frac{dv_{c_2}}{dt}$$

Hence, we have

$$\begin{bmatrix} 1 & 2 \\ 2 & -4 \end{bmatrix}\begin{bmatrix} \dfrac{dv_{c_1}}{dt} \\ \dfrac{dv_{c_2}}{dt} \end{bmatrix} = \begin{bmatrix} -1 & 0 \\ -1 & 1 \end{bmatrix}\begin{bmatrix} v_{c_1} \\ v_{c_2} \end{bmatrix} + \begin{bmatrix} 1 \\ 1 \end{bmatrix} v$$

The state equations are

$$\begin{bmatrix} \dfrac{dv_{c_1}}{dt} \\ \dfrac{dv_{c_2}}{dt} \end{bmatrix} = \begin{bmatrix} -3/4 & 1/4 \\ -1/8 & -1/8 \end{bmatrix} \begin{bmatrix} v_{c_1} \\ v_{c_2} \end{bmatrix} + \begin{bmatrix} 3/4 \\ 1/8 \end{bmatrix} v$$

Let the output be i_s; then

$$i_s = (v - v_{c_1}) + (v - v_{c_1})$$

i.e.

$$i_s = \begin{bmatrix} -2 & 1 \end{bmatrix} \begin{bmatrix} v_{c_1} \\ v_{c_2} \end{bmatrix} + 2v$$

In subsequent sections we formalize the derivation of state equations.

8.3 FORMULATION OF STATE EQUATION

Consider the network shown in Fig. 8.6. This network has two

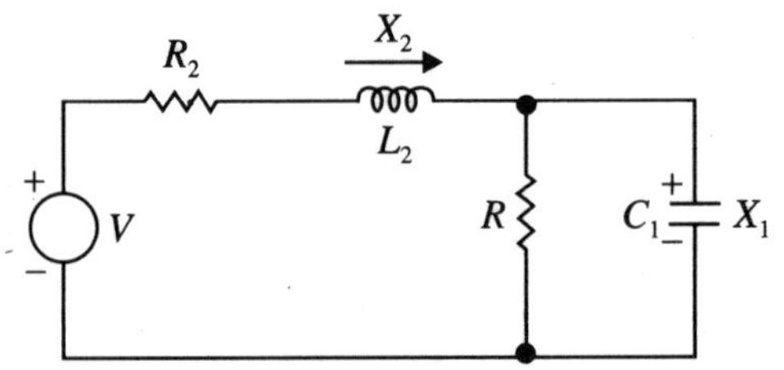

Fig. 8.6

state variables, v_{c_1} (X_1) and v_{L_2} (X_2). To write the state equations we only have to see that $\left(\text{with } \dot{X} = \dfrac{dX}{dt}\right)$.

$$C_1\dot{X}_2 = \text{current through the capacitor} = X_2 - \frac{X_1}{R_1}$$

$$L_2\dot{X}_2 = \text{voltage across the inductor} = v - X_1 - R_2X_2$$

The state equations are

$$\begin{bmatrix} \dot{X}_1 \\ \dot{X}_2 \end{bmatrix} = \begin{bmatrix} -\dfrac{1}{R_1C_1} & \dfrac{1}{C_1} \\ -\dfrac{1}{L_2} & -\dfrac{R_2}{L_2} \end{bmatrix} \begin{bmatrix} X_1 \\ X_2 \end{bmatrix} + \begin{bmatrix} 0 \\ 1/L_2 \end{bmatrix} v$$

Here we have written the state equation by a mere inspection of the network. Obviously, this method of writing these equations (Inspection Method) is limited to very simple networks. For more complicated networks we have to resort to more formal methods.

There are two main methods of writing the state equations—Equivalent Source Method and Graph Theoreti Method.

8.3.1 Equivalent Source Method

In this method, capacitors and inductors are replaced by equivalent voltage or current sources and the current through the capacitors, and the voltage across the inductors are found by application of Superposition theorem.

If the circuit does not contain any degenerate circuits (all capacitor or capacitor-voltage source circuits) or degenerate cut-set (all inductor or inductor-current sources cut-sets), voltages across all the capacitors and currents through all the inductors constitute the states of the network. The capacitors are replaced by voltage sources (v_c) and the inductors by current sources (i_L). Thus, we obtain a resistive network with voltage and current sources. The current through the capacitors $\left(C\dfrac{dv_c}{dt}\right)$ and the voltage across the inductors $\left(L\dfrac{di_L}{dt}\right)$ are found by the application of Superposition theorem to this resistive network. The following examples illustrate the method.

Example 8.2. The network shown in Fig. 8.7(a) does not contain any degenerate cut-sets or circuits. We are required to find the state equations.

Figure 8.7(b) shows the *equivalent source network* obtained by replacing the capacitors by voltage sources and the inductor by a current source. This network

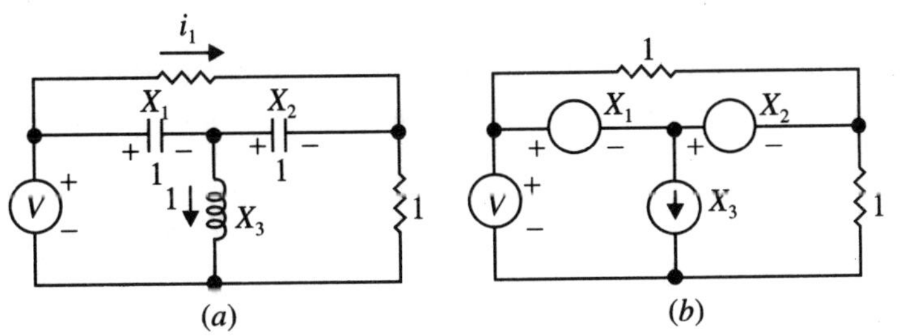

Fig. 8.7

is now to be analysed by superposition. To this end, we remove all sources (short voltage sources and open current sources) except one, and analyse the resulting network for the current through the equivalent voltage sources and the voltage across the equivalent current sources. The final state equations are obtained by combining the results of the analysis of the individual networks.

Let us analyse the network with only the voltage source v in the network. The resulting network is shown in Fig. 8.8(a). It is easily seen that the voltage across the inductor due to the voltage source v is v itself, and the currents through the capacitors are also v. Figures 8.8(b), (c) and (d) show the analyses with respect to the other sources.

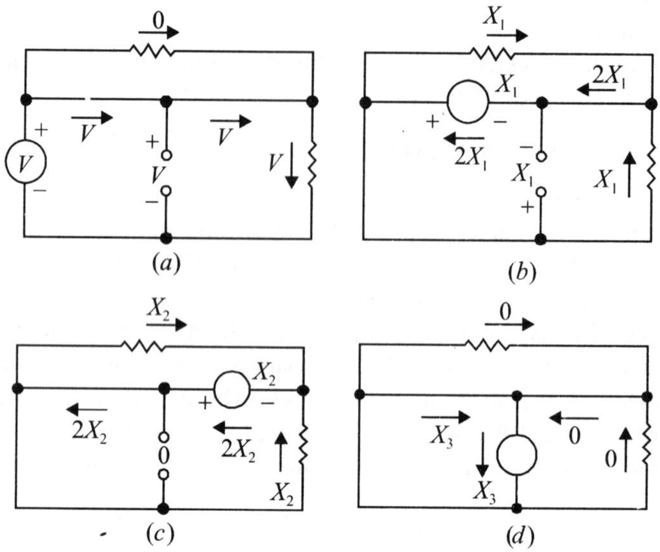

Fig. 8.8

From Figs. 8.7 and 8.8, we have the state equations as

$$\begin{bmatrix} \dot{X}_1 \\ \dot{X}_2 \\ \dot{X}_3 \end{bmatrix} = \begin{bmatrix} -2 & -2 & 1 \\ -2 & -2 & 0 \\ -1 & 0 & 0 \end{bmatrix} \begin{bmatrix} X_1 \\ X_2 \\ X_3 \end{bmatrix} + \begin{bmatrix} 1 \\ 1 \\ 1 \end{bmatrix}$$

Example 8.3. In Fig. 8.9(a) the required output is the source current and the states are X_1, X_2 and X_3. Figure 8.9(b) is the equivalent source network. Let us assume

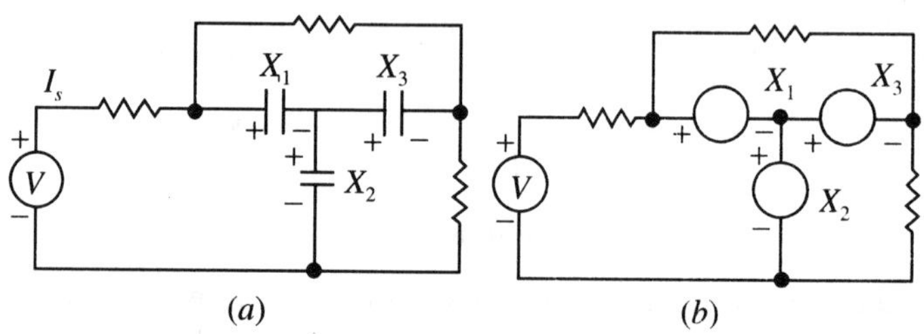

Fig. 8.9

that all the elements are unity valued. Figure 8.10 shows the four networks necessary for the application of superposition theorem.

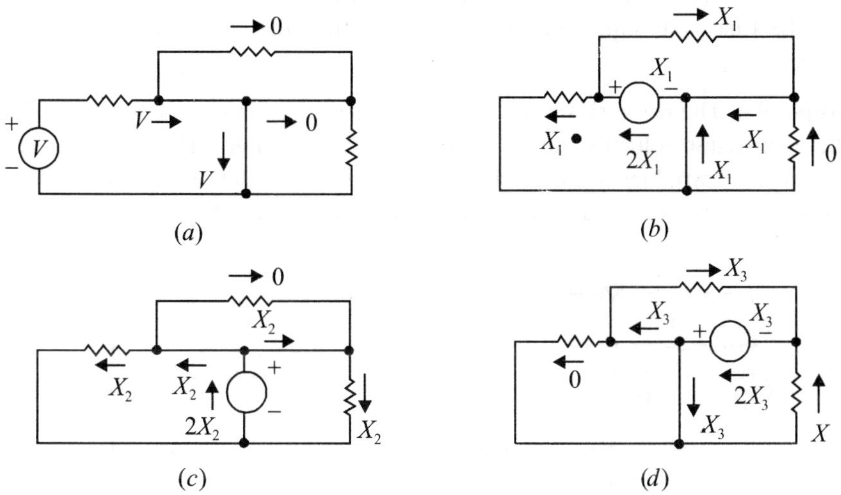

Fig. 8.10

The state and output equations are

$$\begin{bmatrix} \dot{X}_1 \\ \dot{X}_2 \\ \dot{X}_3 \end{bmatrix} = \begin{bmatrix} -2 & -1 & -1 \\ -1 & -2 & 1 \\ -1 & 1 & -2 \end{bmatrix} \begin{bmatrix} X_1 \\ X_2 \\ X_3 \end{bmatrix} + \begin{bmatrix} 1 \\ 1 \\ 0 \end{bmatrix} v$$

$$i_3 = \begin{bmatrix} -1 & -1 & 0 \end{bmatrix} \begin{bmatrix} X_1 \\ X_2 \\ X_3 \end{bmatrix} + v$$

If the network contains degenerate circuits or cut-sets, the method is not directly applicable. In such cases, we have to first select a set of state variables. The voltage of a capacitor is selected as a state, only if it does not form a circuit with a voltage source or capacitors whose voltages are already selected as state variables.The current of an inductor is selected as a state variable, only if it does not form a cut-set with a current source or inductors whose currents are already selected as state variables. The capacitors and the inductors whose voltages and currents are not selected as state variables are called *excess elements.* The voltage across an excess capacitor is the algebraic sum of the state variable voltages and the voltage sources. Similarly, the current in an excess inductor is the algebraic sum of state variable currents and the current sources.

In obtaining the equivalent source network, the excess capacitors are replaced by current sources and excess inductors by voltage sources. The current through the excess capacitor is expressed in terms of state variable voltages and voltage sources. Similarly, the voltage across an excess inductor is expressed in terms of state variable currents and current sources. On eliminating these variables, we can obtain the final state equations. Examples illustrate the state-equations of such networks.

Example 8.4. The network shown in Fig. 8.11(*a*) has a circuit of capacitors. Two of these capacitor voltages can be selected as state-variables. Here we choose v_1 and v_2 as state variables. Figure 8.11(*b*) shows the equivalent source network for this selection. Fig. 8.12 shows the four networks we require for the application of

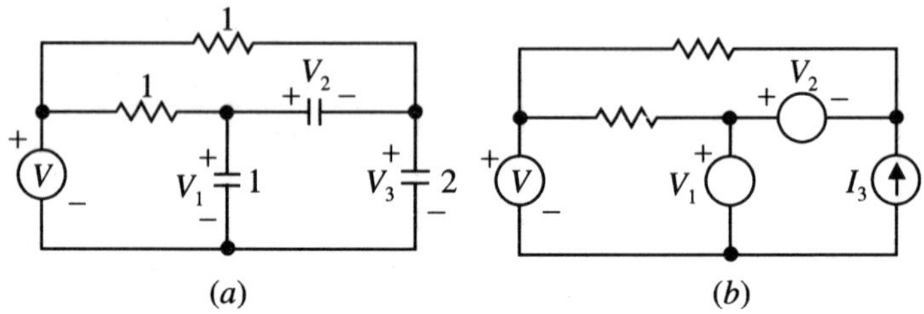

Fig. 8.11

Superposition theorem.

From Fig. 8.12, we have

$$\dot{v}_1 = -2v_1 + v_2 - i_3 + 2v$$

$$2\dot{v}_2 = v_1 - v_2 + i_3 - v$$

But
$$v_3 = v_1 - v_2$$

and
$$i_3 = 2\dot{v}_3 = 2[\dot{v}_1 - \dot{v}_2]$$

Writing these equations in matrix form we have

$$\begin{bmatrix} \dot{v}_1 \\ \dot{v}_2 \end{bmatrix} = \begin{bmatrix} -2 & 1 \\ 1/2 & -1/2 \end{bmatrix} \begin{bmatrix} v_1 \\ v_2 \end{bmatrix} + \begin{bmatrix} -1 \\ 1/2 \end{bmatrix} i_3 + \begin{bmatrix} 2 \\ -1/2 \end{bmatrix} v$$

and
$$i_3 = [2 \ \ -2] \begin{bmatrix} \dot{v}_1 \\ \dot{v}_2 \end{bmatrix}$$

Combining the last two equations, we get

$$\begin{bmatrix} \dot{v}_1 \\ \dot{v}_2 \end{bmatrix} = \begin{bmatrix} -2 & 1 \\ 1/2 & -1/2 \end{bmatrix} \begin{bmatrix} v_1 \\ v_2 \end{bmatrix} + \begin{bmatrix} -1 \\ 1/2 \end{bmatrix} [2 \ -2] \begin{bmatrix} \dot{v}_1 \\ \dot{v}_2 \end{bmatrix} + \begin{bmatrix} 2 \\ -1/2 \end{bmatrix} v$$

i.e.
$$\begin{bmatrix} 3 & -2 \\ -1 & 2 \end{bmatrix} \begin{bmatrix} \dot{v}_1 \\ \dot{v}_2 \end{bmatrix} = \begin{bmatrix} -2 & 1 \\ 1/2 & -1/2 \end{bmatrix} \begin{bmatrix} v_1 \\ v_2 \end{bmatrix} + \begin{bmatrix} 2 \\ -1/2 \end{bmatrix} v$$

or
$$\begin{bmatrix} \dot{v}_1 \\ \dot{v}_2 \end{bmatrix} = \begin{bmatrix} -3/4 & 1/4 \\ -1/8 & -1/8 \end{bmatrix} \begin{bmatrix} v_1 \\ v_2 \end{bmatrix} + \begin{bmatrix} 3/4 \\ 1/8 \end{bmatrix} v$$

which is the required state equation for the network.

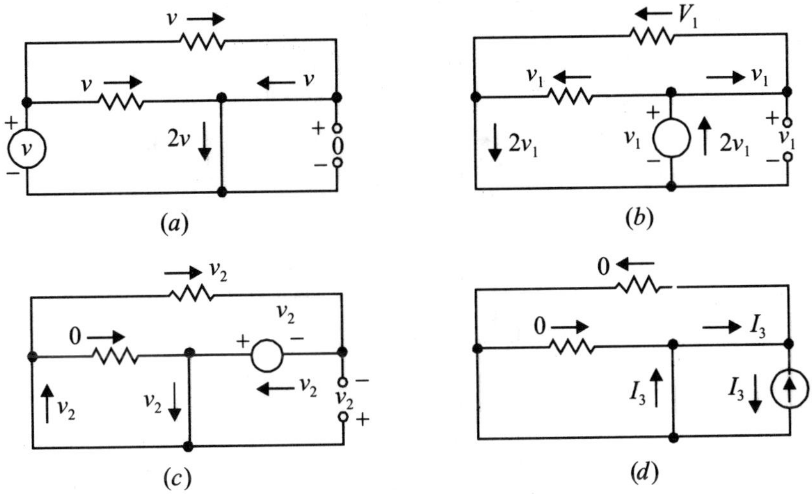

Fig. 8.12

Example 8.5. In the network shown in Fig. 8.13(*a*) there is an excess inductor. Let L_3 be this excess inductor. Figure 8.13(*b*) shows the equivalent source network for this selection. Figure 8.14 shows the four networks required for the application of superposition theorem.

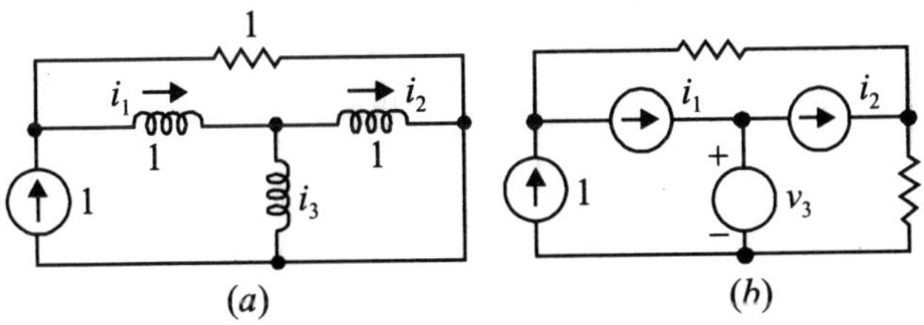

Fig. 8.13

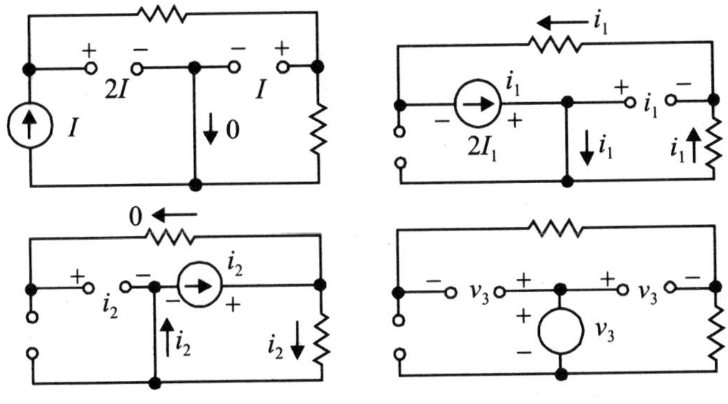

Fig. 8.14

Hence, we have

$$\begin{bmatrix} \dfrac{di_1}{dt} \\ \dfrac{di_2}{dt} \end{bmatrix} = \begin{bmatrix} -2 & 1 \\ 1 & -1 \end{bmatrix} \begin{bmatrix} i_1 \\ i_2 \end{bmatrix} + \begin{bmatrix} -1 \\ 1 \end{bmatrix} v_3 + \begin{bmatrix} 2 \\ -1 \end{bmatrix} I$$

But

$$v_2 = \begin{bmatrix} 1 & -1 \end{bmatrix} \begin{bmatrix} \dfrac{di_1}{dt} \\ \dfrac{di_2}{dt} \end{bmatrix}, \quad (i_3 = i_1 - i_2)$$

Combining the two equations we have

$$\begin{bmatrix} \dfrac{di_1}{dt} \\ \dfrac{di_2}{dt} \end{bmatrix} = \begin{bmatrix} -2 & 1 \\ 1 & -1 \end{bmatrix} \begin{bmatrix} i_1 \\ i_2 \end{bmatrix} + \begin{bmatrix} -1 \\ 1 \end{bmatrix} [1 - 1] \begin{bmatrix} \dfrac{di_1}{dt} \\ \dfrac{di_2}{dt} \end{bmatrix} + \begin{bmatrix} 2 \\ -1 \end{bmatrix} I$$

i.e.

$$\begin{bmatrix} 2 & -1 \\ -1 & 2 \end{bmatrix} \begin{bmatrix} \dfrac{di_1}{dt} \\ \dfrac{di_2}{dt} \end{bmatrix} = \begin{bmatrix} -2 & 1 \\ 1 & -1 \end{bmatrix} \begin{bmatrix} i_1 \\ i_2 \end{bmatrix} + \begin{bmatrix} 2 \\ -1 \end{bmatrix} I$$

or

$$\begin{bmatrix} \dfrac{di_1}{dt} \\ \dfrac{di_2}{dt} \end{bmatrix} = \begin{bmatrix} -1 & 1/3 \\ 0 & -1/3 \end{bmatrix} \begin{bmatrix} i_1 \\ i_2 \end{bmatrix} + \begin{bmatrix} 1 \\ 0 \end{bmatrix} I$$

is the required state equation.

In all the preceding examples, the state equations are of the form

$$\mathbf{X} = A\mathbf{X} + B\mathbf{U}$$

If the network has circuits of capacitors and voltage source and/or cut-sets of inductors and current sources, then, the state equations take the form in eqn. 8.5.

8.3.2 Graph Theoretic Method

The method of equivalent sources is conceptually and algebraically a simple procedure for writing the state equations. If the resistive network is a complicated one, the analysis of the individual network becomes a major problem. Also, the method is not strictly an algorithmic one. The state equations can also be formulated by using graph theoretic concepts. But the method is algebraically quite complex.

Many of the latest computer oriented analysis programs use this method for writing the state equations. We outline this method here.

The formulation of network equations by graph theoretic methods starts with the selection of a tree. To formulate the state equations a proper or modified proper tree is selected. A *proper tree* is a tree which includes all voltage sources and capacitors and excludes all current sources and inductors. A proper tree, if it exists, implies that the network does not have any excess elements. A *modified proper tree* includes all voltage sources, as many as possible capacitors and excludes all current sources and as many as possible inductors. Obviously, every network has at least one modified proper tree. The selection of tree, as described, excludes the possibility of *f*-circuits of capacitors having resistors and/or inductors and *f*-cut-sets of inductors having resistors and/or capacitors.

The notation to be followed in the development is that: $\mathbf{V}_0$ represents voltage source vector, $\mathbf{I}_1$ the current source vector; $\mathbf{V}_{ij}$ and $\mathbf{I}_{ij}$ are the element voltage and current vectors, respectively. The first subscript *i* refers to the tree or chord (*b* or *c*, respectively) and the second subscript *j* to the element type (*r, l* and *c*). For example, $\mathbf{V}_{bc}$ is the branch-capacitor voltage vector and $\mathbf{I}_{cl}$ is the chord inductor current vector. The vector $\mathbf{X} = [\mathbf{V}_{bc}\ \mathbf{I}_{cl}]^t$ constitutes the state vector.

For a modified proper tree KVL and KCL interms of *f*-circuit and *f*-cutset matrices can be written as (see equations 3.1, 3.2 and 2.20)

$$\begin{bmatrix} B_{11} & B_{12} & 0 & 0 & U & 0 & 0 & 0 \\ B_{21} & B_{22} & B_{23} & 0 & 0 & U & 0 & 0 \\ B_{31} & B_{32} & B_{33} & B_{44} & 0 & 0 & U & 0 \\ B_{41} & B_{42} & B_{43} & B_{44} & 0 & 0 & 0 & U \end{bmatrix} \begin{bmatrix} \mathbf{V}_0 \\ \mathbf{V}_{bc} \\ \mathbf{V}_{br} \\ \mathbf{V}_{bl} \\ \mathbf{V}_{cc} \\ \mathbf{V}_{cr} \\ \mathbf{V}_{cl} \\ \mathbf{V}_1 \end{bmatrix} = 0 \qquad \ldots(8.6a)$$

$$\begin{bmatrix} U & 0 & 0 & 0 & -B_{11}^t & -B_{21}^t & -B_{31}^t & -B_{41}^t \\ 0 & U & 0 & 0 & -B_{12}^t & -B_{22}^t & -B_{32}^t & -B_{42}^t \\ 0 & 0 & U & 0 & 0 & -B_{23}^t & -B_{33}^t & -B_{43}^t \\ 0 & 0 & 0 & U & 0 & 0 & -B_{34}^t & -B_{44}^t \end{bmatrix} \begin{bmatrix} \mathbf{I}_0 \\ \mathbf{I}_{bc} \\ \mathbf{I}_{br} \\ \mathbf{I}_{bl} \\ \mathbf{I}_{cc} \\ \mathbf{I}_{cr} \\ \mathbf{I}_{cl} \\ \mathbf{I}_1 \end{bmatrix} = 0 \qquad \ldots(8.6b)$$

where $\mathbf{V}_1$ and $\mathbf{I}_0$ are the voltage across current source and current through voltage source vectors respectively. B_{ij}'s are submatrices of the f-circuit matrix and are of appropriate dimensions. With the selection of a modified proper tree, the submatrices B_{13}, B_{14} and B_{24} are null matrices. In equations 8.6(a) and 8.6(b), U's are unit matrices of commensurate dimensions.

In order to facilitate the easy formulation of state-equations, the network current and voltage variables are grouped as

$$\begin{bmatrix} \mathbf{V}_{bc} \\ \mathbf{I}_{cl} \end{bmatrix}, \begin{bmatrix} \mathbf{I}_{bc} \\ \mathbf{V}_{cl} \end{bmatrix}, \begin{bmatrix} \mathbf{V}_{br} \\ \mathbf{I}_{cr} \end{bmatrix}, \begin{bmatrix} \mathbf{I}_{br} \\ \mathbf{V}_{cr} \end{bmatrix}, \begin{bmatrix} \mathbf{V}_{bl} \\ \mathbf{I}_{cc} \end{bmatrix}, \begin{bmatrix} \mathbf{I}_{bl} \\ \mathbf{V}_{cc} \end{bmatrix} \text{ and } \begin{bmatrix} \mathbf{V}_0 \\ \mathbf{I}_1 \end{bmatrix}$$

The terminal relationships for the elements are written as

$$\begin{bmatrix} C_b & 0 \\ 0 & L_c \end{bmatrix} \begin{bmatrix} \dot{\mathbf{V}}_{bc} \\ \dot{\mathbf{I}}_{cl} \end{bmatrix} = \begin{bmatrix} \mathbf{I}_{bc} \\ \mathbf{V}_{cl} \end{bmatrix} \quad \text{...(8.7}a\text{)}$$

$$\begin{bmatrix} \mathbf{V}_{br} \\ \mathbf{I}_{cr} \end{bmatrix} = \begin{bmatrix} R_b & 0 \\ 0 & G_c \end{bmatrix} \begin{bmatrix} \mathbf{I}_{br} \\ \mathbf{V}_{cr} \end{bmatrix} \quad \text{...(8.7}b\text{)}$$

$$\begin{bmatrix} \mathbf{V}_{bl} \\ \mathbf{I}_{cc} \end{bmatrix} = \begin{bmatrix} L_b & 0 \\ 0 & C_c \end{bmatrix} \begin{bmatrix} \dot{\mathbf{I}}_{bl} \\ \dot{\mathbf{V}}_{cc} \end{bmatrix} \quad \text{...(8.7}c\text{)}$$

where C_b, R_b, L_b and C_c, G_c, L_c are diagonal matrices of branch and chord elements, respectively.

From equations 8.6a and 8.6b we have

$$\begin{bmatrix} \mathbf{I}_{bc} \\ \mathbf{V}_{cl} \end{bmatrix} = \begin{bmatrix} 0 & B_{32}^t \\ -B_{32} & 0 \end{bmatrix} \begin{bmatrix} \mathbf{V}_{bc} \\ \mathbf{I}_{cl} \end{bmatrix} + \begin{bmatrix} 0 & B_{22}^t \\ -B_{33} & 0 \end{bmatrix} \begin{bmatrix} \mathbf{V}_{br} \\ \mathbf{I}_{cr} \end{bmatrix}$$

$$+ \begin{bmatrix} 0 & B_{12}^t \\ -B_{34} & 0 \end{bmatrix} \begin{bmatrix} \mathbf{V}_{bl} \\ \mathbf{I}_{cc} \end{bmatrix} + \begin{bmatrix} 0 & B_{42} \\ -B_{31} & 0 \end{bmatrix} \begin{bmatrix} \mathbf{V}_0 \\ \mathbf{I}_1 \end{bmatrix} \quad \text{...(8.8}a\text{)}$$

$$\begin{bmatrix} \mathbf{I}_{br} \\ \mathbf{V}_{cr} \end{bmatrix} = \begin{bmatrix} 0 & B_{33}^t \\ -B_{22} & 0 \end{bmatrix} \begin{bmatrix} \mathbf{V}_{bc} \\ \mathbf{I}_{cl} \end{bmatrix} + \begin{bmatrix} 0 & B_{23}^t \\ -B_{23} & 0 \end{bmatrix} \begin{bmatrix} \mathbf{V}_{br} \\ \mathbf{I}_{cr} \end{bmatrix}$$

$$+ \begin{bmatrix} 0 & B_{43}^t \\ -B_{21} & 0 \end{bmatrix} \begin{bmatrix} \mathbf{V}_0 \\ \mathbf{I}_1 \end{bmatrix} \quad \text{...(8.8}b\text{)}$$

$$\begin{bmatrix} \mathbf{I}_{bl} \\ \mathbf{V}_{cc} \end{bmatrix} = \begin{bmatrix} 0 & B_{34}^t \\ -B_{12} & 0 \end{bmatrix} \begin{bmatrix} \mathbf{V}_{bc} \\ \mathbf{I}_{cl} \end{bmatrix} + \begin{bmatrix} 0 & B_{44}^t \\ -B_{11} & 0 \end{bmatrix} \begin{bmatrix} \mathbf{V}_0 \\ \mathbf{I}_1 \end{bmatrix} \quad \text{...(8.8}c\text{)}$$

Equations (8.7b) and (8.8b) yield

$$\begin{bmatrix} \mathbf{V}_{br} \\ \mathbf{I}_{cr} \end{bmatrix} = \begin{bmatrix} 0 & R_b B_{33}^t \\ -G_c B_{22} & 0 \end{bmatrix} \begin{bmatrix} \mathbf{V}_{bc} \\ \mathbf{I}_{cl} \end{bmatrix} + \begin{bmatrix} 0 & R_b B_{23}^t \\ -G_c B_{23} & 0 \end{bmatrix} \begin{bmatrix} \mathbf{V}_{br} \\ \mathbf{I}_{cr} \end{bmatrix} + \begin{bmatrix} 0 & R_b B_{43}^t \\ -G_c B_{21} & 0 \end{bmatrix} \begin{bmatrix} \mathbf{V}_0 \\ \mathbf{I}_1 \end{bmatrix}$$

which can be rewritten as

$$\begin{bmatrix} U & -R_b B_{23}^t \\ G_c B_{23} & U \end{bmatrix} \begin{bmatrix} \mathbf{V}_{br} \\ \mathbf{I}_{cr} \end{bmatrix} = \begin{bmatrix} 0 & R_b B_{33}^t \\ -G_c B_{22} & 0 \end{bmatrix} \begin{bmatrix} \mathbf{V}_{bc} \\ \mathbf{I}_{cl} \end{bmatrix} + \begin{bmatrix} 0 & R_b B_{43}^t \\ -G_c B_{21} & 0 \end{bmatrix} \begin{bmatrix} \mathbf{V}_0 \\ \mathbf{I}_1 \end{bmatrix}$$

Let

$$\begin{bmatrix} U & -R_b B_{23}^t \\ G_c B_{23} & U \end{bmatrix}^{-1} = \begin{bmatrix} \theta_1 & \theta_2 \\ \theta_3 & \theta_4 \end{bmatrix}$$

then

$$\theta_1 = (G_b + B_{23}^t\, G_c\, B_{23})^{-1}\, R_b^{-1}$$
$$\theta_4 = (R_b + B_{23}\, R_b\, B_{23}^t)^{-1}\, G_c^{-1}$$
$$\theta_2 = (G_b + B_{23}^t\, G_c\, B_{23})^{-1}\, B_{23}^t$$
$$\theta_2 = -(R_c + B_{23}\, R_b\, B_{23}^t)^{-1}\, B_{23}$$

$Q_1 R_b$ and $\theta_4 G_c$ can be considered as cut-set conductance and loop impedance matrices, respectively. The required inverses for evaluation of θ_1, θ_2, θ_3 and θ_4 exist for a passive network. Finally, we have

$$\begin{bmatrix} \mathbf{V}_{br} \\ \mathbf{I}_{cr} \end{bmatrix} = \begin{bmatrix} -\theta_2 G_c B_{22} & \theta_1 R_b B_{33}^t \\ -\theta_4 G_c B_{12} & \theta_3 R_b B_{33}^t \end{bmatrix} \begin{bmatrix} \mathbf{V}_{bc} \\ \mathbf{I}_{cl} \end{bmatrix} + \begin{bmatrix} -\theta_2 G_c B_{21} & \theta_1 R_b B_{43}^t \\ -\theta_4 G_c B_{21} & \theta_3 R_b B_{43}^t \end{bmatrix} \begin{bmatrix} \mathbf{V}_0 \\ \mathbf{I}_1 \end{bmatrix} \quad \ldots(8.9a)$$

Equations (8.7*c*) and (8.8*c*) yield

$$\begin{bmatrix} \mathbf{V}_{bl} \\ \mathbf{I}_{cc} \end{bmatrix} = \begin{bmatrix} 0 & L_b B_{34}^t \\ -C_c B_{12} & 0 \end{bmatrix} \begin{bmatrix} \dot{\mathbf{V}}_b \\ \dot{\mathbf{I}}_{cl} \end{bmatrix} + \begin{bmatrix} 0 & L_b B_{44}^t \\ -C_c B_{11} & 0 \end{bmatrix} \begin{bmatrix} \mathbf{V}_0 \\ \mathbf{I}_1 \end{bmatrix} \quad \ldots(8.9b)$$

Combining equations (8.7*a*), (8.8*a*), (8.9*a*) and (8.9*b*), we have

$$\begin{bmatrix} C_b + B_{12}^t C_c B_{12} & 0 \\ 0 & L_c + B_{34} L_b B_{34}^t \end{bmatrix} \begin{bmatrix} \dot{\mathbf{V}}_{bc} \\ \dot{\mathbf{I}}_{cl} \end{bmatrix}$$

$$= \begin{bmatrix} -B_{22}^t\theta_4 G_c B_{12} & B_{32}^t + B_{22}^t\theta_3 R_b B_{33}^t \\ -B_{32} + B_{33}\theta_2 G_c B_{22} & -B_{33}\theta_1 R_b B_{33}^t \end{bmatrix} \begin{bmatrix} \mathbf{V}_{bc} \\ \mathbf{I}_{cl} \end{bmatrix}$$

$$+ \begin{bmatrix} -B_{22}^t\theta_4 G_c B_{21} & B_{42}^t + B_{22}^t\theta_3 R_b B_{43}^t \\ -B_{31} + B_{33}\theta_2 G_c B_{21} & -B_{33}\theta_1 R_b B_{43}^t \end{bmatrix} \begin{bmatrix} \mathbf{V}_0 \\ \mathbf{I}_1 \end{bmatrix}$$

$$+ \begin{bmatrix} -B_{12}^t C_c B_{11} & 0 \\ 0 & -B_{34} L_b B_{44}^t \end{bmatrix} \begin{bmatrix} \dot{\mathbf{V}}_0 \\ \dot{\mathbf{I}}_1 \end{bmatrix}$$

Finally, we can write the state equations as

$$\begin{bmatrix} \dot{\mathbf{V}}_{bc} \\ \dot{\mathbf{I}}_{cl} \end{bmatrix} = \begin{bmatrix} A_{11} & A_{12} \\ A_{21} & A_{22} \end{bmatrix} \begin{bmatrix} \mathbf{V}_{bc} \\ \mathbf{I}_{cl} \end{bmatrix} + \begin{bmatrix} B_{11} & B_{12} \\ B_{21} & B_{22} \end{bmatrix} \begin{bmatrix} \mathbf{V}_0 \\ \mathbf{I}_1 \end{bmatrix}$$

$$+ \begin{bmatrix} E_{11} & 0 \\ 0 & E_{22} \end{bmatrix} + \begin{bmatrix} \dot{\mathbf{V}}_0 \\ \dot{\mathbf{I}}_1 \end{bmatrix} \quad \ldots(8.10)$$

where

$$A_{11} = -C^1 B'_{22} (R_c + B_{23} R_b B'_{23})^{-1} B_{22}$$
$$A_{12} = C^1 B^t_{32} - C^1 B'_{22} (R_c + B_{23} R_b B'_{23})^{-1} B_{23} R_b B'_{33}$$
$$A_{21} = -L^1 B_{32} + L^1 B_{33} (G_b + B'_{23} G_c B_{23})^{-1} B'_{23} G_c B_{22}$$
$$A_{22} = -L^1 B_{33} (G_b + B'_{23} G_c B_{23})^{-1} B'_{33}$$
$$B_{11} = -C^1 B'_{22} (R_c + B_{23} R_b B'_{23})^{-1} B_{21}$$
$$B_{12} = C^1 B^t_{42} - C^1 B'_{22} (R_c + B_{23} R_b B'_{23})^{-1} B_{23} R_b B'_{43}$$
$$B_{21} = -L^1 B_{31} + L^1 B_{33} (C_b + B'_{23} G_c B_{23})^{-1} B'_{23} G_c B_{21}$$
$$B_{22} = -L^1 B_{34} (G_4 + B'_{23} G_c B_{23})^{-1} B'_{43}$$
$$E_{11} = -C^1 B'_{12} C_c B_{11}, E_{22} = -L^1 B_{34} L_b B'_{44}$$
$$C_1 = (C_b + B'_{12} C_c B_{12})^{-1} L^1 = (L_c + B_{34} L_b B'_{34})^{-1}$$

Similar to the state-equation of eqn. 8.1 we can derive the output equations by following the procedure indicated. Though the algebraic manipulation involved looks to the horrendous, the actual procedure is algebraic and hence can be easily programmed on a computer.

If the network has a proper tree, *i.e.* there are no excess elements, then matrix E is a null matrix. In case of a time varying network, state equations similar to eqn. 8.10 can be derived. However, the entries of the matrices will be functions of time. Though we have not considered networks with controlled sources, state equations for active networks can be easily obtained.

8.4 STATE EQUATIONS AND TRANSFER FUNCTION

The solution of state equation (say eqns. 8.3 and 8.4) depends on the solution of the matrix differential equation

$$\dot{\Phi}(t) = A\Phi(t) \qquad \text{...(8.11)}$$

The solution to this equation by analogy with scalar differential equations is

$$\Phi(t) = \exp[At]$$

with $\Phi(0) = I$. This is a matrix exponential function. $\Phi(t)$ is called the *state transition matrix* and it can be shown that

$$\mathbf{X}(\mathbf{t}) = \Phi(t - to)\ \mathbf{X}(\mathbf{to}) \qquad \text{...(8.12)}$$

where $\mathbf{X}(to)$ is the state at $t = to$ and this is the *ZIR*.

The Laplace transform of eqn. 8.11 yield

$$s\Phi(s) - \Phi(0) = 4\Phi(s)$$

and hence

$$\Phi(t) = \mathcal{L}^{-1}\left[(sI - A)^{-1}\right] \qquad \text{...(8.13)}$$

which gives a method of evaluating the state transition matrix. Here we have used I for a unit matrix instead of U, in order to differentiate it from input vector.

To find the solution for the state equation 8.3 we resort to Laplace transforms, which yields

$$s\mathbf{X}(s) - \mathbf{X}(0) = A\mathbf{X}(s) + B\mathbf{U}(s)$$

i.e.

$$(sI - A)\ \mathbf{X}(s) = \mathbf{X}(0) + B\mathbf{U}(s)$$

Hence,

$$\mathbf{X}(s) = (sI - A)^{-1}\ \mathbf{X}(0) + (sI - A)^{-1}\ B\mathbf{U}(s)$$

From eqn. 8.13 we have

$$\mathbf{X}(t) = \Phi(t)\mathbf{X}(0) + \int_0^t \Phi(t - \tau) B\mathbf{U}(\tau)\, d\tau \qquad \text{...(8.14)}$$

which is the total response (*i.e. ZIR* + *ZSR*). Equations 8.4 and 8.14 yields

$$\mathbf{Y}(t) = C\Phi(t)\mathbf{X}(0) + \int_0^t C\Phi(t - \tau)B\mathbf{U}(\tau)\, d\tau + D\mathbf{U}(t) \qquad \text{...(8.15)}$$

We recognize the familiar convolution integrals in eqns. 8.14 and 8.15.

From eqn. 8.15

$$ZIR = C\Phi(t)\ \mathbf{X}\ (0)$$

and

$$ZSR = C\Phi(t) * B\mathbf{U}(t) + D\mathbf{U}(t)$$

The impulse response of the system can be found from eqn. 8.15 by setting $\mathbf{X}(0) = 0$ and $U(t) = \delta(t)$

In this case, we have

$$H(t) = C\Phi(t)\ B + D\delta(t)$$

Hence, the transfer function of the network is

$$H(s) = C\ (sI - A)^{-1}B + D \qquad \text{...(8.16)}$$

From eqn. 8.16, $D = H(\infty)$. If

$H(\infty) = 0$, *i.e.* D is a null matrix,

$$H(s) = C(sI - A)^{-1}B \qquad \text{...(8.17)}$$

The poles of the transfer function, the natural frequencies of the network, are the roots of $|sI - A| = 0$ which are the eigenvalues of the matrix (see Appendix A).

In Section 6.1, order of complexity of a network is defined as the number of natural frequencies of the network. As can be seen the order of complexity is the order of the state vector which is the number of independent initial conditions that can be specified for a network. If a network has a proper tree, the order of complexity is the total number of energy storage elements, *i.e.* capacitors and inductors in the network. It can be easily seen that when the network has redundant or excess elements the order of complexity is equal to the number of reactive elements less the number of excess elements.

PROBLEMS

8.1 Write the state equations for the network shown in Fig. P. 8.1. If the currents through the source and the resistor are the outputs, find the output equation and solve for the outputs when $V(t) = u(t)$ and $x(0) = 0$.

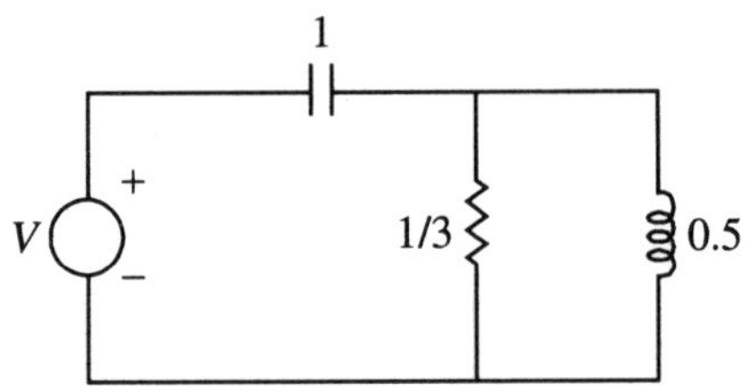

Fig. P. 8.1

8.2 Write the state equations and the output equation (V is the output) for the network of Fig. P. 8.2. Find the input impedance by solving the state equations.

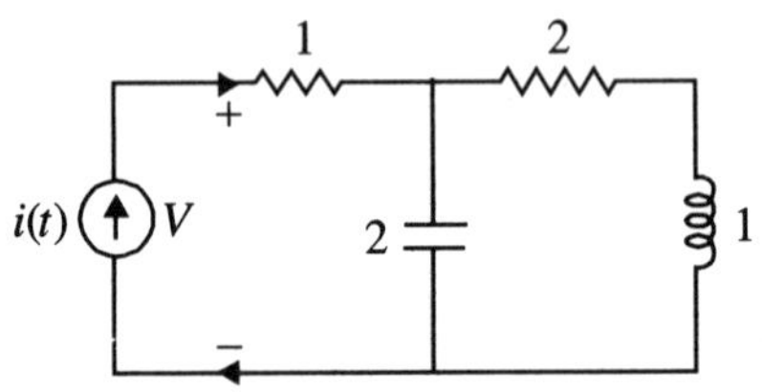

Fig. P. 8.2

8.3 The state equations of the network of Fig. P. 8.3, when V_1 and V_2 are selected as state variables, are

$$\begin{bmatrix} \dot{V}_1 \\ \dot{V}_2 \end{bmatrix} = \begin{bmatrix} -3/4 & 1/4 \\ -1/8 & -1/8 \end{bmatrix} \begin{bmatrix} V_1 \\ V_2 \end{bmatrix} + \begin{bmatrix} 3/4 \\ 1/8 \end{bmatrix} V$$

Determine the state equations if V_1 and V_3 are regarded as state variables.

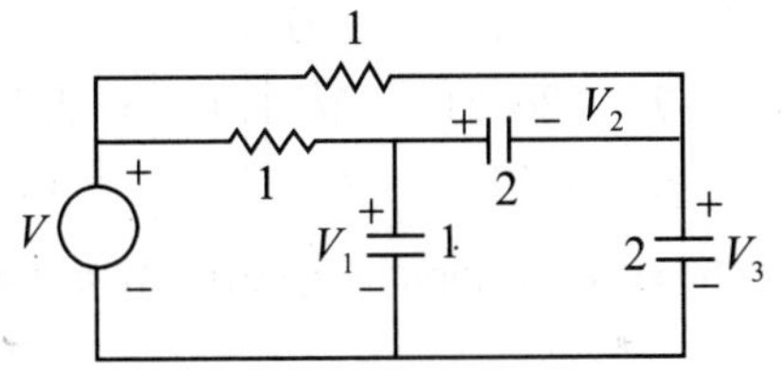

Fig. P. 8.3

8.4 Determine the state equations of the network shown in Fig. P. 8.4 regarding V_2 and V_3 as state variables. If instead V_1 and V_3 are regarded as state variables, what are the state equations?

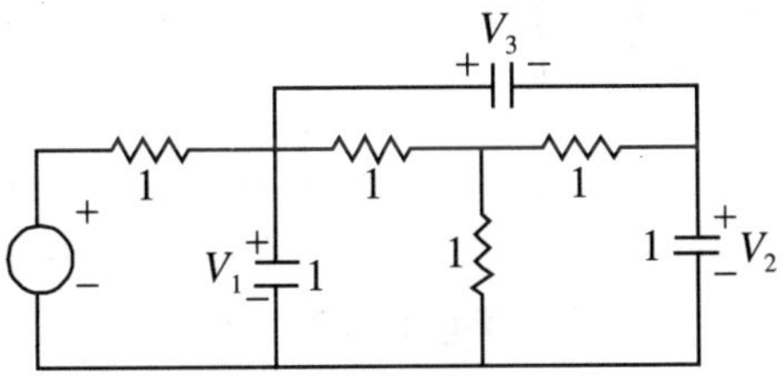

Fig. P. 8.4

8.5 Determine the state equations of the network in Fig. P. 8.5. If the output is taken as the voltage across the current source, determine the output equations. Find the driving point impedance of the network.

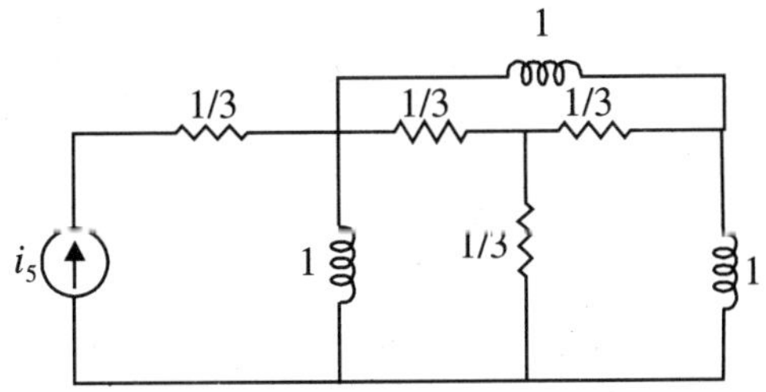

Fig. P. 8.5

8.6 Determine the state equations of the network in Fig. P. 8.6.

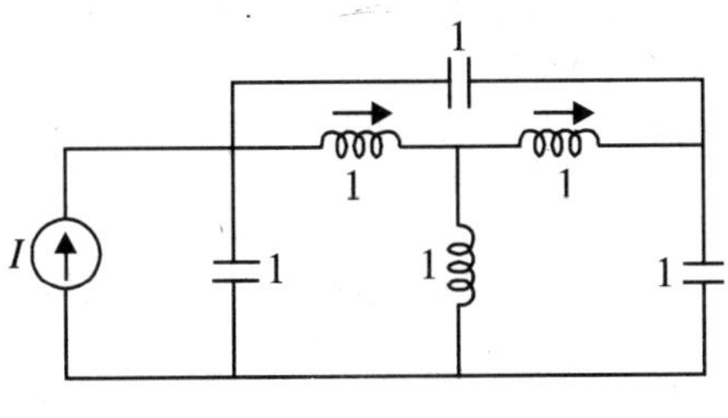

Fig. P. 8.6

8.7 The state equations of the network of Fig. P. 8.7(a).

$$\begin{bmatrix} \dot{V}_1 \\ \dot{V}_2 \\ \dot{I}_3 \end{bmatrix} = \begin{bmatrix} -4/7 & 1/7 & 0 \\ 2/7 & -26/21 & 2/3 \\ 0 & -1/6 & -5/6 \end{bmatrix} \begin{bmatrix} V_1 \\ V_2 \\ I_3 \end{bmatrix} + \begin{bmatrix} 1/7 \\ 2/21 \\ 1/6 \end{bmatrix} V$$

From this determine the state equation of the network of Fig. P. 8.7(*b*).

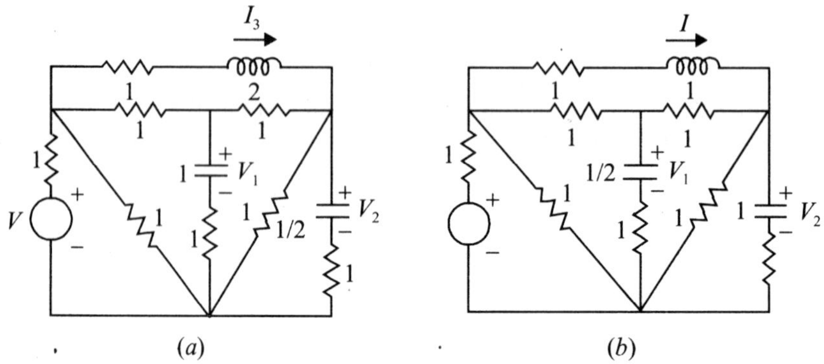

Fig. P. 8.7

8.8 Determine the state equations of the network shown in Fig. P. 8.8 by graph theoretic methods.

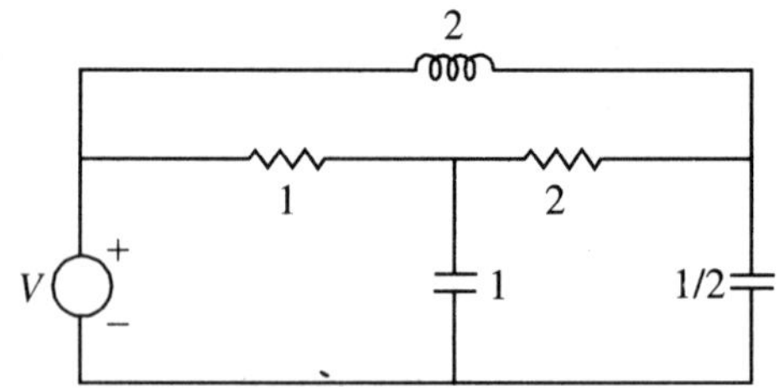

Fig. P. 8.8

NOTES AND REFERENCES

State-variable approach has only found limited application in network analyses and synthesis. An excellent exposition of state variable approach to circuit theory is found in Rohrer's book[1]. Application of state equations to time-varying and non-linear networks are discussed in Balbanian and Bickart's book[2] (Chap. 4). This book also discusses analytical and numerical techniques of evaluating state transition matrix (Chaps. 4 and 5).

Graph theoretic method similar to the one introduced here is applicable to the formulation of state equations of non-electric networks also[3]. State equations and state variable approach finds extensive use in modern control system and the details can be found elsewhere[4,5].

1. R.A. Rohrer, *Circuit Theory—An Introduction to State-Variable Approach*, McGraw-Hill, 1970.
2. N. Balbanian and T. Bickart, *Electrical Network Theory*, John Wiley, 1969.
3. H.E. Koenig, Y. Tokad and H.K. Kesavan, *Analysis of Discrete Physical Systems*, McGraw-Hill, 1967.
4. B.C. Kuo, *Automatic Control Systems*, Prentice-Hall, 1975.
5. L.A. Zadeh and C.A. Desoer, *Linear System Theory*, McGraw-Hill, 1963.

CHAPTER

9 Elements of Network Synthesis

Problems in network theory fall into two categories—analysis and synthesis. Analysis deals with the problem of determining the response when a prescribed input is fed to a given network. The problem of synthesis in contrast deals with the design and fabrication of a network that satisfies a prescribed response specification. The former is comparatively simple and if a solution exists it is generally unique. On the other hand, the synthesis problem is not unique in the sense that we may be able to find more than one network which provides the prescribed response. Indeed, the synthesis problem may not have any solution at all.

Analysis is a prelude to synthesis and we have devoted Chaps. 3–8 to various methods of analysis of a linear passive networks. In this chapter, we consider some aspects of passive network synthesis. We are mainly concerned with the synthesis of driving point immittances and transfer functions.

9.1 POSITIVE REAL FUNCTION

A fundamental theorem in passive network synthesis states that "a rational function of s is the driving point immittance of a passive network if it is positive real."

A rational function $F(s)$ is a *positive real function* (prf) if it satisfies the constraints

$$F(s) \text{ real for real } s$$

$$\text{Re}[F(s)] \geq 0 \text{ for Re } s \geq 0 .$$

If $F(s)$ is a prf, the reciprocal of $F(s)$ (*i.e.* $1/F(s)$) is also a prf. If $F(s)$ is expressed as

$$F(s) = \frac{p(s)}{q(s)} \qquad \ldots(9.1)$$

where $p(s)$ and $q(s)$ are polynomials in s, these polynomials must have real coefficients. This ensures that $F(s)$ is real when $s = \sigma$. p and q being real for $s = \sigma$, complex zeros must appear in conjugate pairs. We first consider methods of testing a function $F(s)$ for positive reality.

We state several theorems, without proof (see notes) and develop the test for *pr* nature of F(s).

Theorem 9.1: A prf $F(s)$ cannot have any poles and zeros in the right half (RH) s-plane. Any zeros of $p(s)$ and $q(s)$ on the $j\omega$-axis must be simple.

Corollary 9.1: The residues of $F(s)$ and $1/F(s)$ at the simple poles on the $j\omega$-axis must be real and positive.

The necessary condition given in Theorem 9.1 can be restated as "the numerator and denominator polynomials of a prf must be *Hurwitz*". A polynomial $f(s)$ is a *strictly Hurwitz polynomial* if all its zeros are in left half (LH) s-plane. It is Hurwitz if its zeros are in LH plane or simple on the $j\omega$-axis.

The analyticity of $F(s)$ and $1/F(s)$ or the Hurwitz nature of $p(s)$ and $q(s)$ can be tested in a number of ways. One of the methods is to construct the Routh-Hurwitz array for the required polynomial.

Let

$$p(s) = a_n s^n + a_{n-1}s^{n-1} + a_{n-2}s^{n-2} + \ldots + a_1 s + a_0$$

For $p(s)$ to be Hurwitz all the coefficients a_i, $i = 0, n$, must be of the same sign. No coefficient should be missing for a strictly Hurwitz polynomial. For a polynomial which has all its zeros on the $j\omega$-axis,' alternate coefficients must be zero, *i.e.* $p(s)$ is either even or odd polynomial. The Routh-Hurwitz array is constructed as follows:

$$\begin{array}{lllll}
s^n & a_n & a_{n-2} & a_{n-4} & \cdots \\
s^{n-1} & a_{n-1} & a_{n-3} & a_{n-5} & \cdots \\
s^{n-2} & b_n & b_{n-1} & b_{n-2} & \cdots \\
s^{n-3} & c_n & c_{n-1} & & \\
\cdot & \cdot & & & \\
\cdot & \cdot & & & \\
\cdot & \cdot & & & \\
s^1 & & & & \\
s^0 & & & &
\end{array}$$

where

$$b_n = \frac{-\begin{vmatrix} a_n & a_{n-2} \\ a_{n-1} & a_{n-3} \end{vmatrix}}{a_{n-1}}, \quad b_{n-1} = \frac{-\begin{vmatrix} a_n & a_{n-4} \\ a_{n-1} & a_{n-5} \end{vmatrix}}{a_{n-1}}$$

$$b_{n-2} = \frac{-\begin{vmatrix} a_n & a_{n-6} \\ a_{n-1} & a_{n-7} \end{vmatrix}}{a_{n-1}}, \ldots$$

$$c_n = \frac{-\begin{vmatrix} a_{n-1} & a_{n-2} \\ b_n & b_{n-1} \end{vmatrix}}{b_n}, \; c_{n-1} = \frac{-\begin{vmatrix} a_{n-1} & a_{n-5} \\ b_n & b_{n-2} \end{vmatrix}}{b_n}$$

and so on.

This array is a triangular one and for a strictly Hurwitz polynomial there should not be any change of sign in the coefficients of the first column, *i.e.* a_n, a_{n-1}, b_n, c_n, ... A vanishing row in the array indicates that the polynomial cannot be strictly Hurwitz. In this case, the vanishing row is replaced by an auxiliary polynomial. Let the *k*th row, the row corresponding to s^k, be the row prior to the vanishing row. Form the polynomial

$$p_k(s) = \alpha_k s^k + \alpha_{k-2} s^{k-2} + ... + \alpha_0$$

where α_i, are the coefficients of the *k*th row. The polynomial $p_k(s)$ will always be an even polynomial. A vanishing row shows the presence of equal and opposite roots and the order of the polynomial $p_k(s)$ indicates the number of equal but opposite roots. An auxiliary polynomial is defined as

$$p_k' = \frac{dp_k(s)}{ds} = k\alpha_k s^{k-1} + (k-2)\alpha_{k-2} s^{k-3} + ...$$

The coefficients of $(k + 1)$th row are replaced by the coefficients of the auxiliary polynomial and the array is now continued. If there is a change of sign in the first column, the polynomial $p(s)$ is not Hurwitz. If there are no changes of sign in the first column, the polynomial $p_k(s)$ is tested for its zeros and if the zeros are simple on the $j\omega$-axis then $p(s)$ is Hurwitz.

Example 9.1. Let $p(s) = s^4 + 16s^3 + 86s^2 + 176s + 105$

s^4	1	86	105
s^3	16	176	
s^2	75	105	
s^1	153.6		
s^0	105		

Hence, $p(s)$ is strictly Hurwitz. The actual factorised form of $p(s)$ is

$$p(s) = (s + 1)\,(s + 2)\,(s + 5)\,(s + 7)$$

(The factors cannot be extracted from Routh's array. Polynomial factorisation methods are necessary for extracting the roots.)

Example 9.2. $p(s) = s^5 + 12s^4 + 45s^3 + 60s^2 + 44s + 48$

s^5	1	45	44	
s^4	12	60	48	
s^3	40	40		

....

s^2	48	48	$p_k(s) = 48s^2 + 48$
s^1	0	(vanishing row)	$p_k(s) = 96s$

....

s^2	48	48	$p_k(s) = 48(s^2 + 1)$
s^1	96		The roots are $s = \pm j1$
s^0	48		

Hence, the polynomial is Hurwitz. The factored form of $p(s)$ is

$$p(s) = (s + 2)\ (s + 4)\ (s + 6)\ (s^2 + 1)$$

A second method of testing for the Hurwitz polynomial is by *continued fraction expansion.* Let $p(s) = m(s) + n(s)$ where m and n are the even and odd parts of $p(s)$. From a rational function $p_1(s)$ given by

$$p_1(s) = \frac{m(s)}{n(s)} \qquad \ldots(9.2)$$

The rational function of (9.2) is expressed in continued fractions as

$$p_1(s) = \beta_1 s + \cfrac{1}{\beta_2 s + \cfrac{1}{\beta_3 s + \cfrac{1}{\beta_4 s + \ldots\ldots\ldots}}}$$

where β_1 may be zero. A necessary and sufficient condition for the Hurwitz nature of $p(s)$ is that the coefficients β_i be real and positive.

Example 9.3. Let $p(s) = s^4 + 16s^3 + 86s^2 + 176s + 105$

Then $m(s) = s^4 + 86s^2 + 105$

and $n(s) = 16s^3 + 176s$

$$p_1(s) = \frac{m}{n} = \frac{s^5 + 86s^2 + 105}{16s^3 + 176s}$$

$$= \frac{1}{16}s + \cfrac{1}{\frac{16}{75}s + \cfrac{1}{75s/153.6 + 1/1.46s}}$$

Hence, $p(s)$ is Hurwitz.

Sometimes the continued fraction expansion may end prematurely. For example consider

$$p(s) = s^5 + 12s^4 + 45s^3 + 60s^2 + 44s + 48$$

Here $m(s) = 12s^4 + 60s^2 + 48$

and $n(s) = s^5 + 45s^3 + 44s$

Then $$p_1(s)=\frac{m(s)}{n(s)}=\frac{12s^4+60s^2+48}{s^5+45s^3+44s}$$

$$=\cfrac{1}{\cfrac{s}{12}+\cfrac{1}{\cfrac{3s}{10}+\cfrac{1}{5s(s^2+1)/6(s^2+1)}}}$$

Because of the cancellation of factor $(s^2 + 1)$ the expansion ends prematurely. A premature termination indicates common factors between m and n. These common factors form an even polynomial and has either roots on the $j\omega$-axis or roots in quadrantal symmetry (equal and opposite real roots are ruled out as this would result in a negative coefficient in the original polynomial). In this particular case, the common factor is due to a pair of $j\omega$-axis roots at $\pm\, j1$. Hence, the polynomial is Hurwitz. In this example

$$m(s) = 12(s^2 + 1)(s^2 + 4)$$

$$z(s) = s(s^2 + 1)(s^2 + 44)$$

and $$p(s) = (s^2 + 1)(s^2 + 12s^2 + 44s + 48)$$

$$= (s^2 + 1)(s + 2)(s + 4)(s + 6)$$

The prf $F(s)$, in addition to having $p(s)$ and $q(s)$ as Hurwitz polynomial must satisfy the Cor. 9.1.

The conditions of the Theorem and Corollary 9.1 are necessary conditions for the positive realness of $F(s)$. The conditions of the theorem imply that $F(s)$ and its reciprocal are analytic in RH s-plane. With this we can restate the condition for a prf in the form of a theorem.

Theorem 9.2: A rational function $F(s)$ with real coefficients is a prf if, $F(s)$ is analytic in RH s-plane with at most simple poles on the $j\omega$-axis.

The residues at any poles on the $j\omega$-axis are real and positive.

$$\text{Re}[F(j\omega)] \geq 0 \text{ for all } \omega.$$

The last condition can be indirectly tested. Let $F(s)=\dfrac{m_1+n_1}{m_2+n_2}$ where m_1, m_2, n_1 and n_2 are the even and odd parts of the numerator and denominator, respectively. It can be easily shown that on the $j\omega$-axis, the Re $F(j\omega)$ is the even part of $F(j\omega)$. The even part of $F(j\omega)$ is

$$\text{Ev.}F(j\omega)=\left.\frac{m_1m_2-n_1n_2}{m_2^2-n_2^2}\right|_{s=j\omega}$$

On the $j\omega$-axis both m_2^2 and $-n_2^2$ being positive, the Re $F(j\omega) \geq 0$ for all ω if, and only if,

$$A(\omega^2) = (m_1 m_2 - n_1 n_2)|_{s=j\omega} \geq 0 \text{ for all } \omega \qquad \ldots(9.3)$$

If all the coefficients of $A(x)$, where $x = \omega^2$, are positive then $A(x)$ is positive for all values of x between 0 and ∞. When all the coefficients of $A(x)$ are not positive we can resort to *Sturm's test.* From $A(x)$ a set of functions $(A_0, A_1, A_2, \ldots A_n)$ are formed so that $A_o = A(x)$, $A_1(s) = dA(x)/dx$. The subsequent functions are formed as follows. The ratio $A_{t-2}(x)$ and $A_{t-1}(x)$ is expressed as

$$\frac{A_{i-2}(x)}{A_{t-1}(x)} = (\alpha_1 x + \alpha_0) - \frac{A_i(x)}{A_{i-1}(x)} \qquad \ldots(9.4)$$

where α_1 and α_0 are constants. This procedure is continued till A_n is a constant. Then Sturm's theorem states that:

Theorem 9.3: The number of zeros of $A(x)$ in the interval $a \le x \le b$ is equal to $S_b - S_a$, where S_a and S_b are the number of sign changes in the set $(A_0, A_1, A_2 \ldots A_n)$ evaluated at $x = a$ and $x = b$, respectively.

By selecting $a = 0$ and $b = \infty$, we can find the number of zeros of $A(x)$ from 0 to ∞. Every time $A(x)$ goes through a zero, the sign of $A(x)$ changes. If there are no zeros of $A(x)$ in the interval $0 \le x \le \infty$ then condition (9.3) is satisfied.

Example 9.4. Let $F(s) = \dfrac{s^3 + 9s^2 + 23s + 15}{s^2 + 6s + 8}$

Then $\quad m_1 = 9s^2 + 15, \qquad n_1 = s^3 + 23s$

$\quad m_2 = s^2 + 8 \qquad n_2 = 6s$

Hence $\quad A(\omega^2) = m_1 m_2 - n_1 n_2$

$\quad = 3\omega^4 + 51\omega^2 + 120.$

Obviously, $A(\omega^2) \ge 0$ for all ω. Since $F(s)$ has all its poles and zeros in the LH and $A(\omega^2) \ge 0$ for all ω, $F(s)$ is a prf.

Example 9.5.

Let $\quad F(s) = \dfrac{2s^4 + 7s^3 + 11s^2 + 12s + 4}{s^4 + 5s^3 + 9s^2 + 11s + 6}$

Then $\quad m_1 = 2s^4 + 11s^2 + 4, \; n_1 = 7s^2 + 12s$

$\quad m_2 = s^4 + 9s^2 + 6, \; n_2 = 5s^3 + 11s$

and $\quad A(\omega^2) = 2\omega^8 + 6\omega^6 - 22\omega^4 + 30\omega^2 + 24$

Now $\quad A_0(x) = 2x^4 + 6x^3 - 22x^2 + 30x + 24$

$$A_1(x) = \frac{dA_0(x)}{dx} = 8x^3 + 18x^2 - 44x + 30$$

$$\frac{A_0(x)}{A_1(x)} = \frac{2x^4 + 6x^3 - 22x^2 + 30x + 24}{8x^3 + 18x^2 - 44x + 30}$$

$$= \left(\frac{1}{4}x + \frac{3}{16}\right) - \frac{115x^2 - 246x - 147}{8A_1(x)}$$

i.e., $A_2(x) = 115x^2 - 246x - 147$

$$\frac{A_1(x)}{A_2(x)} = \frac{8x^3 + 18x^2 - 44x + 30}{115x^2 - 246x - 147}$$

$$= \left(\frac{8}{115}x + 0.305\right) - 41.34\frac{-x - 1.81}{A_2}$$

i.e., $A_3(x) = -x - 1.81$

$$\frac{A_2(x)}{A_3(x)} = \frac{115x^2 - 246x - 147}{-x - 1.81} = (-115x + 454.2) - \frac{675}{A_3}$$

i.e., $A_4(x) = -675.$

Sturm's functions are

$$A_0 = 2x^4 + 6x^3 - 22x^2 + 30x + 24$$
$$A_1 = 8x^3 + 18x^2 - 44x + 30$$
$$A_2 = 115x^2 - 246x - 147$$
$$A_3 = -x - 1.81$$

and $A_4 = -675$

	A_0	A_1	A_2	A_3	A_4	*No. of changes of sign*
$x = 0$	+	+	–	–	–	$S_a = 1$
$x = \infty$	+	+	+	–	–	$S_b = 1$

$$S_a - S_b = 0$$

Hence, Re $F(j\omega) \geq 0$ for all ω.

Let $F(s) - \frac{p(s)}{q(s)}$. Then the continued fraction expansion of $p_1(s)$ and $q_1(s)$ are

$$p_1(s) = \frac{m_1}{n_1} = 0.29s + \cfrac{1}{0.93s + \cfrac{1}{0.91s + \cfrac{1}{2.08s}}}$$

and

$$q_1(s) = \frac{m_2}{n_2} = 0.2s + \cfrac{1}{0.74s + \cfrac{1}{1.03s + \cfrac{1}{1.1s}}}$$

Hence, p and q are strictly Hurwitz and $F(s)$ is a positive real function.

In the preceeding discussions we have presented methods of testing a prf. As already indicated, conditions of positive realness are necessary and sufficient for the realisation of a rational function as the driving point immittance function of a RLCM one-port network. The proofs of the theorems cited here depend on the complex variable theory (see notes). However, a study of energy functions gives a better insight into the *pr* concept.

Let the mesh or the loop equations of a network containing RLCM elements be

$$\mathbf{ZI} = \mathbf{E} \qquad \text{...(9.5)}$$

where
$$Z = \left[R_{ij}\right] + s\left[L_{ij}\right] + \frac{1}{s}\left[S_{ij}\right]$$

and where $[R_{ij}]$, $[L_{ij}]$ and $[S_{ij}]$ are the mesh (loop) resistance, inductance and ellastance matrices respectively. Now, premultiplying eqn. 9.5 by $\mathbf{I}^{t*}$ where * indicates conjugation, we have

$$\mathbf{I}^{t*}\,\mathbf{E} = \mathbf{I}^{t*}\,\mathbf{ZI}$$

which can be written as

$$I^{t*}E = \sum_{i=1}^{l}\sum_{j=1}^{l} R_{ij}\, I_i^*\, I_j + s\sum_{i=1}^{l}\sum_{j=1}^{l} L_{ij}\, I_i^*\, I_j + \frac{1}{s}\sum_{i=1}^{l}\sum_{j=1}^{l} S_{ij}\, I_i^*\, I_j \qquad \text{...(9.6)}$$

where l is the number of loops (meshes). Eqn. 9.6 can be further written as

$$\mathbf{I}^{t*}\mathbf{E} = F_0 + sT_0 + \frac{1}{s}V_0 \qquad \text{...(9.7)}$$

F_0, T_0 and V_0 are quadratic forms. They are related to the dissipated power, magnetic stored energy and electrical stored energy, respectively.

If we consider a one port-network, then the only non-zero source variable in **E** and **I** are E_1 and I_1. In this case

$$E_1 = Z_{11}\, I_1$$

and we have from equation (9.7)

$$F_0 + sT_0 + \frac{1}{s}V_0 = I_1^* Z_{11} I_1 = Z_{11}\left|I_1\right|^2$$

Finally,

$$Z_{11} = \frac{1}{\left|I_1\right|^2}\left(F_0 + sT_0 + \frac{1}{s}V_0\right)$$

without less of generality we set $|\, I_1 \,| = 1$ which yields

$$Z_{11} = F_0 + sT_0 + \frac{1}{s}V_0 \qquad \text{...(9.8)}$$

In eqn. 9.8, the quadratic forms F_0, T_0 and V_0 are called the energy functions though they are only related to the actual energy functions. From the positive semidefinite characteristic of $[R_{ij}]$, $[L_{ij}]$ and $[S_{ij}]$ (see section 4.1) we can conclude that F_0, T_0 and V_0 are real and non-negative for any value of s. Obviously, for real values of s, Z_{11} is real and for $s = \sigma + j\omega$ it can be seen that

$$Z_{11}(s) = \left(F_0 + \sigma T_0 + \frac{\sigma V_0}{\sigma^2 + \omega^2}\right) + j\omega\left(T_0 - \frac{V_0}{\sigma^2 + \omega^2}\right) \qquad \ldots(9.9)$$

From eqn. (9.9), we conclude that

$$\text{Re}\,[Z_{11}(s)] \geq 0 \text{ for } \text{Re}(s) \geq 0$$

We can develop the concept of energy functions starting from node or cut-set equations also. Hence, we conclude:

"The driving point immittance function of a passive RLCM network is positive real."

The sufficiency of the condition will be demonstrated in subsequent sections of this chapter by realising the functions as networks. Table 9.1 summarises several properties of a prf. $F(s)$.

Table 9.1

$$F(s) = \frac{p(s)}{q(s)}$$

1. Coefficients of $p(s)$ and $q(s)$ are real.
2. $p(s)$ and $q(s)$ are Hurwitz polynomials (*i.e.* zeros of p and q are restricted to LHP and/or simple on the $j\omega$-axis).
3. Residues at the $j\omega$-axis poles are real and positive.
4. The degree of p and q can differ at most by one.
5. The degree of the lowest powers of p and q differ utmost by one.
6. $\dfrac{1}{F(s)}$ is also a prf.

In the following sections, we consider the properties and the realisations of two element kind immittances only.

9.2 REACTANCE FUNCTIONS

If a network contains only L's and C's, the network is called a reactance or lossless network. For a reactance network, the energy function F_0 is identically zero. Hence from eqn. 9.8 we have

$$Z_{11} = sT_0 + \frac{V_0}{s} \qquad \ldots(9.9a)$$

Thus, the finite (non-zero) zeros of Z_{11} from eqn. 9.9a are on the $j\omega$-axis, and are given by

$$z_i = \pm j\sqrt{\frac{V_0}{T_0}}$$

If the energy function formation is done on admittance basis, it can be shown that the zeros of Y_{11} (*i.e.* the poles of Z_{11}) are also on the $j\omega$-axis. Hence, we conclude that

"The poles and zeros of the driving point immittance function of an LC-network lie on the $j\omega$-axis".

The real part of the immittance function $F(s)$ on the $j\omega$-axis is zero. The poles and zeros being on the imaginary axis the polynomials $p(s)$ and $q(s)$, where $F(s) = \dfrac{p(s)}{q(s)}$, must be either even or odd. Since the behaviour of the immittance function of an LC network must approach that of an inductor or a capacitor as $s \to 0$ or $s \to \infty$, we conclude that $F(s)$ must have a zero or a pole at the origin and at $s = \infty$. Further it can be shown that the poles and the zeros must alternate on the imaginary axis to satisfy the condition that the residues at the $j\omega$-axis poles be positive real. This forces the slope of a reactance function to be strictly positive for all frequencies. Figure 9.1 shows the plot of a reactance-function as a function of ω.

The preceding discussions are summarized in the following theorem.

Theorem 9.4: Necessary and sufficient conditions for a real rational function $F(s)$ to be the driving point immittance of an *LC*-network are:

(*a*) poles and zeros are simple and interlacing on the $j\omega$-axis.

(*b*) at the origin and at infinity, $F(s)$ must have a pole or a zero.

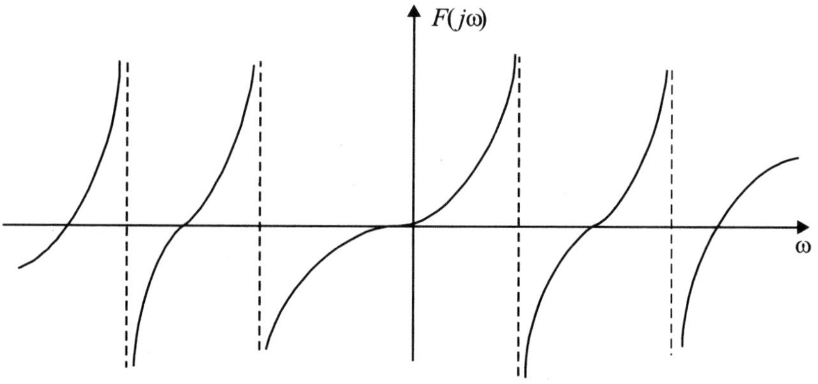

Fig. 9.1

The properties of the LC driving point function are (many of these properties can be derived from the energy function concept):

(*i*) poles and zeros are simple on the imaginary axis.

(*ii*) poles and zeros are interlacing.

(*iii*) at the origin there is a pole or a zero.

(*iv*) at infinity there is a pole or a zero.

(*v*) Re $F(j\omega) = 0$ for all ω.

(*vi*) the slope of $F(j\omega)$ is strictly positive.

(*vii*) $F(s)$ is a ratio of even to odd polynomial or vice-versa.

(*viii*) the residues at the imaginary axis poles are positive and real.

There are a number of methods of realizing a reactance network. We only consider the four basic forms—Foster I, Foster II, Cauer I and Cauer II. The Foster forms are obtained by a partial fraction expansion of $F(s)$ and the Cauer forms by a continued fraction expansion of $F(s)$.

If we combine the terms corresponding to a pair of conjugate poles, the partial fraction expansion of $F(s)$ can be written as

$$F(s) = \frac{p(s)}{q(s)} = \frac{k_0}{s} + \sum_{i=1} \frac{2k_i s}{s^2 + \omega_i^2} + k_\infty s \qquad \ldots(9.10)$$

where k_0, k_i and k_∞ are the residues of $F(s)$ at the poles at the origin, at $j\omega_i$ and at infinity respectively. These residues are given by

$$k_0 = s\, F(s)|_{s=0}$$

$$k_i = (s + j\omega_i) F(s)\Big|_{s=-j\omega_i} = \frac{(s^2 + \omega_i^2) F(s)}{2s}\Bigg|_{s^2=-\omega_i^2} \qquad \ldots(9.11)$$

$$k_\infty = \frac{F(s)}{s}\Bigg|_{t \to \infty}$$

If there is a zero at the origin (at infinity) then $k_0 = 0$ ($k_\infty = 0$).

If $F(s)$ in eqn. 9.10 is an impedance function then it is the series connection of the elemental impedances (Fig. 9.2) and the realized network is shown in Fig. 9.3.

Impedance Function	Network
$\frac{k_0}{s}$	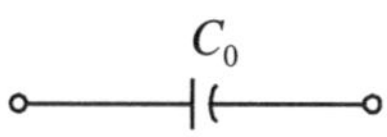 $C_0 = \frac{1}{k_0}$
$\frac{2k_i s}{s^2 + \omega_i^2}$	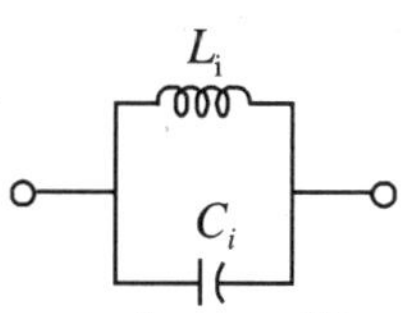 $C_i = \frac{1}{2k_i}$, $L_1 = \frac{2k_i}{\omega_i^2}$
$k_\infty s$	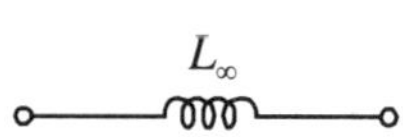 $L_\infty = k_\infty$

Fig. 9.2

If $F(s)$ has a zero at the origin then C_0 will be missing and if it has a zero at infinity L_∞ will be missing. The structure shown in Fig. 9.3 is called *Foster form I.*

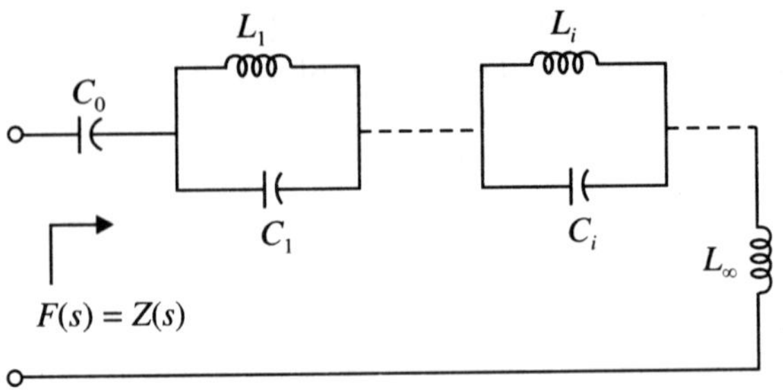

Fig. 9.3

If $F(s)$ in eqn. 9.10 is an admittance function, then it is the parallel combination of the elemental admittances (Fig. 9.4) and the realization is shown in Fig. 9.5. If $F(s)$ has a zero at the origin, L_0 will be missing, and if it has a zero at infinity, C_∞ be missing. The network of Fig. 9.5 is called *Foster form II.*

Admittance Function	Network
$\frac{k_0}{s}$	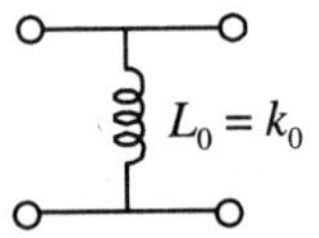
$\frac{2k_i\, s}{s^2 + \omega_i^2}$	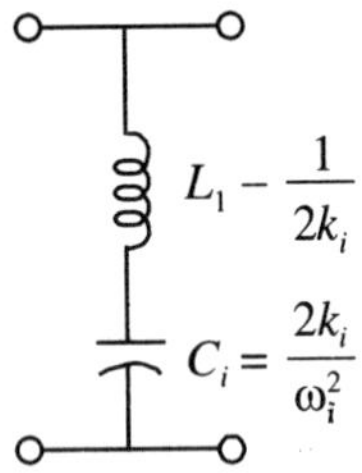
$k_\infty\, s$	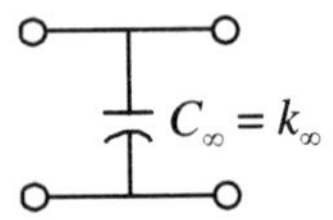

Fig. 9.4

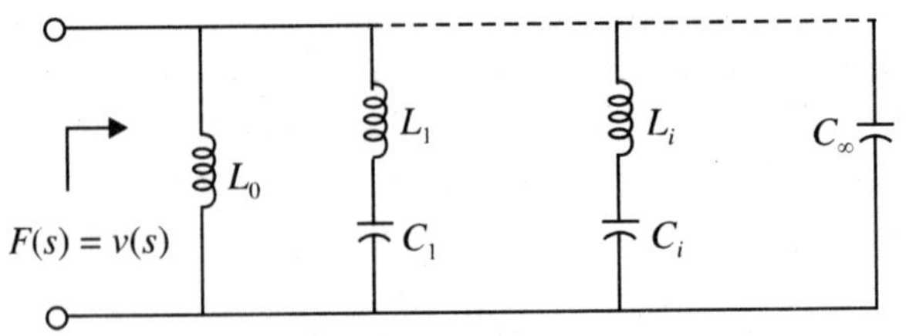

Fig. 9.5

The continued fraction expansion of $F(s)$ is given by

$$F(s) = \alpha_1 s + \cfrac{1}{\alpha_2 s + \cfrac{1}{\alpha_3 s + \cfrac{1}{\dots + \cfrac{1}{\alpha_n s}}}} \quad \dots(9.12)$$

Here $F(s)$ is assumed to have a pole at infinity. The removal of this pole by the factor $\alpha_1 s$ leaves a function with zero at infinity and hence its reciprocal has a pole at infinity which can be removed and the procedure repeated. If $F(s)$ is an impedance function, the realization is shown in Fig. 9.6 and is called *Cauer form I.* If the impedance has no pole at infinity (then the admittance has a pole at infinity) the inductance α_1 will be missing. If the impedance has a zero at the origin the last shunt capacitance will be missing.

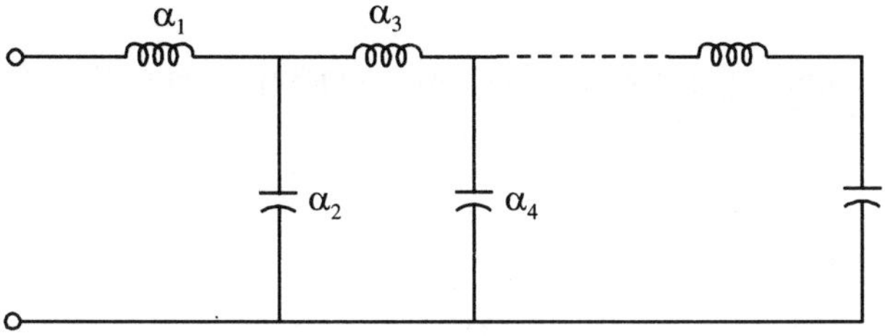

Fig. 9.6

The *Cauer form II* is obtained by removing the pole of $F(s)$ at the origin. In this case, the continued fraction expansion is given by

$$F(s) = \frac{1}{\beta_1 s} + \cfrac{1}{\cfrac{1}{\beta_2 s} + \cfrac{1}{\cfrac{1}{\beta_3 s} + \dots}} \quad \dots(9.13)$$

If $F(s)$ is an impedance, the realization is shown in Fig. 9.7. If the impedance has no pole at infinity the last inductance will be missing.

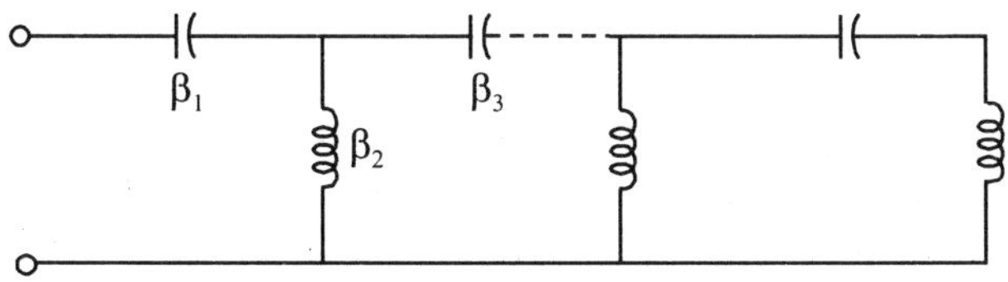

Fig. 9.7

We have shown that a given reactance function $F(s)$ can be realized in four different forms. All these forms have the same number of elements and the number is equal to the number of poles (zeros) of $F(s)$ including any at infinity. We cannot realize an LC network with lesser number of element and as such these forms are called *canonic forms.*

Example 9.6.

Let
$$F(s)=\frac{s^6+5s^4+6.75s^2+2.25}{s^5+3s^3+2s}$$

$F(s)$ can be written as

$$F(s)=\frac{(s^2+0.5)(s^2+1.5)(s^2+3)}{s(s^2+1)(s^2+2)}$$

Poles: $0, \pm j1, \pm\sqrt{2}, \infty$

Zeros: $\pm\sqrt{0.5}, \pm j\sqrt{1.5}, \pm j\sqrt{3}$

$F(s)$ satisfies the conditions of Theorem 9.4 and hence is a reactance function. Let $F(s)$ be an impedance function.

Foster I

The partial fraction expansion of $F(s)$ corresponding to eqn. 9.10 is

$$F(s)=\frac{k_0}{s}+\frac{2k_1 s}{s^2+1}+\frac{2k_2 s}{s^2+2}+k_\infty s$$

From eqn. 9.11

$$k_0 = sF(s)\big|_{s=0} = \frac{(0.5)(1.5)(3)}{(1)(2)} = 1.125$$

$$k_1 = (s^2+1)\frac{F(s)}{2s}\bigg|_{s^2=-1} = \frac{(-1+0.5)(-1+1.5)(-1+3)}{2(-1)(-1+2)} = 0.25$$

$$k_2 = (s^2+2)\frac{F(s)}{2s}\bigg|_{s^2=-2} = \frac{(-2+0.5)(-2+1.5)(-2+3)}{2(-2)(-2+1)} = 0.1875$$

$$k_\infty = \frac{F(s)}{s}\bigg|_{s\to\infty} = 1.$$

The network is given in Fig. 9.8.

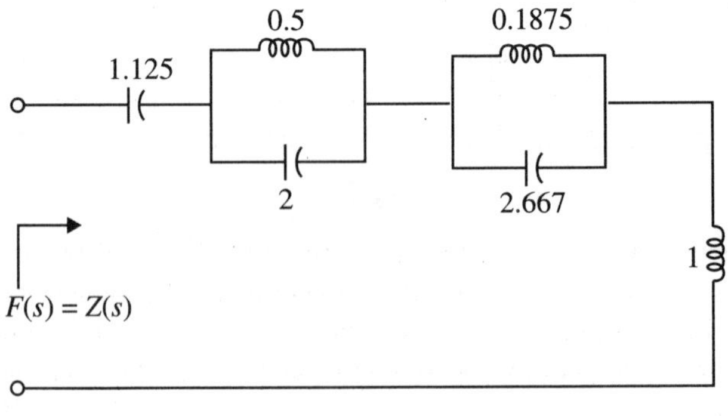

Fig. 9.8

Foster II

The admittance function corresponding to $F(s)$ is

$$F_1(s) = \frac{s(s^2+1)(s^2+2)}{(s^2+0.5)(s^2+1.5)(s^2+3)}$$

The partial fraction expansion is given by

$$F_1(s) = \frac{k_1}{s^2+0.5} + \frac{k_2}{s^2+1.5} + \frac{k_3}{s^2+3}$$

where

$$k_1 = (s^2+0.5)\frac{F_1(s)}{2s}\bigg|_{s^2=-0.5} = \frac{(-0.5+1)(-0.5+2)}{2(-0.5+1.5)(-0.5+3)} = 0.15$$

$$k_2 = (s^2+1.5)\frac{F_1(s)}{2s}\bigg|_{s^2=-1.5} = \frac{(-1.5+1)(-1.5+2)}{2(-1.5+0.5)(-1.5+3)} = 0.083$$

$$k_3 = (s^2+3)\frac{F_1(s)}{2s}\bigg|_{s^2=-3} = \frac{(-3+1)(-3+2)}{2(-3+0.5)(-3+1.5)} = 0.267$$

The network is given in Fig. 9.9.

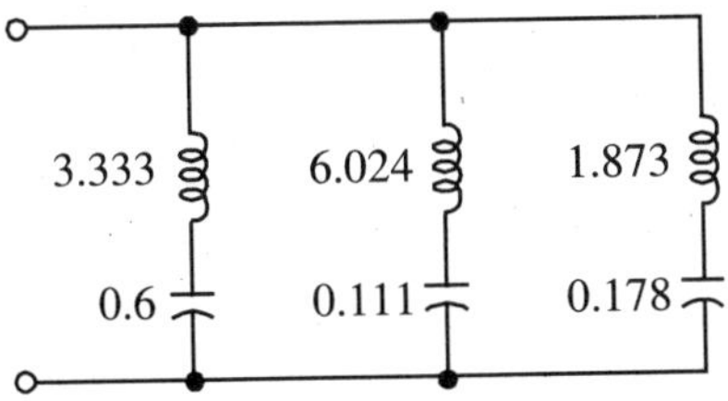

Fig. 9.9

Cauer I

$F(s)$ has a zero at infinity, which can be removed by long hand division

$$F_1(s) = \frac{s^6+5s^4+6.75s^2+2.25}{s^6+3s^3+2s}$$

$$= s + \frac{2s^4+4.75s^2+2.25}{s^6+3s^3+2s}$$

$$= s + F_1(s)$$

$F_1(s)$ has a zero at infinity and hence $1/F_1(s)$ has a pole at infinity which can be removed

$$F_1(s) = \frac{s^5 + 3s^3 + 2s}{2s^4 + 4.75s^2 + 2.25} = 0.5s + \frac{0.625s^3 + 0.875s}{2s^4 + 4.75s^2 + 2.25}$$

$$= 0.5s + F_2(s)$$

$1/F_2(s)$ has a pole at infinity

$$\frac{1}{F_2(s)} = \frac{2s^4 + 4.75s^2 + 2.25}{0.625s^3 + 0.875s} = 3.2 + \frac{1.95s^2 + 2.25}{0.625s^3 + 0.875s}$$

$$= 3.2s + F_3(s)$$

Similarly,

$$\frac{1}{F_3(s)} = \frac{0.625s^3 + 0.875s}{1.95s^2 + 2.25} = 0.321s + \frac{0.153s}{1.95s^2 + 2.25}$$

$$= 0.321s + F_4(s)$$

$$\frac{1}{F_4(s)} = \frac{1.95s^2 + 2.25}{0.153s} = 12.745s + \frac{2.25}{0.153s}$$

$$= 12.745s + F_5(s)$$

$$\frac{1}{F_5(s)} = 0.068s.$$

Hence, the continued fraction expansion for $F(s)$ (eqn. 9.8) is

$$F(s) = s + \cfrac{1}{0.5s + \cfrac{1}{3.2s + \cfrac{1}{0.321s + \cfrac{1}{12.745s + \cfrac{1}{0.068s}}}}}$$

The coefficients of the continued fraction expansion can be obtained in a straightforward manner by *Euclid's algorithm* which employs synthetic division. The algorithm is shown in Table 9.2. The first coefficient is obtained by dividing the numerator by the denominator. This step is indicated by 1. The second coefficient is obtained by dividing the denominator by the remainder of the numerator obtained in step 1. This step is indicated by 2. This procedure of dividing the numerator by the denominator and then the denominator by the numerator is continued till the remainder becomes 0, and steps 3, 4, 5 and 6 are obtained. The coefficient of the continued fraction expansion are given by the corresponding quotients.

Table 9.2

		Numerator F(s)	*Denominator F(s)*		
1	s	$s^6 + 5s^4 + 6.75s^2 + 2.25$ $s^6 + 3s^4 + 2s^2$	$s^5 + 3s^3 + 2s$ $s^5 + 2.2375s^3 + 1.125s$	$0.5s$	2
3	$3.2s$	$2s^4 + 4.75s^2 + 2.25$ $2s^4 + 2.8s^2$	$0.625s^3 + 0.875s$ $0.625s^3 + 0.722s$	$0.321s$	4
5	$12.75s$	$1.95s^2 + 2.25$ $1.95s^2$	$0.153s$ $0.153s$	$0.068s$	6
		2.25	0		

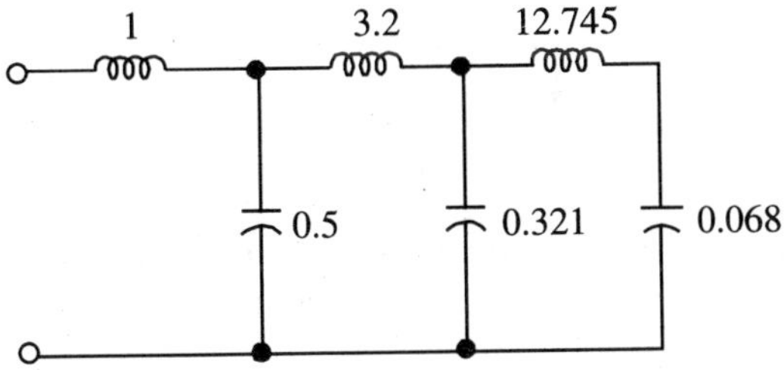

Fig. 9.10

The Cauer I network is given in Fig. 9.10.

Cauer II

In Cauer I the poles at infinity were removed. This is equivalent to dividing the numerator by the denominator to eliminate the highest power in the numerator. In Cauer II, we remove the poles at the origin which is equivalent to dividing the numerator by the denominator to eliminate the lowest power in the numerator. The division is given in Table 9.3.

Table 9.3

		Numerator F(s)	*Denominator F(s)*		
1	$\frac{1}{0.888s}$	$2.25 + 6.75\,s^2 + 5s^4 + s^6$ $2.25 + 3.375s^2 + 1.125s^4$	$2s + 3s^3 + s^5$ $2s + 2.007s^3 + 0.593s^5$	$\frac{1}{1.686s}$	2
3	$\frac{1}{0.294s}$	$3.375s^2 + 3.875s^4 + s^6$ $3.375s^2 + 1.383s^4$	$0.993s^3 + 0.407s^5$ $0.993s^3 + 0.398s^5$	$\frac{1}{2.513s}$	4
5	$\frac{1}{3.6\times10^{-3}s}$	$2.492s^4 + s^6$ $2.492s^4$	$0.009s^6$ $0.009s^5$	$\frac{1}{1.11\times10^{3}s}$	6
		s^6	0		

The continued fraction expansion of $F(s)$ is

$$F(s)=\frac{1}{0.888s}+\cfrac{1}{\cfrac{1}{1.686s}+\cfrac{1}{\cfrac{1}{0.294s}+\cfrac{1}{\cfrac{1}{2.513s}+\cfrac{1}{3.6\times10^{-3}s}+\cfrac{1}{1.11\times10^{3}s}}}}$$

The network is shown in Fig. 9.11.

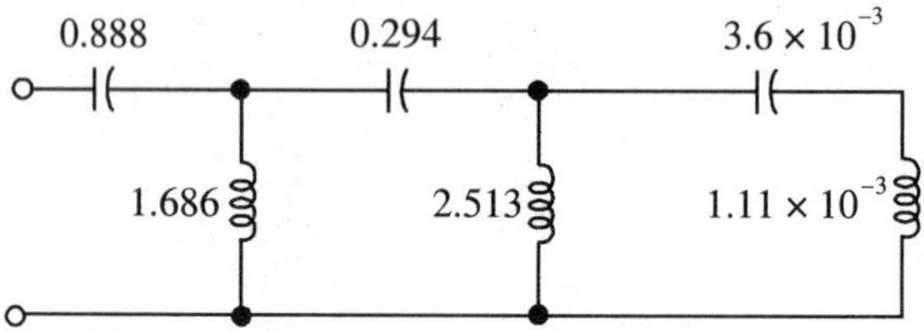

Fig. 9.11

In all the four realizations we have six elements. The number of poles of $F(s)$ is also six, thus demonstrating their canonic nature.

9.3 RC FUNCTIONS

A network with only one type of energy storage element (C or L) cannot produce any ringing and the network can be at best critically damped. As such, an RC network cannot be designed with Q greater than 1/2. Yet, the study of RC network is of fundamental importance as they form the building blocks in active filters. High Q realization can be obtained by embedding an active element in an RC network in feedback configuration (see Chap. 12).

For an RC network, the energy function T_0 is identically zero and hence from eqn. 9.8, we have

$$Z_{11}=F_0+\frac{V_0}{s} \qquad \text{...(9.14)}$$

The non-zero finite zeros of Z_{11} are given by

$$z_i=-\frac{V_0}{F_0}$$

Zeros of Y_{11} also satisfy identical relationship. Hence, we conclude **"poles and zeros of an RC immittance function lie on negative real axis".** Further properties of RC immittances can be derived from eqn. 9.14. It can also be seen that

$$Z_{RC}(s)=\frac{1}{p}Z_{LC}(p)\bigg|_{p^2=s} \qquad \text{...(9.15)}$$

The properties of an RC impedance function can be derived from the properties of an LC impedance function and eqn. 9.15 (prob. 9.5). The properties of an RC impedance function $Z_{RC}(s)$ are:

(*i*) The poles and zeros of $Z_{RC}(s)$ are on the negative real axis of the s-plane and are simple.

(*ii*) Poles and zeros of $Z_{RC}(s)$ are interlacing.

(*iii*) The critical frequency nearest to the origin is a pole. There can be a pole at the origin (this indicates a series capacitor). There cannot be a zero at the origin as this indicates an inductor.

(*iv*) The critical frequency farthest from the origin is a zero. This zero can be at infinity (this would indicate a shunt capacitor). There cannot be a pole at infinity.

(*v*) The slope of $Z_{RC}(\sigma)$ is strictly negative.

(*vi*) If $Z_{RC}(s) = p(s)/q(s)$, the order of $q(s)$ must be one greater than or equal to that of $p(s)$.

(*vii*) The residues of $Z_{RC}(s)$ at its poles are all positive real.

(*viii*) $Z_{RC}(0) > Z_{RC}(\infty)$.

A transformation similar to the one for eqn. 9.11 between the RC admittance and LC admittance is given by

$$Y_{RC(s)} = pY_{LC}(p)\Big|_{p^2=s} \qquad \ldots(9.16)$$

The properties of $Y_{RC}(s)$ can be summarized as:

(*i*) The poles and zeros of $Y_{RC}(s)$ are all on the negative real axis of s-plane and are simple.

(*ii*) Poles and zeros of $Y_{RC}(s)$ are interlacing.

(*iii*) The critical frequency nearest to the origin is zero and this can be at the origin. There cannot be a pole at the origin.

(*iv*) The critical frequency farthest from the origin is a pole and this can be at infinity. There cannot be a zero at infinity.

(*v*) The slope of $Y(\sigma)$ is strictly positive.

(*vi*) If $Y_{RC}(s) = p(s)/q(s)$, the order of $p(s)$ must be one greater than or equal to that of $q(s)$.

(*vii*) The residues of $Y_{RC}(s)$ at its poles are real and negative and those of $Y_{RC}(s)/s$ are real and positive.

(*viii*) $Y_{RC}(\infty) > Y_{RC}(0)$.

Figures 9.12 and 9.13 show the plots of an RC-impedance and RC-admittance respectively, as functions of σ.

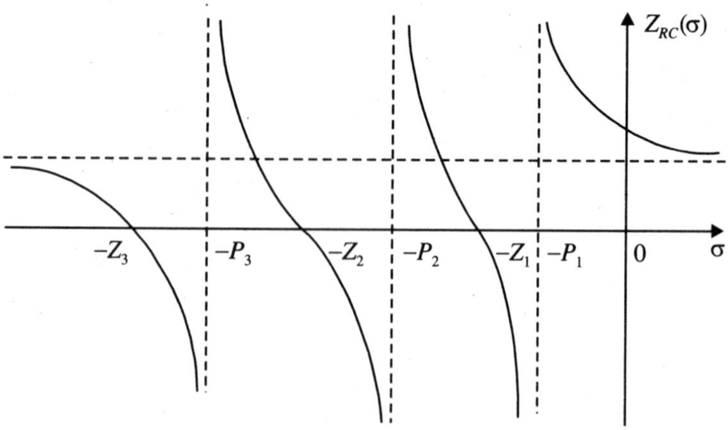

$$Z_{RC}(s) = k\frac{(s+Z_1)(s+Z_2)(s+Z_3)}{(s+P_1)(s+P_2)(s+P_3)}$$

$$0 \le P_1 < Z_1 < P_2 < Z_2 < P_3 < Z_3$$

Fig. 9.12

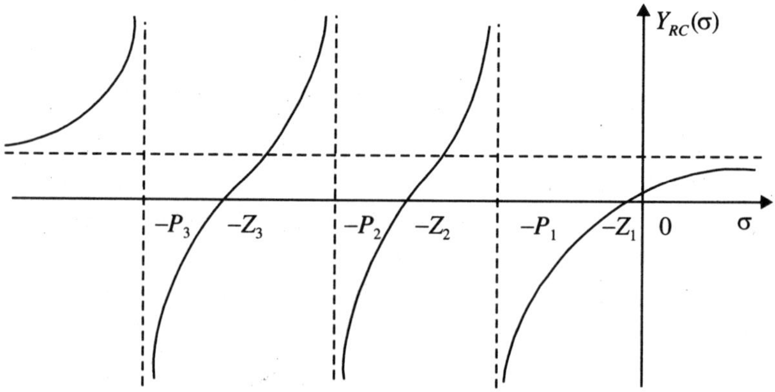

$$Y_{RC}(s) = k\frac{(s+Z_1)(s+Z_2)(s+Z_3)}{(s+P_1)(s+P_2)(s+P_3)}$$

$$0 \le Z_1 < P_1 < Z_2 < P_3 < Z_3 < P_3$$

Fig. 9.13

As in the case of reactance functions, RC functions can be realized in four different ways. The Foster form I is obtained by a partial fraction expansion of $Z_{RC}(s)$. Let $F(s)$ be a function satisfying the properties of an RC impedance. Then $F(s)$ can be expanded as

$$F(s) = \frac{k_0}{s} + \sum_{i=1} \frac{k_i}{s+p_i} + k_\infty \qquad \text{...(9.17)}$$

where k_0, k_i and k_∞ are the residues of the poles at the origin, at $s = -pi$ and at infinity respectively. These constants are given by

$$\left.\begin{aligned} k_0 &= sF(s)\big|_{s=0} \\ k_i &= (s+p_i)F(s)\big|_{s=-pi} \\ \text{and} \qquad k_\infty &= \lim_{s\to\infty} F(s) \end{aligned}\right\} \qquad \ldots(9.18)$$

All the constants are real and positive, If there is no pole and the origin then $k_0 = 0$ and if there is a zero at infinity then $k_\infty = 0$. The impedance function of eqn. 9.17 is realized by a series of connections of the elemental impedance shown in Fig. 9.14. The final network is shown in Fig. 9.15.

Elemental Impedance	Network
$\dfrac{k_0}{s}$	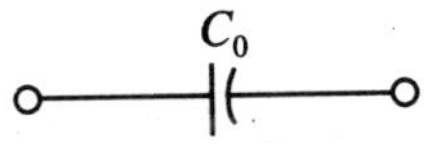 $C_0 = \dfrac{1}{k_0}$
$\dfrac{k_i}{s+p_i}$	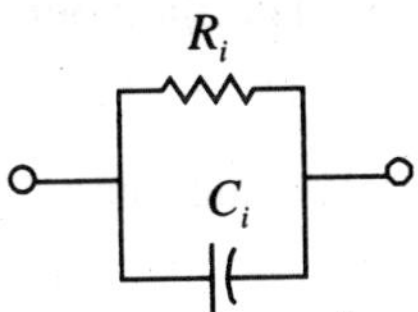 $C_i = \dfrac{1}{k_i}$, $R_i = \dfrac{k_i}{p_i}$
k_∞	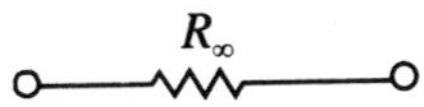 $R_\infty = k_\infty$

Fig. 9.14

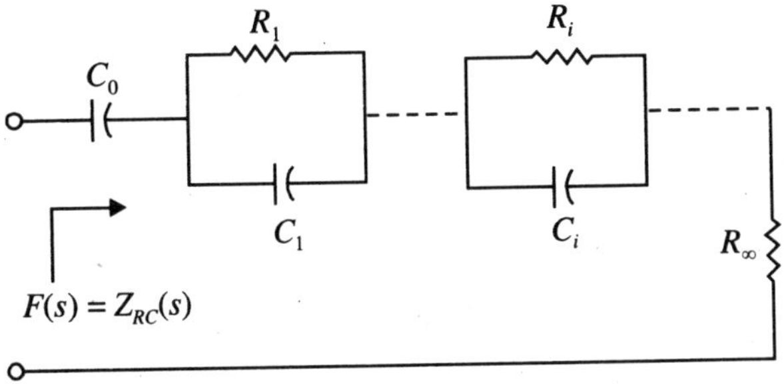

Fig. 9.15

The Foster form II is obtained by expanding the RC admittance in partial fractions. Let $F(s)$ be the specified RC admittance. The partial fraction expansion of $F(s)/s$, which is an RC impedance, is similar to that in eqn. 9.17 and is given by

$$\frac{F(s)}{s} = \frac{\alpha_0}{s} + \sum_{i=1} \frac{\alpha_i}{s + p_i} + \alpha_\infty \qquad \ldots(9.19)$$

where, as before, α_0 and α_i are the residues of $F(s)/s$ at the pole and at the origin, and at $s = -pi$ and and $\alpha_\infty = \lim_{s \to \infty} F(s)/s$. (Here we have to expand $F(s)/s$ instead of $F(s)$ as an RC impedance has real and negative residues at its poles while the RC impedance has real and positive residues at its poles). If $F(s)$ has a zero at the origin, $F(s)/s$ does not have a pole at the origin and α_0 will be 0. If $F(s)$ does not have a zero at infinity, $F(s)/s$ has a zero at infinity and α_∞ will be 0. From eqn. 9.19 we have

$$F(s) = \alpha_0 + \sum_{i=1} \frac{\alpha_i s}{s + p_i} + \alpha_\infty s \qquad \ldots(9.20)$$

The admittances function is realized by a parallel connection of the elemental admittances (Fig. 9.16). The final network is shown in Fig. 9.17.

Elemental Admittance	Network
α_0	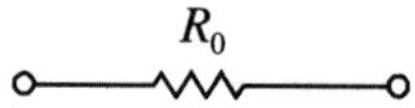 $R_0 = \frac{1}{\alpha_0}$
$\frac{\alpha_i s}{s + p_i}$	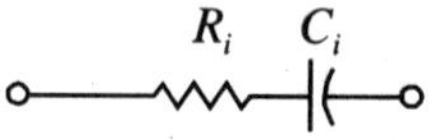 $R_i = \frac{1}{\alpha_i}$, $C_i = \frac{\alpha_i}{p_i}$
$\alpha_\infty s$	 $C_\infty = \alpha_\infty$

Fig. 9.16

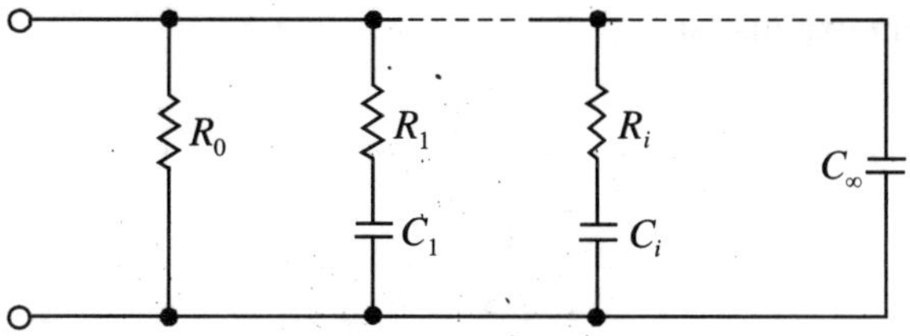

Fig. 9.17

The Cauer form I is obtained by expanding the RC impedance function in continued-fractions about infinity. From the property of RC impedance function the maximum constant that can be removed is $Z_{RC}(\infty)$. The removal of this constant is equivalent to dividing the numerator by the denominator to eliminate the highest power of s in the numerator. If the impedance function has a zero at infinity obviously we cannot remove any constant. In any case, the first step leaves an impedance function with a zero at infinity. The pole at infinity of the reciprocal admittance function is then removed. The cycle repeats by the removal of a constant from the impedance function and the pole at infinity of the admittance function. The final continued fraction expansion is given by

$$F(s) = \beta_1 + \cfrac{1}{\beta_2 s + \cfrac{1}{\beta_3 s + \cfrac{1}{\beta_4 s + \cfrac{1}{\dots + \beta_n s}}}} \qquad \dots(9.21)$$

The network is shown in Fig. 9.18. If $F(s)$ has a zero at infinity then β_1 is zero. The final element may be either a resistor or a capacitor.

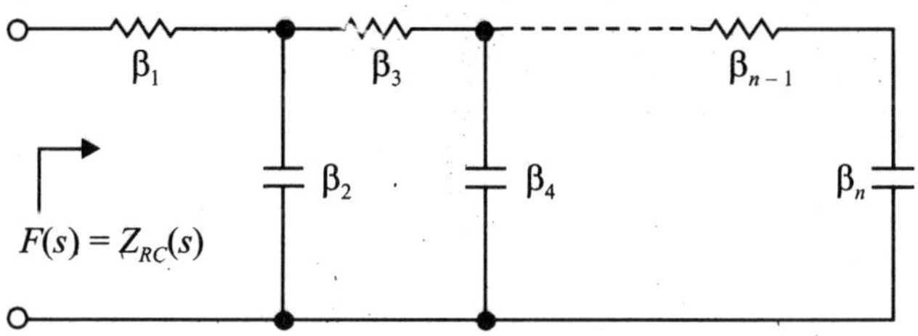

Fig. 9.18

The Cauer form II is obtained by expanding the given function into continued fraction about the origin. If the given impedance function has a pole at the origin then it is removed by dividing the numerator by the denominator to eliminate the constant term in the numerator. The reciprocal of the remainder function being on RC-admittance function has minimum value at $s = 0$ which can be removed by creating zero at the origin. If the original impedance function does not have a pole at the origin, the first cycle is the removal of the constant at $s = 0$ in the corresponding admittance function. The procedure of removal of a pole at $s = 0$ of the impedance function and then the constant at $s = 0$ of the admittance function leads to the continued fraction expansion.

$$F(s) = \frac{1}{\gamma_1 s} + \cfrac{1}{\cfrac{1}{\gamma_2} + \cfrac{1}{\cfrac{1}{\gamma_3 s} + \cfrac{1}{\cfrac{1}{\gamma_4} + \cfrac{1}{\gamma_n}}}} \qquad \ldots(9.22)$$

and the network is shown in Fig. 9.19. If the given function has no pole at the origin then γ_1 is zero. The last element may be either a capacitance or a resistance.

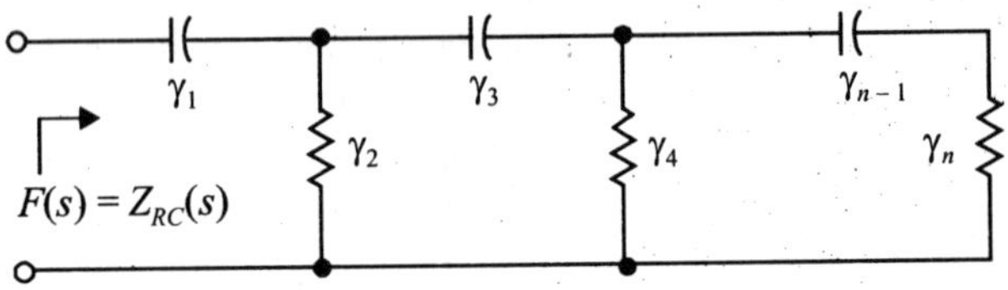

Fig. 9.19

The two Cauer forms can be obtained by the synthetic division algorithm introduced earlier.

In all the four forms of realisation, the number of elements is the same and this is one more than the number of internal critical frequencies (*i.e.* finite non-zero poles and zeros). Such realisations are called canonic forms and no realisation can have fewer than this number of elements.

Example 9.7.

Let $$F(s) = \frac{s^3 + 9s^2 + 23s + 15}{s(s^3 + 12s^2 + 44s + 48)}$$

Poles: 0, –2, –4, –6

Zeros: –1, –3, –5, ∞

$F(s)$ is an RC–impedance (Z_{RC}).

(i) Foster I

Expanding $Z_{RC}(s)$ in the form of eqn. 9.17

$$Z_{RC}(s) = \frac{k_0}{s} + \frac{k_1}{s+2} + \frac{k_2}{s+4} + \frac{k_3}{s+6}$$

$$k_0 = sZ_{RC}\big|_{s=0} = 0.3125$$

$$k_1 = (s+2)Z_{RC}\big|_{s=-2} = \frac{(-2+1)(-2+3)(-2+5)}{(-2)(-2+4)(-2+6)} = 0.1875$$

$$k_2 = (s+4)Z_{RC}\big|_{s=-4} = \frac{(-4+1)(-4+3)(-4+5)}{(-4)(-4+2)(-4+6)} = 0.1875$$

$$k_3 = (s+6)Z_{RC}\Big|_{s=-6} = \frac{(-6+1)(-6+3)(-6+5)}{(-6)(-6+2)(-6+4)} = 0.3125$$

(Here $k_\infty = 0$ as Z_{RC} has a zero at infinity)

The network is shown in Fig. 9.20.

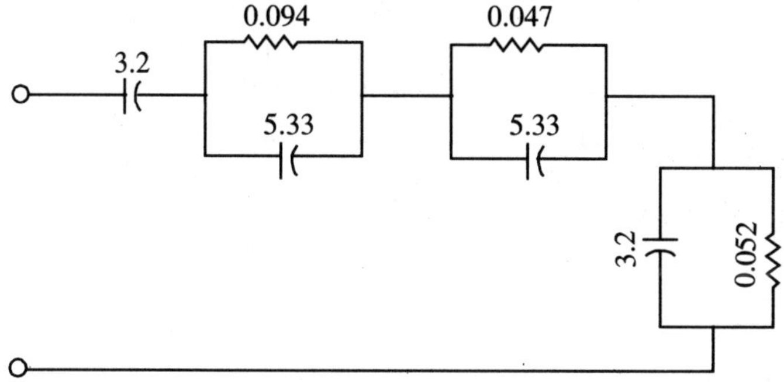

Fig. 9.20

(*ii*) *Foster II*

$$Y_{RC}(s) = \frac{1}{Z_{RC}(s)} = \frac{s(s^3+12s^2+44s+48)}{s^3+9s^2+23s+15}$$

$$\frac{Y_{RC}(s)}{s} = \frac{s^3+12s^2+44s+48}{s^3+9s^2+23s+15}$$

The partial fraction expansion corresponding to eqn. 9.19 is given by

$$\frac{Y_{RC}(s)}{s} = \frac{\alpha_1}{s+1} + \frac{\alpha_2}{s+3} + \frac{\alpha_3}{s+5} + \alpha_\infty$$

$$\alpha_1 = (s+1)\frac{Y_{RC}(s)}{s}\Bigg|_{s=-1} = \frac{(-1+2)(-1+4)(-1+6)}{(-1+3)(-1+5)} = 2.5$$

$$\alpha_2 = (s+3)\frac{Y_{RC}(s)}{s}\Bigg|_{s=-3} = \frac{(-3+2)(-3+4)(-3+6)}{(-3+1)(-3+5)} = 0.75$$

$$\alpha_3 = (s+5)\frac{Y_{RC}(s)}{s}\Bigg|_{s=-5} = \frac{(-5+2)(-5+4)(-5+6)}{(-5+1)(-5+3)} = 0.375$$

$$\alpha_\infty = \lim_{s\to\infty}\frac{Y_{RC}(s)}{s} = 1$$

Here α_0 is zero as $Y_{RC}(s)/s$ does not have pole at the origin. Finally, the network is shown in Fig. 9.21.

$$Y_{RC}(s) = \alpha_\infty s + \frac{\alpha_1 s}{s+1} + \frac{\alpha_2 s}{s+3} + \frac{\alpha_3 s}{s+5}$$

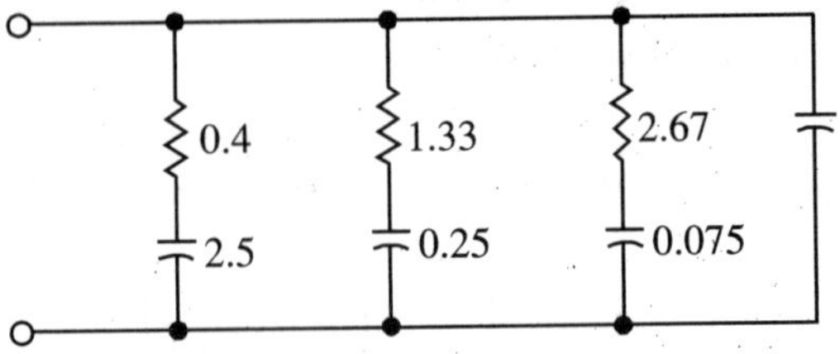

Fig. 9.21

(*iii*) *Cauer I*

		Numerator	Denominator		
1	0	$s^3 + 9s^2 + 23s + 15$ $0 + 0 + 0 + 0$	$s^4 + 12s^2 + 44s + 48s$ $s^4 + 9s^2 + 23s^3 + 15s$	s	2
3	0.3	$s^3 + 9s^2 + 23s + 15$ $s^3 + 7s^2 + 11s$	$3s^3 + 21s^2 + 33s$ $3s^8 + 18s^2 + 22.5s$	$1.5s$	4
5	0.67	$2s^2 + 12s + 15$ $2s^2 + 7s$	$3s^2 + 10.5s$ $3s^2 + 9s$	$0.6s$	6
7	3.33	$5s + 15$ $5s$	$1.5s$ $1.5s$	$0.1s$	8
		15	0		

The continued-fraction expansion is

$$Z_{RC}(s) = \cfrac{1}{s + \cfrac{1}{0.33 + \cfrac{1}{1.5s + \cfrac{1}{0.67 + \cfrac{1}{0.6s + \cfrac{1}{3.33 + \cfrac{1}{0.1s}}}}}}}$$

The network is given in Fig. 9.22.

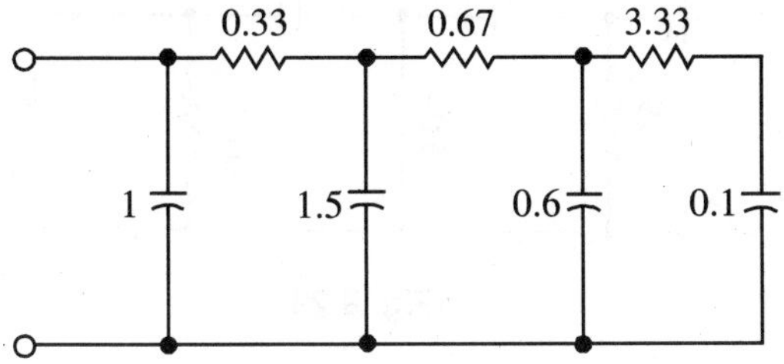

Fig. 9.22

(iv) Cauer II

	Numerator	*Denominator*	
$\frac{0.3125}{s}$	$15 + 2.3 + 9s^2 + s^3$	$48s + 44s^2 + 12s^3 + s^4$	5.1892
	$15 + 13.75s + 3.75s^2 + 0.3125s^3$	$41s + 27.2432s^3 + 3.5675s^3$	
$\frac{0.5520}{s}$	$9.25s + 5.25s^2 + 0.6875s^3$	$16.7567s^2 + 8.4324s^3 + s^4$	28.1483
	$9.25s + 4.6547s^2 + 0.5520s^3$	$16.7567s^2 + 3.8141s^3$	
$\frac{0.1289}{s}$	$0.5953s^2 + 0.1355s^3$	$4.6183s^3 + s^4$	699.742
	$0.5953s^2 + 0.1289s^3$	$4.6183s^3$	
$\frac{0.0066}{s}$	$0.0066s^3$	s^4	
	$0.0066s^3$		
	0		

The continued-fraction expansion corresponding to eqn. 9.22

$$Z_{RC}(s) = \frac{1}{3.2} + \cfrac{1}{\cfrac{1}{0.19} + \cfrac{1}{\cfrac{1}{1.81s} + \cfrac{1}{\cfrac{1}{0.04} + \cfrac{1}{\cfrac{1}{7.76s} + \cfrac{1}{\cfrac{1}{0.001} + \cfrac{1}{\cfrac{1}{151.62s}}}}}}}$$

The network is shown in Fig. 9.23.

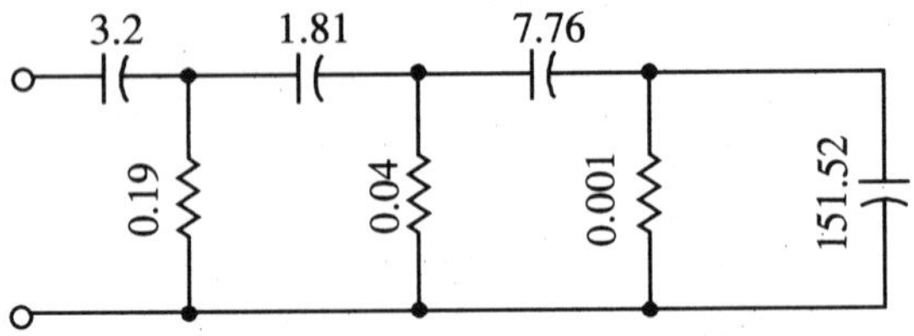

Fig. 9.23

All the four networks have seven elements (one + number of internal critical frequencies) of which four are capacitors (degree of the highest power of s in $F(s)$).

In the continued-fraction expansion realizations we have to be quite careful about the numerical accuracy. If we do not express the coefficients with sufficient accuracy there may be premature termination of the expansion leading to erroneous conclusions.

9.4 RL NETWORK

The admittance of an inductor is similar to the impedance of a capacitor; we conclude that the properties of an RL admittance are identical to those of RC impedance and vice-versa, *i.e.*,

$$Z_{RC}(s) = Y_{RL}(s)$$

$$Z_{RL}(s) = Y_{RC}(s)$$

(An RL-admittance can be considered as the dual of an RC-impedance and vice-versa). The properties and realizations of RL-functions follow from those of RC-functions. The four realizations for an RL-functions are shown in Fig. 9.24.

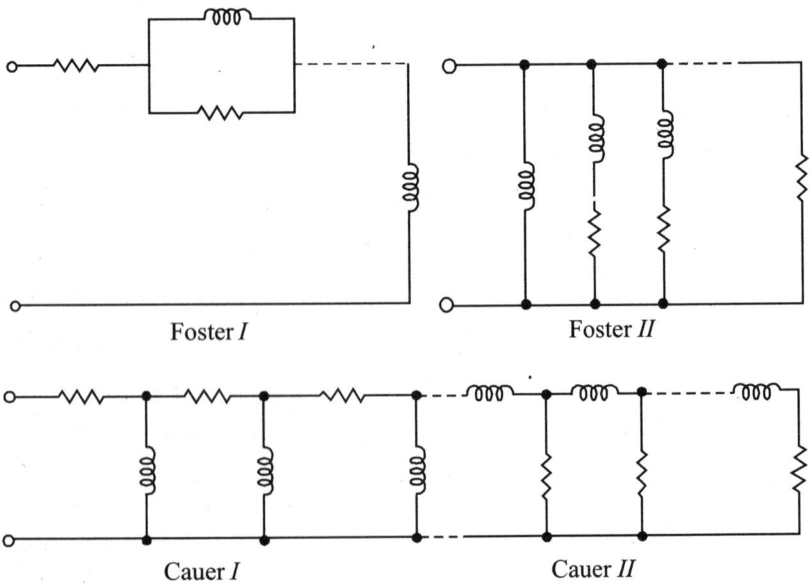

Fig. 9.24

9.5 TWO-PORT FUNCTIONS

A one-port network is completely characterized by its driving-point impedance or admittance function. In contrast, a two-port network requires, in general, the specification of four parameters (see Chap. 7). In the realization of filter networks an often specified quantity is the *open circuit voltage ratio* $T(s)$ ($= V_2/V_1$, V_2 is the output or the load end voltage and V_1 the input or the source end voltage). From Table 7.1, we know that

$$T(s) = \frac{1}{A(s)} = \frac{-y_{12}}{y_{22}} = \frac{z_{12}}{z_{11}} \qquad \ldots(9.23)$$

Hence, the realized network must simultaneously satisfy the specifications on y_{12} (z_{12}) and y_{22} (z_{11}).

In a two-port network, we say that there is transmission through the network when finite input results in finite output. If for a finite input, zero output results we say that there is zero transmission. The frequencies at which zero transmission occurs are called *zeros of transmission* and these can be identified with zeros of $T(s)$ (see section 6.3).

In Chapter 6, we showed that the transfer function (from now on when we say transfer function we imply open-circuit voltage transfer ratio) of a lumped passive network is a rational function of the variable s.

Let
$$T(s) = \frac{p(s)}{q(s)} \qquad \ldots(9.24)$$

The poles of the transfer function are restricted to the left half s-plane (any pole on the $j\omega$-axis must be simple) from stability considerations. The zeros of transmission can be anywhere in the complex plane. The synthesis of the two-port from a specification of its transfer function proceeds with the identification of a driving point and a transfer impedance or admittance. $T(s)$ of eqn. 9.24 can be rewritten as

$$T(s) = \frac{p(s)/Q(s)}{q(s)/Q(s)} \qquad \ldots(9.25)$$

where $Q(s)$ is a real polynomial of s. On identifying eqn. 9.25 with 9.23 we have

$$-y_{12}(z_{12}) = \frac{p(s)}{Q(s)}, \; y_{22}(z_{11}) = \frac{q(s)}{Q(s)} \qquad \ldots(9.26)$$

Q(s) must be so selected that $q(s)/Q(s)$ is a *pr* function and hence a driving point immittance function of a passive network. This driving point function is realized in such a way that it simultaneously realizes the required transmission zeros.

In what follows we consider only the realization of RC and LC ladders. Consider the ladder network shown in Fig. 9.25. A transmission zero is created when the series arm is open or a shunt arm is is a short (see Chap. 6 for further restrictions). The series arm is open at the pole frequencies of the series impedances and the shunt arm is short at the pole frequencies of the shunt admittances. The series arm impedance and shunt arm admittance functions being *pr*, we conclude that "the zeros of transmission for a ladder network are restricted to the left half s-plane".

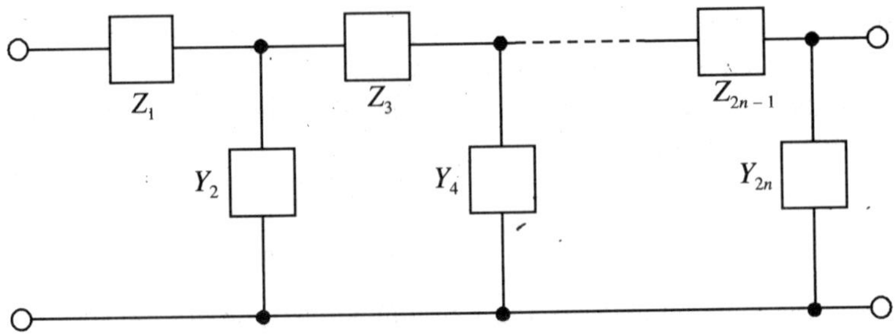

Fig. 9.25

9.5.1 RC Ladders

For an RC ladder z_i ($i = 1, 3,\ldots 2n - 1$) and y_i ($i = 2, 4,\ldots 2n$) are all RC functions. From eqn. 9.25 and 9.26 the poles of the transfer function are restricted to the negative real axis. Since at high frequencies an RC network degenerates to a resistive network, the output voltage must be finite and $T(s)$ cannot have a pole at infinity. It can also be shown that $T(s)$ cannot have pole at the origin. (This is more involved as it depends upon the relation between z_{12} and z_{11} or y_{12} and y_{11}). The zeros of transmission are restricted to the negative real axis. In summary:

"The poles of the transfer function of an RC-ladder network are simple and restricted to the negative real axis excluding the origin and infinity. The transmission zeros are all restricted to the negative real axis". An important point to note is that an RC ladder is incapable of generating zeros in RHP complex or imaginary transmission zeros. If we require such transmission zeros we have to resort to either parallel ladders or Twin-T networks.

Let the transfer function of an RC ladder be as given in eqn. 9.25, *i.e.*,

$$T(s) = \frac{p(s)}{q(s)}$$

where both $p(s)$ and $q(s)$ are real polynomials with zeros on the negative real axis. The degree of the denominator polynomial is at least as great as that of the numerator and "s" is not a factor of the denominator. Identifying z_{11} and z_{12} as in eqn. 9.26 we have

$$z_{11} = \frac{q(s)}{Q(s)}, \quad z_{12} = \frac{p(s)}{Q(s)}$$

Here $Q(s)$ is selected in such a way that its zeros are on the negative real axis and interlacing with those of $q(s)$. $Q(s)$ can have a zero at the origin, and the degree of the highest power of s in $Q(s)$ cannot be less than that of $q(s)$. With such a selection z_{11} satisfies the requirements of an RC-impedance function.

To simplify the synthesis problem we consider the case of an all pole transfer function, *i.e.*, $p(s) = 1$ indicating that all the transmission zeros are at infinity. In this case, the transfer function is

$$T(s) = \frac{1}{q(s)} \qquad \ldots(9.27)$$

This represents transfer function of a low-pass filter. The network is realized by developing $q(s)/Q(s)$ into a ladder in Cauer I form by continued fraction expansion. The final network has the form shown in Fig. 9.26. In this RC-ladder, the transmission zeros are created only when the shunt arm offers zero impedance, *i.e.*, at $s = \infty$.

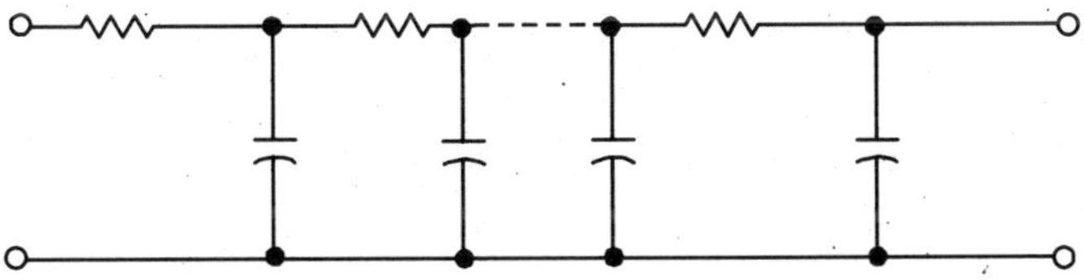

Fig. 9.26

This method only assures that the transfer function will have the proper poles and zeros. z_{11} is realized exactly and z_{12} with a constant multiplier. Hence, the transfer function is only realized with a constant multiplier. In order to evaluate this constant multiplier we need not perform a complete analysis; it suffices to determine the behaviour of the network and of the transfer function as $s \to 0$ or $s \to \infty$. This procedure is illustrated by examples.

Example 9.8. Let $T(s) = \dfrac{1}{(s+2)(s+5)}$

i.e., $\qquad p(s) = 1$

$$q(s) = (s + 2)(s + 5)$$

We develop two separate networks to realize this $T(s)$. Remember that synthesis problems may have multiple solution.

(*a*) Choose $Q(s)$ as $(s + 1)(s + 3)$. Then

$$z_{11} = \frac{(s+2)(s+5)}{(s+1)(s+3)} = \frac{s^2+7s+10}{s^2+4s+3}$$

The continued fraction expansion of z_{11} is

$$z_{11} = 1 + \cfrac{1}{\cfrac{s}{3} + \cfrac{1}{\cfrac{9}{5} + \cfrac{1}{\cfrac{25s}{24} + \cfrac{1}{\cfrac{8}{15}}}}}$$

and the network is given in Fig. 9.27(*a*)

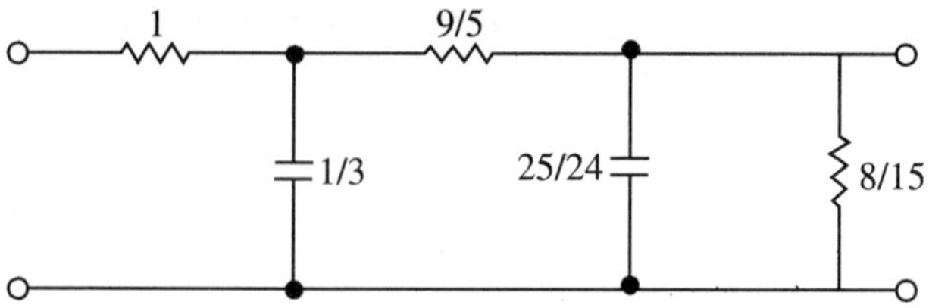

Fig. 9.27(*a*)

To determine the constant multiplier we study the behaviour of the network and of the transfer function as $s \to 0$. The transfer function reduces to:

$$T(0) = \frac{K}{10}$$

(Here it is assumed that the transfer function is realized with a constant multiplier *K*). The network reduces to that shown in Fig. 9.27(*b*). The transfer function of the network is

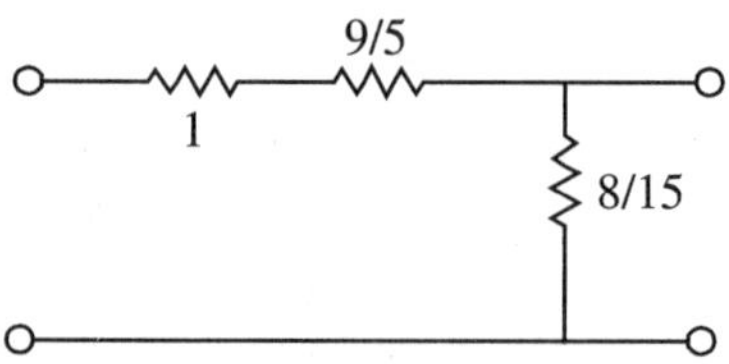

Fig. 9.27(*b*)

$$T(0) = \frac{8/15}{1 + 9/5 + 8/15} = \frac{8}{50}$$

i.e.,
$$\frac{K}{10} = \frac{8}{50} \to K = \frac{8}{5}$$

Hence, Fig. 9.27(*a*) realizes the transfer function with a constant multiplier of 8/5.

(*b*) Choose $Q(s)$ as $s(s + 4)(s + 6)$, then

$$z_{11} = \frac{(s+2)(s+5)}{s(s+4)(s+6)} = \frac{s^2 + 7s + 10}{s^3 + 10s^2 + 24s}$$

The continued fraction expansion of z_{11} is

$$z_{11} = \cfrac{1}{s + \cfrac{1}{\cfrac{1}{3} + \cfrac{1}{\cfrac{9s}{7} + \cfrac{1}{\cfrac{49}{25} + \cfrac{1}{\cfrac{8}{70}s}}}}}$$

and the network is shown in Fig. 9.28. To evaluate the constant multiplier we study the behaviour of the network and of the function as $s \to 0$.

$$T(0) = \frac{K}{10}$$

For the network $T(0) = 1$, yielding $K = 10$. In Fig. 9.28, the first capacitor is not required in the network to realize the given transfer function. This capacitor is a result of selecting a factor in $Q(s)$.

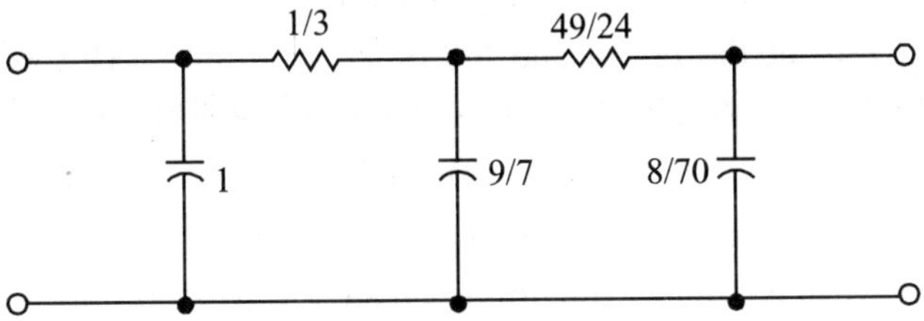

Fig. 9.28

Networks in Fig. 9.27(*a*) and Fig. 9.28 both realize the same transfer function.

If all the transmission zeros are at the origin instead of infinity as in eqn. 9.27, then the synthesis proceeds by developing z_{11} in Cauer II form, instead of Cauer I form.

Example 9.9. Let $T(s) = \dfrac{s^2}{(s+2)(s+5)}$

Choose $Q(s)$ as $(s + 1)(s + 3)$ and expand z_{11} in into continued fraction about the origin instead of about infinity. The corresponding expansion is

$$z_{11} = \cfrac{1}{\cfrac{1}{3.33} + \cfrac{1}{\cfrac{1}{0.19s} + \cfrac{1}{\cfrac{1}{1.83} + \cfrac{1}{\cfrac{1}{0.04s} + \cfrac{1}{\cfrac{1}{6.54}}}}}}$$

and the network is given in Fig. 9.29. It is high-pass network. In this example, the transfer function is realized exactly (*i.e.*, $K = 1$). The constant can be evaluated by considering the behaviour of the network and the function as $s \to \infty$.

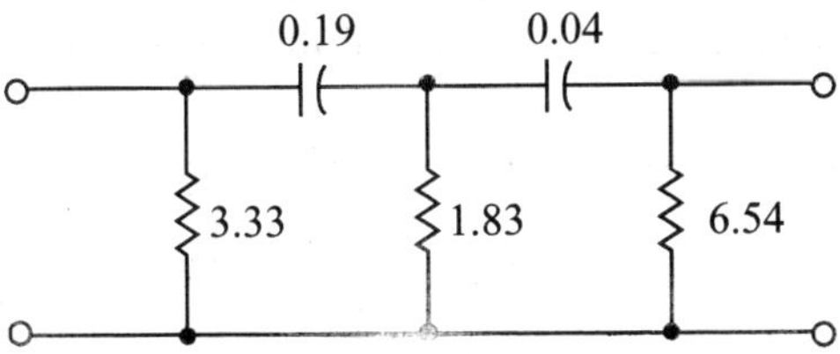

Fig. 9.29

Though, insofar as the driving point functions are concerned the two Cauer forms yield the same function. We have seen that Cauer I leads to a low-pass network and Cauer II to a high-pass network. If the zeros of transfer function are both at the origin and at infinity, the associated driving point function is to expand in continued fraction partially about $s = 0$ and partially about $s = \infty$.

For example consider the transfer function

$$T(s) = \frac{s}{(s+2)(s+5)}$$

This has a zero at the origin and the other at infinity. (This is a bandpass transfer function). Let $Q(s)$ be chosen as in example 9.5(a). Then

$$z_{11} = \frac{(s+2)(s+5)}{(s+1)(s+3)}$$

We expand z_{11} in continued fraction to create a zero at infinity, *i.e.*,

$$z_{11} = 1 + \cfrac{1}{\cfrac{s}{3} + \cfrac{\frac{5}{3}s+3}{3s+7}}$$

Now, the admittance function

$$y^1 = \frac{\frac{5}{3}s+3}{3s+7}$$

is expanded about the origin yielding

$$y^1 = \frac{3}{7} + \cfrac{1}{\cfrac{147}{8s} + \cfrac{1}{8/63}}$$

The overall expansion of z_{11} is

$$z_{11} = 1 + \cfrac{1}{\cfrac{s}{3} + \cfrac{3}{7} + \cfrac{1}{\cfrac{147}{8s} + \cfrac{1}{8/63}}}$$

The corresponding network is shown in Fig. 9.30. The constant multiplier can be evaluated easily. Here we have removed the zero at infinity first and then the zero at the origin. In general, any mixing of the order of realization is possible.

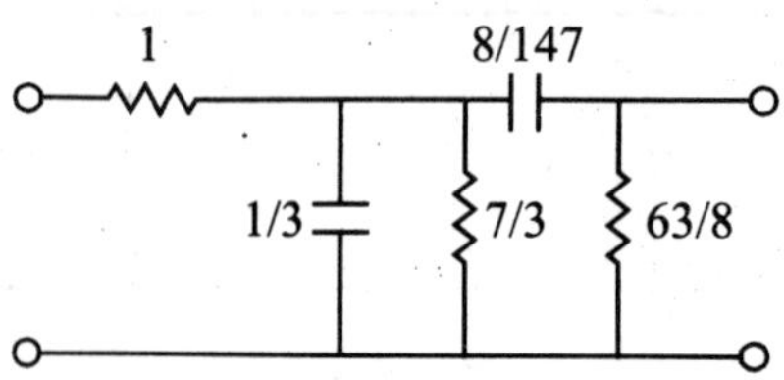

Fig. 9.30

The creation of finite –ve real axis transmission zeros not at the origin leads to a more complicated network. In the all-pole transfer function, the synthesis proceeds by the removal of a constant at infinity. This constant is $z_{11}(\infty)$. The removal of this constant creates a zero at infinity in the remainder of the impedance function. For the example 9.8(*a*), we have

$$\hat{z}_{11} = z_{11} - z_{11}(\infty) = \frac{s^2+7s+10}{s^2+4s+3} - 1$$

$$= \frac{3s+7}{9s^2+4s+3}$$

$\hat{z}_{11}$ has a zero, and hence $\hat{y}_{11} = 1/\hat{z}_{11}$ has a pole, at infinity. This pole is removed in the next cycle to yield a shunt capacitance and a transmission zero at infinity. If instead of removing $z_{11}(\infty)$ we only remove $kz_{11}(\infty)$, where $k < 1$, we can create a 'finite zero' in z_{11}. For example let $k = 0.5$. Then

$$\hat{z}_{11} = z_{11} - 0.5z_{11}(\infty) = \frac{(s+2)(s+5)}{(s+1)(s+3)} - 0.5$$

$$= \frac{0.5s^2+5s+8.5}{s^2+4s+3} = \frac{(0.5s+3.91)(s+2.17)}{(s+1)(s+3)} \qquad \ldots(9.28)$$

The removal of $k\, z_{11}(\infty)$ shifts the zeros of z_{11} to the left. In the first case when $k = 1$, the zero at –5 and –2 have moved to infinity and –7/3, respectively. In the second case, when $k = 0.5$. the zeros have moved to –7.82 and –2.17 respectively.

The pole of $1/\hat{z}_{11}$ at –7.82 is now removed.

i.e.,

$$\hat{y}_{11} = \frac{(s+1)(s+3)}{(0.5s+3.91)(s+2.17)}$$

$$\frac{\hat{y}_{11}}{s} = \frac{(s+1)(s+3)}{s\,(0.5s+3.91)(s+2.17)} = \frac{1.488}{s+7.82} + \frac{0.512(s+1.5)}{s(s+2.17)}$$

i.e.,

$$\hat{y}_{11} = \frac{1.488s}{s+7.82} + \frac{0.512(s+1.5)}{s+2.17} \qquad \ldots(9.29)$$

Now, we can write z_{11} as

$$z_{11} = 0.5 + \cfrac{1}{\cfrac{1.488}{s+7.82} + \cfrac{1}{\cfrac{s+2.17}{0.512(s+1.5)}}} \qquad \ldots(9.30)$$

The network realized so far is shown in Fig. 9.34. Hence, we have created a transmission zero at $s = -7.82$.

We have shown that a partial removal of the constant $z_{11}(\infty)$ has created a finite transmission zero. By varying the constant k between 0 and 1 we can create a other finite transmission zero from –5 to ∞.

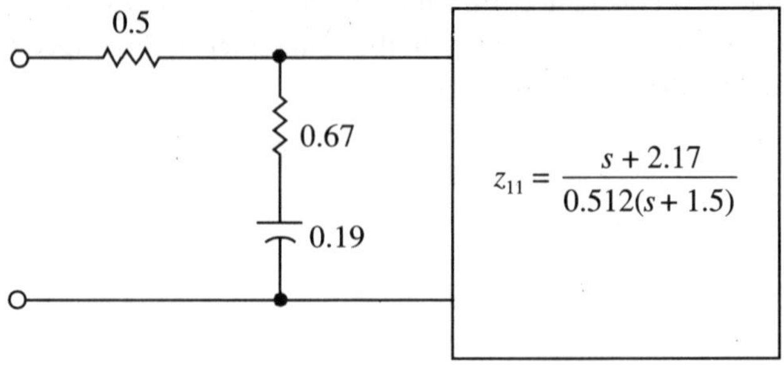

Fig. 9.31

The method we have illustrated by an example is called *zero shifting* technique. By this technique we can create finite transmission zeros not at the origin or at infinity. To select the constant k, we are only to observe that from eqn. 9.28.

$$\hat{z}_{11}\Big|_{s=-7.82} = 0 = z_{11}\Big|_{s=-7.82} - 0.5$$

i.e.,
$$k = z_{11}\Big|_{s=-7.82}$$

Hence, we conclude that to create a transmission zero at $s = -t_i$ we select the constant k as

$$k = z_{11}\Big|_{s=-i}$$

The final form of the network realized by this method is as shown in Fig. 9.32. A series RC branch in the shunt arm creates a finite transmission zero, excluding

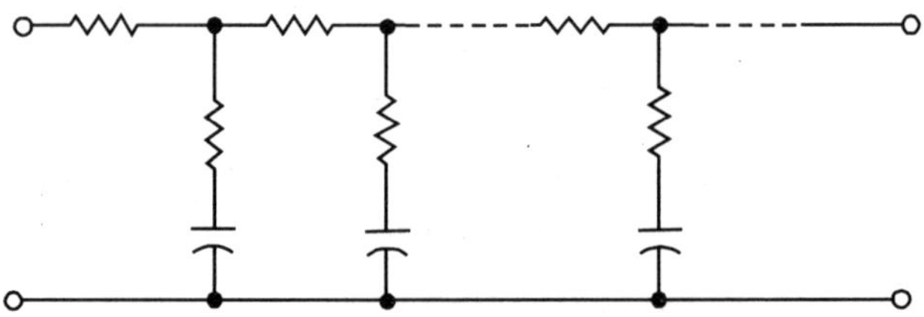

Fig. 9.32

the origin, and a shunt capacitance creates a transmission zero at $s = \infty$. Transmission zeros can also be created in the series arm, in which case the network will be shown in Fig. 9.33.

Here, we have removed a constant $kz_{11}(\infty)$ from z_{11} to produce a zero shifting sequence. If we start with y_{22} instead we remove ky_{22} (0), $0 < k \le 1$, to produce a zero shifting sequence. In general we can partially remove a pole at $s = -p_i$ to produce a zero shifting sequence. In this case, the factor to be removed from z_{11} is $\dfrac{kr(p_i)}{(s+p_i)}$ where as before $0 < k \le 1$ and $r\,(p_i)$ is the residue of z_{11} at the

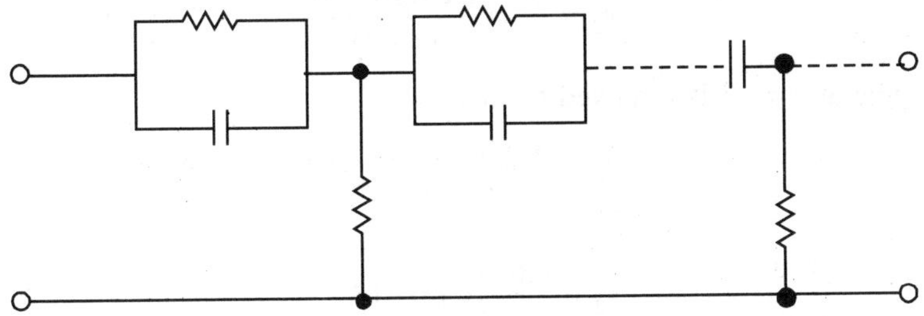

Fig. 9.33

pole p_i. This shifts the zeros towards that pole, the amount of shift depending on the constant k. In all cases, the zero shifts towards the quantity being removed. However, in no case can the zero be shifted beyond an adjacent pole. The complete removal of the constant at the origin (infinity) shifts the zero to the origin (infinity) and in the case of complete removal of a pole, the zero shifts to the position of the pole.

The following example illustrates ladder realization with zero shifting technique.

Example 9.10. $$T(s) = \frac{(s+3.0)(s+7.82)}{(s+2)(s+5)}$$

Transmission zeros: $-3, -7.82$

Let $$Q(s) = (s+1)(s+3)$$

$$z_{11} = \frac{(s+2)(s+5)}{(s+1)(s+3)}$$

(Here the example chosen is the same as that was earlier used to demonstrate the zero shifting technique, except that the transmission zeros at –7.82 and –3 have been specified). z_{11} has zeros at –5 and –2. The zero at –5 is shifted to –7.82 by removing a constant of 0.5 as explained earlier. The results of zero shifting and the ensuing pole removal are shown in eqns. 9.28 to 9.30.

With z_{11}^1

$$z_{11}^1 = \frac{(s+2.17)}{0.512(s+1.5)}$$

In order to create a transmission zero at $s = -3$, we remove a constant k_1 which is given by

$$k_1 = z_{11}^1\Big|_{s=-3} = \frac{0.553}{0.512}$$

Now

$$z_2 = z_{11}^1 - k_1 = 0.808\frac{(s+3)}{s+1.5}$$

placing in evidence the zero created at $s = -3$.

$$y_2 = \frac{1}{z_2} = \frac{1.23(s+1.5)}{(s+3)}$$

The pole at $s = -3$ is removed from y_2

$$\frac{y_2}{s} = \frac{1.23(s+1.5)}{s(s+3)} = \frac{0.615}{s+3} + \frac{0.615}{s}$$

i.e.,

$$y_2 = \frac{0.615s}{s+3} + 0.615$$

The final network realizing $T(s)$ is shown in Fig. 9.34.

The constant of realization can be evaluated, as indicated earlier.

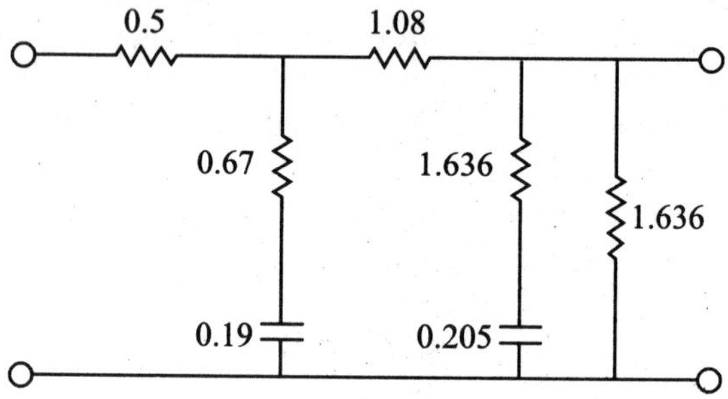

Fig. 9.34

Example 9.11. $T(s) = \dfrac{(s+2)(s+7)}{(s+3)(s+6)}$

Tr. zeros: $-2, -7$

Let $Q(s) = (s + 1)(s + 4)$, then

$$z_{11} = \frac{(s+3)(s+6)}{(s+1)(s+4)}$$

we first create the transmission zero at -7 by removing a constant from z_{11}. This constant is

$$z_{11}\Big|_{s=-7} = \frac{(-4)(-1)}{(-6)(-3)} = \frac{2}{9}$$

Now, we form a new function z_2

$$z_2 = z_{11} - \frac{2}{9}$$

$$= \frac{7}{9}\frac{(s+7)(s+22/7)}{(s+1)(s+4)}$$

The constant 2/9 is removed as a series resistor and z_2 has the required zero at -7. This zero is produced by removing the pole of $y_2(= 1/z_2)$ at $s = -7$

$$y_2 = \frac{9}{7}\frac{(s+1)(s+4)}{(s+7)(s+22/7)}$$

$$\frac{y_2}{s} = \frac{9(s+1)(s+4)}{7s(s+7)(s+22/7)} = \frac{6/7}{s+7} + \frac{3/7(s+12/7)}{s(s+22/7)}$$

$$y_2 = \frac{6/7\,s}{s+7} + \frac{3/7(s+12/7)}{(s+22/7)}$$

The first factor is removed as a shunt branch of series RC, leaving

$$y_3 = \frac{3}{7}\frac{(s+12/7)}{(s+22/7)}$$

Neither y_3 nor $z_3(1/y_3)$ has the required transmission zero at $s = -2$. z_3, a pole at $-12/7$ and a zero at $-22/7$. The transmission zero at $s = -2$ is created by partially removing the pole at $-12/7$.

Let $$z_4 = z_3 - \frac{k}{(s+12/7)}$$

$$k = \left(s + \frac{12}{7}\right) z_3 \Big|_{s=-2} = 8/3$$

Hence, $$z_4 = \frac{7}{3}\frac{(s+22/7)}{(s+12/7)} - \frac{8/3}{s+12/7} = \frac{7}{3}\frac{(s+2)}{(s+12/7)}$$

z_4 has the required zero. The factor $\dfrac{8/3}{s+12/7}$ is removed as a series branch of parallel RC combination. The transmission zero at $s = -2$ is produced by removing the pole of y_4 $(=1/z_4)$ at -2.

$$\frac{y_4}{s} = \frac{3}{7s}\frac{(s+12/7)}{(s+2)} = \frac{3/49}{s+2} + \frac{18/49}{s}$$

i.e., $$y_4 = \frac{3}{49}\frac{s}{s+2} + \frac{18}{49}$$

The first factor yields a shunt element of series RC and the second a shunt element R.

The final network is shown in Fig. 9.35.

In Fig. 9.35, it is to be observed that parallel RC combination (the component due to weakening of a pole) does not produce a transmission zero at $-12/7$ as the network to the left of the branch also produces a pole at $-12/7$ resulting in a cancellation. The constant multiplier can be evaluated as earlier.

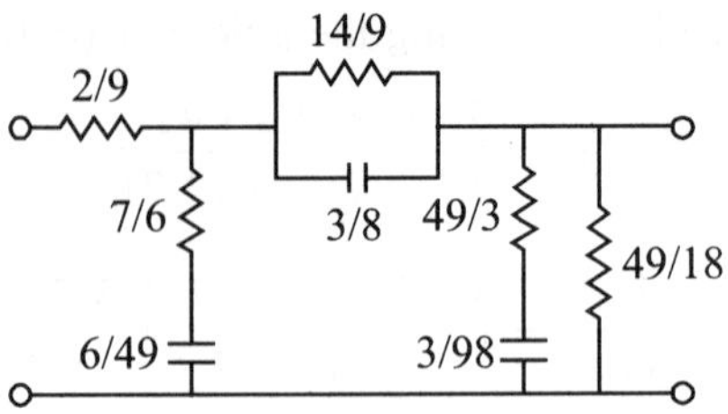

Fig. 9.35

Following the method delineated here finite non-zero transmission zeros can be created in RC ladders.

9.5.2 LC Ladders

For an LC ladder $z_i(1, 3, \ldots, 2n - 1)$ and $y_i(i = 2, 4, \ldots, 2n)$ are all reactance functions. The poles of the transfer function of such a ladder are simple and restricted to the $j\omega$-axis excluding the origin and infinity. The transmission zeros are all on the $j\omega$-axis. An LC ladder is incapable of producing transmission zeros anywhere other than the $j\omega$-axis.

Let the transfer function of an LC ladder be as given in eqn. 9.25

i.e.,
$$T(s) = \frac{p(s)}{q(s)}$$

where $p(s)$ and $q(s)$ are both real polynomials with zeros on the $j\omega$-axis. The degree of the denominator polynomial is at least as great as that of the numerator and s is not a factor of the denominator (compare this with the RC ladder transfer function). Identifying z_{11} and z_{12} as in eqn. 9.26, we have

$$z_{11} = \frac{q(s)}{Q(s)}, \; z_{12} = \frac{p(s)}{Q(s)}$$

Here $Q(s)$ is selected so that its zeros are on the $j\omega$-axis and interlacing with those of $q(s)$. $Q(s)$ can have a zero at the origin and the degree of the highest power of s in $Q(s)$ can be one more, equal to or one less than that of $q(s)$. Such a selection makes z_{11} reactance function.

As in the case of an RC ladder we first consider an all-pole transfer function

$$T(s) = \frac{1}{q(s)}$$

The network is realized by developing $q(s)/Q(s)$ into a ladder in Cauer I form by continued fraction expansion about infinity. The final network is shown in Fig. 9.36.

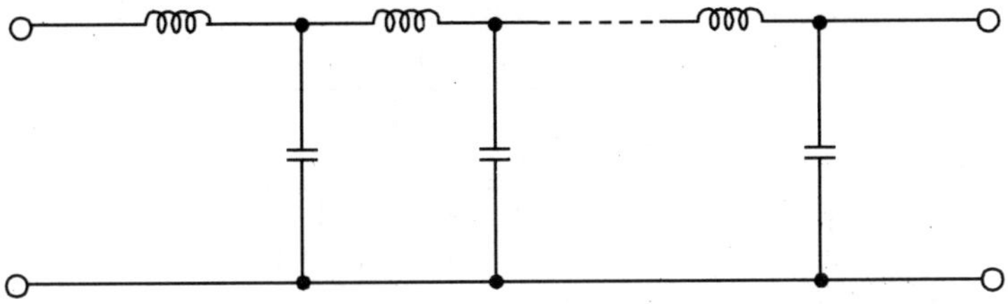

Fig. 9.36

Example 9.12. Let $T(s) = \dfrac{1}{(s^2+1)(s^2+1)}$

i.e., $p(s) = 1$, and $q(s) = (s^2 + 1)(s^2 + 3)$

(*a*) Choose $Q(s)$ as $s(s^2 + 2)$. Then

$$z_{11} = \frac{(s^2+1)(s^2+3)}{s(s^2+2)}$$

The continued-fraction expansion of z_{11} about infinity is

$$z_{11} = s + \cfrac{1}{\cfrac{s}{2} + \cfrac{1}{4s + \cfrac{1}{\cfrac{1}{6}s}}}$$

and the network is shown in Fig. 9.37. The transfer function is realized with a constant multiplier of 3. This is computed by considering the network and the transfer function behaviour as $s \to 0$.

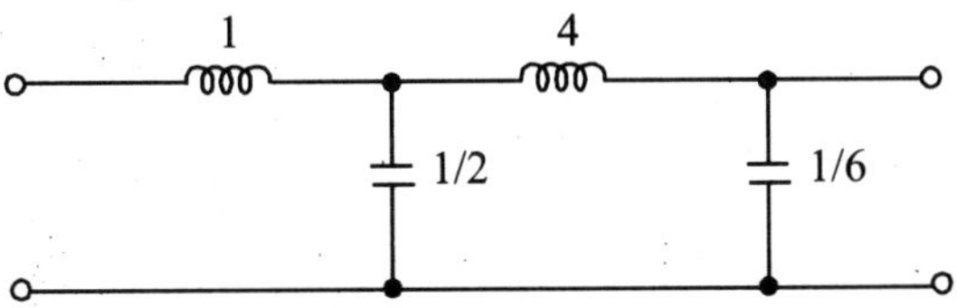

Fig. 9.37

(*b*) Choose $Q(s)$ as $s(s^2 + 2)(s^2 + 4)$. Then

$$z_{11} = \frac{(s^2+1)(s^2+3)}{s(s^2+2)(s^2+4)}$$

The continued-fraction expansion about infinity is

$$z_{11} = \cfrac{1}{s + \cfrac{1}{\cfrac{s}{2} + \cfrac{1}{\cfrac{4}{3}s + \cfrac{1}{\cfrac{3}{2}s + \cfrac{1}{\cfrac{s}{3}}}}}}$$

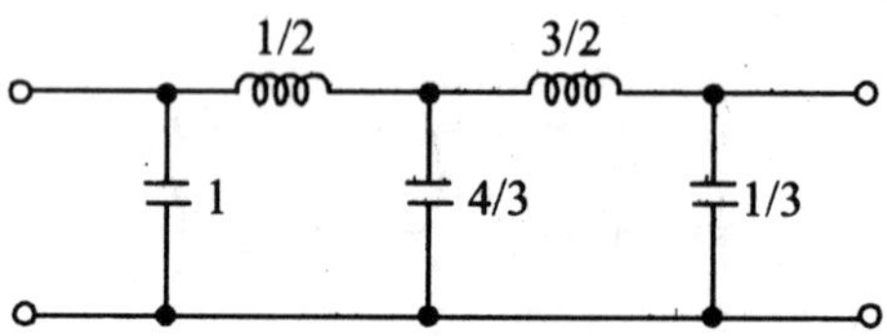

Fig. 9.38

and the network is given in Fig. 9.38. The value of the constant-multiplier for this realization is 3. Here again, as in the case of example 9.8(b), the first capacitor does not affect the realization of $T(s)$. Part (b) is included to illustrate the generality of the selection of $Q(s)$. Obviously, realization of Fig. 9 has a redundant element.

If all the transmission zeros are at the origin, then the synthesis proceeds by expanding z_{11} in Cauer-form II, *i.e.* z_{11} is expanded in continued-fraction expansion about the origin.

Example 9.13. Let $T(s) = \dfrac{s^4}{(s^2+1)(s^2+3)}$

Choose $Q(s)$ as $s(s^2 + 2)$ and expand z_{11} about the origin. The expansion is given by

$$z_{11} = \frac{3}{2s} + \cfrac{1}{\cfrac{4}{5s} + \cfrac{1}{\cfrac{25}{2s} + \cfrac{1}{\cfrac{1}{5s}}}}$$

and the network is given in Fig. 9.39. The constant multiplier can be evaluated by considering the behaviour as $s \to \infty$. The multiplier value is 1.

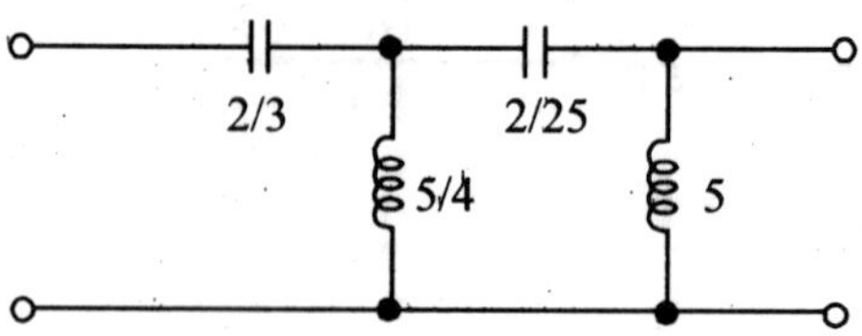

Fig. 9.39

If the transmission zeros are both at the origin and at infinity we can resort to the method similar to the one indicated for RC ladders.

Example 9.14. $$T(s) = \frac{s^2}{(s^2+1)(s^2+3)}$$

Let $Q(s) = s\,(s^2 + 2)$ then

$$z_{11} = \frac{(s^2+1)(s^2+3)}{s(s^2+2)}$$

We expand z_{11} in continued fractions about infinity till we realize two transmission zero at infinity and then expand, the remainder of the function about the origin. The final expansion is given by

$$z_{11} = s + \cfrac{1}{\cfrac{s}{2} + \cfrac{1}{\cfrac{6}{s} + \cfrac{1}{\cfrac{1}{4s}}}}$$

The network realized is shown in Fig. 9.40. Here $T(s)$ is realized with a constant multiplier of 2.

Finite transmission zeros, not at the origin, can be created by zero shifting (or pole weakening) as in the case of RC networks.

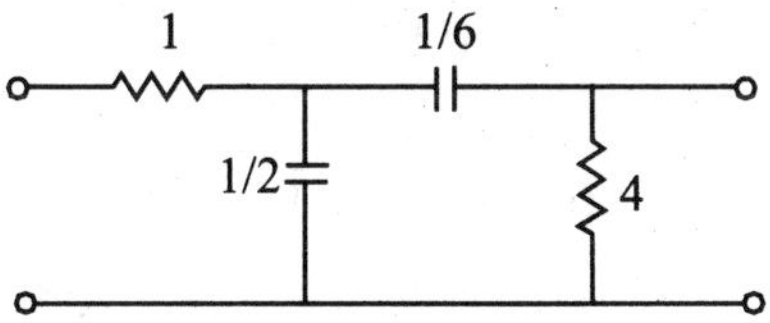

Fig. 9.40

In the pole weakening (or zero shifting) technique, the zeros move towards the weakened pole. However, a zero cannot cross an intermediate pole. The creation of such transmission zeros in shunt arm leads to the network of Fig. 9.41(a) or (b). If the finite transmission zeros are created in series arm, the structures are

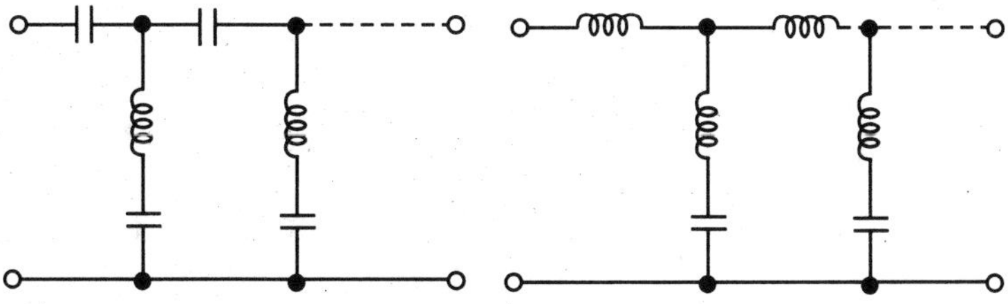

(a) (b)

Fig. 9.41

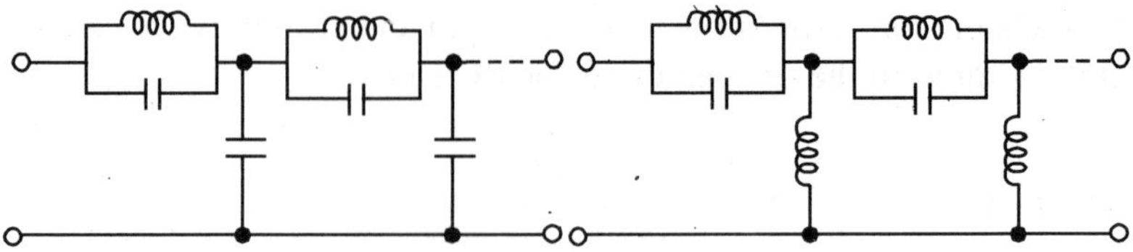

Fig. 9.42

shown in Fig. 9.42. These structures are called *mid-series* and *mid-shunt* ladders, respectively.

Example 9.15. $T(s) = \dfrac{(s^2+1)(s^2+4)}{(s^2+9)(s^2+25)}$

Transmission zeros: $\pm j1, \pm j2$

Let $Q(s) = s\,(s^2 + 16)$

i.e.,
$$z_{11} = \frac{(s^2+9)(s^2+25)}{s(s^2+16)}$$

z_{11} has a pole at the origin. The removal of this pole shifts the zero at $j3$ to the origin. By partial removal of this pole the zero can be shifted anywhere between $j3$ and the origin. Here we shift it to $j1$.

To create a zero at $j1$ we remove $\dfrac{k}{s}$ from z_{11}.

$$k = \left[s\,z_{11}(s)\right]_{s=j1}$$

$$= \frac{8\times 24}{15} = \frac{64}{5}$$

Observe that we have removed a constant $k = k\ r(0)$ where $0 < k \infty 1$ and $r(0)$ is the residue of z_{11} at its pole at the origin. Here

$$r(0) = \frac{9\times 25}{16}$$

The new function is

$$z_1 = z_{11} - \frac{k}{s}$$

$$= \frac{(s^2+1)(s^2+20.2)}{s(s^2+16)}$$

z_1 has a zero at $j1$ as expected. The pole of y_1 at $j1$ is removed to produce the zero at $j1$.

i.e.,
$$y_1 = \frac{s(s^2+16)}{(s^2+1)(s^2+20.2)} = \frac{\frac{15}{19.2}s}{s^2+1} + \frac{\frac{4.2}{19.2}s}{s^2+20.2}$$

The admittance function $\dfrac{\dfrac{15}{19.2}s}{s^2+1}$ produces the required transmission zero. Removal of this leaves an admittance function

$$y_2 = \frac{\dfrac{4.2}{19.2}s}{s^2+20.2}$$

Now $z_2 = \dfrac{s^2+20.2}{\dfrac{4.2}{19.2}s}$, has a zero at $j(\sqrt{20.2})$. By partially removing the pole of z_2 at the origin we can shift the zero to $j2$, the second transmission zero. As before we remove $\dfrac{k}{s}$ from z_2 where

$$k = [s\, z_2]_{s\,=\,j2}$$

i.e.,

$$k = \frac{16.2}{\left(\dfrac{4.2}{19.2}\right)}$$

we have

$$z_3 = z_2 - \frac{k}{s} = \frac{s^2+4}{\dfrac{4.2}{19.2}s}$$

and thus creating a transmission zero at $s = j2$. To produce this we consider the admittance function.

$$y_3 - \frac{\dfrac{4.2}{19.2}s}{s^2+4}$$

The network realizing the transfer function can be obtained by referring to Figs. 9.2 and 9.4. The network is shown in Fig. 9.30. The constant multiplier can be evaluated as indicated earlier.

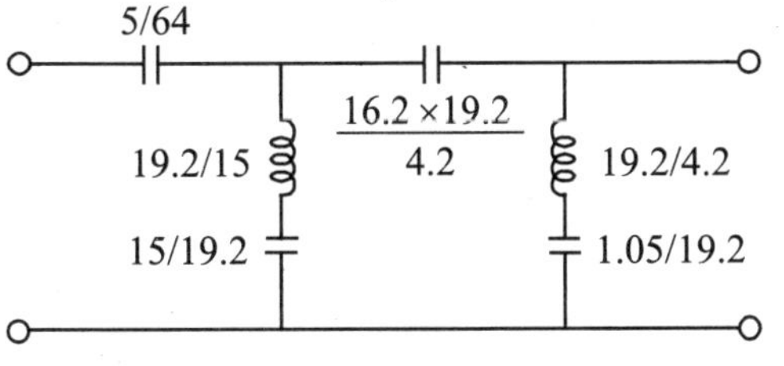

Fig. 9.43

From the examples of this section it is clear that a given transfer function has many realizations. In zero shifting technique one can obtain alternate realizations by changing the order of transmission zeros created and removed. The selection of the final network depends on factors like the number of elements and the elemental spread. Before leaving this topic we summarize some of the properties of the realizable LC and RC ladder transfer functions.

1. The poles of LC transfer functions must be on $j\omega$-axis while those of RC transfer functions are on the negative real axis.
2. The transmission zeros of LC ladders must be on the $j\omega$-axis and must appear in conjugate pair. Those of RC ladders must appear on negative real axis.
3. There can be no pole at the origin or at infinity.
4. The degree of the numerator is less than or equal to that of the denominator.

9.6 MINIMUM-PHASE NETWORKS

The zeros of transmission of an RLC network can occur anywhere in the complex s-plane. However, a ladder can realize transmission zeros in the LH s-plane including the imaginary axis. A network whose transmission zeros are restricted to the LH plane is called a *minimum-phase network* and its transfer function a *minimum-phase transfer function.* Hence, we conclude that a *passive ladder network is a minimum-phase network.*

A non-minimum phase network must necessarily have zeros of transmission in the RH s-plane. Let $T(s)$ be such a transfer function and let

$$T(s) = \frac{p_1(s)p_2(s)}{q(s)} \qquad \ldots(9.31)$$

where the zeros of $p_1(s)$ are in LH plane and those of $p_2(s)$ in RH plane. Further, let $p_2(s)$ be written in the factorized form as

$$p_2(s) = \prod_{t=1}^{\gamma}(s - z_i)$$

Now, $T(s)$ can be written as

$$T(s) = \frac{p_1(s)}{q(s)} \cdot p_2(s) \cdot \frac{q_1(s)}{q_1(s)} \qquad \ldots(9.32)$$

where $q_1(s) = \prod_{i=1}^{\gamma}(s + z_i)$. On regrouping eqn. 9.32, we have

$$T(s) = \left\{\frac{p_2(s)q_1(s)}{q(s)}\right\}\left\{\frac{p_2(s)}{q_1(s)}\right\}$$

$$= T_1(s)T_2(s) \qquad \ldots(9.33)$$

where $T_1(s)$ is a minimum-phase transfer function and $T_2(s)$ is given as

$$T_2(s) = \frac{\prod_{i=1}^{\gamma}(s - z_i)}{\prod_{i=1}^{\gamma}(s + z_i)} \qquad \ldots(9.34)$$

The poles of $T_2(s)$ are in the LHP and the zeros are the RHP images of these poles. Such a function is called an *all-pass* function and has unity magnitude for all frequencies.

The first and second order *all-pass* function are given by

$$\frac{s-a}{s+a}, \frac{s^2 - as + b}{s^2 + as + b}$$

From eqn. 9.33 we conclude that "a non-minimum-phase function can be expressed as a product of a minimum-phase function and an *all-pass* function."

Non-minimum phase and all pass networks can be realized by lattice, twin-T or bridge T-networks. The splitting of a non-minimum phase function itself indicates a method of realizing the same as a cascade combination of a minimum phase network and an all-pass network. But the loading effect of passive network makes the method non-applicable without further modifications. A method of avoiding the loading effect is to make the networks constant resistance networks. A two-port constant resistance network has an input impedance of R at one port when the other port is terminated in a resistance R. If both the networks of the cascade combination are realized as constant resistance networks, they can be cascaded without any loading effect. Lattice and bridge-T networks can be realized as constant resistance networks. The details of design and application of constant resistance networks are found elsewhere (see Ref. 2).

PROBLEMS

9.1 Construct the Routh's array for the following and test for the Hurwitz nature of the polynomials:

(*a*) $s^5 + 20s^4 + 147s^3 + 484s^2 + 612s + 336$

(*b*) $s^5 + 5s^4 + 6s^3 + s^2 + 5s + 6$

(*c*) $s^4 + 6s^3 + 9s^2 + 24s + 20$

(*d*) $s^6 + 3s^5 + 3s^4 + 3s^3 + 3s^2 + 3s + 2$

9.2 Test for the Hurwitz nature of the polynomials by continued fraction expansion:

(*a*) $s^5 + 17s^4 + 107s^3 + 307s^2 + 396s + 180$

(*b*) $s^5 + 5.5s^4 + 9s^3 + 6.5s^2 + 8s + 6$

(*c*) $s^4 + 5s^3 + 11s^2 + 25s + 6$

(*d*) $s^5 + 2s^4 + 5s^3 + 10s^2 + 9s + 18$

9.3 Test for the positive reality of the following rational functions:

(*a*) $\dfrac{s^2+4s+3}{s^2+6s+8}$ (*b*) $\dfrac{s^4+2.5s^2+1}{s^5+4.5s^3+4.5s}$

(*c*) $\dfrac{s^4+2s^3+8s^2+s+1}{s^4+s^3+3s^2+2s+1}$ (*d*) $\dfrac{s^3+10s^2+27s+18}{s^2+7s+11.25}$

9.4 Classify the following RC immittances:

(*a*) $\dfrac{s^3+10s^2+24s}{s^2+7s+10}$ (*b*) $\dfrac{s^2+10s+24}{s^2+7s+10}$

(*c*) $\dfrac{s^2+3s+1.25}{s^2+7.5s^2+9s}$ (*d*) $\dfrac{s^2+12s+35}{s^3+15s^2+62s+48}$.

9.5 The properties of a reactance function are given in sec. 9.2. Using LC → RC transformation of eqns. 9.15, 9.16 obtain the properties of RC-immittance functions.

9.6 Realize the following reactance functions in the four canonic forms:

(*a*) $\dfrac{(s^2+1)(s^2+6.25)(s^2+25)}{s(s^2+4.5)(s^2+9)}$ (*b*) $\dfrac{s(s^2+3.24)(s^2+36)}{(s^2+2)(s^2+4)(s^2+40)}$.

9.7 Realize the RC functions of prob. 9.4 in the four canonic forms.

9.8 Realize the following transfer functions:

(*a*) $\dfrac{1}{s^3+15s^2+62s+48}$ (*b*) $\dfrac{1}{s^3+9s^2+23s+15}$

(*c*) $\dfrac{1}{s^6+5s^4+6.75s^2+2.25}$ (*d*) $\dfrac{1}{s^4+15.25s^2+56.25}$

(*e*) $\dfrac{s^3}{s^3+15s^2+62s+48}$ (*f*) $\dfrac{s^2}{s^3+9s^2+23s+15}$

(*g*) $\dfrac{s^6}{s^6+5s^4+6.75s^2+2.25}$ (*h*) $\dfrac{s^2}{s^4+15.25s^2+56.25}$.

9.9 Realize the following transfer functions:

(*a*) $\dfrac{s+1}{s^2+5s+2}$ (*b*) $\dfrac{s(s+0.5)}{(s+2)(s+4)}$

(*c*) $\dfrac{(s+2)(s+3)}{(s+1)(s+8)}$ (*d*) $\dfrac{s(s+3/4)}{(s+1/2)(s+2)}$.

9.10 Realize the following transfer functions:

(*a*) $\dfrac{s(s^2+9)}{(s^2+1)(s^2+25)}$ (*b*) $\dfrac{s(s^2+4)}{(s^2+1)(s^2+25)}$

(*c*) $\dfrac{s(s^2+1)}{(s^2+3)(s^2+4)}$ (*d*) $\dfrac{(s^2+1)(s^2+4)}{(s^2+9)(s^2+23)}$.

NOTES AND REFERENCES

The subject of network synthesis is too vast to be covered properly as a part of a book. The various topics of this chapter have been selected to provide sufficient background material for the synthesis of two element kind one-port networks, RC and LC 2-port ladder networks. Driving point function synthesis and ladder network realization are essential parts of insertionloss synthesis to be discussed in Chapter 11. Many active filter realizations use RC ladder in feed-back configuration with an operational amplifier.

Fundamental works of Brune and Darlington pertaining to the realization of general *pr* functions is not considered here for lack of space. The classic books on network synthesis by Guillemin[1], Balbanian[2] and Valkenberg[3] cover these topics in detail. These books also provide proofs of theorems on *pr* function besides providing further information on the same.

A *pr* function can be determined either from its real part or its imaginary part. Also the real and imaginary parts can also be interrelated by the familiar Hilbert transform relationship. The determination of system function is discussed in detail in Guillemin's book (Chap. 8). A comprehensive presentation is found in Volkenberg (Chap. 8) and Weinberg's[4] (Chap. 6) books.

Among many methods of realization of two-port networks, we have only discussed the realization of ladders from a specification of the voltage transfer functions. All books referred here provide details of two-port synthesis along with their realizability conditions. The realization of Butterworth and Chebyshev filters discussed in Chapter 11 is based on realization of all-pole transfer functions while elliptic filter realization require zero shifting technique.

Balbanian and Bickarh's book[5] gives a comprehensive coverage of network functions and synthesis (Chaps. 6 and 7).

1. E.A. Guillemin, *Synthesis of Passive Networks*, John Wiley, 1957.
2. N. Balbanian, *Network Synthesis*, Prentice-Hall, 1958.
3. M.E. Van Valkenberg, *Introduction to Modern Network Synthesis*, John Wiley, 1960.
4. Weinberg, *Network Analysis and Synthesis*, McGraw-Hill, 1962.
5. N. Balbanian and T.A. Bikart, *Electrical Network Theory*, John Wiley, 1969.

CHAPTER

10 Approximation

Filters can be considered as frequency selective networks. A filter is required to separate an unwanted signal from a mixture of unwanted and wanted signals. In general, the overlapping of these signals is such that a complete separation of these signals is not feasible, if not impossible. In such cases, the transfer characteristic of the filter is so shaped that the ratio of unwanted to wanted signal at the output of the filter is minimized. The filter specifications are generally given in terms of *cut-off frequencies* or *pass-band* and *stop-band regions*. *Pass-band* is the frequency band of wanted signals and *stop-band* is the frequency band of unwanted signals. An ideal filter should pass the wanted signals with no attenuation and provide infinite attenuation for the unwanted signals.

Electric filters are ubiquitous and it is difficult to conceive a modern electronic device or a system which does not employ an electric filter. The theory of filters originated with the works of Campbell and Wagner on electric-wave filters. Campbell and Zobel considered the design of filters based on '*image parameters*' which fundamentally depended on image or iterative impedances. This classical filter design has outlasted its usefulness. A brief outline of this method is given in Appendix F.

In contrast to the image parameter theory, modern filter design is based on the selection of the filter transfer function to satisfy the specification, and then the realization of this function by synthesis techniques.

The steps involved in the modern filter design can be summarized as:

(*i*) selection of the filter specification,
(*ii*) selection of a realizable rational function which satisfies this specification.
(*iii*) realization of the transfer function and calculation of the component values of the chosen filter structure, and
(*iv*) construction and testing of the filter.

The second step is generally described as the *approximation problem.* The filter specification—cut-off frequencies, pass-band gain, transition, bandwidth, stop-band attenuation—are generally specified in the frequency domain. In the approximation problem a realizable rational function is selected so that it meets the specifications with prescribed tolerance. This procedure is mainly algebraic, and there are many standard functions available. In this chapter, we consider the frequency domain approximation procedures. The approximation can be achieved in the time domain also. The realization of the transfer function approximating the given specifications is the subject of the next chapter.

10.1 FILTER SPECIFICATIONS

The characteristics of an ideal low-pass filter are shown in Fig. 10.1. The filter network is required to pass all frequencies below ω_p with no attenuation and frequencies above ω_p with infinite attenuation. The phase of the filter network is to be a linear function of ω. These specifications cannot be achieved by a physical network. If we consider the magnitude as the Fourior transform, the corresponding time function is of the form sin α_t/α_t and exists for all t. This makes the impulse response of the filter network non-causal and hence non-realizable. Indeed, it can be shown that any band limited frequency response results in a non-causal time response.

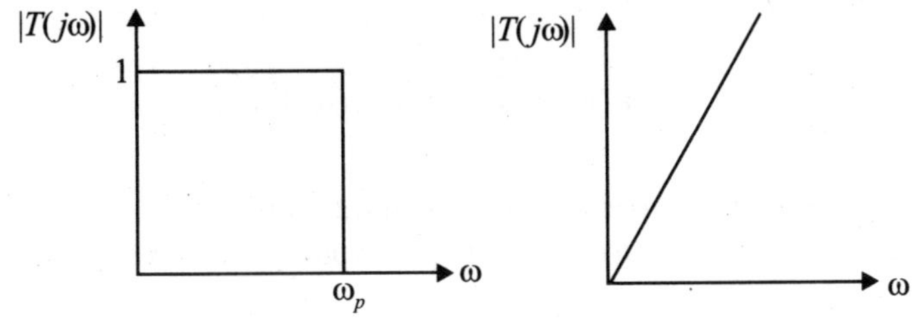

Fig. 10.1

In order to circumvent this problem the filter specifications are modified. Fig. 10.2(a) shows the specifications for a realizable low-pass filter.

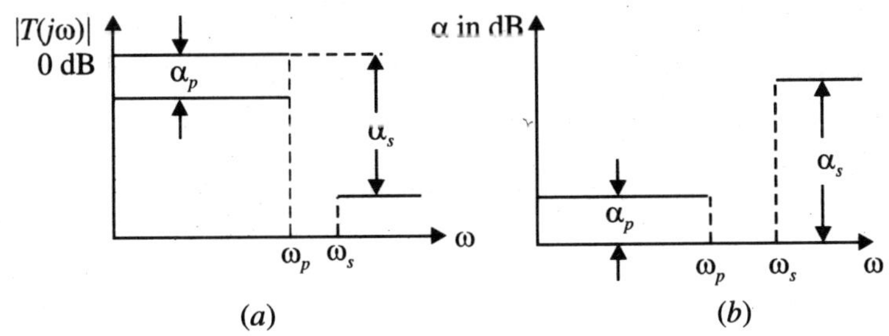

Fig. 10.2

$\omega_p \rightarrow$ pass-band edge

$\omega_s \rightarrow$ stop-band edge

$\omega_p < \omega < \omega_s \rightarrow$ transition band

Maximum pass-band attenuation α_p dB

Maximum stop-band attenuation α_s dB

i.e., $\alpha \le \alpha_p \qquad \omega \le \omega_p$

$\alpha \ge \alpha_s \qquad \omega \ge \omega_s$

where α is the attenuation of the filter in dB.

Strictly speaking, α being attenuation should be expressed as negative quantity. When we specify α_p and α_s we should interpret it as α_pdB loss at $\omega = \omega_p$ and α_sdB loss at $\omega = \omega_s$. All the quantities in Fig. 10.29 are referred to 0 dB (unity gain) axis and $|T(j\omega)|$ in dB is negative for all values of $\omega > \omega_p$. If we plot the loss or attenuation in dB versus ω then α_p and α_s will always be positive (Fig. 10.2(*b*)).

The approximation methods solve the problem of selecting a realizable rational function whose frequency response approximates that in Fig. 10.2.

Let $f(\omega)$ be the required frequency response function and $t(\omega)$ the function which approximates $f(\omega)$. We may require that

$$\varepsilon(\omega) = \underset{\omega_a \le \omega \le \omega_b}{\text{Max}} |f(\omega - t(\omega)|$$

be minimized. Here the criterion is to minimize the maximum error over the frequency band $\omega_a \le \omega \le \omega_b$. We may also pick a mean-square error criterion, *i.e.*, $t(\omega)$ is selected such that it minimizes,

$$\varepsilon(\omega) = \int_{\omega_a}^{\omega_b} |f(\omega) - t(\omega)|^2 \, d\omega$$

Another method of approximating is to have $t(\omega)$ exactly equal to $f(\omega)$ at a prescribed point $\omega = \omega_o$. Let $f(\omega)$ and $t(\omega)$ be expanded in Taylor-series about ω_o, *i.e.*,

$$f(\omega) = f(\omega_o) + f'(\omega_o)(\omega - \omega_o) + \frac{f''(\omega_o)}{2!}(\omega - \omega_o)^2 + \ldots$$

and $$t(\omega) = t(\omega_o) + t'(\omega_o)(\omega - \omega_o) + \frac{t''(\omega_o)}{2!}(\omega - \omega_o)^2 + \ldots$$

(where *o* indicates differentiation with respect to ω). If $f(\omega_o) = t(\omega_o)$ and as many as possible derivatives at $\omega = \omega_o$ are set to zero we have a *Taylor series approximation.*

In all the approximation the transfer function is selected as

$$|T(j\omega)|^2 = \frac{1}{1 + \varepsilon^2 K_n^2(\omega)} \qquad \ldots(10.1a)$$

with $$0 \le K_n^2(\omega) \ll 1, \; 0 \le \omega \le \omega_p$$

and $$K_n^2(\omega) \gg 1, \; \omega > \omega_s$$

With this selection, $|T(j\omega)|^2 \approx 1$ in the pass-band and approximately zero in the stop-band. From eqn. 10.1*a* the attenuation function is

$$\alpha(\omega) = 10 \log [1 + \varepsilon^2 K_n^2(\omega)] \qquad \ldots(10.1b)$$

and the constant ε determines the pass-band and/or stop band attenuation.

In what follows we consider two well-known methods—Butterworth and Chebyshev approximations—in detail, and briefly discuss some other methods of approximations.

10.2 BUTTERWORTH APPROXIMATION

Butterworth approximation is a special form of Taylor series approximation in which the approximating function $t(\omega)$ and the specified function $f(\omega)$ are identical at $\omega = 0$. For this approximation $K_n(\omega)$ is selected as

$$K_n(\omega) = \beta_0 + \beta_1\omega + \beta_2\omega^2 + \ldots + \beta_n\omega^n$$

For a Taylor series approximation the function $K_n(\omega)$ must be *maximally flat* at the origin (*i.e.*, $\omega = 0$). Hence, as many derivatives of $K_n(\omega)$ as possible must vanish at $\omega = 0$. Therefore, for Butterworth approximation

$$K_n(\omega) = \omega^n$$

The magnitude function $|T(j\omega)|^2$ and the attenuation function $\alpha(\omega)$ are then given by

$$|T(\omega)|^2 = \frac{1}{1+c^2\omega^{2n}}$$

$$\alpha(\omega) = 10 \log[1 + c^2 \omega^{2n}] \quad \ldots(10.2)$$

(Here we have used c in place of ε in eqn. 10.1). In eqns. 10.1 and 10.2 $\boldsymbol{\omega}$ should be interpreted at the frequency normalized with respect to the pass-band edge ω_p, *i.e.*,

$$|T(j\omega)|^2 = \frac{1}{1+c^2\left(\frac{\omega}{\omega_p}\right)^{2n}} \quad \ldots(10.3a)$$

and

$$\alpha(\omega) = 10\log\left[1+c^2\left(\frac{\omega}{\omega_p}\right)^{2n}\right] \quad \ldots(10.3b)$$

The frequency response of a Butterworth filter for various values of n is shown in Fig. 10.3. All the curves pass through the same point at $\omega = \omega_p$ and this point is determined by α_p.

If the specifications are barely satisfied, then from eqn. 10.3b we have

$$\alpha_p = 10\log\left[1+c^2\left(\frac{\omega_p}{\omega_p}\right)^{2n}\right]$$

$$\alpha_s = 10\log\left[1+c^2\left(\frac{\omega_s}{\omega_p}\right)\right] \quad \ldots(10.4)$$

The factor ω_p/ω_s is called the *selectivity parameter* and is represented by k

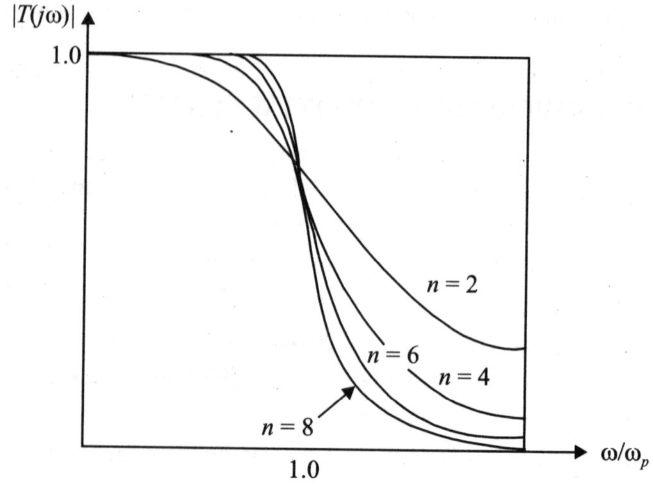

Fig. 10.3

i.e.,

$$k = \frac{\omega_p}{\omega_s} \quad \text{...(10.5)}$$

Equation 10.4 can be rewritten as

$$c^2 = 10^{0.1\alpha_p} - 1$$

and

$$\frac{c^2}{k^{2n}} = 10^{0.1\alpha_s} - 1 \quad \text{...(10.6)}$$

Dividing one part by the other we have

$$k^{2n} = \frac{10^{0.1\alpha_p} - 1}{10^{0.1\alpha_s} - 1}$$

If we define

$$\left[\frac{10^{0.1\alpha_p} - 1}{10^{0.1\alpha_s} - 1}\right] = k_1 \quad \text{...(10.7)}$$

as the *discrimination parameter,* we have

$$n = \frac{\log k_1}{\log k}$$

The order of the filter n should be selected so that

$$n \geq \frac{\log k_1}{\log k}, \text{ } n \text{ an integer.} \quad \text{...(10.8)}$$

If n happens to be equal to log k_1/log k then the value of c obtained from both the equations in 10.6 are the same. If $n \neq \log k_1/\log k$, then c can be selected to satisfy either the passband edge or the stop-band edge requirements exactly.

In the first case: $c = \sqrt{10^{0.1\alpha_p} - 1}$ ...(10.9*a*)

In the second case: $c = \sqrt{k^{2n}\left(10^{0.1\alpha_s} - 1\right)}$ ...(10.9*b*)

Example 10.1. The specifications for a LP filter are

$$\alpha \le 1 \text{ dB, for } f \le 3 \text{ MHz}$$

$$\alpha \ge 60 \text{ dB, for } f \ge 12 \text{ MHz}$$

For this filter the selectivity and the discrimination parameters are

$$k = \frac{3}{12} = 0.25,$$

$$k_1 = \left[\frac{10^{0.1} - 1}{10^6 - 1}\right]^{\frac{1}{2}} = 0.5089 \times 10^{-3}$$

The order of the Butterworth polynomial from eqn. 10.8 is

$$n \ge \frac{\log k_1}{\log k} = \frac{-3.2934}{-0.6021} = 5.4702$$

i.e., $n = 6$

(*i*) If c is to satisfy pass-band requirement (eqn. 10.9*a*), then

$$c = \sqrt{10^{0.1} - 1} = 0.5089$$

(*ii*) If c is to satisfy stop-band requirement (eqn. 10.9*b*), then

$$c = \sqrt{(0.25)^{12}(10^6 - 1)} = 0.2441$$

(If n is chosen as 5.4702, then from eqn. 10.9 (*a*) and (*b*) we have $c = 0.5089$ as indicated earlier.)

Filter specification often includes the cut-off frequency which is defined as the—3dB point. If ω_c is the cut-off frequency then eqn. 10.3 yields

$$\frac{1}{\sqrt{1 + c^2\left(\frac{\omega_c}{\omega_p}\right)^{2n}}} = \frac{1}{\sqrt{2}}$$

i.e.,

$$c^2\left(\frac{\omega_c}{\omega_p}\right)^{2n} = 1$$

and

$$\omega_c = \frac{\omega_p}{\sqrt[n]{c}} \quad \text{...(10.10)}$$

For the previous example, if c is selected to satisfy the pass-band requirement

$$\omega_c = \frac{\omega_p}{\sqrt[6]{0.5089}} = 1.12\omega_p$$

If the pass-band edge is the cut-off frequency, *i.e.*, $\alpha_p = 3\text{dB}$, then $c = 1$.

From now on for convenience we consider a normalized Butterworth filter, *i.e.*, $\omega_p = 1$ and $c = 1$.

In order to realize a filter we have to determine its transfer function. From realizability conditions, the poles of the transfer function, here an all pole transfer function, must be in the left half of s-plane. The poles of the transfer function are determined by factorizing the denominator of $|T(j\omega)|^2$ and properly allocating its roots. To this end let

$$T(j\omega)^2 = \frac{1}{|H(j\omega)|^2} \quad \text{...(see eqn. 11.17)}$$

Here $|H(j\omega)|^2 = 1 + \omega^{2n}$, and by analytical continuation we can write

$$\left|H(j\omega)^2\right| = H(s)H(-s)\Big|_{s=j\omega} = 1 + (-s^2)^n\Big|_{s^2=-\omega^2}$$

The $2n$ zeros of $H(s)\ H(-s)$ are obtained by solving

$$1 + (-1)^n s^{2n} = 0 \quad \text{...(10.11)}$$

As can be easily seen, the roots of $H(s)\ H(-s)$ appear in quadrantal symmetry and lie on a circle, Butterworth circle, if radis 1_2 (n roots of unity). Hence, the roots in the LH plane can be separated and assigned as the poles of $T(s)$. Two cases arise depending upon whether n is even or odd.

(*i*) *If n is even:* Equation 10.11 reduces to

$$s^{2n} = -1 = e^{j(2k-1)\pi}$$

The $2n$-roots are

$$p_k = \exp\left[j\frac{2k-1}{2n}\pi\right],\ k = 1, 2, ..., 2n \quad \text{...(10.12)}$$

i.e.,

$$p_k = \cos\left[\frac{2k-1}{2n}\pi\right] + j\sin\left[\frac{2k-1}{2n}\pi\right], k = 1, 2, ..., 2n$$

(*ii*) *If n is odd*: Equation 10.11 reduces to

$$s^{2n} = 1 = e^{j2k\pi}$$

The $2n$ roots are

$$p_k = \exp\left[j\frac{k}{n}\pi\right],\ k = 0, 1, ..., (2n-1), \quad \text{...(10.13)}$$

i.e.,

$$p_k = \cos\left[\frac{k}{n}\pi\right] + j\sin\left[\frac{k}{n}\pi\right], k = 0, 1, ..., (2n-1).$$

The transfer function is constructed from the LH plane roots and is given by (see prob. 10.3)

$$n\ even: T(s) = \frac{1}{\prod_{k=1}^{n/2}(s^2 + 2\cos\theta_k s + 1)} \quad \text{...(10.14a)}$$

with $$\theta_k = \frac{2k-1}{2n}\pi$$

$$n\ odd: T(s) = \frac{1}{(s+1)\prod_{k=1}^{(n-1)/2}(s^2 + 2\cos\theta_k s + 1)} \qquad \ldots(10.14b)$$

with $$\theta_k = \frac{\pi}{n}$$

For example for $n = 2$

$$T(s) = \frac{1}{s^2 + 2\cos\frac{\pi}{4}s + 1} = \frac{1}{s^2 + \sqrt{2s} + 1}$$

and for $n = 3$

$$T(s) = \frac{1}{(s+1)\left(s^2 + 2\cos\frac{\pi}{3}s + 1\right)}$$

$$= \frac{1}{s^3 + 2s^2 + 2s + 1}$$

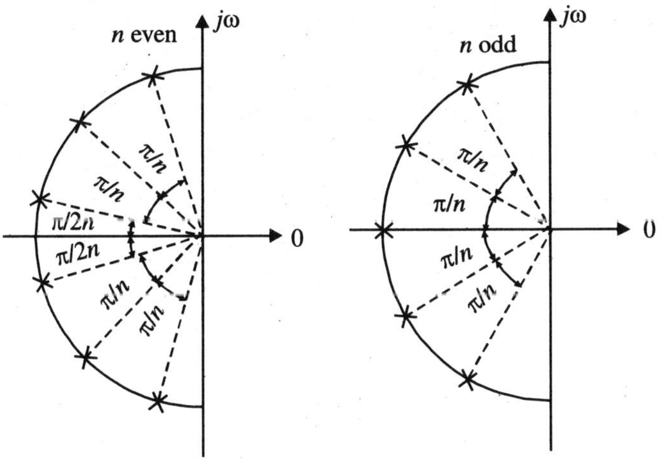

Fig. 10.4

The poles of $T(s)$ for even and odd n arc shown in thc Fig. 10.4. It can be seen that for odd n, there is always a pole on the negative real axis and for both the cases there cannot be a pole on the $j\omega$ = axis. Table 10.1 gives the denominator polynomial of $T(s)$ for various n.

In the preceding discussion c was assumed to be 1. If it was not the case, *i.e.*, when the pass-band edge is not the 3 dB point, the radius of the *Butterworth circle,*

Table 10.1

a	*Roots*	*Factors*	*Polynomial*
1	-1	$s + 1$	$s + 1$
2	$-\cos \pi/4 \pm j \sin \pi/4$	$s^2 + \sqrt{2s} + 1$	$s^2 + \sqrt{2s} + 1$
3	$-1, -\cos \pi/3 \pm j \sin \pi/3$	$(s + 1), (s^2 + s + 1)$	$s^3 + 2s^2 + 2s + 1$
4	$-\cos k\frac{\pi}{8} \pm j\sin k\frac{\pi}{8} (k = 1, 3)$	$(s^2 + 0.765s + 1), (s^2 + 1.848s + 1)$	$s^4 + 2.613s^3 + 3.414s + 2.613s + 1$
5	$-1, -\cos\frac{k\pi}{5} \pm j\sin\frac{k\pi}{5} (k = 1, 2)$	$(s + 1), (s^2 + 0.618s + 1), (s^2 + 1.618s + 1)$	$s^5 + 3.236s^4 + 5.236s^3 + 5.236s^2 + 3.236s + 1$
6	$-\cos k\frac{\pi}{12} \pm j\sin k\frac{\pi}{12} (k = 1, 3, 5)$	$(s^2 + 0.518 + 1), (s^2 + \sqrt{2s} + 1),$ $(s^2 + 1.932s + 1)$	$s^6 + 3.864s^5 + 7.464s^4 + 9.142s^3 + 7.464^2 +$ $3.864s + 1$
7	$-1, -\cos k\frac{\pi}{7} \pm j\sin k\frac{\pi}{7} (k = 1, 2, 3)$	$(s + 1), (s^2 + 0.44s + 1), (s^2 + 1.247s + 1),$ $(s^2 + 1.802s + 1)$	$s^7 + 4.494s^6 + 10.098s^5 + 14.592s^4 +$ $14.592s^3 + 10.098s^2 + 4.494s + 1$
8	$-\cos k\frac{\pi}{16} \pm j\sin k\frac{\pi}{16} (k = 1, 3, 5, 7)$	$(s^2 + 0.3s + 1), (s^2 + 1.111s + 1),$ $(s^2 + 1.166s + 1), (s^2 + 1.962s + 1)$	$s^8 + 5.126s^7 + 13.137s^6 + 21.846s^5 +$ $25.688s^4 + 21.846s^3 + 13.137s^5 + 5.126s + 1$
9	$-1, -\cos k\frac{\pi}{9} \pm j\sin k\frac{\pi}{9} (k = 1, 2, 3, 4)$	$(s + 1), (s^2 + 0.347s + 1), (s^2 + s + 1),$ $(s^2 + 1.532s + 1), (s^2 + 1.879s + 1),$	$s^9 + 5.759s^8 + 16.582s^7 + 31.163s^6 +$ $41.986s^5 + 41.986s^4 + 31.163s^3 +$ $16.582s^2 + 5.759s + 1$
10	$-\cos k\frac{\pi}{20} \pm j\sin k\frac{\pi}{20} (k = 1, 3, 5, 7, 9)$	$(s^2 + 0.313s + 1), (s^2 + 0.908s + 1),$ $(s^2 + \sqrt{2s} + 1), (s^2 + 1.792s + 1),$ $(s^2 + 1.975s + 1)$	$s^{10} + 6.393s^9 + 20.432s^8 + 42.802s^7 +$ $64.882s^6 + 74.233s^5 + 64.882s^4 + 42.802s^3$ $+ 20.432s^2 + 6.393s + 1$

is $\frac{1}{\sqrt[n]{c}}$ and the roots of $H(s)$ are given by

$$n \text{ even: } p_k = \frac{1}{\sqrt[n]{c}} \exp\left[j\frac{2k-1}{2n}\pi \right], \quad k = 1, 2, ..., 2n \qquad ...(10.15)$$

$$n \text{ odd: } p_k = \frac{1}{\sqrt[n]{c}} \exp\left[i\frac{k}{n}\pi \right], \; k = 0, 1, ..., 2n-1$$

and the transfer function $T(s)$ is obtained from the *LH* plane roots as indicated earlier.

10.3 CHEBYSHEV APPROXIMATION

From the standpoint of obtaining the best approximation to the ideal filter characteristic from a polynomial of a given degree, the Butterworth function does not do as well as may be done because it concentrates all of the approximating ability of the polynomial at $\omega = 0$, instead of distributing it over the range $0 < \omega < 1$. A better result in this regard may be obtained if we look for a rational function that approximates the constant value unity throughout this range in an oscillatory manner, rather than a monotonic manner. *Chebyshev Approximation* does exactly this.

Chebyshev polynomials are defined as linearly independent solutions of the differential equation

$$(1-\omega^2)\ddot{y} - \omega\dot{y} - n^2 y = 0$$

One of the solution is

$$y = T_n(\omega) = \cos\,(n \cos^{-1} \omega),\; |\omega| < 1$$

$$= \cosh\,(n \cosh^{-1} \omega),\; |\omega| \geq 1 \qquad ...(10.16)$$

$T_n(\omega)$ is called the Chebyshev polynomial of the first kind. A power series expansion for $T_n(\omega)$ can be obtained by rewriting eqn. 10.16 as

$$T_n(\omega) = \text{Re}\, |e^{jn}\, \phi| = \text{Re}\, |\cos\phi + j\sin\phi\,|^n \qquad ...(10.17)$$

where $\phi = \cos^{-1}\omega$

i.e., $\cos\phi = \omega,\; \sin\phi = \sqrt{1-\omega^2}.$

Binomial expansion of eqn. 10.17 yields

$$T_n(\omega) = \text{Re}\, [\omega + j(1 - \omega^2)]^n$$

$$= \omega^n - \frac{n(n-1)}{2!}\omega^{n-2}(1-\omega^2) + \frac{n(n-1)(n-2)(n-3)}{4!}\omega^{n-4}(1-\omega^2)^2 \; ...(10.18)$$

From which

$$T_0(\omega) = 1$$

$$T_1(\omega) = \omega$$

Rather than using eqn. 10.18 we can obtain a recurrence relation for $T_n(\omega)$. From eqn. 10.16 we have

$$\begin{aligned} T_{n+1}(\omega) &= \cos[(n+1)\phi] \\ &= \cos\phi \cos n\phi - \sin\phi \sin n\phi \\ &= 2\cos\phi \cos n\phi - (\cos n\phi + \sin\phi \sin\phi) \\ &= 2\omega \cos n\phi - \cos[(n-1)\phi] \\ &= 2\omega\, T_n(\omega) - T_{n-1}(\omega) \end{aligned}$$

i.e., $$T_{n+1}(\omega) = 2\omega T_n(\omega) - T_{n-1}(\omega) \qquad \text{...(10.19)}$$

with $T_0(\omega) = 1$ and $T_1(\omega) = \omega$

Table 10.2 gives $T_n(\omega)$ for various values of n (observe that the coefficient of the highest power is 2^{n-1}).

Table 10.2

n	$T_n(\omega)$
0	1
1	ω
2	$2\omega^2 - 1$
3	$4\omega^3 - 3\omega$
4	$8\omega^4 - 8\omega^2 + 1$
5	$16\omega^5 - 20\omega^3 + 5\omega$
6	$32\omega^6 - 48\omega^4 + 18\omega^2 - 1$
7	$64\omega^7 - 112\omega^5 + 56\omega^3\ 7\omega$
8	$128\omega^8 - 256\omega^6 + 160\omega^4 - 32\omega^2 + 1$
9	$256\omega^9 - 576\omega^7 + 432\omega^5 - 120\omega^3 + 9\omega$
10	$512\omega^{10} - 1280\omega^8 + 1120\omega^6 - 400\omega^4 + 50\omega^2 - 1$

Some properties of these polynomials are:

(*i*) The zeros of the polynomials are all located in the interval $|\omega| \le 1$.

(*ii*) $T_n(0) = (-1)^{n/2}$ and $T_n(+1) = 1$ for n even

$T_n(0) = 0$ and $|T_n(\pm 1)| = 1$ for n odd.

i.e., $T_n(\omega) \le 1$ for $|\omega|\ \omega\ 1$.

(*iii*) Beyond the interval $|\omega| \le 1$, $T_n(\omega)$ increases rapidly for increasing values of $|\omega|$.

For Chebyshev approximation $K_n(\omega)$ is written as $T_n(\omega)$ in equation 10.1(*a*). Thus,

$$|T(j\omega)|^2 = \frac{1}{1+\varepsilon^2 T_n^2\left(\dfrac{\omega}{\omega_p}\right)} \qquad \text{...(10.20)}$$

and

$$\alpha(\omega) = 10\log\left[1+\varepsilon^2 T_n^2\left(\frac{\omega}{\omega_p}\right)\right] \quad \ldots(10.21)$$

With the specification for a low-pass filter as given in Fig. 10.2, eqn. 10.21 yields

$$\alpha_p = 10\log\left[1+\varepsilon^2 T_n^2\left(\frac{\omega_p}{\omega_p}\right)\right]$$
$$\alpha_s = 10\log\left[1+\varepsilon^2 T_n^2\left(\frac{\omega_s}{\omega_p}\right)\right] \quad \ldots(10.22)$$

These equations can be solved to yield ε and n. From the first eqn. in 10.22

$$\varepsilon^2 = 10^{0.1\alpha p} - 1 \quad \ldots(10.23)$$

as T_n^2 (1) = 1. From the same equation it can be shown that the order of the filter is given by (prob. 10.4).

$$n \geq \frac{\cosh^{-1}(1/k_1)}{\cosh^{-1}(1/k)} \quad \ldots(10.24)$$

where k and k_1 are given as earlier by eqn. 10.5 and 10.7, respectively.

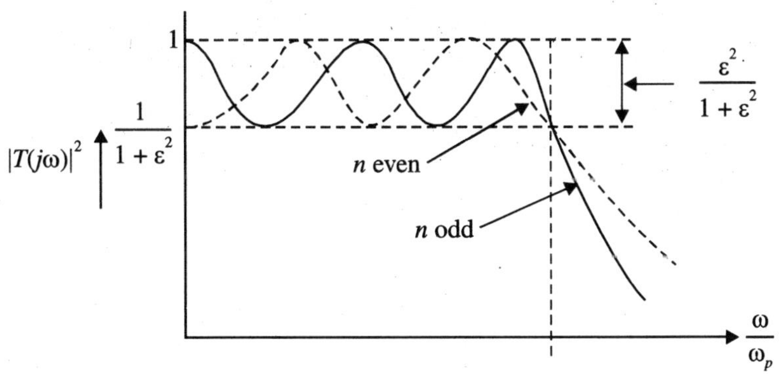

Fig. 10.5

Figure 10.5 shows a plot of | $T(j\omega)$ |^2 for two different values of n(n = 4 and n = 5).

From eqn. 10.20 it can be seen that

$$|T(j\omega)|^2 \text{ max} = 1 \text{ when } T_n(\omega) = 0$$

$$|T(j\omega)|^2 \min = \frac{1}{1+\varepsilon^2} \text{ when } T_n(\omega) = \pm 1.$$

Hence, the peak to peak ripple is $\dfrac{\varepsilon^2}{1+\varepsilon^2}$ and the approximating function ripples about the value $\dfrac{1+\varepsilon^2/2}{1+\varepsilon^2}$. Typically, the allowable ripple α_p is specified in terms of decibels. The ripple in dB is given by

$$\text{dB (ripple)} = 10 \log (1 + \varepsilon^2) \qquad \ldots(10.25)$$

(This is only an alternate way of writing eqn. 10.23). Figure 10.6 shows the relationship between dB (ripple) and ε.

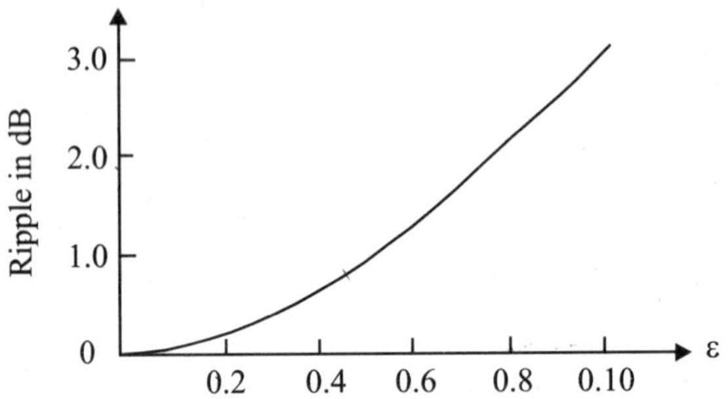

Fig. 10.6

The transfer function can obtained as in the case of Butterworth approximation by determining its poles. To this end, we define $|H(j\omega)|^2$ as

$$|H(j\omega)|^2 = 1 + \varepsilon^2 T^2(\omega) \qquad \ldots(10.26)$$

The zeros of $H(s)\,H(-s)_{s=j\omega} = |H(\omega)|^2$ appear, as already indicated, in quadrantal symmetry and can be separated into *LH* and *RH* plane roots. The *LH* roots are assigned as the poles of $T(s)$ and $T(s)$ is obtained from these poles.

It can be shown that the poles $T(s)$ are given by (prob. 10.6).

$$s_k = \sigma_k + j\omega_k$$

$$\sigma_k = \sinh v \sin\left(\frac{2k+1}{2\pi}\pi\right),\ k = 0, 1, 2, \ldots, 2n-1 \qquad \ldots(10.27)$$

$$\omega_k = \cosh v \cos\left(\frac{2k+1}{2n}\pi\right)$$

where $v = 1/n \sinh^{-1} 1/\varepsilon$.

In Butterworth approximation, the poles were located on a circle of radius $\dfrac{1}{n\sqrt{c}}$. The locus of poles of Chebyshev approximation can also be similarly determined. From eqn. 10.27 we have

$$\frac{\sigma_k}{\sinh v} = \sin\left(\frac{2k+1}{2n}\pi\right)$$

and

$$\frac{\omega_k}{\cosh v} = \cos\left(\frac{2k+1}{2n}\pi\right)$$

and finally

$$\frac{\sigma_k^2}{\sinh^2 v} + \frac{\omega_k^2}{\cosh^2 v} = 1$$

indicating that the poles of $T(s)$ lie on an ellipse with the major axis along the $j\omega$-axis and the minor axis along the σ-axis.

It is not necessary to evaluate eqn. 10.27 to find the poles of $T(s)$. The poles can also be found graphically (Fig. 10.7). The semi-major and minor axes of the ellipse are $b(b = \cosh v)$ and $a(a = \sinh v)$ respectively. (Here the pass-band edge

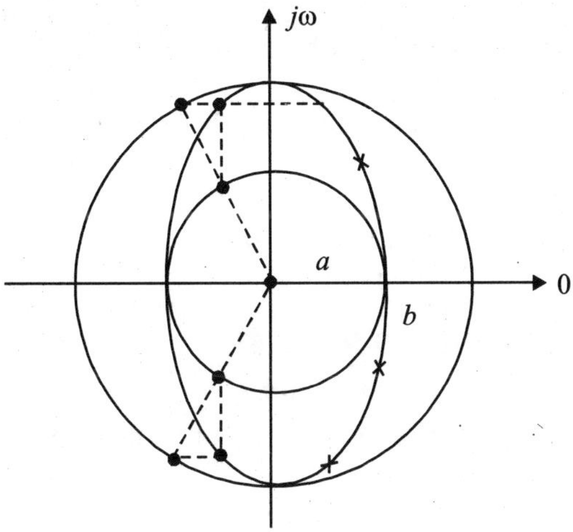

Fig. 10.7

is assumed to be 1. For a pass-band edge of ω_p the semi-major and semi-minor axes are $h\omega_p$, and $b\omega_p$, respectively.) It can be shown that (prob. 10.7)

$$a - \frac{1}{2}\left(\Delta^{1/n} - \Delta^{-1/n}\right),\ b = \frac{1}{2}\left(\Delta^{1/n} + \Delta^{-1/n}\right) \qquad \text{...(10.28)}$$

where $$\Delta = \frac{1}{\varepsilon} + \sqrt{\frac{1}{\varepsilon^2} + 1}$$

ε and n are found from the pass and the stop band specifications as in eqns. 10.25 and 10.24. To locate the poles of the Chebyshev filter, we first identify the points on the major and minor circles. These points, which are symmetrically located with respect to $j\omega$ axis, are equally spaced along the circle with a spacing of π/n radians. No point falls on the imaginary axis and for odd n a point occurs on the real axis. The points on the circles can be considered to correspond to the poles of the Butterworth filter. The poles of the Chebyshev filter are on the ellipse. The ordinate of the poles is specified by the point on the major circle and abscissa by the point on the minor circle, *i.e.*, the real part of the Chebyshev pole is the same as the real part of the Butterworth pole on the minor circle, and the imaginary part is the same as the imaginary part of the Butterworth pole on the major circle.

Example 10.2. The specifications for a low-pass filter are:

Pass band ripple	1 dB
Pass band	0 to 1.75 MHz
Stop band loss	at least 20 dB at 2.5 MHz

From eqn. 10.25

$$1 = 10 \log (1 + \varepsilon^2)$$

i.e., $\varepsilon^2 = 0.2589$ or $\varepsilon = 0.5088$.

From eqn. 10.24

$$n \geq \frac{\cosh^{-1}(1/k_1)}{\cosh^{-1}(1/k)}$$

$$\frac{1}{k} = \frac{2.5}{1.75} = 1.43$$

$$\frac{1}{k_1} = \left[\frac{10^{20/10} - 1}{10^{1/10} - 1}\right] = 19.56$$

$$n \geq \frac{\cosh^{-1} 19.56}{\cosh^{-1} 1.43} = \frac{3.67}{0.897} = 4.0914$$

Let $n = 5$, n being odd, there will be a root on the real axis.

From eqn. 10.28

$$\Delta = \frac{1}{0.5088} + \sqrt{1 + \frac{1}{0.2589}} = 4.1705$$

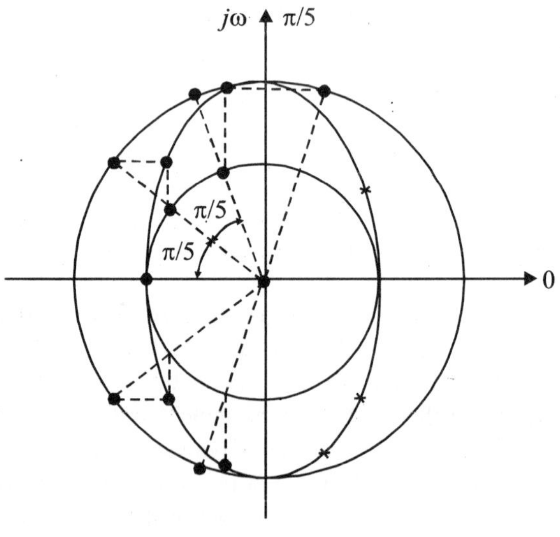

Fig. 10.8

$$a = \frac{1}{2}\left(4.1705^{1/5} - 4.1705^{-1/5}\right) = 0.2895$$

$$b = \frac{1}{2}(4.1705^{1/5} + 4.1705^{-1/5}) = 1.0411$$

The spacing between the points on the circle is $\pi/5$ (Fig. 10.8).

The poles of the Chebyshev filter are at

$$-a,\ -a\cos\pi/5 \pm jb\sin\pi/5,\ -a\cos 2\pi/5 \pm jb\sin 2\pi/5$$

i.e., the poles are at

$$-0.2895,\ -0.2342 \pm j0.612,\ -0.0895 \pm j0.9902$$

Hence,

$$H(s) = (s + 0.2895)\ \{(s + 0.2342)^2 + (0.612)^2\}\ \{(s + 0.0895)^2 + (0.9902)^2\}$$
$$= (s + 0.2895)\ (s^2 + 0.4684s + 0.4294)\ (s^2 + 0.179s + 0.9902).$$

10.4 COMPARISON BETWEEN BUTTERWORTH AND CHEBYSHEV TRANSFER FUNCTIONS

The Butterworth and Chebyshev approximations discussed in Secs. 10.2 and 10.3 are both polynomial approximations for ideal frequency responses of Fig. 10.2. These lead to an all pole transfer function.

i.e.,
$$T(s) = \frac{1}{s^n + a_{n-1}s^{n-1} + \ldots + a_1 s + a_0}$$

The order of the Chebyshev filter to satisfy a set of specifications is always lower than that of the corresponding Butterworth filter.

Example 10.3. Let the specifications for a low-pass filter be:

$\alpha_p \le 1$ dB $\qquad \omega \le 150\ k$ rad/sec

$\alpha_s \ge 60$ dB $\qquad \omega \ge 200\ k$ rad/sec.

The selectivity and discrimination parameters for this filter are

$$k = \frac{\omega_p}{\omega_s} = \frac{150}{200} = 0.75$$

$$k_1 = \left[\frac{10^{0.1} - 1}{10^6 - 1}\right]^{1/2} = 0.5089 \times 10^{-3}$$

(*i*) *Butterworth filter:*

From eqn. 10.9(*a*)

$$c = \sqrt{10^{0.1} - 1} = 0.5089$$

From eqn. 10.8

$$n \geq \frac{\log(0.5089 \times 10^{-3})}{\log(0.75)} = 26.36$$

i.e., $n = 27.$

(*ii*) *Chebyshev filter:*

From eqn. 10.23

$$\varepsilon = \sqrt{10^{0.1} - 1} = 0.5089.$$

From eqn. 10.24

$$n \geq \frac{\cosh^{-1}\left(\dfrac{1}{0.5089 \times 10^{-3}}\right)}{\cosh^{-1}\left(\dfrac{1}{0.75}\right)} = 10.4$$

$$n = 11.$$

From Fig. 10.7 it is evident that the quality factor of the Chebyshev roots are higher than those of the Butterworth roots.

Hence, for a given order of the filter transfer function the transition band of the former is narrower than the latter.

The two approximations can also be compared by their behavior at high frequencies. Let $|T_B(j\omega)|^2$ and $|T_C(j\omega)|^2$ be the Butterworth and Chebyshev magnitude functions. As ω tends to ∞.

$$|T_B(j\omega)|^2 \to \frac{1}{c^2 \omega^{2n}}$$

and
$$|T_C(j\omega)|^2 \to \frac{1}{\varepsilon^2 (2^{2n-2} \omega^{2n})}$$

i.e.,
$$10 \log |T_B(j\omega)|^2 = -(20 \log c + 20n \log \omega)\,\text{dB}$$

and
$$10 \log |T_C(j\omega)|^2 = -20[\log \varepsilon + 20n \log \omega + 3(2n-2)]\,\text{dB} \qquad \ldots(10.29)$$

Both the responses have the same slope ($-20n$ dB decode). Butterworth is offset by an amount depending on c while the offset in Chebyshev is dependent on both n and ε. Because of the factor $3(2n - 2)$ the Chebyshev response has a smaller value in the stop band for all values of n and for $c = \varepsilon_1$ and consequently is a better approximation.

Figure 10.9 shows the phase responses of Butterworth and Chebyshev filters. As can be observed from the graph a Butterworth filter gives a fairly linear phase response and is superior to Chebyshev phase response.

As the order of the Chebyshev response becomes higher, the phase response becomes more non-linear. If linear phase response is a requirement, neither Butterworth nor Chebyshev is suitable (see Sec. 10.7).

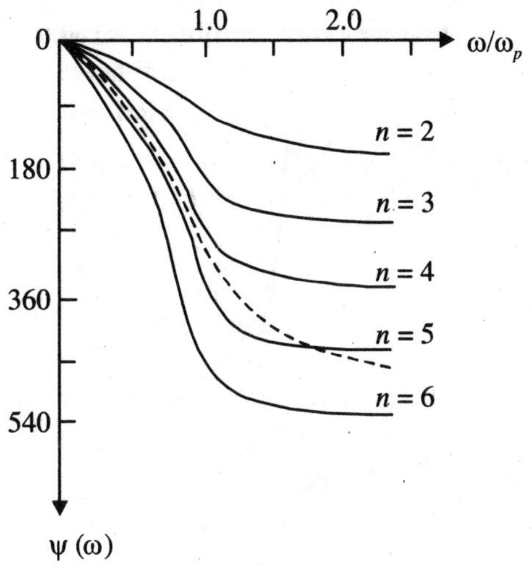

Fig. 10.9

From eqn. 10.29 it is evident that for a given order of the Chebyshev polynomial *n,* the increase of ε, *i.e.,* the increase of the pass-band ripple, will improve the stop-band performance. The effect of variation of ε (or α_p) is shown in Fig. 10.10. It is also evident from eqn. 10.29 that the increase of the order of the polynomial will increase the loss in the stop band (Fig. 10.11). From Figs. 10.10 and 10.11 it can be concluded that a smaller pass-band ripple and sharper cut-off requirements cannot be simultaneously achieved; if one is decreased and other increases.

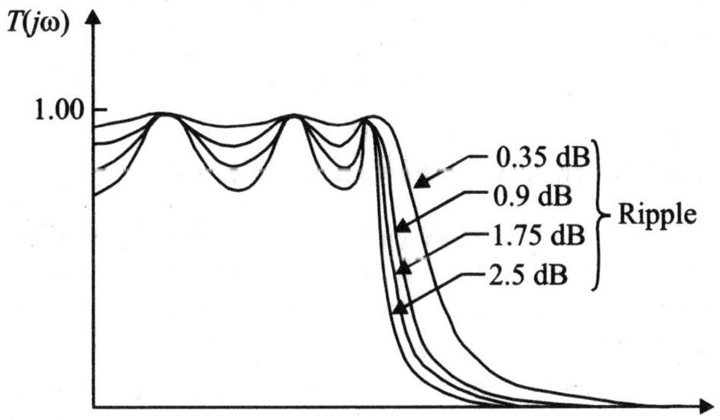

Fig. 10.10

A final comparison of Fig. 10.12 shows the response curves of a 6th order Butterworth filter (α_p = 3dB) and two 6th order Chebyshev filters (α_p = 0.1 dB and α_p = 0.969 dB). The conclusions we have arrived at in the preceding discussion are confirmed by these curves. There is no better *n*th order all-pole filter than Chebyshev filter with equal or better performance in both the pass and stop band.

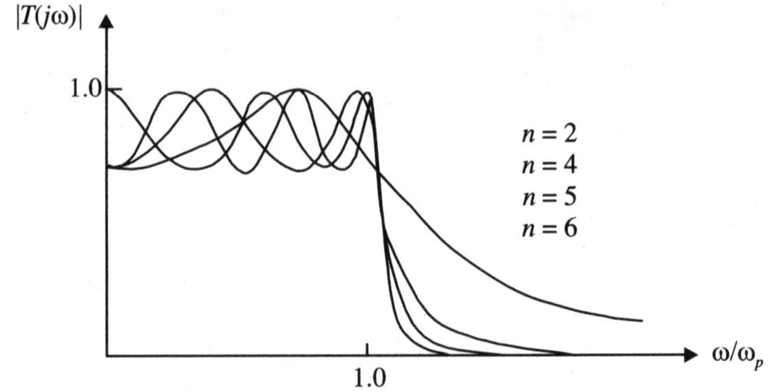

Fig. 10.11

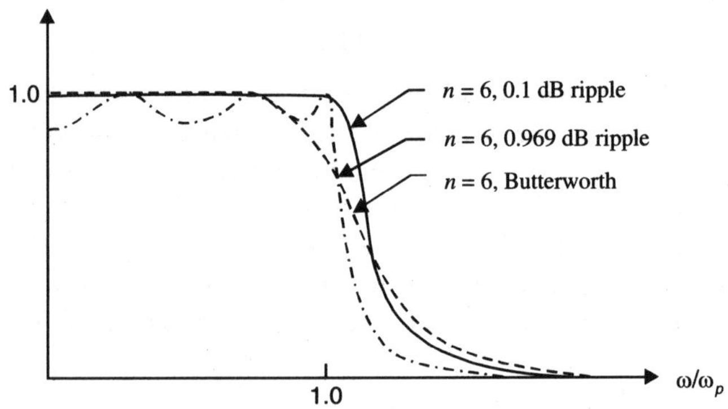

Fig. 10.12

10.5 FREQUENCY TRANSFORMATION

In the earlier sections of this chapter, we have only discussed the design of low-pass (LP) filters. There are many other types of filters which find use in signal processing. Figure 10.13 shows the frequency characteristics of ideal *high-pass* (HP) and *band-pass* (BP) and *band elimination* (BE) filters. Once the low-pass prototype filter has been designed, the other filters can be obtained by a simple *frequency transformation.* The frequency variable of the low-pass filter is transformed by a *reactance transformation* to obtain high-pass, band-pass and band elimination characteristics. In the reactance transformation, the low-pass frequency variable Ω is replaced by a realizable, rational function $X(\omega)$ of another frequency variable ω. By a proper choice of this function we can transform the low-pass characteristic to other characteristics. The transformation $X(\omega)$ is selected as a *reactance function* (see Chap. 9).

Let $$\Omega = X(\omega)$$

be the required transformation where

$$X(\omega) = A\frac{\omega(\omega_2^2 - \omega^2)(\omega_4^2 - \omega^2)}{(\omega_1^2 - \omega^2)(\omega_3^2 - \omega^2)...}$$

and $$0 < \omega_1 < \omega_2 < \omega_3 < ...$$

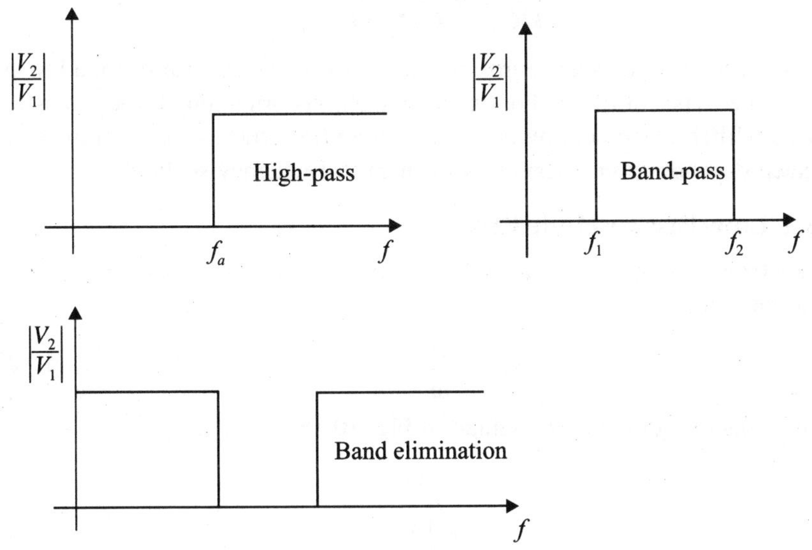

Fig. 10.13

This transformation can be graphically represented, as shown in Fig. 10.14. Let the low-pass cut-off frequency be $-\Omega_c$ and + Ω_e. (Here we have assumed that the low-pass is symmetric about the 0 frequency). The pass-band $-\Omega_c \le \Omega \le \Omega_c$ of the prototype is transformed to a set of pass-bands centered at 0, $\pm\ \omega_2$, $\pm\ \omega_4$ ... Thus, we have obtained a multiple pass and stop-band filter by transforming a single pass-band filter. The elements of the transformed filter can be obtained by

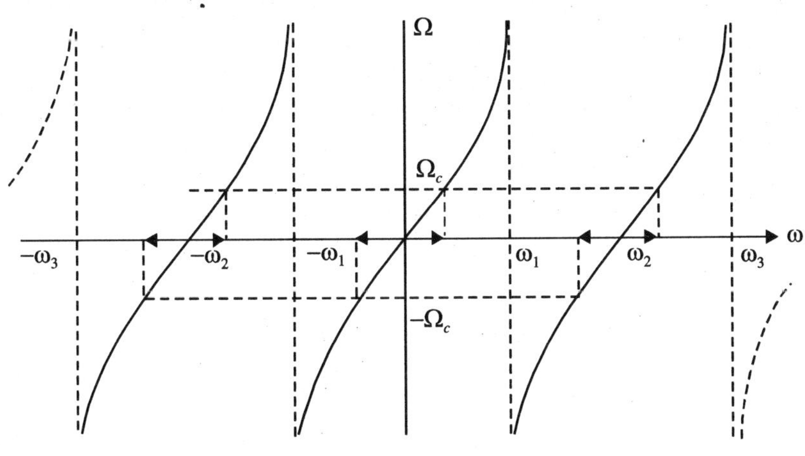

Fig. 10.14

applying the transformation to the reactive elements of the prototype. An inductor of the prototype L_p is transformed by the rule

$$\Omega L_p \to L_p X(\omega)$$

and a capacitance C_p by the rule

$$\frac{1}{\Omega C_p} \to \frac{1}{C_p X(\omega)}$$

By selecting the poles and zeros of $X(\omega)$ properly we can transform a low-pass filter to other types of filters. For example, the transformation $\Omega = A\omega$ transforms a low-pass filter to another low-pass filter whose frequency scale is compressed or elongated depending on A (this is equivalent to frequency scaling).

10.5.1 Low-Pass to High-Pass

To transform a low-pass to a high-pass filter, select $X(\omega)$ as $-A/\omega$, *i.e.*, the transformation is

$$\Omega = -\frac{A}{\omega} \qquad \text{...(10.29}a\text{)}$$

This transformation is represented in Fig. 10.15.

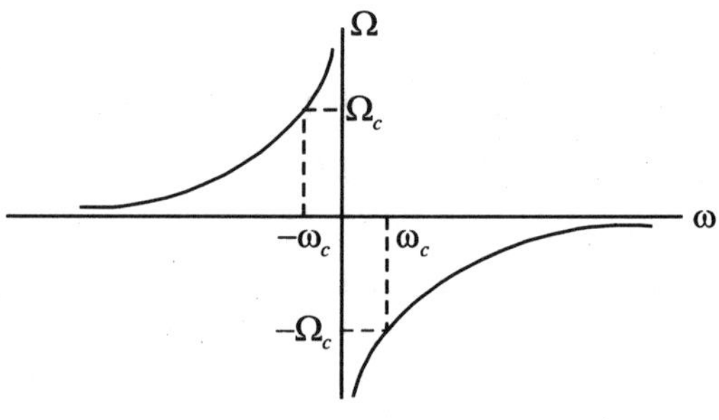

Fig. 10.15

From the figure it can be observed that the pass-band $(-\Omega_c, \Omega_c)$ of the low-pass is transformed to two pass-bands (ω_c, ∞) and $(-\omega_c, -\infty)$. Thus, a low-pass is transformed to a HP. The transformation of eqn. 10.29*a* is simple frequency inversion. The inductance (capacitance) of the low-pass is transformed to a capacitance (inductance) in the HP.

$$j\Omega L_p \to jL_p\left(-\frac{A}{\omega}\right) = \frac{1}{jC_\omega}$$

$$\frac{1}{j\Omega C_p} \to \frac{1}{jC_p\left(-\frac{A}{\omega}\right)} = jL\omega$$

where $$C = \frac{1}{AL_p}$$

and $$L = \frac{1}{AC_p}$$

are the capacitance and inductance of the high-pass. Figure 10.16 shows a *LP* prototype and the transformed HP network.

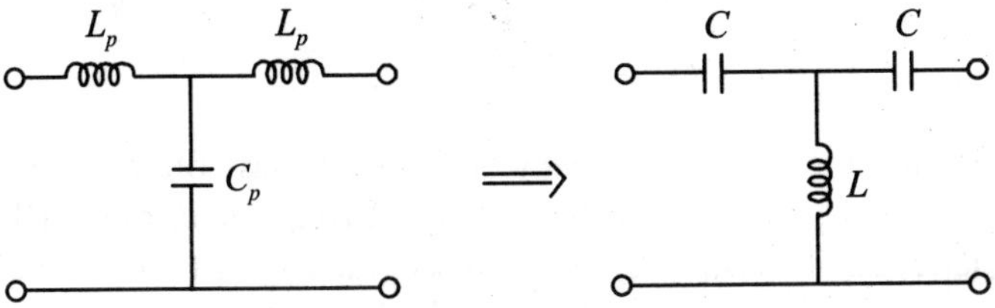

Fig. 10.16

10.5.2 Low-Pass to Band-Pass

The transformation which transforms a low-pass filter to a band-pass filter has to be a second order function, as the band-pass filter has two cut-off frequencies ω_1 and ω_2. Consider the transformation

$$\Omega = -A\frac{\omega_0^2 - \omega^2}{\omega} \quad \text{...(10.30)}$$

This transformation is represented graphically in Fig. 10.17. From the figure it can be observed that the passband $(-\Omega_c, \Omega_c)$ of the LP is transformed to two passbands $(\omega_1, -\omega_2)$ and $(-\omega_1, \omega_2)$. Thus, a LP is transformed to a BP. Also, it can be seen that the transformed frequencies corresponding to Ω_c are $-\omega_1$ and ω_2 and those to $-\Omega_c$ are ω_1 and $-\omega_2$.

If we select Ω equal to Ω_c, then eqn. 10.30 yields

$$\Omega_c = -A\frac{\omega_0^2 - \omega^2}{\omega}$$

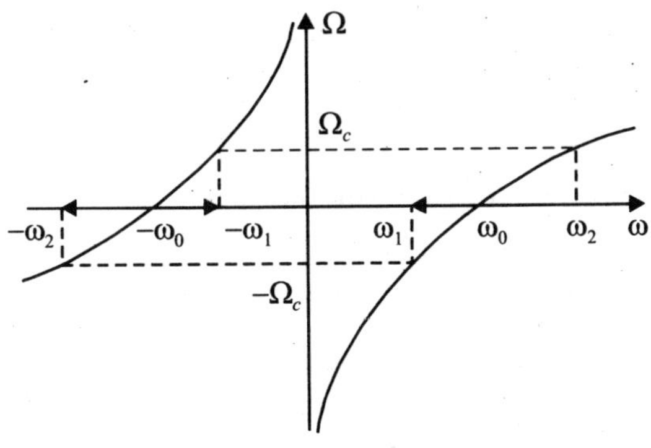

Fig. 10.17

i.e.,
$$-\omega^2 = -\frac{\Omega_c \omega}{A} - \omega_0^2 \qquad \text{...(10.31)}$$

The roots of these equations are known to be $-\omega_1$, and ω_2 an hence we have

$$\omega_1 \omega_2 = \omega_0^2$$

and
$$\omega_2 - \omega_1 = \frac{\Omega_c}{A}$$

We see that ω_0, the *centre frequency* of the band-pass filter, is the geometric mean of the *upper* (ω_2) and the *lower* (ω_1) cut-off frequencies of the BP. If the LP cut-off frequency Ω_c is normalized to 1 rad/sec, then $1/A$ is the *bandwidth* of the BP filter. Equation 10.31 can be solved explicitly for ω_2 and ω_1, assuming Ω_c to be 1 rad/sec.

This yields

$$\omega_1 = -\frac{1}{2A} + \sqrt{\frac{1}{4A^2} + \omega_0^2} \qquad \text{...(10.32)}$$

$$\omega_2 = \frac{1}{2A} + \sqrt{\frac{1}{4A^2} + \omega_0^2}$$

The elements of the transformed network can be found as before

$$\textit{Inductance } L_p : j\Omega L_p \to jL_p \left\{ A\left(\omega - \frac{\omega_0^2}{\omega} \right) \right\}$$

i.e.,
$$j\Omega L_p \to j\omega A L_p + \frac{1}{j\omega 1/AL_p\omega_0^2}$$

An inductor gets transformed to a series connection of an inductor L^1 and a capacitor C^1 where

$$L^1 = AL_p$$

$$C^1 = \frac{1}{L^1\omega_0^2} \qquad \text{...(10.33)}$$

$$\textit{Capacitance } C_p : \frac{1}{j\Omega C_p} \; \frac{1}{C_p A\left(\omega - \omega_0^2/\omega\right)}$$

i.e.,
$$\frac{1}{j\Omega C_h} \to \frac{1}{j\omega A C_p + A\omega_0^2 C_p / j\omega}$$

i.e., A capacitor gets transformed to a parallel combination of a capacitor C^2 and an inductor L^2 where

$$C^1 = AC_p$$

$$L^2 = \frac{1}{C^2\omega_0^2} \qquad \text{...(10.34)}$$

Figure 10.18 shows a LP prototype and the transformed BP network.

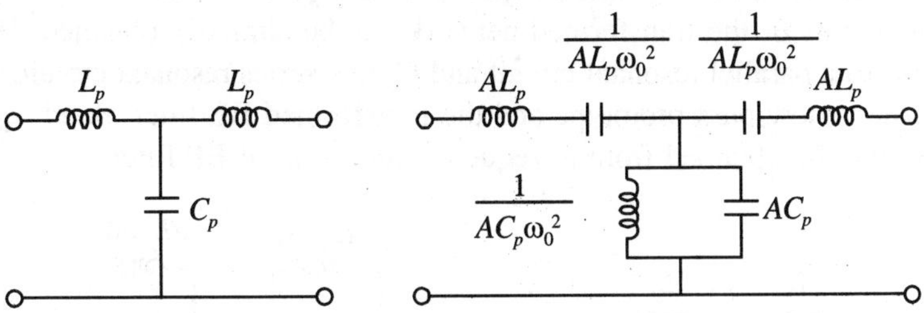

Fig. 10.18

10.5.3 Low-Pass to Band Elimination

Consider the transformation

$$\Omega = \frac{\omega}{\omega_0^2 - \omega^2} \qquad \ldots(10.35)$$

Figure 10.19 gives the graphical representation of this transformation. The pass band $(-\Omega_c, \Omega_c)$ of the LP is transformed to the passbands $(0, \omega_1)$, (ω_2, ∞) and $(0, -\omega_1)$, $(-\omega_2, -\infty)$. Obviously, the transformed filter is a *band elimination* or rejection filter with the two cut-off frequencies ω_1 and ω_2. The cut-off frequency Ω_c yields the two frequencies ω_1 and $-\omega_2$. From eqn. 10.35, we have

$$\omega^2 + \frac{A}{\Omega}\omega - \omega_0^2 = 0$$

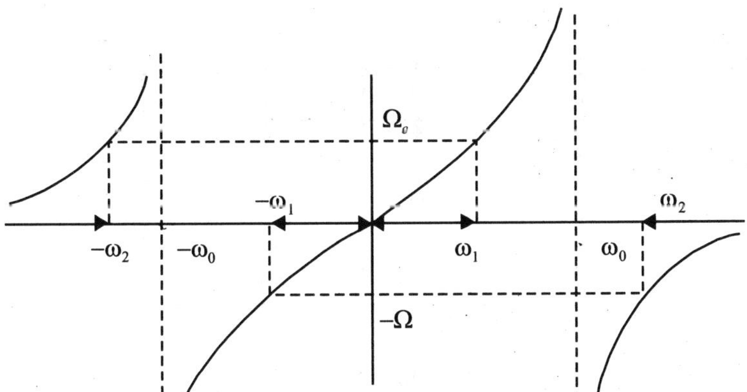

Fig. 10.19

This yields

$$\omega_2 - \omega_1 = \frac{A}{\Omega_c}$$

$$\omega_1 \, \omega_2 = \omega_0^2$$

$(\omega_2 - \omega_1)$ is the band of frequencies that is rejected. If we are considering a normalized LP with $\Omega_c = 1$, then A specifies the rejection band.

The elements of the transformed network can be similarly obtained. Here L_p transforms to a parallel resonant circuit and C_p to a series resonant circuit. Figure 10.20 shows a low-pass prototype and the transformed BE filter. Band rejection filter can also be obtained from a frequency inversion of BP filter.

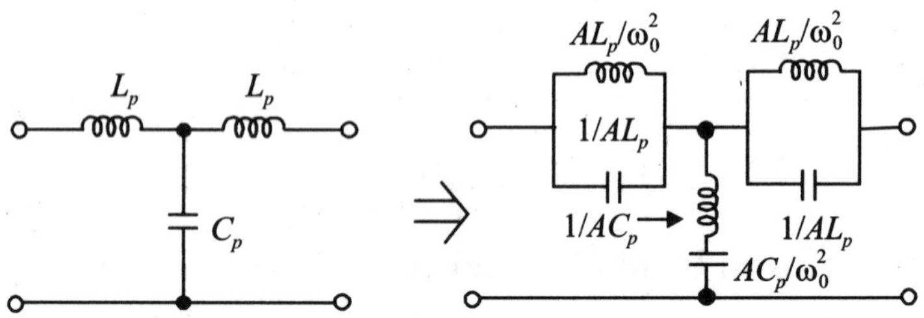

Fig. 10.20

10.6 BAND-PASS FILTER APPROXIMATION

We have discussed the basic frequency transformation that transform a LP to HP or BP filters. It is instructive to study the application of the same to obtain the approximation for a BP filter.

Figure 10.21 shows the specifications (or the frequency response) of a band-pass filter, ω_1 and ω_2 are the lower and the upper pass-band edges respectively, and ω_{s_1} and ω_{s_2} are the lower and the upper stop-band edges. ω_0 is the centre frequency of the filter.

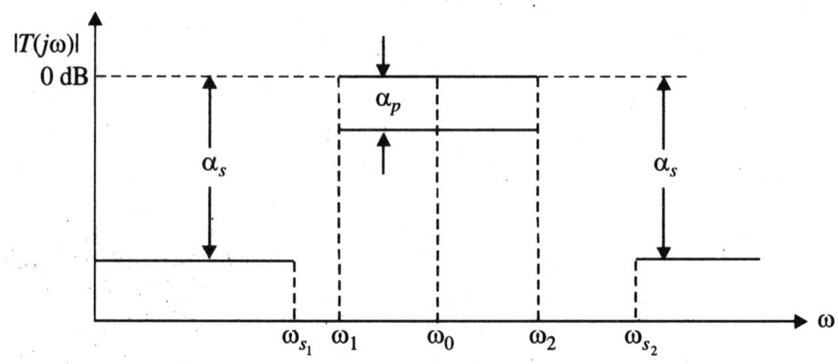

Fig. 10.21

From eqn. 10.31, we know that

$$\omega_0 = \sqrt{\omega_1 \omega_2}\,.$$

If the pass-band and the stop-band edge frequencies are such that

$$\frac{\omega_1}{\omega_{s_1}} = \frac{\omega_{s_2}}{\omega_2} \qquad \ldots(10.36)$$

(*i.e.*, $\omega_0 = \sqrt{\omega_1 \omega_2} = \sqrt{\omega_{s_1}, \omega_{s_2}}$), the band-pass filter is called *a geometrically symmetric band-pass filter.* The frequency or the spectral transformation from low-pass to band-pass given by

$$\Omega = -A \frac{\omega_0^2 - \omega^2}{\omega} \text{ with } \frac{\Omega_p}{\omega_2 - \omega_1} = A \qquad \ldots(10.37)$$

always results in a *geometrically symmetric filter.* Here Ω_p is the pass-band edge of the low-pass prototype.

If we are to design a band-pass filter our first task is to convert the band-pass filter specification to a corresponding low-pass filter specification and then perform the approximation on this filter specification. This yields a low-pass filter transfer function which is then transformed to a band-pass filter transfer function by the transformation.

$$s \to A \frac{s^2 + \omega_0^2}{s} \qquad \ldots(10.38)$$

The equation is obtained from 10.37 by substituting *s/j* for Ω and ω.

If the specified band-pass filter is not a geometrically symmetric filter, then lower of two ratios in eqn. 10.36 is used for the design. This improves the other transition band. An all-pole low-pass transfer function cannot be transformed to a geometrically asymmetric filter. To generate the latter the corresponding low-pass filter must necessarily have transmission zeros in the stop-band (see Sec. 10.7).

Let the specifications for a geometrically symmetric band-pass filter be

$$\alpha \le \alpha_p, \qquad \omega_1 \le \omega \le \omega_2$$

$$\alpha \ge \alpha_s, \qquad \omega \le \omega_{s_1}, \; \omega \ge \omega_{s_2}.$$

Let the normalized pass-band edge of the low-pass prototype be $\Omega_p = 1$ rad/sec (there is no loss of generality here). Then the constant *A* is given by

$$A = \frac{1}{\omega_2 - \omega_1}.$$

The stop-band edge of the low-pass prototype is now given by

$$\Omega_s = -A \cdot \frac{\omega_2 \omega_1 - \omega_{s_2}^2}{\omega_{s_2}} \qquad \ldots(10.39)$$

The specifications for the required low-pass filters is

$$\alpha \le \alpha_p, \qquad \Omega \le 1$$

$$\alpha \ge \alpha_s, \qquad \Omega \ge \alpha_s$$

The low-pass filter approximation is performed with these specifications.

Example 10.4. The specifications for a band-pass filter are:

$\alpha_p \le 30$ dB $\qquad$ 50 k rad < ω < 72 k rad,

$\alpha_s \ge 40$ dB $\qquad$ ω < 30 k rad; ω > 120 k rad

The specifications indicate that

$$\omega_1 = 50 \text{ k rad/sec}, \qquad \omega_2 = 72 \text{ k rad/sec}$$

Hence, $\omega_{s_1} = 30 \text{ k rad/sec}, \qquad \omega_{s_2} = 120 \text{ k rad/sec},$

$$\omega_0 = \sqrt{\omega_1 \omega_2} = \sqrt{\omega_{s_1} \omega_{s_2}} = 160 \text{ k rad/sec}.$$

and the filter is a geometrically symmetric filter. Let the pass-band edge of the low-pass prototype be 1 rad/sec. Then

$$A = \frac{1}{BW} = \frac{1}{\omega_2 - \omega_1} = \frac{1}{22k}$$

$$= 0.0455 \times 10^{-3}.$$

The stop-band edge of the low-pass prototype from eqn. 10.39 is

$$\Omega_s = -0.0455 \times 10^{-3} \times \frac{(60 \times 10^3)^2 - (120 \times 10^3)^2}{(120 \times 10^3)}$$

$$= 4.095$$

Hence, the low-pass filter specifications are

$$\alpha_p \leq 3 \text{ dB}, \qquad \Omega \leq 1 \text{ rad/sec}$$

$$\alpha_s \geq 40 \text{ dB}, \qquad \Omega \geq 4.095$$

We can now approximate this by a Butterworth filter. We have

$$k = \frac{1}{4.095} = 0.2442$$

$$k_1 = \left[\frac{10^{0.3} - 1}{10^4 - 1}\right] = 0.0100$$

and $$n \geq \frac{\log k_1}{\log k} = 3.27. \text{ Let } n = 4.$$

Since the attenuation at the pass-band edge is 3 dB, $c = 1$. From Table 10.1 the low-pass transfer function (with $\omega_p = 1$ rad/sec) is

$$T_1(s) = \frac{1}{s^4 + 2.613s^3 + 3.414s^2 + 2.613s + 1}$$

The transformation from LP $\rightarrow$ BP from eqn. 10.38 is

$$s \rightarrow 0.0455 \times 10^{-3} \frac{s^2 + (60 \times 10^3)^2}{s}.$$

yielding the band-pass transfer function as

$$T_{B_p}(s) = \frac{2.332 \times 100^{17} s^4}{\begin{aligned}(s^8 &+ 2.1978 \times 10^4 s^7 + 1.4883 \times 10^{10} s^6 + 2.4795 \times 10^{14} s^5 \\ &+ 8.1238 \times 10^{19} s^4 + 8.9272 \times 10^{23} s^3 + 1.9259 \times 10^{29} s^2 \\ &+ 1.0255 \times 10^{33} s + 1.6798 \times 10^{38})\end{aligned}}$$

10.7 OTHER METHODS OF APPROXIMATION

In both Butterworth and Chebyshev approximations, the filter response in stop-band is monotonic. This, as we have seen, lead to an all-pole transfer function as

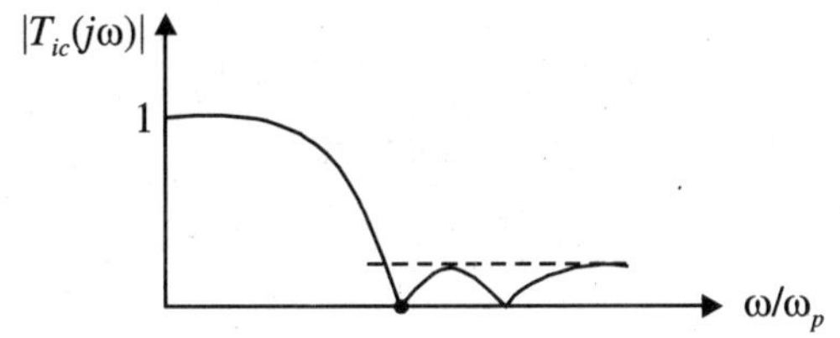

Fig. 10.22

the function $K_n^2\ (\omega)$ is a polynomial in ω. It is possible to approximate the ideal filter characteristic by a function which has equiripple response in stop-band. For this case the function $K_n^2\ (\omega)$ is a rational function and the transfer function has finite zeros.

10.7.1 Inverse Chebyshev Approximation

By manipulation of the Chebyshev approximation of eqn. 10.20 it is possible to obtain an approximation which has monotonic response in the pass-band and equiripple in the stop-band

Let
$$|T'(j\omega)|^2 = 1 - |T_C(j\omega)|^2$$

i.e.,
$$|T'(j\omega)|^2 = 1 - \frac{1}{1+\varepsilon^2 T_n^2(\omega)}$$

(Here we have used the subscript C to indicate Chebyshev approximation of eqn. 10.20)

i.e.,
$$|T'(j\omega)|^2 = \frac{\varepsilon^2 T_n^2(\omega)}{1+\varepsilon^2 T_n^2(\omega)} \quad \ldots(10.40)$$

Obviously, if $|\ T_C\ (j\omega)\ |^2$ is a low-pass function, then $|\ T'(j\omega)\ |^2$ is a high-pass function with equiripple stop-band extending from $0 \le \omega \le 1$ and monotonic pass-band beyond $\omega = 1$. Now, if we perform a HP to LP transformation on eqn. 10.40, *i.e.,* by replacing ω by $1/\omega$, we have

$$|T_{IC}(j\omega)|^2 = |T'(j/\omega)|^2 = \frac{\varepsilon^2 T_n^2(1/\omega)}{1+\varepsilon^2 T_n^2(1/\omega)} \quad \ldots(10.41)$$

$|T_{IC}(j\omega)|^2$ is a low-pass function with monotonic pass-band and equiripple stop-band (stop-band edge being $\omega = 1$). This called '*Inverse Chebyshev*' approximation (hence the subscript *IC*) and the response is shown in Fig. 10.22. In order to determine the transfer function, let

$$T_{IC}(s) = \frac{p(s)}{q(s)} \quad \ldots(10.42)$$

$q(s)$ can be determined by first finding $q_C(s)$ for the Chebyshev filter. Then, as can be seen from eqn. 10.41,

$$q(s) = s^n q_C(1/s)$$

To determine $p(s)$ we observe that

$$p(s)p(-s) = \omega^{2n} T_n^2(1/\omega)\big|_{\omega^2=-s^2} \qquad \text{...(10.43)}$$

An example illustrates the approximation.

Example 10.5. Consider the Chebyshev filter of example 10.2 with $n = 5$ and $\varepsilon = 0.5088$.

$$q_C(s) = (s + 0.2895)(s^2 + 0.4684s + 0.4294)(s^2 + 0.179s + 0.9902)$$

Hence,

$$q(s) = s^5\left(\frac{1}{s} + 0.2895\left(\frac{1}{s^2} + \frac{0.4684}{s} + 0.4294\right)\left(\frac{1}{s^2} + 0.179s + 0.9902\right)\right)$$

$$= 0.1231(s + 3.454)(s^2 + 1.093s + 2.329)(s^2 + 0.18s + 1.009)$$

From eqn. 10.43 and Table 10.2

$$p(s)p(-s) = \omega^{10}\left(\frac{16}{\omega^5} - \frac{20}{\omega^3} + \frac{5}{\omega}\right)^2\Bigg|_{\omega^2=-s^2}$$

$$= (5\omega^4 - 20\omega^2 + 16^2)\big|_{\omega^2=-s^2}$$

$$= (5s^4 + 20s^2 + 16)^2$$

Hence, $\qquad p(s) = 5s^4 + 20s^2 + 16.$

In inverse Chebyshev approximation we have to observe that the poles of $T_{IC}(s)$ are in the pass-band and the zeros in the stop-band.

10.7.2 Elliptic Filter Approximation

If we are to improve the performance of a filter beyond that which is achieved by the Chebyshev filter we have to allow equiripple response in both the pass– and the stop-band. This leads to narrower transition band. This is achieved by choosing $K_n^2(\omega)$ in eqn. 10.1 as

$$K_n(\omega) = R_n(\omega) \qquad \text{...(10.43a)}$$

where $R_n(\omega)$ is the Chebyshev rational function.

This function introduced to network theory by Cauer is given by

$$R_n(\omega) = \frac{\omega(\omega_1^2 - \omega^2)(\omega_2^2 - \omega^2)...(\omega_k^2 - \omega^2)}{(1-\omega_1^2\omega^2)(1-\omega_2^2\omega^2)...(1-\omega_k^2\omega^2)} \quad n \text{ odd} \qquad \text{...(10.44)}$$

$$= \frac{(\omega_1^2 - \omega^2)(\omega_2^2 - \omega^2)...(\omega_k^2 - \omega^2)}{(1-\omega_1^2\omega^2)(1-\omega_2^2\omega^2)...(1-\omega_k^2\omega^2)} \quad n \text{ even} \qquad \text{...(10.45)}$$

It can be observed from eqns. 10.44 or 10.45 that

$$R_n\left(\frac{1}{\omega}\right) = \frac{1}{R_n(\omega)}$$

and that its poles and zeros and reciprocal of each other. This ensures that the value of $R_n(\omega)$ at a frequency ω_i in $0 \le \omega < 1$ is the reciprocal of its value at $1/\omega_i$, $1 \le \omega \le \infty$. The ω_k^2 are related to Jacobi Elliptic functions and hence the filter is called "Elliptic Filter". Figure 10.23 shows the response of such a filter.

The algebra involved in the development of elliptic filters is rather involved and can be found elsewhere (see notes). In passing we state that elliptic filters are optimal in the sense that no filter of lower order will satisfy the specifications of the filter. Charts and nomographs are available to facilitate the design of these filters.

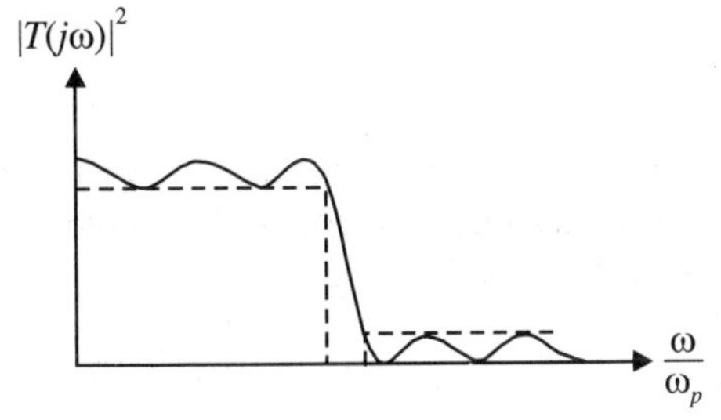

Fig. 10.23

10.7.3 Bessel Approximation

In all the approximation techniques discussed so far, we have concentrated on approximating the magnitude of the transfer function. In many signal processing applications, linearity of the phase or constant phase delay (phase delay is defined as $\phi(\omega)/\omega$, where $\phi(\omega)$ is the phase of the transfer function) is an important factor. As indicated earlier, the phase distortion (deviation from linearity of phase) is more in the Chebyshev filter than the Butterworth filter. In fact it can be shown that a sudden change in the amplitude is accompanied by a similar change in the phase. An equiripple filter has greater amount of phase distortion than the maximally flat filter.

A linear phase filter is also a constant time delay (time delay is defined as— $d\phi(\omega)/d\omega$) filter. A constant delay response can be approximated by a maximally flat delay at $\omega = 0$ by using *Bessel polynomials.*

An ideal-delay network with a delay time t_0 has an output of $x(t - t_0)$ for an input of $x(t)$, *i.e.*, the output is a delayed replica of the input. If the frequency is normalized with respect to $\omega_0(= 1/t_0)$, then transfer function of such a filter is e^{-s}. A rational function approximation for such a filter can be obtained by expansion of phase angle of e^{-s} in continued fractions and truncate it after n terms.

We know that

$$T(s) = e^{-s} = \frac{1}{\cosh s + \sinh s} \qquad \text{...(10.46)}$$

Here cosh s and sinh s can be associated with even and odd polynomials M and N given by

$$\cosh s = M = 1 + \frac{s^2}{2!} + \frac{s^4}{4!} + \ldots$$

$$\sinh s = N = s + \frac{s^8}{3!} + \frac{s^5}{5!} + \ldots$$

Then

$$\frac{M}{N} = \frac{1}{s} + \cfrac{1}{\frac{3}{s} + \cfrac{1}{\frac{5}{s} + \cfrac{1}{\frac{2n-1}{s}}}} \qquad \ldots(10.47)$$

where the continued fraction expansion is terminated after n terms. Here we observe that M/N has the form of a reactance function with its poles and zeros simple and interlacing on the $j\omega$-axis. For $n = 3$ we have

$$\frac{M}{N} = \frac{1}{s} + \cfrac{1}{\frac{3}{s} + \frac{1/5}{s}} = \frac{6s^2 + 15}{s^2 + 15s}$$

and the transfer function is

$$T(s) = \frac{15}{s^3 + 6s^2 + 15s + 15}$$

From the expansion of eqn. 10.47, the transfer function can be written as

$$T(s) = \frac{k}{q(s)} \qquad \ldots(10.48)$$

where k is selected equal to the constant term in $q(s)$ to render $T(0) = 1$ as is required by equation 10.46 and

$$q(s) = M(s) + N(s)$$

It can be shown that

$$q(s) = s^n B_n\left(\frac{1}{s}\right) \qquad \ldots(10.49)$$

when $B_n(1/s)$ is the Bessel polynomial; hence the name of the filter, given by the recurrence relationship

$$B_n\left(\frac{1}{s}\right) = \frac{2n-1}{s} B_{n-1}\left(\frac{1}{s}\right) + B_{n-2}\left(\frac{1}{s}\right) \qquad \ldots(10.50)$$

with $B_0 = 1$ and $B_{-1} = 1$. In view of eqn. 10.50, $q(s)$ can also be expressed as

$$q(s) = \sum_{i=1}^{n} b_i s^i \qquad \dots(10.51)$$

where

$$b_i = \frac{(2n-i)!}{2^{n-i}\, i!(n-i)!}$$

Table 10.3 lists the coefficients of $q(s)$ for n up to 6. The constant k is given by b_0.

Table 10.3

n	b_0	b_1	b_2	b_3	b_4	b_5	b_6
0	1						
1	1	1					
2	3	3	1				
3	15	15	16	1			
4	105	105	45	10	1		
5	945	945	420	105	15	1	
6	10,395	10,395	4,725	1,260	210	21	1

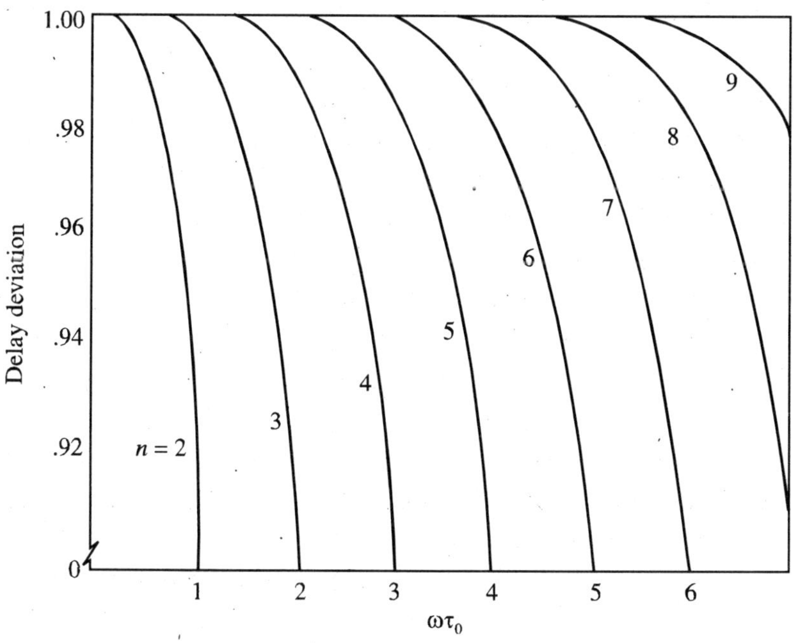

Fig. 10.24

The delay $\tau_n(\omega)$ is defined by

$$\tau_n(\omega) = -\frac{d}{d\omega}\phi_n(\omega)$$

where $\phi_n(\omega)$ is the phase of $T(j\omega)$ and it can be shown that

$$\tau_n(\omega) = 1 - \frac{\omega^{2n}}{b_0}|T(j\omega)|^2 \quad \text{...(10.52)}$$

Hence,

$$\tau_n(0) = 1 \text{ (normalized delay).}$$

Furthermore, the first $(n - 1)$ derivatives of τ_n with respect to ω are zero at $\omega = 0$ indicating that the delay is maximally flat at the origin. The deviation from this flat delay for other values of ω is generally expressed as per cent of the ideal delay $\tau(0)$.

The only design parameter for a Bessel filter is its order n. The specific value of n is to be selected to satisfy both the delay-error specification and the magnitude deviation specification of $|T(j\omega)|$. Figure 10.24 and Fig. 10.25 give graphs of delay error vs $\omega\tau$ and magnitude error vs $\omega\tau$.

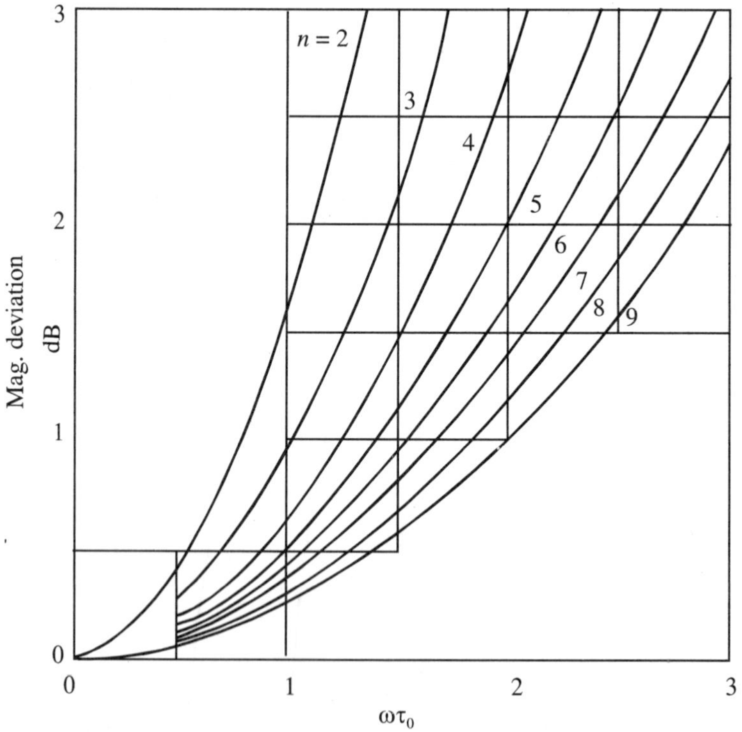

Fig. 10.25

Example 10.6. Design a Bessel filter to satisfy the following specifications:

$\tau(0) = 8$ ms, $\tau_{error} = 1\%$ for $\omega \le 200$ rad/s.

Magnitude deviation: 1.5 dB for $\omega \le 200$ rad/s

$$\tau(0)\ \omega = 1.6$$

From graph of Fig. 10.24 $n = 4$

From graph of Fig. 10.25 $n = 5$

Hence $n = 5$

$T(s)$ can be constructed from eqn. 10.51.

In this section we have briefly discussed three additional approximation techniques-two for magnitude and one for delay. The standard approximations discussed here are facilitated by filter design charts and graphs, details of which are found elsewhere.

PROBLEMS

10.1 The specifications for a Butterworth LP filter are

Pass-band	0–2 M rad/sec
Pass-band loss	≤ 2 dB
Stop-band loss	≤ 60 dB at 6 M rad/sec

Find n and c.

10.2 The specifications for an LP filter are

$$\alpha \le 1 \text{ dB for } f \le 2 \text{ MHz}$$

$$\alpha \le 60 \text{ dB for } f \le 8 \text{ MHz}$$

Find the transfer function of the Butterworth filter which satisfies these specifications. Select c to satisfy the pass-band edge characteristic exactly.

10.3 Show that the transfer function for an nth order Butterworth filter is

$$T = \frac{1}{\prod_{k=1}^{\pi/2}(s^2 + 2s\cos\theta_k + 1)}, \quad \theta_k = \frac{2k-1}{2n}\pi, \ n \text{ even}$$

and

$$= \frac{1}{(s+1)\prod_{k=1}^{(n-1)/2}(s^2 + 2s\cos\theta_k + 1)}, \quad \theta_k = \frac{\pi}{n}, \ n \text{ odd.}$$

Here c is assumed to be 1. If c is not equal to 1 what modifications are to be made to the preceding equations.

10.4 Show that for a Chebyshev filter

$$n \ge \frac{\cosh^{-1}(1/k_1)}{\cosh^{-1}(1/k)}$$

where k and k_1 are given by eqn. 10.5 and 10.7.

10.5 Find n and c for a Chebyshev filter with the following specifications

Pass-band ripple	0.5 dB
Pass-band	0–2 M rad/sec
Stop-band loss	60 dB at 4 M rad/sec

10.6 Show that the poles of the transfer function for a Chebyshev filter are given by

$$s_k = \sigma_k + j_{\omega_k}$$

where σ_k and ω_k are given in eqn. 10.27.

10.7 The poles of a Chebyshev filter transfer function lie on an ellipse. Show that the semi-minor and semi-major axes for this ellipse are $a\omega_p$ and $b\omega_p$ where

$$a = 1/2\ (\Delta^{1/n} - \Delta^{-1/n}),\ b = 1/2(\Delta^{1/n} + \Delta^{-1/n}),$$

$\Delta = 1/\varepsilon + \sqrt{1 + 1/\varepsilon^2}$ and ω_p is the pass-band edge. Hence, justify the determination of the poles of the filter from the two Butterworth circles.

10.8 The specification for a LP filter are as given in prob. 10.2. Find the transfer function of the corresponding Chebyshev filter. Compare this with the Butterworth filter of prob. 10.2.

10.9 The specification for a band-pass filter are:

$$\alpha_p < 2{,}0 \text{ dB for } 30 \text{ k rad} \le \omega < 30 \text{ k rad}$$

$$\alpha_k \ge 50 \text{ dB for } \omega \le 10 \text{ k rad and } \omega\ 40 \text{ k rad.}$$

Find its transfer function.

NOTES AND REFERENCES

The approximation problem is a highly algebraic one and is, at times, quite involved as in the case of Elliptic filters. The references cited in Chapter 9 provide additional information on the approximation, especially Guillemin's (Chap. 14) and Weinberg's (Chap. 11). Of late some books have been written to specifically deal with approximation. Daniel's book[1] is once such discussing may aspects of approximation in detail.

Modern filter design is based on tables and charts. Zverev's book[2] provides excellent and detailed study filter design from charts.

1. R.W. Daniels, *Approximation Methods for Electronic Filter Design*, McGraw-Hill, 1974.
2. A.I. Zverev, *Handbook of Filter Synthesis*, McGraw-Hill, 1976.

CHAPTER

11 Insertion Loss Synthesis

Among many methods of two-port network and filter design, Darlington's insertion loss synthesis is a popular one and has stood the test of time. The design is based on a LC 2-port network terminated in a resistances.

The approximation techniques discussed in Chapter 10 result in a rational function for transfer voltage ratio. In this chapter, we utilize the synthesis methods discussed in Chapter 9 to realize the approximated transfer function. In section 11.1, for completeness sake we introduce coefficient matching technique which is popular in active filter design.

11.1 COEFFICIENT MATCHING TECHNIQUE

The coefficient matching technique is a simple, efficient method of realizing transfer functions. In this method, a filter structure or a network is assumed and its transfer function is evaluated. A set of equations is obtained by comparing the coefficients of this transfer function and the filter function is obtained by approximation techniques or otherwise. The coefficients in the transfer function being functions of the network elements, the elemental values can be obtained by solving these equations.

The voltage transfer function of the doubly terminated network shown in Fig. 11.1 is

$$T(s) = \frac{V_2}{V_s} = K\frac{1/LCK}{s^2 + s\left(1/CR_2 + R_1/L\right) + 1/LCK} \quad \text{...(11.1)}$$

where $$K = \frac{R_2}{R_1 + R_2}$$

Here, R_1 is considered to be the source impedance and R_2 the load impedance.

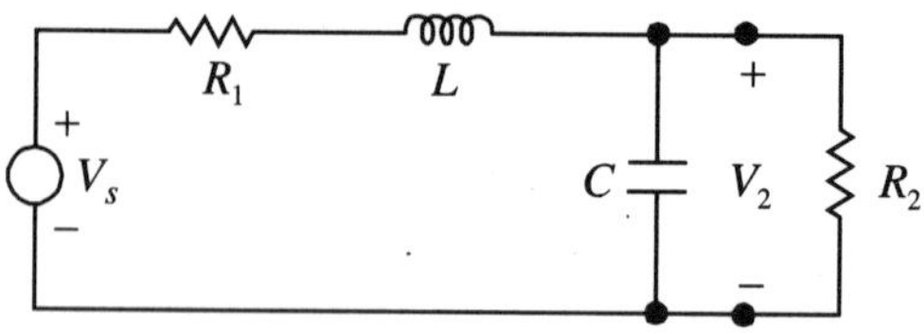

Fig. 11.1

Let the filter specifications lead to a second order Butterworth filter transfer function,

$$T(s) = K\frac{1}{s^2 + \sqrt{2}s + 1} \quad \text{...(11.2)}$$

where the cut-off frequency is normalized to 1 rad/sec. If we consider equal termination, *i.e.*, $R_1 = R_2$, and normalize the impedance to 1 Ω, we have $R_1 = R_2 = 1$ and $K = 1/2$. From eqns. 11.1 and 11.2, on comparing coefficients, we have

$$\frac{2}{LC} = 1, \frac{1}{C} + \frac{1}{L} = \sqrt{2}$$

Solving these equations for C and L leads to

$$L = \sqrt{2} \text{ and } C = \sqrt{2}$$

Hence, the elemental values of a normalized second order Butterworth filter (ω_c = 1 rad/sec, R = 1) are

$$R_1 = R_2 = 1, L = \sqrt{2}h, C = \sqrt{2}f$$

with the *dc* gain of the filter (K) as 1/2.

Consider the third order network shown in Fig. 11.2. The voltage transfer function is

$$T(s) = \frac{K \times 1/LC_1C_2R_p}{s^3 + s^2(1/R_2C_2 + 1/R_1C_1) + s[1/L(1/C_1 + 1/C_2) + 1/C_1C_2R_1R_2] + 1/LC_1C_2R_p} \quad \text{...(11.3)}$$

where $$K = \frac{R_2}{K_1 + R_2} \text{ and } R_p = \frac{R_1R_2}{R_1 + R_2}$$

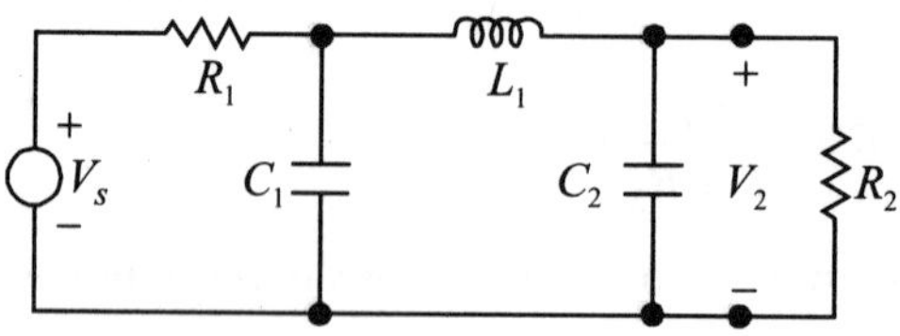

Fig. 11.2

The transfer function of a third-order Butterworth filter with $\omega_c = 1$ rad/sec is

$$T(s) = \frac{1}{s^3 + 2s^2 + 2s + 1} \qquad \text{...(11.4)}$$

On comparing coefficients in eqns. 11.3 and 11.4 (assuming equal termination of 1 Ω) we have

$$R_p = \frac{1}{2},\ K = \frac{1}{2},\ \frac{1}{LC_1C_2R_p} = 1$$

$$\frac{1}{C_1} + \frac{1}{C_2} = 2,\ \frac{1}{L}\left(\frac{1}{C_1} + \frac{1}{C_2}\right) + \frac{1}{C_1C_2} = 2$$

and the solution is

$$C_1 = C_2 = 1f,\ L = 2h.$$

The elements of the filters obtained can now be denormalized to the required cut-off frequency and resistance terminations.

Example 11.1. We are required to realize a third-order Butterworth filter to cut-off at 5 k rad/sec with an equal termination of $R_1 = R_2 = 600\ \Omega$.

The normalized filter network is shown in Fig. 11.2. The denormalizing factors are

$$\gamma_n = 600$$

$$\Omega_n = 5 \times 10^3$$

Hence, the elemental values of the filter are

$$R_1 = R_2 = 600\ \Omega$$

$$L_1 = \frac{2 \times 600}{5 \times 10^3} = 240 \text{ mh}$$

$$C_1 = C_2 = \frac{1}{600 \times 5 \times 10^3} = 3.33\ \mu f$$

Though the coefficient matching technique is a simple and elegant method of filter design, the design of an nth order filter requires the solution of a set of n equations which, in itself, is an involved problem. Besides this, the design method does not demonstrate the use of sophisticated network synthesis methods. We refer to coefficient matching technique again when we discuss active filter design (Chap. 12).

11.2 DARLINGTON'S METHOD

As demonstrated in the earlier chapters, a low-pass characteristic can be obtained by RC, RL or LC networks. Low cost, low sensitivity to component variation and simplicity of design make the LC two-port networks the most widely used filter networks. Here, we consider the overall filter to be a doubly terminated LC two-port (Fig. 11.3). As in the case of coefficient matching technique, we consider R_1 to be the source resistance and R_2 to be the load resistance.

A powerful method of designing doubly terminated LC networks is the well-known Darlington's method. In the Darlington method of filter design the specification of the insertion loss of the filter is converted to the reflection coefficient of the filter and from the latter the driving-point impedance of the terminated network is obtained. Then this driving-point impedance is developed into a resistively terminated LC-Ladder network. Some of the concepts used in this method are introduced next.

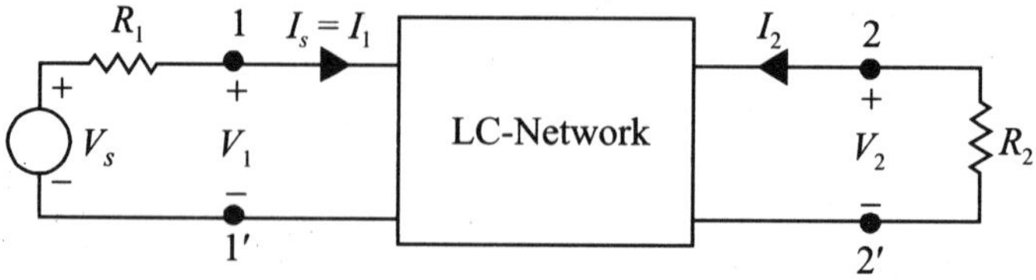

Fig. 11.3

11.2.1 Insertion Loss and Reflection Coefficient

To define and to interrelate insertion loss and reflection coefficient we consider three different modes of operation of the network. In the first mode, the voltage source is directly connected to the load as shown in Fig. 11.4. We represent by V_{20} and P_{20} the voltage at the load terminals and the power absorbed by the load,

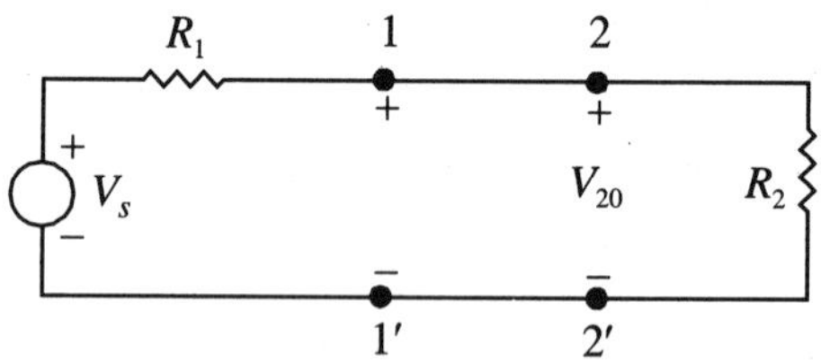

Fig. 11.4

respectively. In the second mode of operation an LC network is inserted between the source and the load as shown in Fig. 11.3. The load voltage and power are represented by V_2 and P_2 respectively. The first mode of operation can be considered as the degenerate form of the second mode in which the lossless LC-network is shorted or removed. In the third mode of operation, the source is connected to the load through a transformer of turns ratio $\sqrt{R_1} : \sqrt{R_2}$ (Fig. 11.5).

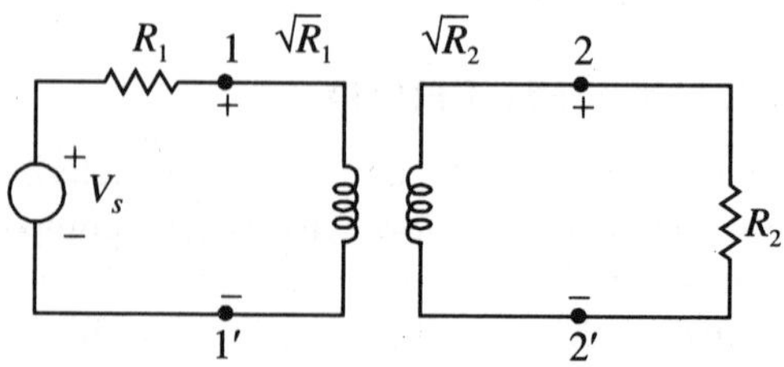

Fig. 11.5

In the first mode of operation, from Fig. 11.4,

$$V_{20} = \frac{R_2}{R_1 + R_2} V_s \quad \text{...(11.5)}$$

$$P_{20} = \frac{R_2}{(R_1 + R_2)} |V_s|^2 \quad \text{...(11.6)}$$

In the second mode of operation,

$$P_2 = \frac{V_2^2}{R_2} \quad \text{...(11.7)}$$

where P_2 and V_2 are the actual power absorbed and the actual voltage at the load terminals, respectively. From eqn. 11.5 to 11.7 we have

$$P_I = \frac{P_{20}}{P_2} = e^{2\theta} = \left[\frac{R_2}{R_1 + R_2} \left| \frac{V_s}{V_2} \right| \right]^2 \quad \text{...(11.8)}$$

$$V_I = \frac{V_{20}}{V_2} = e^{\gamma} = \frac{R_2}{R_1 + R_2} \frac{V_s}{V_2} \quad \text{...(11.9)}$$

P_I is called the *insertion power ratio* and V_I the *insertion voltage ratio* for obvious reasons. V_I being complex is represented by e^{γ} when γ is a complex number. The *insertion loss* is defined as

$$IL = 10 \log P_I = 20 \log \left[\frac{R_2}{R_1 + R_2} \left| \frac{V_s}{V_2} \right| \right] \quad \text{...(11.10)}$$

From eqns. 11.8 and 11.9 it can be seen that the real part of γ is θ. γ can be written as

$$\gamma = \theta + j\beta \quad \text{...(11.11)}$$

and β is called the *insertion phase.*

In the third mode of operation the transformer ratio is so selected that there is maximum power transfer to the load. The maximum power available from the source under matched conditions is (see sec. 5.7)

$$P_{2\max} = \frac{|V_s|^2}{4R_1} \quad \text{...(11.12)}$$

From eqns. 11.7 and 11.12 we have

$$|t(j\omega)|^2 = \frac{P_2}{P_{2\max}} = \frac{4R_1}{R_2} \left| \frac{V_2}{V_s} \right|^2 \quad \text{...(11.13)}$$

where $t(s)$ is called the *transmission coefficient.* As $P_{2\max}$ is always greater than P_2 we can write

$$\frac{P_2}{P_{2\max}} = \frac{1}{1 + |K(j\omega)|^2} \quad \text{...(11.14)}$$

Here, $K(s)$ is called the *loss* or *characteristic function* (compare this with eqn. 10.1).

Under matched conditions (Fig. 11.5) the load voltage is

$$V_2' = \sqrt{\frac{R_2}{R_1}} \cdot \left(\frac{V_s}{2}\right) \quad \text{...(11.15)}$$

This is the largest possible load voltage. The ratio V_2'/V_2 is sometimes called the *transducer function*, is given by

$$H'(s) = \frac{1}{2}\sqrt{\frac{R_2}{R_1}}\left(\frac{V_s}{V_2}\right) \quad \text{...(11.16)}$$

(observe that $H'(s) = 1/t(s)$)

From eqns. 11.13, 11.14 and 11.16 we have

$$|H'(j\omega)|^2 = 1 + |K(j\omega)|^2 \quad \text{...(11.17)}$$

For a lumped, LTI network $t(s)$ and $k(s)$ are rational functions of s with real coefficients and can be represented as

$$t(s) = \frac{q(s)}{p(s)},\; k(s) = \frac{f(s)}{q(s)}$$

Hence, we can write from eqn. 11.17

$$p(s)\, p(-s) = q(s)\, q(-s) + f(s)\, f(-s) \quad \text{...(11.18)}$$

This is also called the *Feldtkellers condition.* From eqn. 11.16 it can be observed that the transmission coefficient differs from the overall voltage transfer ratio by only a multiplicative constant. Hence, the zeros of $p(s)$ and the zeros of $q(s)$ are the natural frequencies and the transmission zeros of the network, respectively. The zeros of $f(s)$ are the zero-loss frequencies. Equation 11.18 gives the relation between these zeros. The specification of either the transducer (or transfer) function uniquely specifies the zero-loss frequencies.

The LC-network of Fig. 11.3 can also be characterized by the reflection coefficient which is defined as

$$\rho_1(s) = \frac{Z_1(s) - R_1}{Z_1(s) + R_1} \quad \text{...(11.19)}$$

where Z_1 is the input impedance at 1-1′ of the LC-network with the load R_2. (The concept of reflection coefficient is quite familiar in the study of transmission lines. Here, ρ_1 is the source-end reflection coefficient. We can also define a load-end reflection coefficient).

Though we have defined a number of new network functions we can interrelate all of them and obtain some bounds on these functions. From eqns. 11.6 and 11.12 we have

$$\frac{P_{20}}{P_{2\max}} = \frac{4R_1R_2}{(R_1 + R_2)^2} \quad \text{...(11.20)}$$

and from eqn. 11.8

$$\frac{P_2}{P_{2\max}} = |t(j\omega)|^2 = \frac{4R_1R_2}{(R_1+R_2)^2}e^{-2\theta} \qquad \ldots(11.21)$$

Under equal termination, *i.e.*, $R_1 = R_2$, P_2 is equal to $P_{2\max}$ and

$$|t(j\omega)|^2 = e^{-2\theta}$$

The bound on P_I also depends on the termination. As $\frac{P_{20}}{P_2} \geq \frac{P_{20}}{P_{2\max}}$ from eqns. 11.8 and 11.20 we have

$$P_I \geq \frac{4R_1R_2}{(R_1+R_2)^2} \qquad \ldots(11.22)$$

Under equal termination, then

$$P_{20} \geq P_2$$

The transducer function, loss function and the reflection coefficient can be similarly related. The power delivered to the LC-network can be written as

$$P_1 = \left|\frac{V_s}{R_1 + Z_1}\right|^2 R$$

where $Z_1(j\omega) = R + jX$. Since the network is lossless this is also the power delivered to the load

i.e., $P_1 = P_2$.

Hence from eqns. 11.6 and 11.8 we have

$$P_I = e^{2\theta} = \frac{|R_1 + Z_1|^2}{(R_1+R_2)^2}\frac{R_2}{R} \qquad \ldots(11.23)$$

Here, the insertion power ratio is expressed in terms of the network parameters only. From eqns. 11.19 and 11.22 it can be easily shown that

$$1 - \frac{4R_1R_2}{(R_1+R_2)^2}e^{-2\theta} = \left|\frac{Z_1 - R}{Z_1 + R}\right|^2 = \rho_1(j\omega)\rho_1(-j\omega) \qquad \ldots(11.24)$$

and using eqn. 11.21 we have

$$1 - \frac{P_2}{P_{2\max}} = \rho_1(j\omega)\rho_1(-j\omega)$$

i.e., $|\rho_1(j\omega)|^2 = 1 - |t(j\omega)|^2$

or $|\rho_1(j\omega)|^2 + |t(j\omega)|^2 = 1 \qquad \ldots(11.25)$

Equation 11.25 indicates that the sum of the reflected power and power delivered to the load is equal to the total power available, as it should be. If $Z_1 = R_1$ then $\rho_1 = 0$ indicating that $P_2 = P_{1\max}$. In this case, eqn. 11.22 holds with an equality.

In Chap. 10 we defined the attenuation of a filter as

$$\alpha(\omega) = 10 \log \left[1 + |K(j\omega)|^2 \right]$$

i.e.,

$$\alpha(\omega) = 10 \log \frac{P_{2\,\max}}{P_2} \qquad \ldots(11.26)$$

From eqns. 11.8, 11.10 and 11.26 it can be shown (see prob. 10.11) that

$$\alpha(\omega) = IL + 10 \log \frac{(R_1 + R_2)^2}{4R_1R_2} \qquad \ldots(11.27)$$

With equal termination, the attenuation and insertion loss are the same. For an LC low-pass filter the low-frequency insertion loss approaches zero because inductors become short and capacitors open. With this condition

$$\alpha(0) = 10 \log \frac{(R_1 + R_2)^2}{4R_1R_2}$$

implying that the attenuation at $\omega = 0$ cannot be zero unless the load and source resistances are equal.

Though we have considered the case of $R_1 \neq R_2$, in practice filter networks are more often designed with equal termination.

11.2.2 Realization Procedure

As pointed out in Sec. 11.2.1, a doubly terminated filter network can be specified in a number of ways. In the Darlington realization procedure these specifications have to be converted finally to yield the squared transmission coefficient $|t(j\omega)|^2$. The specified function can be

$$\left| \frac{V_2}{V_s}(j\omega) \right|^2, \left| \frac{V_2}{I_s}(j\omega) \right|^2 \text{ or } P_I$$

If

$$\left| \frac{V_2}{V_s}(j\omega) \right|^2$$

is the specified function then eqn. 11.13 yields $|t(j\omega)|^2$.

If $\left| \frac{V_2}{I_s}(j\omega) \right|^2$ is the specified function then

$$|t(j\omega)|^2 = \frac{4}{R_1R_2} \left| \frac{V_2}{I_s}(j\omega) \right|^2 \qquad \ldots(11.28)$$

(This equation is obtained from eqn. 11.13 by substituting $V_s = R_1 I_s$). If P_I (or $e^{2\theta}$) is specified then eqn. 11.21 yields $|t(j\omega)|^2$.

We assume, without loss of generality, that the insertion loss characteristic of the filter network is specified. In this case, we can write from eqn. 11.24, that

$$\rho(s)\,\rho(-s) = 1 - \frac{4R_1R_2}{(R_1+R_2)^2}F(s) \qquad \ldots(11.29)$$

where $F(s)$ is the rational function obtained from the approximation methods of Chap. 10. If the specified insertion loss characteristic is approximated by Butterworth polynomials then

$$F(s) = \frac{K}{1+c^2(-1)^n s^{2n}} \qquad \ldots(11.30)$$

and if approximated by Chebyshev it is

$$F(s) = \frac{K}{1+\varepsilon^2 T_n^2(s)} \qquad \ldots(11.31)$$

For Butterworth and odd-order Chebyshev the constant K is 1 and for even order Chebyshev it is $(1 + \varepsilon^2)$. This can be shown by observing the behaviour of $F(0)$ and recognizing the fact that the insertion losses for a low-pass LC ladder is zero at $s = 0$. This coupled with equation 11.22 leads to the restriction on termination as

$$1 \geq \frac{4R_1R_2}{(R_1+R_2)^2} \text{ Butterworth odd order Chebyshev} \qquad \ldots(11.32)$$

$$\frac{1}{1+\varepsilon^2} \geq \frac{4R_1R_2}{(R_1+R_2)^2} \text{ even order Chebyshev}$$

Hence, we conclude that we can realize Butterworth and odd order Chebyshev filters, filter functions which have a reflection zero at $\omega = 0$, with equal terminations, *i.e.*, $R_1 = R_2$. But we cannot realize even order Chebyshev, a filter function with no reflection zero at $\omega = 0$, with equal termination.

The realization procedure is summarized as:

(*i*) From the given pass-band and stop-band requirements determine $F(s)$ (see Chap. 10). The normalized pass-band is assumed to be 1 rad/sec.

(*ii*) From $F(s)$ obtain the reflection coefficient ρ (s). Here we have used ρ in place of ρ_1. From eqn. 11.29, we have

$$\rho(s)\rho(-s) = \frac{p(s)}{q(s)} = 1 - \frac{4R_1R_2}{(R_1+R_2)^2}F(s) \qquad \ldots(11.33)$$

R_1 and R_2 must satisfy the restriction imposed by eqn. 11.32. $\rho(s)$ can be determined from this equation by observing that the zeros of $p(s)$ and $q(s)$ appear in quadratral symmetry. Let $p(s) = p_1(s)/q_1(s)$. The zeros of $q_1(s)$ are the left half plane zeros of $q(s)$. The zeros of $p(s)$ are equally distributed between the zeros of $\rho(s)$ and $\rho(-s)$, with the restriction that if z_i is a zero of $\rho(s)$ then $-z_i$ is a zero of $\rho(-s)$. The complex zeros must be selected in conjugate pairs.

(*iii*) From $\rho(s)$ determine $Z(s)$. From equation 11.19 we have

$$\rho(s) = \pm\frac{Z_1(s) - R_1}{Z_1(s) + R_1} = \pm\frac{Z(s) - 1}{Z(s) + 1}$$

where $Z(s)$ is normalized with respect to

$$R_1 = 1 \left(i.e., \frac{Z_1}{R_1} = Z \right).$$

For the positive sign, we have

$$Z(s) = \frac{1 + \rho(s)}{1 - \rho(s)}$$

and for the negative sign

$$Z(s) = \frac{1 - \rho(s)}{1 + \rho(s)}$$

(*iv*) Expand $Z(s)$ into continued fraction expansion about infinity and obtain the ladder. The two impedances defined in (*iii*) lead to two (dual) ladders, one with termination R_2 and the other with termination $1/R_2$.

(*v*) Denormalize the network to yield the required cut-off frequency and terminations.

We demonstrate the realization procedure by examples.

Example 11.2. Let the insertion-loss characteristic be approximated by a third-order Butterworth filter with a cut-off frequency of 1 rad/sec. Let the terminating resistances R_1 and R_2 be 1 Ω. We are required to design an LP filter to satisfy this insertion loss characteristic.

(*i*) The cut-off frequency is also the pass-band edge. Hence c = 1 and eqn. 11.30 yields

$$F(s) = \frac{1}{1 + (-1)^3 s^6} = \frac{1}{1 - s^6}$$

The specified terminations satisfy eqn. 11.32

(*ii*) From eqn. 11.33

$$\rho(s)\rho(-s) = \frac{p(s)}{q(s)} = 1 - \frac{1}{1 - s^6}$$

$$= \frac{s^6}{s^6 - 1}$$

The zeros of $q(s)$ are given by (see eqn. 10.12)

$$q_k = e^{jk\pi/3}, \; k = 0, 1, 2, ..., 5$$

The poles of $\rho(s)$ are the left half plane zeros of $q(s)$ and they are $e^{j2\pi/3}$, $e^{j\pi}$ and $e^{j4\pi/3}$, *i.e.*, $p_0 = -1$, $p_{1,2} = -0.5 \pm j\, 0.866$. $p(s)$ has a sixth-order zero at $s = 0$ and hence $\rho(s)$ has a third order zero at $s = 0$. Now $\rho(s)$ is given by

$$\rho(s) = \frac{s^3}{(s-p_0)(s-p_1)(s-p_2)} = \frac{s^3}{s^3+s^2+2s+1}$$

(*iii*) From eqn. 11.34

$$Z(s) = \frac{1+\{s^3/(s^3+2s^2+2s+1)\}}{1-\{s^3/(s^3+2s^2+2s+1)\}}$$

i.e.,

$$Z(s) = \frac{2s^3+2s^2+2s+1}{2s^2+2s+1}$$

(*iv*) The continued fraction expansion of *Z*(*s*) is given by

$$Z(s) = s + \cfrac{1}{2s + \cfrac{1}{s + \cfrac{1}{1}}}$$

the filter network is given in Fig. 11.6.

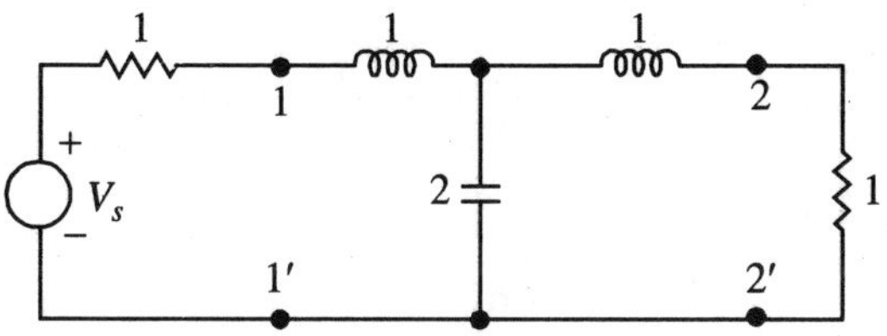

Fig. 11.6

Example 11.3. The insertion-loss specification for a LP filter are

Pass-band	:	0–25 KHz
Maximum pass-band loss	:	3 dB at 25 KHz
Minimum stop-band loss	:	20 dB at 50 KHz

$$R_1 = R_2 = 1k$$

We design a Butterworth filter for the preceding specifications.

(*i*) Let the pass-band edge be normalized to 1 rad/sec. As the attenuation at the pass-band edge is 3 dB, $c = 1$. The order of the Butterworth polynomial is

$$n \geq \frac{\log k}{\log k_1}$$

$$k = \frac{2\pi(25\times 10^3)}{2\pi(50\times 10^3)} = 0.5$$

$$k_1 = \left[\frac{10^{0.3}-1}{10^2-1}\right]^{1/2} = 0.1005$$

$$n \geq \frac{\log(0.1005)}{\log(0.5)} = 3.315$$

Let $$n = 4$$

Hence, $$F(s) = \frac{1}{1+(-1)^4 s^8} = \frac{1}{1+s^8}$$

The specified terminations satisfy eqn. 11.32

(*ii*) From eqn. 11.33

$$\rho(s)\rho(-s) = \frac{p(s)}{q(s)} = 1 - \frac{1}{1+s^8} = \frac{s^8}{1+s^8}$$

The zeros of $q(s)$ are $q_k = e^{j(2k-1)\pi/8}$, $k = 1, 2, \ldots, 8$. Hence, the poles of $\rho(s)$ are $e^{j5\pi/8}$, $e^{j7\pi/8}$, $e^{j9\pi/8}$ and $e^{j11\pi/8}$

i.e., the poles are $p_{1,2} = -0.924 \pm j\,0.383$ and $p_{3,4} = -0.383 \pm j\,0.924$.

$\rho(s)$ has a fourth order zero at $s = 0$. Hence,

$$\rho(s) = \frac{s^4}{(s-p_1)(s-p_2)(s-p_3)(s-p_4)}$$

$$= \frac{s^4}{s^4 + 2.613s^3 + 3.414s^2 + 2.613s + 1}$$

(*iii*) From eqn. 11.34

$$Z(s) = \frac{2s^4 + 2.613s^3 + 3.414s^2 + 2.613s + 1}{2.613s^3 + 3.414s^2 + 2.613s + 1}$$

(*iv*) The continued fraction expansion is

$$Z(s) = 0.765s + \cfrac{1}{1.848s + \cfrac{1}{1.848s + \cfrac{1}{0.765s + \cfrac{1}{1}}}}$$

The normalized network is shown in Fig. 11.7.

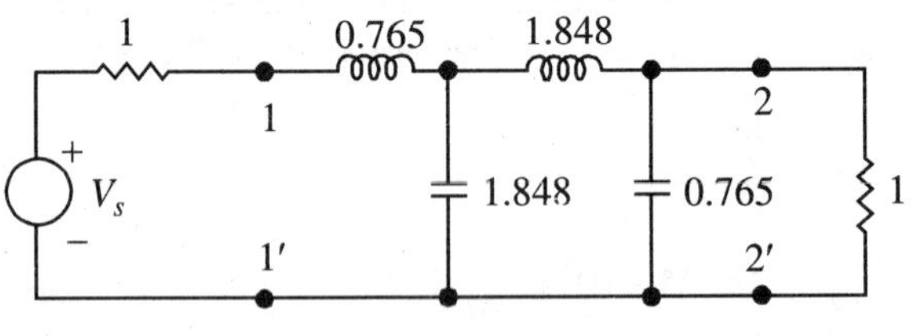

Fig. 11.7

(*v*) The frequency and impedance denormalizing factors are

$$\Omega_n = \frac{2\pi(25\times10^3)}{1} = 157.08\times10^3$$

and $$\gamma_n = \frac{1\times10^3}{1}$$

The final values of the elements are obtained by

$$L = \frac{L_n\gamma_n}{\Omega_n} \text{ and } C = \frac{C_n}{\gamma_n\Omega_n}$$

The final network is shown in Fig. 11.8.

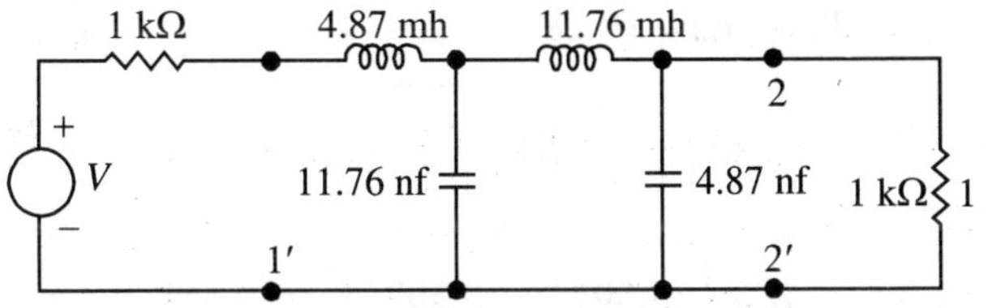

Fig. 11.8

Example 11.4. The insertion loss specifications for a LP filter are

Pass-band	:	0.3 MHz
Pass-band ripple	:	0.5 dB
High frequency discrimination	:	40 dB at 6 MHz

$$R_1 = R_2 = 100\ \Omega$$

(*i*) The specifications suggest a Chebyshev approximation.
From eqn. 10.25

$0.5 = 10 \log (1 + \varepsilon^2)$

Hence, $\varepsilon^2 = 0.122$ and $\varepsilon = 0.349$.

From equation 10.24

$$n \geq \frac{\cosh^{-1}(1/k_1)}{\cosh(1/k)}$$

Here, $$1/k = 6/3 = 2$$

and $$1/k_1 = \left[\frac{10^{40/10}-1}{10^{0.5/10}-1}\right]^{1/2} = 286.263$$

and $$n \geq \frac{\cosh^{-1} 286.263}{\cosh^{-1} 2} = 4.78$$

Let $n = 5$. Then

$$F(s) = \frac{1}{1+\varepsilon^2 T_5^2(s)}$$

From Table 10.2

$$T_5^2(s) = -(16s^5 + 20s^3 + 5s)^2 = T_5'^2(s)$$

(*ii*) From eqn. 11.33

$$\rho(s)\rho(-s) = \frac{p(s)}{q(s)} = 1 - \frac{1}{1-\varepsilon^2 T_5'^2(s)}$$

$$= \frac{(16s^5 + 20s^3 + 5s)^2}{(16s^5 + 20s^3 + 5s)^2 - 1/\varepsilon^2}$$

The zeros of $q(s)$ are the roots of the Chebyshev function filter with $n = 5$ and $\varepsilon = 0.349$. From eqn. 10.28

$$\Delta = \frac{1}{0.349} + \sqrt{1 + \frac{1}{0.122}} = 5.597,$$

$$a = \frac{1}{2}(5.597^{1/5} - 5.597^{-1/5}) = 0.351$$

and $$b = \frac{1}{2}(5.597^{1/5} + 5.597^{-1/5}) = 1.06$$

The poles of $\rho(s)$ are at

$-a, -a\cos\pi/5 \pm jb\sin\pi/5, -a\cos 2\pi/5 \pm jb\sin 2\pi/5$

i.e., the poles are at

$p_0 = -0.351$, $p_{1,2} = -0.284 \pm j0.623$, $p_{3,4} = 0.109 \pm j1.008$.

The zeros of $p(s)$ are the roots of the equation

$$(16s^5 + 20s^3 + 5s)^2 = 0$$

These can be obtained by solving for the roots of $T_5(\omega) = 0$ and substituting $\omega = s/j$. $T_5(\omega) = 0$ can be written as

$$\cos(5\cos^{-1}\omega) = 0$$

i.e., $$5\cos^{-1}\omega = \frac{k\pi}{2}, \; k = 1, 3, 5, 7, 9$$

or $$\omega = \cos\frac{k\pi}{10}, \; k = 1, 3, 5, 7, 9$$

Hence, $$\omega = 0, \pm 0.588, \pm 0.951$$

and finally the zeros of $\rho(s)$ are at

$$z_0 = 0, \; z_{1,4} = \pm j0.588, \pm j0.951$$

Hence, $$\rho(s) = \frac{(s-z_0)(s-z_1)(s-z_2)(s-z_3)(s-z_4)}{(s-p_0)(s-p_1)(s-p_2)(s-p_3)(s-p_4)}$$

i.e., $$\rho(s) = \frac{s^5 + 1.251s^3 + 0.313s}{s^5 + 1.171s^4 + 1.935s^3 + 1.307s^2 + 0.75s + 0.167}$$

(*iii*) From eqn. 11.34

$$Z(s) = \frac{2s^5 + 1.171s^4 + 3.186s^3 + 1.307s^2 + 1.063s + 0.167}{1.171s^4 + 0.684s^3 + 1.307s^2 + 0.437s + 0.167}$$

(*iv*) The continued fraction expansion of *Z*(*s*) is

$$Z(s) = 1.707s + \cfrac{1}{1.298s + \cfrac{1}{2.69s + \cfrac{1}{1.09s + \cfrac{1}{1.944s + \cfrac{1}{1}}}}}$$

The normalized filter is shown in Fig. 11.9.

(*v*) The frequency and impedance scaling factors are

$$\Omega_n = \frac{2\pi(3 \times 10^6)}{1}, \; \gamma_n = \frac{100}{1}$$

The denormalized network is shown in Fig. 11.10

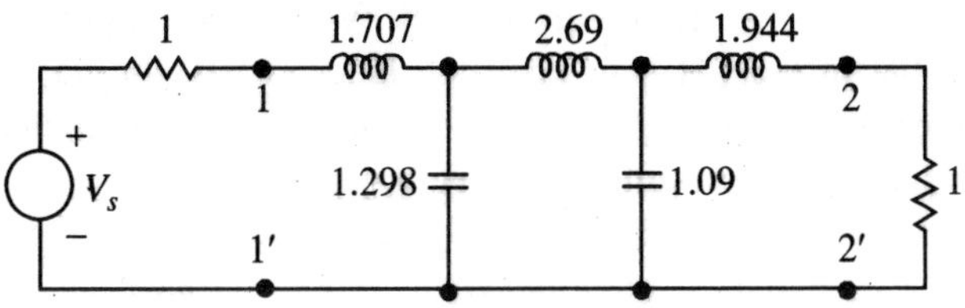

Fig. 11.9

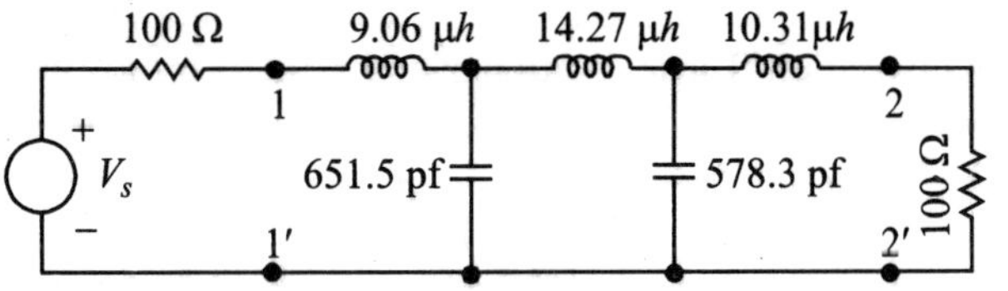

Fig. 11.10

11.3 DISCUSSIONS

We have discussed the design of LP filters only. As already described, we can use frequency transformation technique (see Sec. 10.5 and 10.6) to design high pass, bandpass and band elimination filters. The insertion loss specification of HP, BP or BE filters is first converted to that of an LP filter and the LP ladder is realized as indicated. Then the LC elements of the LP filters are transformed by the transformation given in Sec. 10.5. In the design of BP filters we once again stress the point that the application of the spectral transformation technique leads to geometrically symmetric filter.

Examples 10.2 to 10.4 bring out some interesting features of the Darlington design procedure. The Butterworth approximation of the insertion loss leads to a reflection coefficient with all its zeros at the origin. If Chebyshev approximation is used instead, the zeros of the reflection coefficient are all on the imaginary axis.

The transfer functions of the filters generated are all-pole functions. If we need finite transmission zeros elsewhere—other than the origin—we have to resort to approximation techniques other than the Butterworth and Chebyshev as indicated earlier.

The continued fraction expansion of $Z(s)$ by Euclids algorithm may run into numerical problems and to alleviate this the coefficients must be represented with the sufficient accuracy. The last term in the expansion must be a constant equal to the load resistance.

Though we have developed $Z(s)$ into a ladder network, any other method of realizing the input impedance is equally suitable. In the examples, we have considered all the transmission zeros to be at infinity and hence a ladder was the most convenient network to realize. We may resort to other realization methods (other than continued fraction expansion) as long as the network obtained has specifically the required transmission zeros and the terminating resistances. If the specifications lead to an insertion loss function with finite transmission zeros elsewhere other than the origin (Inverse Chebyshev and elliptic approximations for example) but in the LH plan, then the ladder has to be developed by zero shifting or pole weakening technique. If we consider a non-minimum phase network obviously we cannot resort to ladder structure.

A large part of Chap. 9 was devoted to the realization of a driving point function. This is often the case in most network synthesis books. As we have seen from Sec. 9.5 and the Darlington procedure, the realization of the transfer or insertion loss function fundamentally depends on the realization of a driving point function. The coefficient matching technique on the other hand does not depend on the driving point function realization. The realization of active filters mostly utilizes coefficient matching technique and this forms the topic for the next chapter.

PROBLEMS

11.1 A doubly terminated, two-port is shown in Fig. P. 11.1. Find the insertion-voltage ratio and power ratio. Also, find the insertion phase characteristic. Plot $\beta(\omega)$ vs ω.

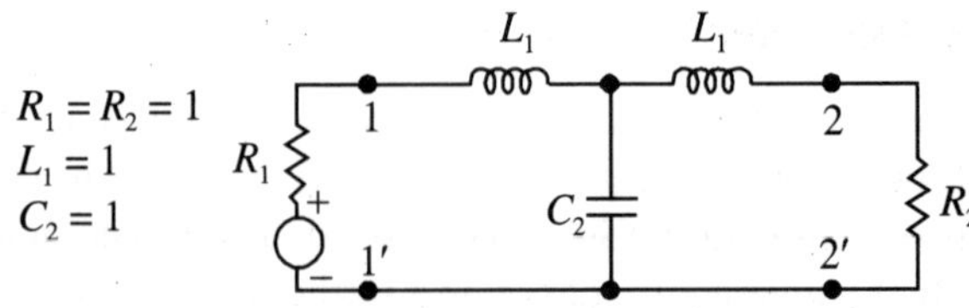

Fig. P. 11.1

11.2 Find $\rho(s)$ and $Z(s)$ for the following filter specifications:

(*a*) $e^{2\theta} = K(1 + \omega^{2n})$

(i) $n = 2$, $K = 1$, $R_1/R_2 = 3$

(ii) $n = 3$, $K = 1$, $R_1/R_2 = 1$

(b) $e^{2\theta} = K[1 + \varepsilon^2 T_n^2(\omega)]$

(i) $n = 3$, $\varepsilon = 0.125$, $K = 1$, $R_1/R_2 = 1$

(ii) $n = 3$, $\varepsilon = 1$. $K1$, $R_1/R_2 = 1$

(iii) $n = 2$, $\varepsilon = 0.125$, $R_1/R_2 = 1/2$.

11.3 Design low-pass Butterworth filter for $n = 3$, $\omega_c = 3$ k rad/sec. $R_1 = R_2 = 1$ kΩ by coefficient matching technique.

11.4 If $R_2 = 2R_1$, and $R_1 = 1$ kΩ is prob. 11.3, find the elemental values of the Butterworth filter.

11.5 Design a low-pass filter for the specifications of prob. 10.1 with $R_1/R_2 = 600$ Ω.

11.6 Design a low-pass filter for the specifications of prob. 10.2 with $R_1/R_2 = 3$, $R_2 = 100$ Ω.

11.7 Design a Chebyshev filter for the specifications of prob. 10.5 with $R_1 = 1$ kΩ. Select the proper value of R_2.

11.8 Design a Chebyshev filter for the specifications of prob. 10.8 with $R_1 = 600$ Ω. Select the proper value of R_2.

11.9 Design a third order band-pass Butterworth filter with bandwidth = 20 kHz, centre frequency = 1 MHz and $R_1 = R_2 = 600$.

11.10 Design a band-pass filter for the following specifications. The upper and lower 3 dB points are at 940 kHz and 1060 kHz.
A loss of at least 20 dB at 910 kC and 1090 kC $R_1 = R_2 = 600$.

11.11 A doubly terminated LC network with R_1 and R_2 as terminating resistances is specified by the insertion-loss IL and attenuation $\alpha(\omega)$. Show that

$$\alpha(\omega) = IL + 10\log\frac{(R_1 + R_2)^2}{4R_1R_2}$$

11.12 Design a third order Butterworth filter with 3 dB pass-band loss and $R_2 = 3R_1$.

11.13 The insertion loss specifications for a low-pass Butterworth filter are the following:

Pass-band loss	:	1.5 dB
Stop-band loss	:	25 dB at 3.3 M rad/sec
Pass-band	:	0 to 1M rad/sec

$$R_1 = 150\ \Omega,\ \frac{R_2}{R_1} = 2$$

Design a LC ladder for these specifications.

11.14 A Chebyshev filter is required to satisfy the following insertion loss requirements:

Pass-band	:	0 to 25 k rad/sec
Pass-band ripple	:	1.50 dB
Stop-band	:	30 dB at 50 k rad/sec

The filter is to be driven by voltage source with an internal impedance of 150 Ω. Design a filter to satisfy these requirements.

NOTES AND REFERENCES

The concept of insertion loss introduces here for a lumped linear passive network, plays an import role in the design of VHF/UHF and microwave circuit also. Here, we have shown the application of the concept in filter design. Chen's book[1] discusses this topic in considerable detail. Further details can be had from references of Chapter 9.

Insertion loss synthesis was introduced by Darlington in his celebrated paper on synthesis of reactance 4-poles and has to date stayed as the most powerful method of designing doubly terminated filter networks.

1. W.H. Chen, *Linear Network Design and Synthesis*, McGraw-Hill, 1976.
2. S. Darlington: *Synthesis of Reactance 4-pole which produce Prescribed Insetium, Loss Characteristics*, J. Math and Physics, Vol. 18, 1939, pp. 257–353.

CHAPTER

12 Active Networks and Filters

A wide class of networks and filters that has received considerable attention in the last two decades is the active RC networks. Some of the limitations of the passive filters, viz., the size of the inductors at low frequencies, the necessity of buffer or isolation amplifiers to prevent loading while cascading sections of filters, and the need for an external amplifier to adjust the required gain have led to intensive research and development work in the field of active filters. The present trend to microminiaturize the circuit in order to increase the package density and the reliability of the circuit has also been an important factor. Unfortunately, it has not been possible to integrate inductors with practical value of inductances and reasonable Q-factors. The availability of high quality inexpensive monolithic operational amplifier as an off-the-shelf unit has made the active filter design an attractive one.

As indicated earlier, RC filters can, at best, realize a Q of 0.5 and hence cannot be used to design a high selectivity filter. Most of the active filter structures are realized by embedding an operational amplifier in RC network in a feedback configuration, providing high Q networks.

In this chapter, we discuss the design of active RC filters. We introduce the commonly employed active elements. The concept of sensitivity which plays a major role in the active filter design is also discussed. Several single amplifier and a few multiamplifier filters are also presented. Inductance and ladder simulation for higher order filters is also introduced.

12.1 ACTIVE ELEMENTS

Network elements can be classified as active and passive. An element is considered active if it can supply energy to the external world. If $v(t)$ and $i(t)$ are the instantaneous voltage and current respectively, at the terminals of the element, the element is active if, $\{W(t_0, t) + E(t_0)\}| \le 0$ for some t (see Sec. 1.3). From this definition it is clear that the lumped R, L and C elements defined in chapter one are all passive. In contrast a transistor is an active element.

Of the many active elements used in filters, the most important one is the operational amplifier. Of late, several active elements like gyrators, generalised impedance convertors are becoming an integral part of active filters. However, a controlled source (see Sec. 1.1.2) may be considered as the basic active element. Other active elements, to be discussed later in this section, can be realized by using controlled sources. Controlled sources are classified in four different ways. They are

(*i*) *Current-controlled current source* (CCT), e.g., a transistor.
(*ii*) *Voltage-controlled current source* (VCT), e.g., field effect transistor.
(*iii*) *Voltage-controlled voltage source* (VVT), e.g., an operational amplifier. (Though an operational amplifier is realized by using transistor or FETs, we consider operational amplifier as a basic element.)
(*iv*) *Current-controlled voltage source* (CVT).

Figure 12.1 shows the diagramatic representation of the ideal controlled sources. The transmission parameter matrices and the input and output impedances (R_i and R_o, respectively) of the four controlled sources are

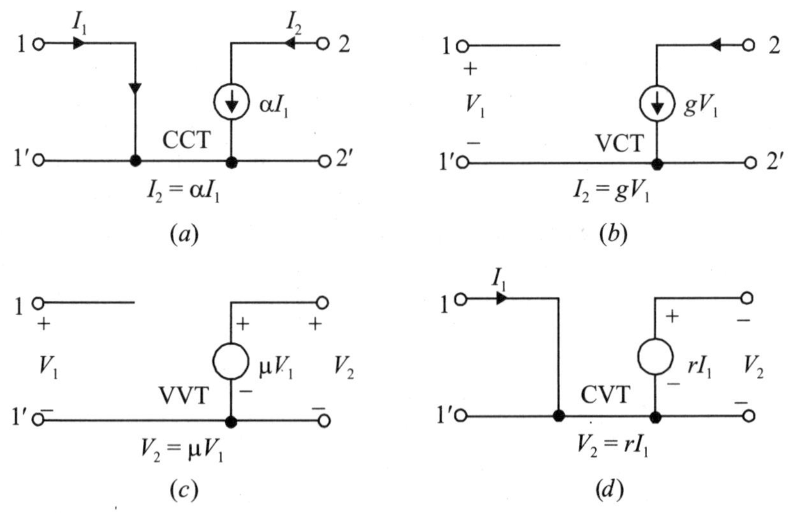

Fig. 12.1

CCT: $\begin{bmatrix} 0 & 0 \\ 0 & -\dfrac{1}{\alpha} \end{bmatrix}$, $\quad R_i = 0$, $R_o = \infty$

VCT: $\begin{bmatrix} 0 & -\dfrac{1}{g} \\ 0 & 0 \end{bmatrix}$, $\quad R_i = 0$, $R_o = \infty$

VVT: $\begin{bmatrix} \frac{1}{\mu} & 0 \\ 0 & 0 \end{bmatrix}$, $R_i = \infty$, $R_o = 0$

CVT: $\begin{bmatrix} 0 & 0 \\ \frac{1}{r} & 0 \end{bmatrix}$, $R_i = 0$, $R_o = 0$

It can be observed from the transmission parameter matrices that we can generate a VVT by cascading a VCT and a CVT. We can also generate a CCT by cascading a CVT and a VCT.

12.1.1 Operational Amplifier

An operational amplifier (op. amp) is a high gain differential input dc amplifier. Ideally, it is considered as a dual input infinite gain VVT. The symbol and the equivalent circuit of an op. amp are shown in Fig. 12.2. Output voltage V_0 is given by

$$V_0 = -AV = -A(V_1 - V_2)$$

i.e.,
$$V_0 = A(V_2 - V_1) \quad \ldots(12.1)$$

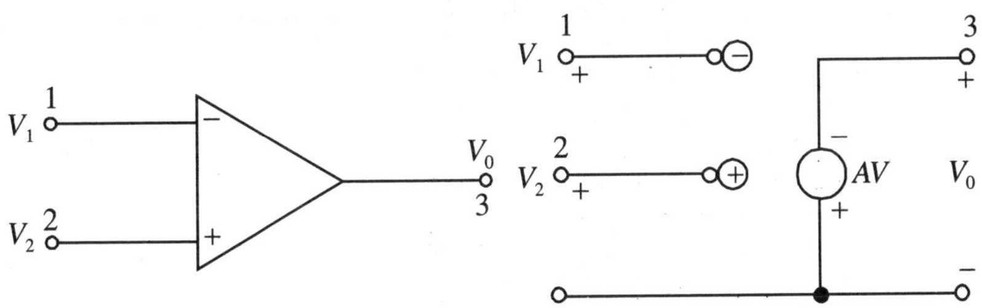

Fig. 12.2

In an ideal op. amp A is considered to be infinity and the amplifier is assumed to have infinite bandwidth. Also, the input impedance and the output admittance are assumed to be infinity. Because of infinite gain A, the differential voltage V is zero and if one of the terminals of an op. amp (*i.e.*, either 1 or 2) in feedback configuration is grounded then the other is a virtual ground. It is also assumed that an op. amp has zero off set, *i.e.*, the output voltage V_0 is zero when the differential voltage $(V_1 - V_2)$ is made zero. The current drawn by the two input terminals of an op. amp is zero as the input impedance is assumed to be infinity.

Op. amps are rarely operated in open loop mode. Some of the feedback mode operations and the corresponding terminal relations are shown in Fig. 12.3. Op. amps find extensive use in linear and nonlinear networks.

A practical op. amp deviates considerably from its ideal behaviour and we discuss this in a later section.

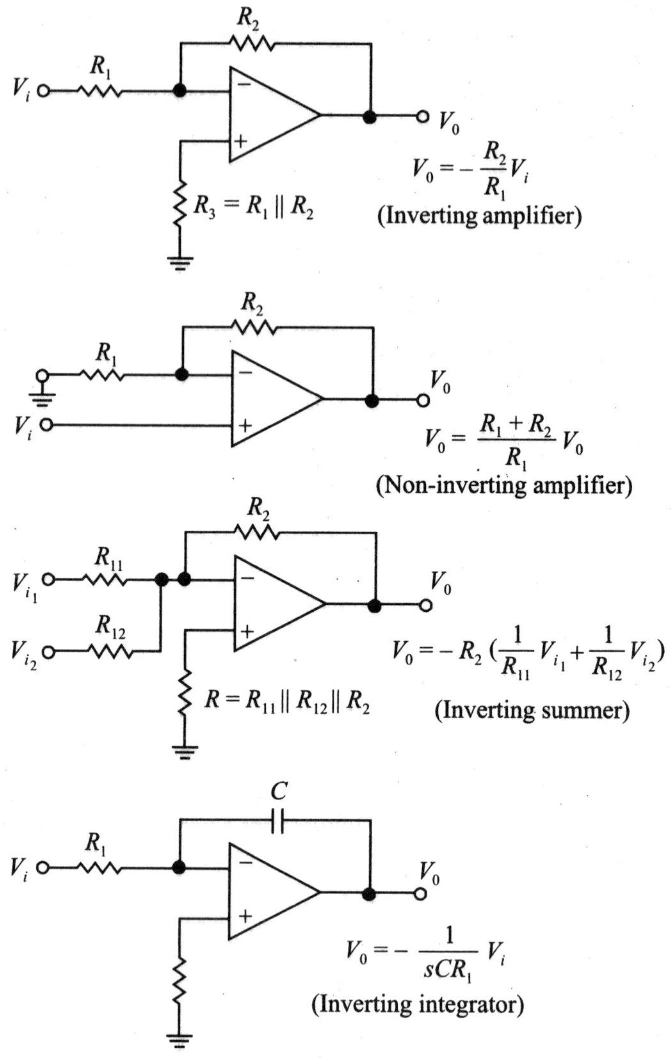

Fig. 12.3

12.1.2 Negative Impedance Converter

A *negative impedance converter* (NIC) is a two-port device whose input impedance, when terminated in an impedance Z_L, *is* $-kZ_L$ where k is called the *conversion ratio*. From Fig. 12.4

$$Z_{in} = \frac{V_1}{I_1}, \quad Z_L = -\frac{V_2}{I_2}$$

Hence, a NIC is characterized by the relation

$$\frac{V_1}{I_1} = k\frac{V_2}{I_2} \qquad \text{...(12.2)}$$

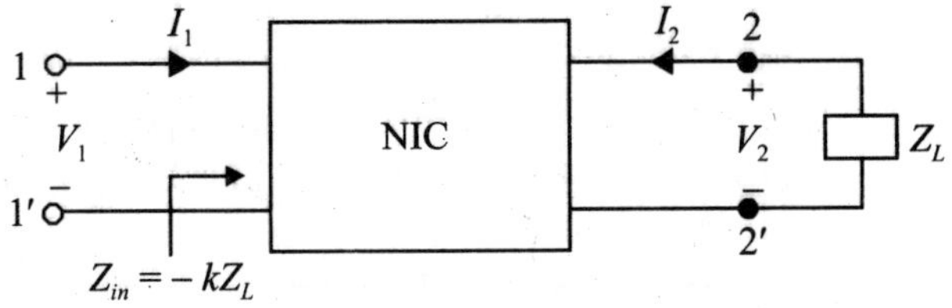

Fig. 12.4

We have two different possibilities

(*i*) $V_1 = -k_1 V_2, \quad I_1 = -\dfrac{I_2}{k_2}$ and $k = k_1 k_2$

The transmission matrix for such a NIC is

$$\begin{bmatrix} -k_1 & 0 \\ 0 & 1/k_2 \end{bmatrix}$$

and this is a *voltage inversion* type NIC or VNIC.

(*ii*) $V_1 = k_1 V_2, \quad I_1 = \dfrac{I_2}{k_2}$ and $k = k_1 k_2$

The transmission matrix is

$$\begin{bmatrix} k_1 & 0 \\ 0 & -1/k_2 \end{bmatrix}$$

and such a NIC is of *current inversion type* or CNIC.

To see the usefulness of NICs, apart from being able to generate negative valued components, consider the network shown in Fig. 12.5. The input impedance of the network is

$$Z_{in} = 1 - \frac{1}{s+1} = \frac{s}{s+1}$$

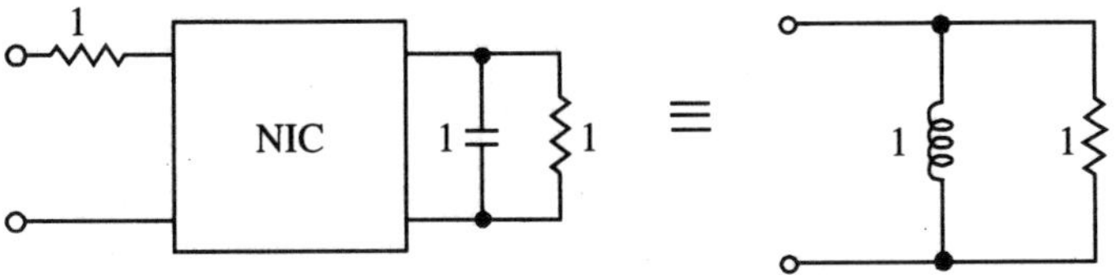

Fig. 12.5

and hence a NIC can be used to simulate the effect of an inductance.

NICs can be realized by using controlled sources. Figure 12.6 shows realizations of a VNIC and a CNIC. Figure 12.7 shows an op. amp realization of a CNIC and a VNIC. (Observe that in the Circuit *b* the output is ungrounded).

NICs are nonreciprocal devices and are potentially unstable networks. We will not study the stability of NICs here.

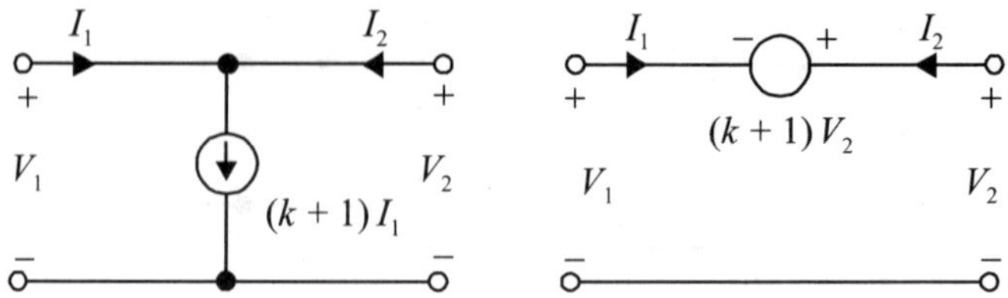

Fig. 12.6

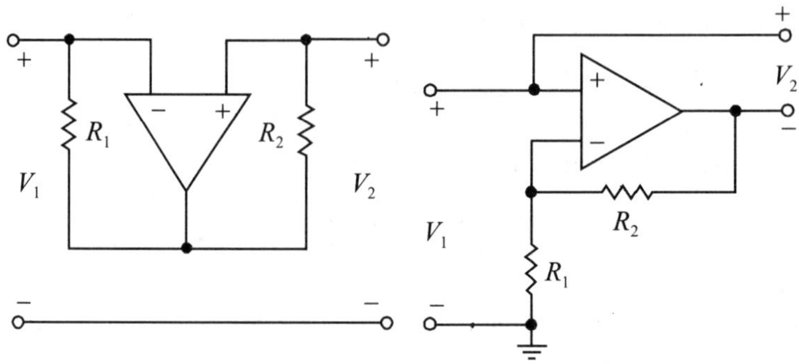

Fig. 12.7

12.1.3 Gyrators

A two-port device, which is becoming an integral part of active filter design and inductor simulation, is the *gyrator.* A gyrator is characterized by the terminal relationships

$$V_1 = -r_1 I_2$$
$$V_2 = r_2 I_1$$

and its symbolic representation is shown in Fig. 12.8. The total power delivered to the gyrator is

$$V_1 I_1 + V_2 I_2 = (r_2 - r_1)\, I_1 I_2$$

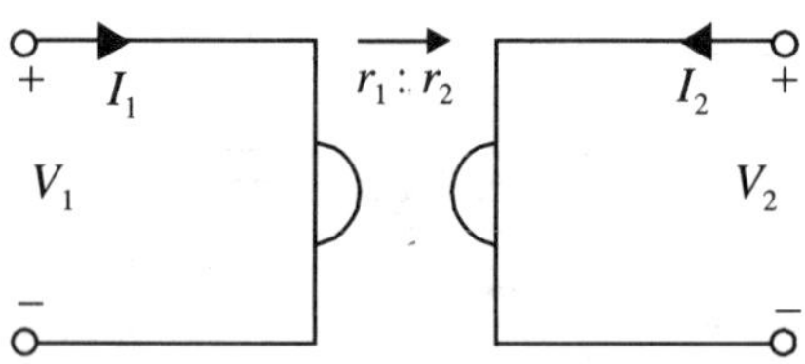

Fig. 12.8

If $r_1 = r_2$, the power delivered to the gyrator is zero. Such a device is a passive lossless one and is called an ideal gyrator. If $r_2 \neq r_1$, the power can be made negative and then the device is called an active gyrator. The two-port parameters of a gyrator are

$$Z = \begin{bmatrix} 0 & -r_1 \\ r_2 & 0 \end{bmatrix}, \; Y = \begin{bmatrix} 0 & g_2 \\ -g_1 & 0 \end{bmatrix}, \; T = \begin{bmatrix} 0 & r_1 \\ g_2 & 0 \end{bmatrix} \qquad \text{...(12.3)}$$

where $g_i = 1/r_i$. It can be seen from eqn. 12.3 that a gyrator is a nonreciprocal device.

If the gyrator is terminated in an impedance Z_L, the input impedance is

$$Z_{in} = \frac{V_1}{I_1} = \frac{-r_1 I_2}{V_2/r_2} = r_1 r_2\left(-\frac{I_2}{V_3}\right)$$

i.e.,
$$Z_{in} = \frac{k}{Z_L} \text{ where } k = r_1 r_2 \qquad \ldots(12.4)$$

A gyrator can thus be considered as an impedance inverter. If the load is a capacitance, *i.e.*, $Z_L = 1/sC$, then Z_{in} assuming $r_1 = r_2 = r$, is $s\, r^2C$. Hence, a capacitance terminated gyrator is equivalent to an inductor of value r^2C (Fig. 12.9)

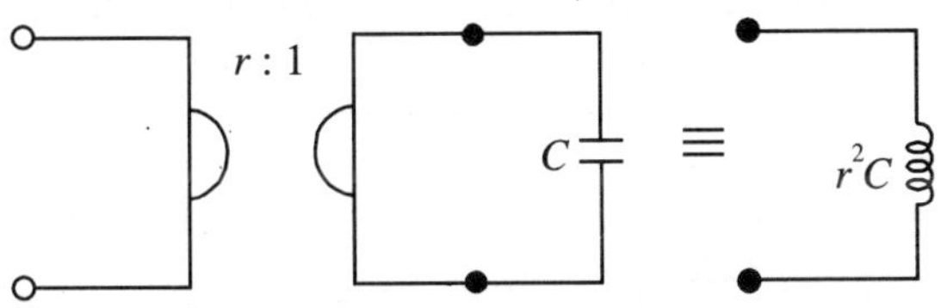

Fig. 12.9

Figure 12.10 shows a realization of an ideal gyrator using two controlled sources and the gyration ratio is $r = 1/G$. Figure 12.11 shows a realization of an ideal gyrator using two op. amps. This is commonly referred to as *"Antoniou Gyrator"*. (Observe that the capacitor C_4 is being gyrated to an inductance).

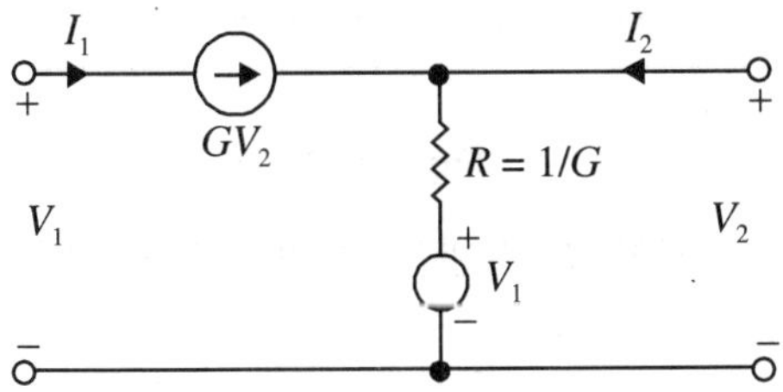

Fig. 12.10

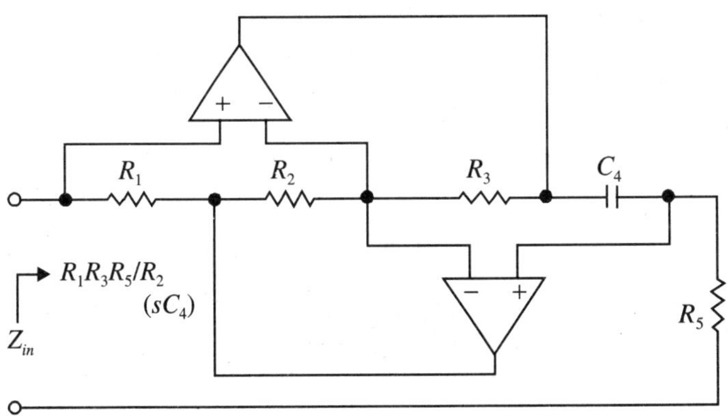

Fig. 12.11

12.1.4 Generalised Impedance Converter

Let a two-port network be terminated in an impedance Z_L. If the input impedance of the terminated two-port is

$$Z_{in} = K(s)Z_1^n, \quad n \pm 1 \qquad \text{...(12.5)}$$

where $K(s)$ is real rational function of s, then the two-port is called a *generalised impedance converter* (GIC). $K(s)$ may also be a constant (The definition here differs somewhat from elsewhere in the literature.) If $n = 1$ it is called a converter and $n = -1$ it is an impedance inverter. If $K(s) = -1$ and $n = 1$, the GIC is the already familiar NIC and thus can be realized by a single operational amplifier. If $K(s) = 1$ and $n = 1$, then we have a positive impedance converter or a transformer. If $K(s) = k$, and $n = -1$, the GIC is a gyrator. On the other hand, if $K(s) = -k$ and $n = -1$, the GIC acts like a *negative impedance inverter* (see Sec. 12.2).

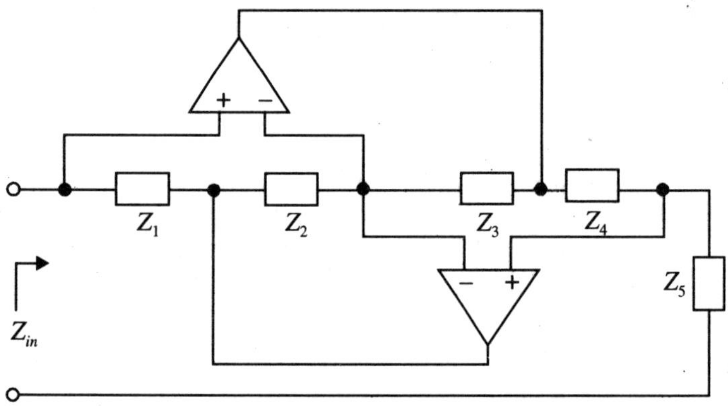

Fig. 12.12

Figure 12.12 shows a realization of a two amplifier GIC. The input impedance of this network is

$$Z_{in} = \frac{Z_1 Z_3 Z_5}{Z_2 Z_4} \qquad \text{...(12.6)}$$

If in Fig. 12.12 (or in eqn. 12.6) we select Z_i, (i =1, 2, 3, 5) as resistors and Z_4 as a capacitor, we obtain the familiar Antoniou gyrator. In fact, we can generate a second two amplifier gyrator circuit by selecting Z_2 as a capacitor and the rest of the impedances as resistors.

If Z_1 and Z_5 are capacitors and Z_2, Z_3 and Z_4 resistors, the input impedance is

$$Z_{in}(s) = \frac{R_3}{R_2 R_4} \frac{1}{s^2 C_1 C_5} = \frac{1}{Ds^2} \qquad \text{...(12.7)}$$

where $$D = \frac{R_2 R_4 C_1 C_5}{R_3}$$

Then $$Z_{in}(j\omega) = -\frac{1}{D\omega^2}$$

and hence the input impedance behaves like a *frequency dependent negative resistance* (FDNR). The circuit realization an FDNR is shown in Fig. 12.13. This is known as *Bruton's* FDNR. As can be seen from Fig. 12.12 we can generate two more FDNR circuits by selecting (Z_1, Z_3) or (Z_3, Z_5) as capacitors and the rest of the impedances as resistors. FDNRs are becoming popular active components in the design of active filters.

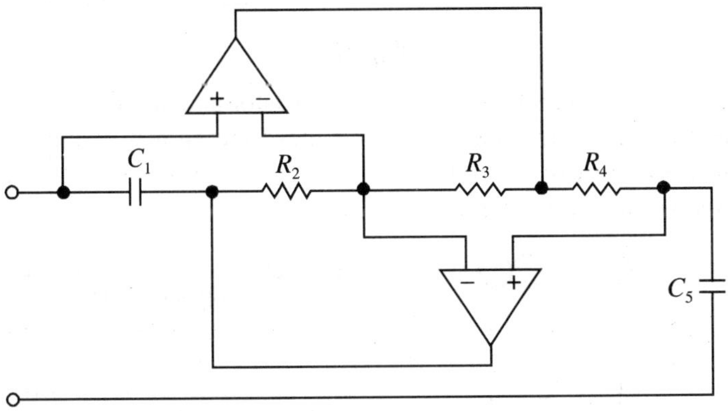

Fig. 12.13

If in Fig. 12.12 we select (Z_2, Z_4) as capacitors and the rest as resistors we have

$$Z_{in}(s) = B\, s^2 \text{ or } Z_{in}(j\omega) = -B\omega^2$$

where $B = R_1R_3R_5C_2C_4$. This component is called a *frequency dependent negative conductance* (FDNC).

We see that from the basic topology of the two-amplifier GIC we can generate two gyrators, three FDNRs and an FDNC. We cannot generate an NIC or a NIV from this topology.

12.2 NULLATOR-NORATOR MODELS

In Chap. 1 we introduced two pathological or singular network elements—nullator and norator. The ideal behaviour of a transistor or an op. amp can be represented by a nullator-norator (or nullor) models. Figures 12.14 and 12.15 show such a representation for a transistor and an op. amp, respectively. Ideally, the current drawn by the base terminal and the voltage across the base-emitter junction of a transistor are zero. The current drawn by the emitter terminal is independent of

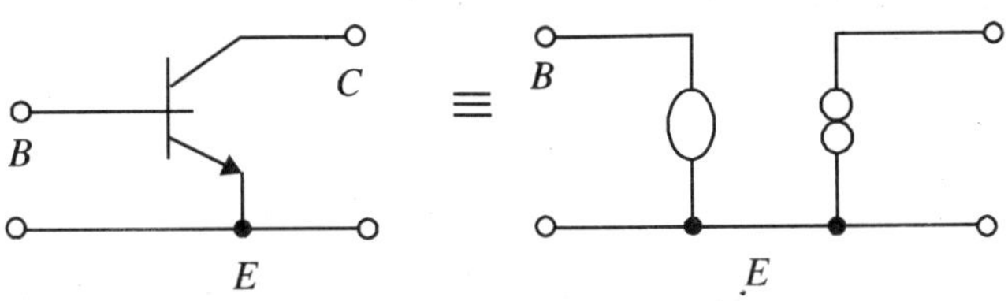

Fig. 12.14

the collector to emitter voltage thus justifying the nullator-norator model for a transistor. In an ideal op. amp the current drawn by and voltage between the negative and positive input terminals are zero. The output voltage is independent of the current drawn by the output terminals and hence the model of Fig. 12.15.

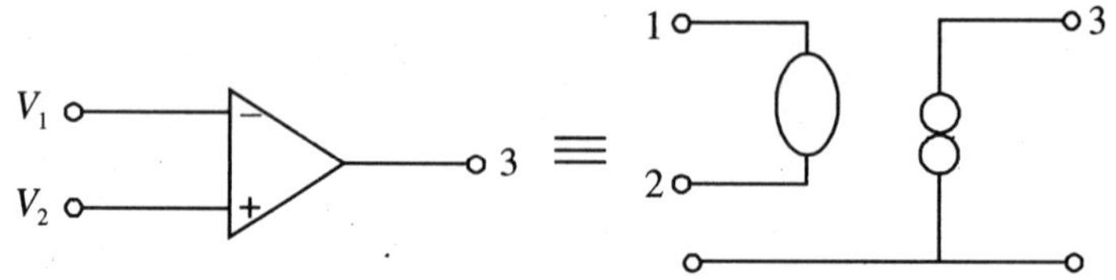

Fig. 12.15

In addition to the nullator-norator of a transistor and an op. amp. two nullator-norator equivalences are very useful and these are shown in Fig. 12.16. From these nullator-norator models, we can generate various active element configurations as will be demonstrated now.

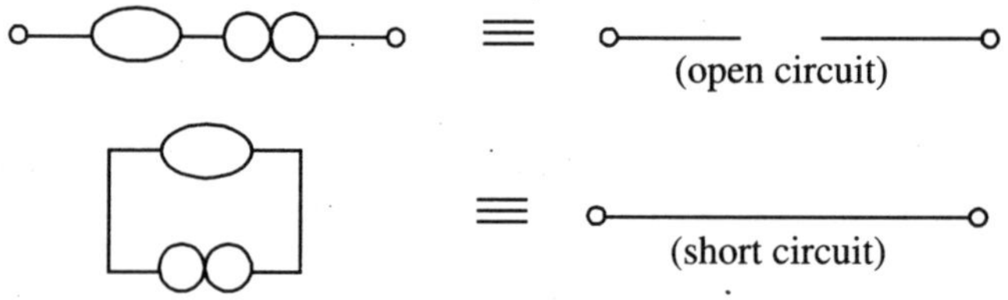

Fig. 12.16

The concept of nullator-norator models is very useful in developing transistor or op. amp circuits for the active elements discussed in Sec. 12.1.

Nullator-norator equivalent circuits for the four different types of controlled sources are shown in Fig. 12.17. Of course, we can generate many more circuits by judicially adding nullators and norators. It is to be observed that the nullator and norator should always occur in pair. By properly combining a nullator with a norator, one can generate transistor or op-amp circuits for these controlled sources.

The nullator-norator model shown in Fig. 12.18 represents a CNIC. As there is a nullator between terminals 1 and 2, the two voltages V_1 and V_2 are the same. Also, the voltage drops across R_1 and R_2 are the same and hence

$$I_2 = \frac{R_1}{R_2} I_1$$

The equations describing Fig. 12.18 are then

$$V_1 = V_2$$

$$I_2 = kI_1, \; k = \frac{R_1}{R_2}$$

which is a CNIC with $k_1 = 1$ and $k_2 = k$. It is not possible to generate a transistor realization of Fig. 12.18 directly as there is no common point between the nullator

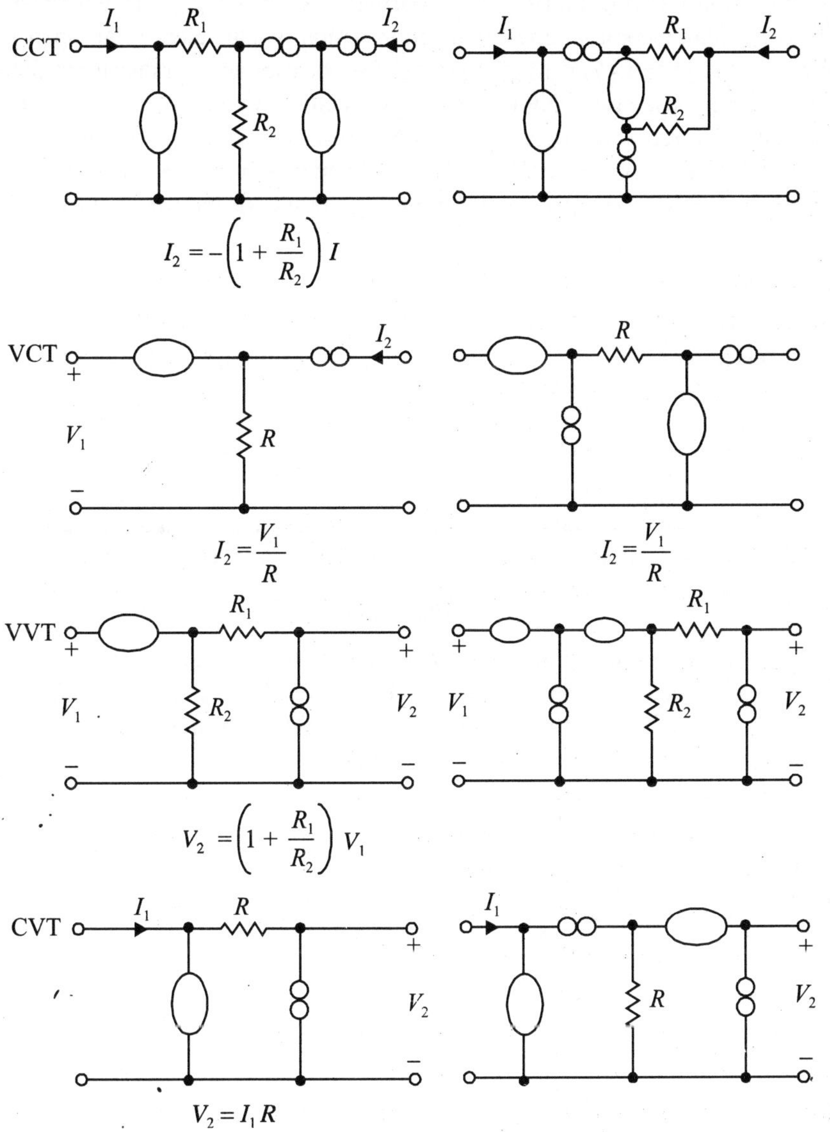

Fig. 12.17

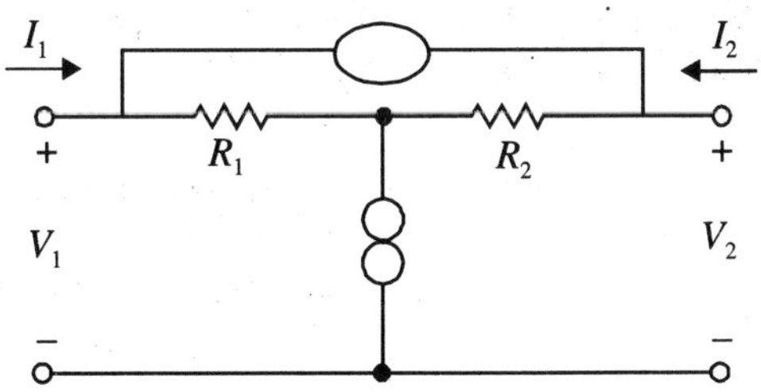

Fig. 12.18

and the norator. Figure 12.19(*a*) is a modified version of Fig. 12.18 where we have added a nullator-norator series connection which is equivalent to an open circuit (Fig. 12.16), whereas Fig. 12.19(*b*) shows the corresponding transistor realization. Here, we have used two PNP transistors. We can use two NPN transistors as well. The only difference between using a PNP or a NPN transistor is in the biasing circuit which is not shown in Fig. 12.19(*b*). From now on without loss of generality we only use PNP transistors and the biasing circuit for the transitors will not be shown.

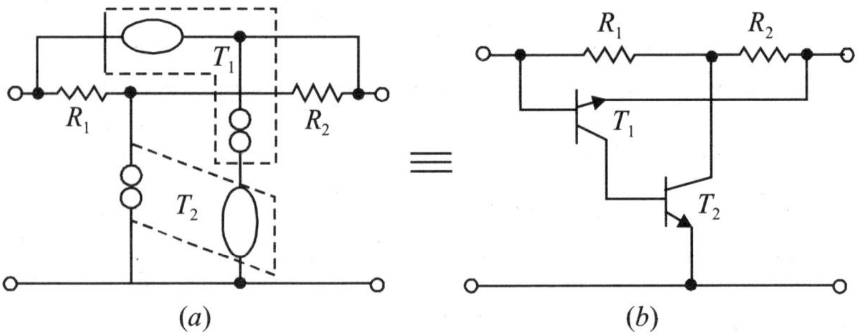

Fig. 12.19

By connecting a series nullator-norator combination at different places we can generate various CNIC transistor circuits. Figure 12.20 shows another CNIC circuit. The selection of the final circuit depends on stability, biasing and other practical considerations.

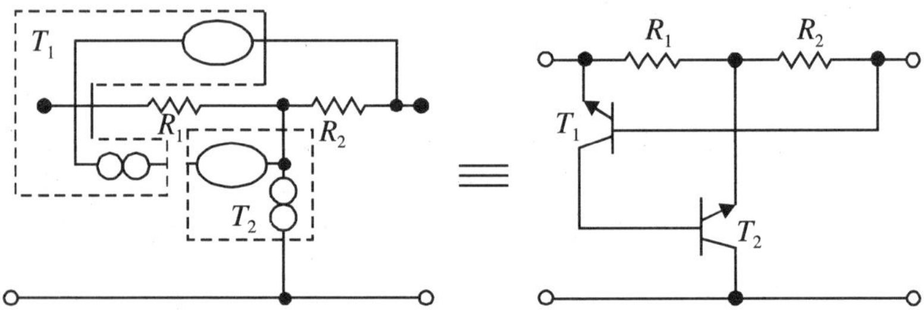

Fig. 12.20

The nullator-norator circuit of Fig. 12.21 represents a VNIC. The modified nullator-norator equivalent and the corresponding transistor realization are shown in Fig. 12.22.

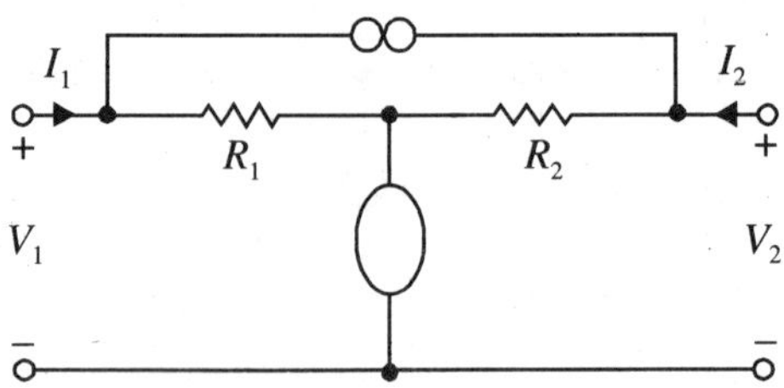

Fig. 12.21

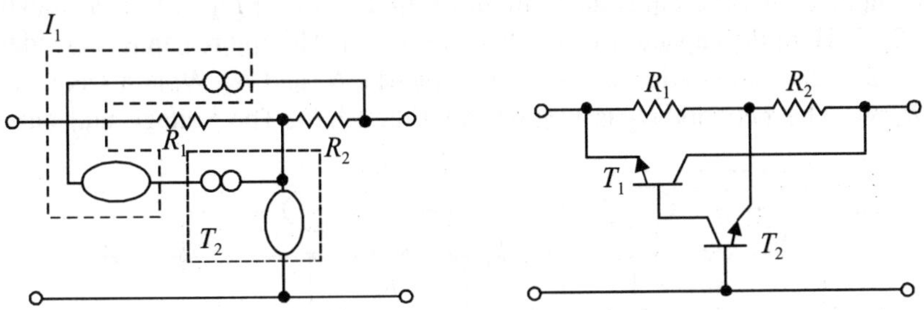

Fig. 12.22

Figure 12.23 shows a nullator-norator equivalent circuit of the two-controlled source representation of a gyrator of Fig. 12.10. Here, n_i and o_i represent nullators and norators, respectively. In order to analyse the circuit we observe that the voltage at point *c* is V_1 because of n_1: the voltage across *ab* is V_2 because of n_2 and n_3 and the current from *a* to *b* is I_1. The current from *a* to *b* is also given by GV_2

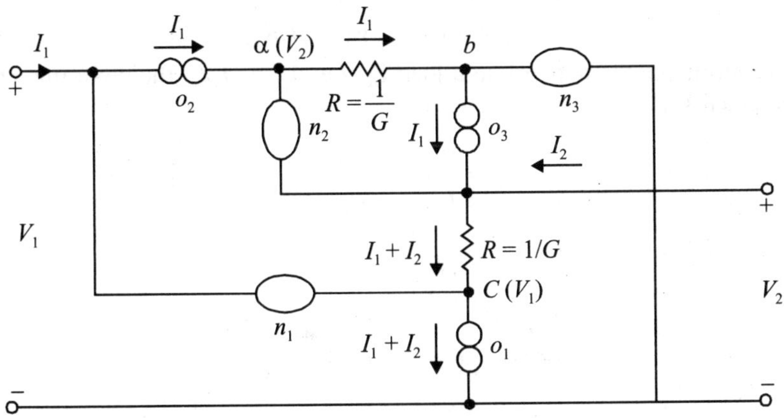

Fig. 12.23

as is required by Fig. 12.10. A transistor realization of this gyrator is obtained by grouping (n_1, o_1), (n_2, o_2) and (n_3, o_3). This is shown in Fig. 12.24 and is known as *Shenoi Gyrator.*.

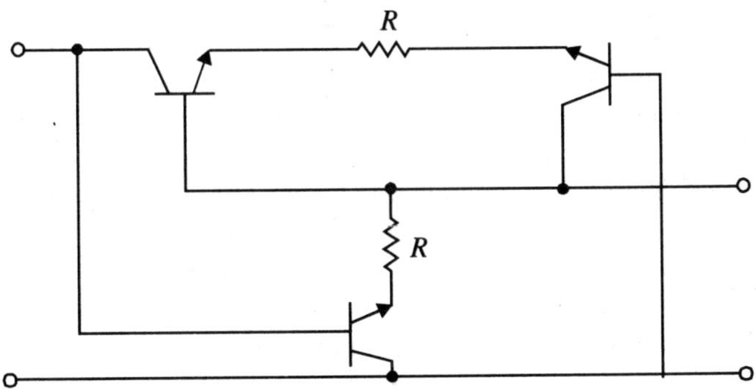

Fig. 12.24

The nullator-norator equivalent circuit of the gyrator of Fig. 12.11 is shown in Fig. 12.25. Here the capacitor C_4 has been removed. The figure can be considered as the cascade combination of two sub-networks N_1 and N_2. N_1 is a CNIC with $k = R_1/R_2$. The network N_2 is redrawn in Fig. 12.26. The voltage and current

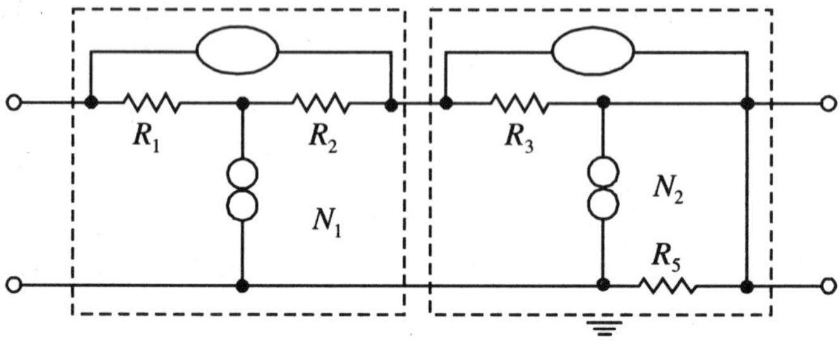

Fig. 12.25

distributions are also shown. From the figure we have

$$V_1 = -R_5 I_2, \quad V_2 = -I_1 R_3$$

If we terminate the network in a load Z_L, *i.e.*, $V_2 = -I_2 Z_L$, the input impedance of the network is

$$Z_{in} = \frac{V_1}{I_1} = -\frac{R_3 R_5}{Z_L}$$

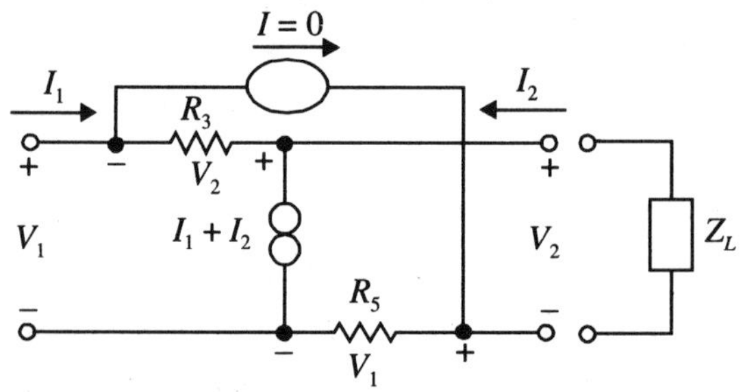

Fig. 12.26

and the network is the equivalent of a *negative impedance inverter* (*NIV*). A gyrator can be realized as the cascade combination of a NIC and a NIV.

The nullator-norator equivalent circuit of a GIC is similar to the one shown in Fig. 12.25. From the same nullor model by grouping nullators-norators differently we can generate alternate op. amp circuits for the GIC or the gyrator. The nullor equivalent circuits are useful for the initial analysis of the circuit. As already indicated, the final transistor or the op. amp circuit realization for the particular active element can only be selected by a detailed study of the biasing circuits and the stability of the circuits for various terminations.

12.3 SENSITIVITY

In the design of filters considered so far we have assumed the parameters of the elements (*i.e., R, L* and *C*) to be constant. The parameters or the characteristics of the elements depend on many factors like temperature, humidity and ageing. As such, we cannot consider the parameters to be constants. This deviation of the parameters, active or passive, from their nominal value manifests itself as a change in the response of the network. In fact, for a highly selective network this may induce oscillations and hence instability. The variation of the network response due to an incremental change in the value of the parameter is expressed in terms of sensitivity of the particular network function with respect to the parameter (see Secs. 5.3 and 5.6).

Classically, the sensitivity of a network function $F(s)$ with respect to a parameter x is defined as

$$S_x^{F(s)} = \frac{\partial \ln F(s)}{\partial \ln x} = \frac{x}{F}\frac{\partial F}{\partial x} \quad \text{...(12.8)}$$

Let the transfer function (open-circuit voltage transfer ratio) of a network N be $T(s) = p(s)/q(s)$. Then the sensitivity of $T(s)$ with respect to the parameter x can be shown to be

$$S_x^T = S_x^P - S_x^q$$

$$= x\left[\frac{1}{p}\frac{\partial p}{\partial x} - \frac{1}{q}\frac{\partial q}{\partial x}\right] \quad \text{...(12.9)}$$

In general, S_x^T is a complex quantity. If we write $T(j\omega)$ as $|\,T(j\omega)\,|\, e^{j\phi(\omega)}$ then

$$S_x^{T(j\omega)} = S_x^{\alpha(\omega)} + j\phi S_x^{\phi(\omega)}$$

where $\alpha(\omega) = |\,T(j\omega)\,|$. The real part of $S_x^{T(j\omega)}$ is called the *gain sensitivity* and the imaginary part the *phase sensitivity*.

Example 12.1. Consider an op. amp in the inverting mode and let the gain of the amplifier by μ. Then

$$T(s) = \frac{-\mu R_2}{((R_1 + R_2) + \mu R_1)} = \frac{p}{q}$$

(If $\mu \to \infty$, $s) \to -R_2/R_1$ as previously stated).

Then from eqn. 12.9

$$S_{R_1}^T = R_1\left[\frac{1}{p}\frac{\partial p}{\partial R_1} - \frac{1}{q}\frac{\partial q}{\partial R_1}\right]$$

$$= R_1\left[\frac{1}{p}0 - \frac{1+\mu}{q}\right]$$

i.e.,

$$S_{R_1}^T = \frac{-(1+\mu)R_1}{(R_1 + R_2) + \mu R_1}$$

Similarly
$$S_{R_2}^T = R_2\left[\frac{1}{p}(-\mu) - \frac{1}{q}\right]$$

$$= R_2\left[\frac{1}{R_2} - \frac{1}{(R_1 + R_2) + \mu R_1}\right]$$

i.e.,
$$S_{R_2}^T = \frac{R_1(1+\mu)}{(R_1 + R_2) + \mu R_1}$$

$$S_{\mu}^T = \mu\left[\frac{1}{(-\mu R_2)}(-R_2) - \frac{R_1}{(R_1 + R_2) + \mu R_1}\right]$$

i.e.,
$$S_{\mu}^T = \mu\frac{R_1 + R_2}{(R_1 + R_2) + \mu R_1}$$

As $\mu \to \infty$, $S_{R_1}^T \to -1$, $S_{R_2}^T \to 1$ and $S_{\mu}^T \to 0$.

Here, we have indicated the basic form of 'sensitivity of a network function' with respect to a parameter. The topic of sensitivity is an involved and an important one. Here our intention is to introduce the term and stress its importance in circuit design (see notes). The computation of sensitivity has been considered by several investigators and the "Adjoint Network" introduced in Section 5.6 is one such.

As indicated in Chapters 6 and 10, the network functions and the filter transfer functions are often specified by their poles and zeros. The poles and zeros themselves being functions of network parameters get perturbed due to variations in these parameters. The change in pole-zero location changes the frequency response characteristics of a filter and may indeed indicate potential instability of a filter. In order to study this aspect the sensitivity of a root (a pole or a zero) is found to be useful.

Let s_i be a root of either the numerator or denominator. Then the *root sensitivity* $S_x^{s_i}$ is defined as

$$S_x^{s_i} = x\frac{ds_i}{dx}\bigg|_{s-s_i} \qquad \ldots(12.10)$$

The root sensitivity in contrast to the transfer-function sensitivity is a complex constant.

Let a polynomial $p(s)$ be written as

$$p(s) = a(s) + xb(s)$$

then it can be shown that

$$s_x^{s_i} = -x\frac{b(s_i)}{p'(s_i)} \qquad \ldots(12.11)$$

where s_i is a root of $p(s)$ and $p'(s_i) = \frac{dp}{dx}\bigg|_{s=s_i}$

Example 12.2. Consider the low-pass filter shown Fig. 6.7. The transfer function of the filter was shown to be

$$T(s) = \frac{k}{s^2 + (3-k)s + 1} = \frac{k}{q(s)}$$

Let the nominal value of k be 1.5. Then

$$p(s) = s^2 + 1.5s + 1$$

The roots of $p(s)$ are

$$s_1,\ s_2 = -0.7500 \pm j\ 0.6614$$

$p(s)$ can be written as $p(s) = a(s) + k\ b(s)$, where $a(s) = s^2 + 3s + 1$ and $b(s) = -s$. Then, from eqn. 12.11, we have

$$S_k^{s_1} = -k\frac{b(s_1)}{p'(s_1)}$$

$$= -k\frac{(-s_1)}{2s_1 + (3-k)}$$

At the nominal value of k

$$S_k^{s_1} = 0.7500 + j\ 0.8505$$

Similarly, $$S_k^{s_2} = 0.7500 + j\ 0.8505$$

(Note that the sensitivity of the root at s_2 is the conjugate of the sensitivity of the root at s_1).

Let the value of x change from x to $x + \Delta x$. Then from eqn. 12.10

$$\Delta s_i = \frac{\Delta x}{x} S_x^{s_i}$$

and the new location of the root is

$$s_i^* = s_i + \Delta s_i$$

Let in the example 15.2, k change from 1.5 to 1.6, *i.e.*, $\Delta k = 1$.

Then

$$\Delta s_1 = \frac{0.1}{1.5}(0.75 + j0.8505) = 0.0500 + j0.0567$$

and $$\Delta s_2 = 0.0500 - j0.0567$$

Hence, the new locations of the roots are

$$s_1^* = -0.7000 + j\ 0.7181$$

$$s_2^* = -0.7000 + j\ 0.7181$$

The classical sensitivity, as defined by eqn. 12.8, can be related to the pole-zero sensitivity of the transfer function. If the transfer function $T(s)$ is expressed in the factorized form as

$$T(s) = K \frac{\prod_{i=1}^{n} (s - z_i)}{\prod_{i=1}^{m} (s - p_i)}$$

then it can be shown that

$$S_x^T = S_x^K + \sum_{i=1}^{m} \frac{S_x^{p_i}}{(s - p_i)} - \sum_{j=1}^{n} \frac{S_x^{z_j}}{s - z_j} \qquad \text{...(12.12)}$$

which shows that classical sensitivity is a weighted sum of the root sensitivities.

A filter is quite often defined by its Q-factor and centre frequency (or bandwidth). The transfer function of a second order filter is generally written as

$$T(s) = \frac{f(s)}{s^2 + \omega_0 s/Q + \omega_0^2} \qquad \text{...(12.13)}$$

where ω_0 is the resonant frequency (in the case of a BP filter it is the centre frequency) and Q is the quality factor. For this transfer function we can define ω_0 and Q sensitivities as

$$S_x^Q = \frac{x}{Q}\frac{dQ}{dx} \quad \text{and} \quad S_x^{\omega_0} = \frac{x}{\omega_0}\frac{d\omega_0}{dx} \qquad \text{...(12.14)}$$

The ω_0 and Q sensitivities are closely related to the transfer function sensitivity itself. For the transfer function of eqn. 12.13 with $f(s) = s^k\,\omega_0^2$, where $k = 0$, 1 or 2, it can be shown that

$$S_x^T = S_{\omega_0}^T \left[S_x^{\omega_0} + \lambda S_x^Q \right] \qquad \text{...(12.15)}$$

where

$$\lambda = \frac{\dfrac{\omega_0}{2Q}}{s + \dfrac{\omega_0}{2Q}}$$

A couple of examples (one for passive filter and the other for an active filter) illustrate this relationship.

Example 12.3. Consider the passive low-pass RC filter shown in Fig. 12.27. The transfer function is

$$T(s) = \frac{1/R_1R_2C^2}{s^2 + s(2R_1 + R_2)/CR_1R_2 + \dfrac{1}{R_1R_2C^2}} \qquad \text{...(12.16)}$$

which is of the form of eqn. 12.13 with $f(s) = \omega_0^2$. Here

$$\omega_0 = \frac{1}{C\sqrt{R_1R_2}}, \qquad Q = \frac{\sqrt{R_1R_2}}{2R_1 + R_2} \qquad \text{...(12.17)}$$

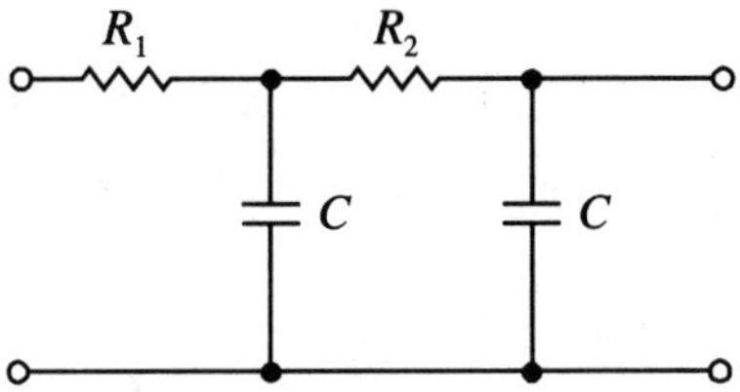

Fig. 12.27

From eqn. 12.14

$$S_{R_1}^{\omega_0} = \frac{R_1}{\omega_0} \frac{d}{dR_1}\left[\frac{1}{C\sqrt{R_1 R_2}}\right]$$

$$= \frac{R_1}{\omega_0} \frac{1}{C\sqrt{R_2}}\left(-\frac{1}{2R_1\sqrt{R_1}}\right)$$

i.e.,

$$S_{R_1}^{\omega_0} = -\frac{1}{2} \qquad \text{...(12.18)}$$

and

$$S_{R_1}^{Q} = \frac{R_1}{Q} \frac{d}{dR_1}\left[\frac{R_1 R_2}{2R_1 + R_2}\right]$$

After some algebra this yields

$$S_{R_1}^{Q} = Q \cdot \frac{R_2 - 2R_1}{2\sqrt{R_1 R_2}} \qquad \text{...(12.19)}$$

From eqns. 12.9 and 12.13 with $f(s) = \omega_0^2$

$$\left(\text{here } p(s) = \omega_0^2 \text{ and } q(s) = s^2 + \frac{\omega_0}{Q} s + \omega_0^2\right),$$

$$S_{\omega_0}^{T} = \omega_0\left[\frac{1}{\omega_0^2} 2\omega_0 - \frac{1}{s^2 + s\omega_0/Q + \omega_0^2}\left(\frac{s}{Q} + 2\omega_0\right)\right]$$

$$= \frac{2s(s + \omega_0/2Q)}{s^2 + \omega_0 s/Q + \omega_0^2} \qquad \text{...(12.20)}$$

From eqn. 12.9 with $p(s) = 1/R_1R_2C^2$ and

$$q(s) = s^2 + s\left(\frac{2R_1 + R_2}{CR_1R_2} + \frac{1}{R_1R_2C^2}\right)$$

We have

$$S_{R_1}^T = R_1\left\{R_1R_2C^2\frac{1}{R_2C^2}\left(-\frac{1}{R_1^2}\right)-\frac{s\cdot 1/CR_1^2 - 1/R_2C^2\cdot R_1^2}{s^2+s(2R_1+R_2)/CR_1R_2+1/R_1R_2C^2}\right\}$$

$$= -1+\frac{(s/CR_1)1/R_1R_2C^2}{s^2+(2R_1+R_2)s/CR_1R_2+1/R_1R_2C^2}$$

$$= -\frac{-s(s+2/CR_2)}{s^2+(2R_1+R_2)s/CR_1R_2+1/R_1R_2C^2}$$

From eqn. 12.17, we have

$$S_{R_1}^T = \frac{-s(s+2/CR_2)}{s^2+\omega_0 s/Q+\omega_0^2}$$

From eqns. 12.15 and 12.18 to 12.20, we have

$$S_{R_1}^T = \frac{2s(s+\omega_0/2Q)}{s^2+\omega_0 s/Q+\omega_0^2}\left[-\frac{1}{2}+\frac{\omega_0/2Q}{s+\omega_0/2Q}\cdot Q\frac{(R_2-2R_1)}{2\sqrt{R_1R_2}}\right]$$

$$= \frac{s}{s^2+\omega_0 s/Q+\omega_0^2}\left\{-s-\left[\frac{\omega_0}{2Q}\frac{\omega_0}{2\sqrt{R_1R_2}}(R_2-2R_1)\right]\right\}$$

But

$$\frac{\omega_0}{2Q}-\frac{\omega_0}{2\sqrt{R_1R_2}}(R_2-2R_1) = \frac{2R_1+R_2}{2CR_1R_2}-\frac{R_2-2R_1}{2C\,R_1R_2}$$

$$= \frac{4R_1}{2C\,R_1R_2} = \frac{2}{CR_2}$$

Hence $\quad S_{R_1}^T = -\dfrac{s(s+2/CR_2)}{s^2+\omega_0 s/Q+\omega_0^2}$

and this confirms the relation between S_x^T, $S_x^{\omega_0}$ and S_x^Q for the passive network.

Example 12.4. Consider the active low-pass filter shown in Fig. 12.28. The transfer function is

$$T(s) = \frac{1/R_1R_2C^2}{s^2+\frac{1}{C}\frac{R_1+R_2}{R_1R_2}s+\frac{1}{C^2R_1R_2}}$$

Hence, $\quad \omega_0 = \dfrac{1}{C\sqrt{R_1R_2}},\ Q = \dfrac{\sqrt{R_1R_2}}{R_1+R_2}$

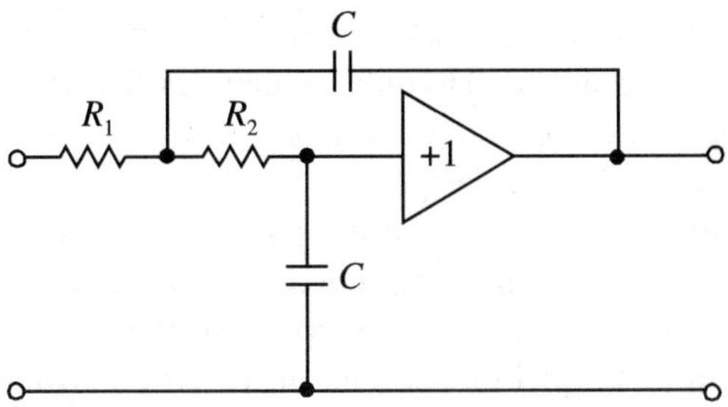

Fig. 12.28

From eqns. 12.9 and 12.14, we have

$$S_{R_1}^{\omega_0} = -\frac{1}{2},\ S_{R_1}^{0} = Q\frac{R_2 - R_1}{2\sqrt{R_1R_2}}$$

$$S_{\omega_0}^{T} = \frac{2s\left(s+\dfrac{\omega_0}{2Q}\right)}{s^2+\dfrac{\omega_0}{2Q}s+\omega_0^2},\ S_{R_1}^{T} = -\frac{s\left(s+\dfrac{1}{CR_2}\right)}{s^2+\dfrac{\omega_0}{Q}s+\omega_0^2}$$

Again, after a little algebra we see that eqn. 12.15 is satisfied.

The preceding two examples demonstrate the generality of the expression in 12.15 and show the correlation between the transfer function, ω_0 and Q sensitivities for a second-order LP filter. It can be similarly shown that the relationship is true for HP and BP filters.

We have defined several sensitivity functions and have interrelated them. However, in the final analysis, we are interested in minimizing the deviation of the filter response due to incremental variation of some network parameter. In highly selective networks (*i.e.,* high Q networks), the pole sensitivity is an important factor to consider because of the stability of the network. In the design of active filters, the sensitivity of the filter transfer function to the variation of the active parameters is a major consideration.

The study of sensitivity is an integral part of the design of active filters. In this section, we have just introduced some of the fundamental definitions and interrelation that occur in sensitivity studies.

12.4 SINGLE AMPLIFIER FILTERS

In Chapter 11, we considered the realisation of passive LC filters terminated in a load resistance. These filters still find extensive use in communication systems. However, the present tendency is to design active RC filters to replace the passive LC filters wherever possible. (Over the last decade and half, digital filters are finding wide use. Such filters are not considered in this book). This, as already indicated, is motivated by a desire to miniaturise the filter circuits.

Most active filters are generated by embedding an operational amplifier in a passive RC network. Active network design has received considerable attention from many authors (see notes). Though we can develop filter realizations based on network synthesis methods, here we resort to coefficient matching technique as an efficient method of filter design.

The synthesis problem starts, as in the passive case, from the approximation technique. The transfer function so obtained is realised by an active RC network. The active element used is, in general, an operational amplifier. In this section, we consider only single amplifier second order filter structures. The general configuration is shown in Fig. 12.29. Here, N is a three port RC network (ports are

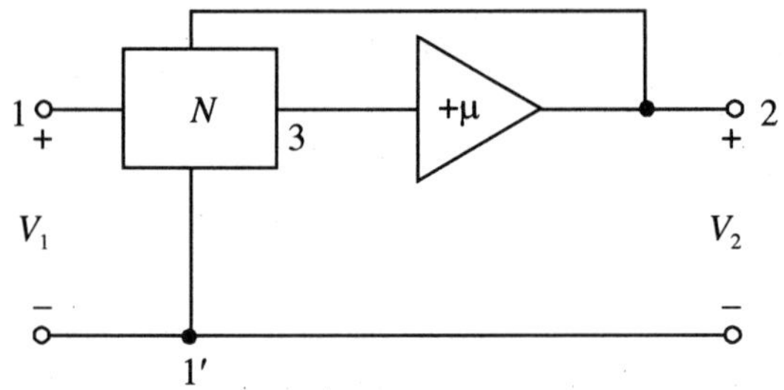

Fig. 12.29

11′, 21′ and 31′) and μ is a finite gain amplifier (see Fig. 12.3). General analysis of such networks are considered in books on active networks. Here, we consider the design LP, HP, BP and BE filters only.

12.4.1 Low-Pass Filter

A circuit realizing a second order low-pass filter with a positive gain amplifier is shown in Fig. 12.30. The transfer function of the circuit is (see Example 12.4).

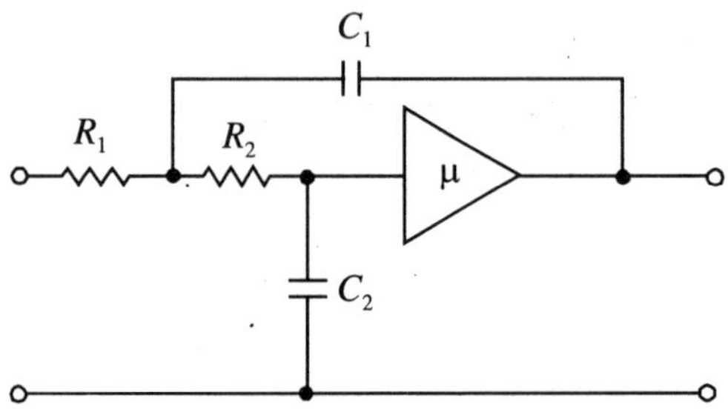

Fig. 12.30

$$T(s) = \frac{\mu\left(\dfrac{1}{R_1R_2C_1C_2}\right)}{s^2 + \left[\dfrac{1}{R_1C_1} + \dfrac{1}{R_2C_1} + \dfrac{1}{R_2C_2}(1-\mu)\right]s + \dfrac{1}{R_1R_2C_1C_2}} \qquad \ldots(12.21)$$

The general transfer function of a second order LP filter is

$$T(s) = \frac{K\omega_0^2}{s^2 + \omega_0 s/Q + \omega_0^2} \qquad \ldots(12.22)$$

where ω_0, Q and K are the resonant frequency (peak response frequency), quality factor and *dc* gain respectively of the filter. On comparing the coefficients in eqns. 12.21 and 12.22.

$$K = \mu,\ \omega_0 = \frac{1}{\sqrt{R_1R_2C_1C_2}},\ Q = \frac{1}{\sqrt{\dfrac{R_2C_2}{R_1C_1}} + \sqrt{\dfrac{R_1C_2}{R_2C_1}} + (1-\mu)\sqrt{\dfrac{R_1C_1}{R_2C_2}}}$$

The ω_0 and Q sensitivities of this filter with respect to the passive and the active elements are

$$S_{R_1}^{\omega_0} = S_{R_2}^{\omega_0} = S_{C_1}^{\omega_0} = S_{C_2}^{\omega_0} = -1/2$$

$$S_{\mu}^{\omega_0} = 0\ (\omega_0 \text{ is not a function of } \mu)$$

and

$$S_{R_1}^{Q} = -S_{R_2}^{Q} = Q\sqrt{\frac{R_2C_2}{R_1C_1}} - \frac{1}{2},$$

$$S_{C_1}^{Q} = -S_{C_2}^{Q} = Q\sqrt{\frac{C_2}{C_1}}\left(\frac{R_1+R_2}{\sqrt{R_1R_2}}\right) - \frac{1}{2}$$

$$S_{\mu}^{Q} = Q\left[\sqrt{\frac{R_1C_1}{R_2C_2}} + \sqrt{\frac{R_2C_2}{R_1C_1}} + \sqrt{\frac{R_2C_2}{R_2C_1}}\right] - 1$$

Though the ω_0 sensitivities are low, the Q sensitivities are proportional to Q. As can be observed from the transfer function, the structure will become unstable for values of μ greater than

$$\left[1 + \frac{C_2}{C_1}\left(1 + \frac{R_2}{R_1}\right)\right]$$

In active filters built in thin film form it is customary to keep the elemental spread (the ratio of maximum to minimum value) as small as possible. In the structure under consideration, if we select equal valued elements, *i.e.*, $R_1 = R_2 = R$ and $C_1 = C_2 = C$, we have

$$\omega_0 = \frac{1}{RC},\ Q = \frac{1}{3-\mu}$$

The design equations for this second order filter circuit are:

$$RC = \frac{1}{\omega_0},\ \mu = 3 - \frac{1}{Q}$$

Example 12.5. We are required to design a second-order LP Butterworth filter with a cut-off frequency of 2.5 kHz.

From Table 10.1, the transfer function of Butterworth filter with a cut-off frequency of 1 rad/sec and a dc gain of μ is

$$T(s) = \frac{\mu}{s^2 + 2s + 1}$$

Denormalizing the cut-off frequency to $2\pi(2.5 \times 10^3)$ we have the desired transfer function

$$T(s) = \frac{\mu(2\pi \times 2.5 \times 10^3)^2}{s^2 + s(2\pi \times 2.5 \times 10^3)\sqrt{2} + (2\pi \times 2.5 \times 10^3)^2}$$

From the design equations $(Q = 1/\sqrt{2},\ \omega_0 = 2\pi \times 2.5 \times 10^3)$

$$\mu = 3 - \sqrt{2} = 1.586$$

$$RC = \frac{1}{2.467 \times 10^8}$$

If we select R = 10 kΩ, then C = 0.405 pf.

Instead of choosing equal valued capacitors, as indicated, we can select a unity gain amplifier (voltage follower) and $R_1 = R_2 = R$. This selection yields

$$\omega_0 = \frac{1}{R\sqrt{C_1C_2}},\ Q = \sqrt{\frac{C_1}{C_2}}$$

Then design equations are

$$RC_1 = \frac{2Q}{\omega_0} \text{ and } RC_2 = \frac{1}{2\omega_0 Q}$$

and the Q sensitivities are

$$S_{R_1}^Q = S_{R_2}^Q = 0,\ S_{C_1}^Q = -S_{C_2}^Q = \frac{1}{2},\ S_\mu^Q = 2Q^2.$$

A positive gain realization leads to a low-gain amplifier but the Q sensitivities are large. This is a general feature of all positive gain structures.

A circuit realizing a second-order LP filter with a negative gain amplifier is shown in Fig. 12.31.

The transfer function is

$$\frac{-\mu\left(\frac{1}{R_1R_2C_1C_2}\right)}{s^2 + \left[\frac{1}{C_1}\left(\frac{1}{R_1} + \frac{1}{R_2} + \frac{1}{R_3}\right) + \frac{1}{C_2}\left(\frac{1}{R_2} + \frac{1}{R_4}\right)\right]s + \frac{1}{R_1R_2C_1C_2}\left[1 + \frac{R_1}{R_3}(1+\mu) + \frac{R_1}{R_3}\left(1 + \frac{R_2}{R_3} + \frac{R_2}{R_1}\right)\right]}$$

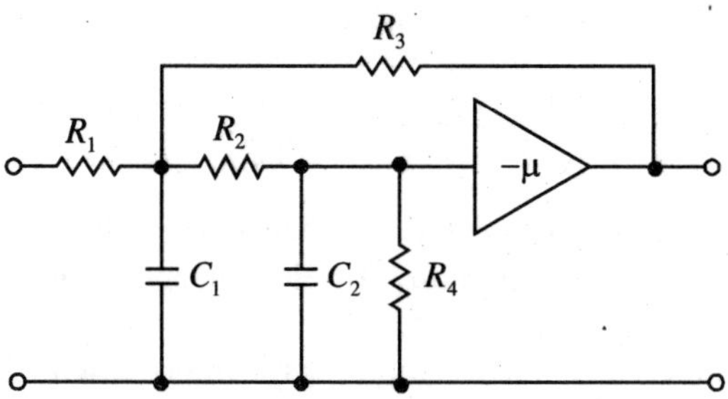

Fig. 12.31

The resistor R_4 can be considered as a part of the gain producing network. Here again we consider equal elemental valued realization, *i.e.*, $R_1 = R_2 = R_3 = R_4 = R$ and $C_1 = C_2 = C$. The resonant frequency and Q-factors are then

$$\omega_0 = \frac{\sqrt{5+\mu}}{RC}, \; Q = \frac{\sqrt{5+\mu}}{5}$$

Hence, the design equations are

$$\mu = 25Q^2 - 5, \; RC = \frac{5Q}{\omega_0}$$

It can be shown that the magnitude of ω_0 and Q sensitivities for $Q > 1/2$ are all less than or equal to 1/2. The closed loop gain of the amplifier for a reasonable Q is very large (for example $Q = 10$ yields $\mu = 2495$). This is a common feature of all negative gain realizations.

A second-order LP filter can also be realized by using multiple feedback infinite gain amplifier structure of Fig. 12.32. The transfer function is

$$T(s) = \frac{\dfrac{1}{R_1R_2C_1C_2}}{s^2 + s\dfrac{1}{C_1}\left(\dfrac{1}{R_1} + \dfrac{1}{R_2} + \dfrac{1}{R_3}\right) + \dfrac{1}{R_2R_3C_1C_2}}$$

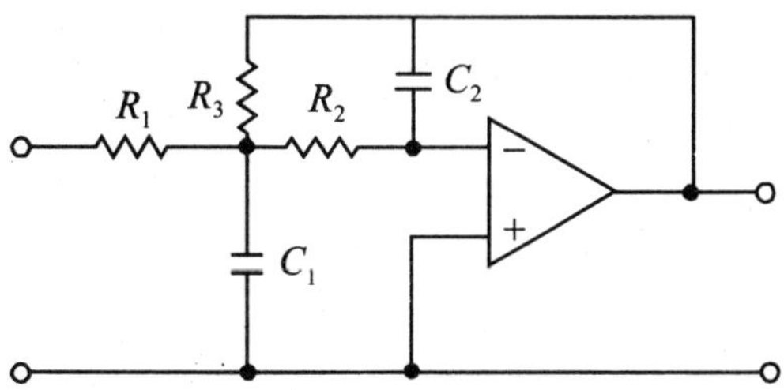

Fig. 12.32

If we select $R_1 = R_2 = R_3 = R$, we have

$$\omega_0 = \frac{1}{R\sqrt{C_1C_2}} \text{ and } Q = \frac{1}{3}\sqrt{\frac{C_1}{C_2}}$$

The magnitude of the ω_0 and the Q sensitivities with respect to the passive elements are less than or equal to half. The element spread in this case is $9\,Q^2$ (*i.e.*, $C_1/C_2 = 9Q^2$).

We have discussed three single amplifier LP filters. On the basis of sensitivity alone the infinite gain and negative gain amplifier structures are superior to the positive gain structure. The latter uses a low gain amplifier. If we consider the gain-sensitivity product (*i.e.*, μ, S_x^Q) as a measure, the positive gain structure is superior to the other two. In many practical filter design situations the positive gain realization is preferred to other two.

12.4.2 HP, BP and BE Filters

The method of spectral transformations developed in Chap. 11 is not applicable for designing active RC HP, BP and BE filters as this would necessitate the inclusion of an inductor in the circuit. The design of these filters is similar to that of the LP filters. The structures of various filters, their transfer functions, ω_0 and $\underline{C}$ sensitivities and design equations are summarized in Table 12.1.

The transfer function of these structures is given by

$$T(s) = \frac{Kf(s)}{s^2 + a_1 s + a_0}$$

$$= \frac{Kf(s)}{s^2 + \dfrac{\omega_0}{Q}s + \omega_0^2}$$

Hence, $\omega_0 = \sqrt{a_0}$, $Q = \dfrac{\sqrt{a_0}}{a_1}$ and $f(s)$ determines the zeros.

The transfer function for a notch (or band elimination filter) is

$$T(s) = \frac{K(s^2 + \alpha^2)}{s^2 + \dfrac{\omega_0}{Q}s + \omega_0^2} = \frac{K(s^2 + b)}{s^2 + a_1 s + a_0}.$$

These different situations may arise depending on α,

(*i*) *High-pass notch*: $\alpha < \omega_0$

(*ii*) *Low-pass notch*: $\alpha > \omega_0$

(*iii*) *Active notch*: $\alpha = \omega_0$.

(Observe that low-pass and high-pass notch, transfer functions are the familiar low-pass and high-pass functions with a finite zero of transmission.) Table 12.2 shows the structures and other particulars of these three filters. In HP notch and LP notch the element spread is controlled by k and can be arbitrarily selected.

Table 12.1

Structure	*Transfer function*	*Sensitivities*	*Design equations*
1. HP: Positive gain 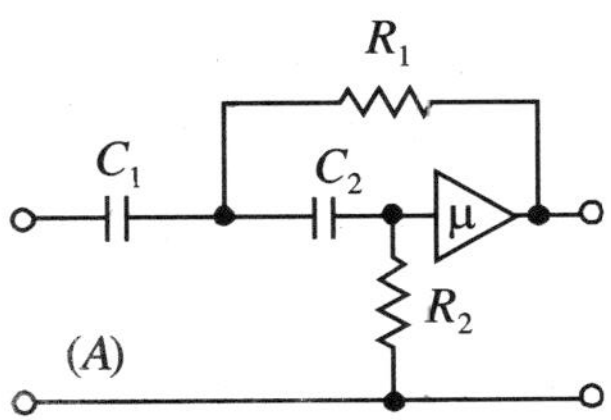	$K=\mu,\ f(s)=s^2$ $a_0=\dfrac{1}{R_1R_2C_1C_2}$ $a_1=\dfrac{1}{R_2}\left(\dfrac{1}{C_1}+\dfrac{1}{C_2}\right)+\dfrac{1-\mu}{R_1C_1}$ $\omega_0=\dfrac{1}{\sqrt{R_1R_2C_1C_2}}$ $Q=\dfrac{1}{\sqrt{\dfrac{R_1}{R_2}}\dfrac{C_1+C_2}{\sqrt{C_1C_2}}+\sqrt{\dfrac{R_2C_2}{R_1C_1}}(1-\mu)}$	$S_{R_1}^{\omega_0}=S_{R_2}^{\omega_0}=S_{C_1}^{\omega_0}=S_{C_2}^{\omega_0}=-\dfrac{1}{2}$ $S_{\mu}^{\omega_0}=0$ $S_{R_1}^{Q}=-S_{R_2}^{Q}=\dfrac{1}{2}-2Q$ $S_{C_1}^{Q}=-S_{C_2}^{Q}=\dfrac{1}{2}-Q$ $S_{\mu}^{Q}=3Q-1$	$R_1=R_2=R$ $C_1=C_2=C$ $\mu=3-\dfrac{1}{Q}$ $RC=\dfrac{1}{\omega_0}$

(Contd.)

Table 12.1 *(Continued)*

Structure	*Transfer function*	*Sensitivities*	*Design equations*
2. HP: Negative gain 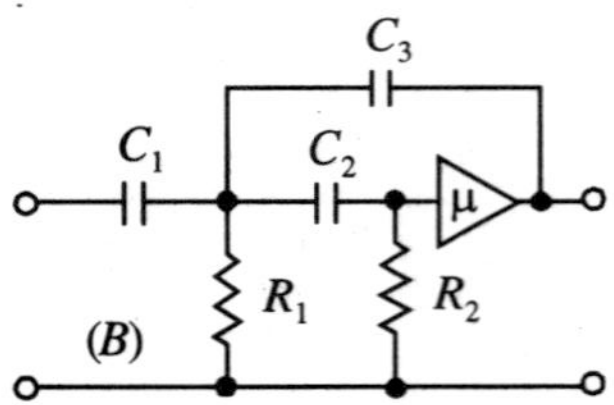	$K = \frac{-\mu}{1+\frac{C_3}{C_1}+\mu}$, $f(s) = s^2$ $a_0 = \frac{1}{R_1R_2C_2\left[C_1 + C_3(1+\mu)\right]}$ $a_1 = \frac{(R_1+R_2)C_2 + R_1(C_1+C_3)}{R_1R_2C_2\left[C_1 + C_3(1+\mu)\right]}$ $\omega_0 = \frac{1}{\sqrt{R_1R_2C_2\left[C_1 + C_3(1+\mu)\right]}}$ $Q = \frac{R_1R_2C_2C_1 + C_3(1+\mu)}{(R_1+R_2)C_2 + R_1(C_1+C_3)}$	$S_{R_1}^{\omega_0} = S_{R_2}^{\omega_0} = S_{C_2}^{\omega_0} = -\frac{1}{2}$ $S_{C_1}^{\omega_0} = -\frac{1}{32Q^2}$ $S_{C_3}^{\omega_0} = -\frac{1}{2} + \frac{1}{16Q^2}$ $S_{R_2}^{Q} = -S_{R_2}^{Q} = \frac{1}{4}$ $S_{C_2}^{Q} = 0$, $S_{C_1}^{Q} = \frac{1}{32Q^2} - \frac{1}{4}$ $= -S_{C_3}^{Q}$ $S_{\mu}^{Q} = \frac{1}{2}\cdot\frac{1}{16Q^2}$	$R_1 = R_2 = R$ $C_1 = C_2 = C_3 = C$ $\mu = 16Q^2 - 2$ $RC = \frac{1}{4Q\omega_0}$

Table 12.1 *(Continued)*

Structure	*Transfer function*	*Sensitivities*	*Design equations*
3. HP: Infinite gain 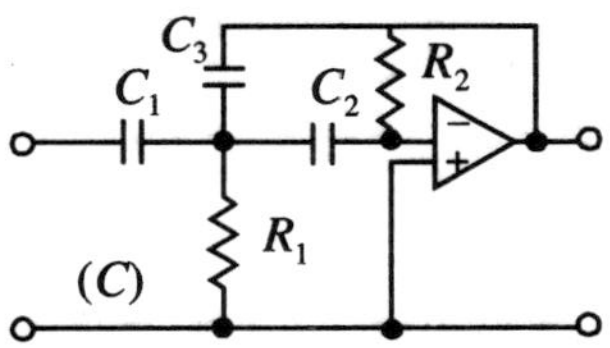	$K = -\frac{C_3}{C_1},\ f(s) = s^2$ $a_0 = \frac{1}{R_1 R_2 C_1 C_2}$ $a_1 = \frac{1}{R_2}\left(\frac{1}{C_1} + \frac{1}{C_2} + \frac{C_3}{C_1 C_2}\right)$ $\omega_0 = \frac{1}{\sqrt{R_1 R_2 C_1 C_2}}$ $Q = \frac{\sqrt{R_2/R_1}}{\sqrt{\frac{C_1}{C_2}} + \sqrt{\frac{C_2}{C_1}} + \frac{C_3}{\sqrt{C_1 C_2}}}$	$S_{R_1}^{\omega_0} = S_{R_2}^{\omega_0} = S_{C_1}^{\omega_0} = S_{C_2}^{\omega_0} = -\frac{1}{2}$ $S_{C_3}^{\omega_0} = 0$ $S_{R_2}^{Q} = -S_{R_1}^{Q} = \frac{1}{2}$ $S_{C_1}^{Q} = S_{C_3}^{Q} = \frac{1}{6}$ $S_{C_3}^{Q} = -\frac{1}{3}$	$C_1 = C_2 = C_3 = C$ $\sqrt{\frac{R_2}{R_1}} = 3Q$ $C\sqrt{R_1 R_2} = \frac{1}{\omega_0}$

Table 12.1 *(Continued)*

Structure	*Transfer function*	*Sensitivities*	*Design equations*
4. BP: Positive gain 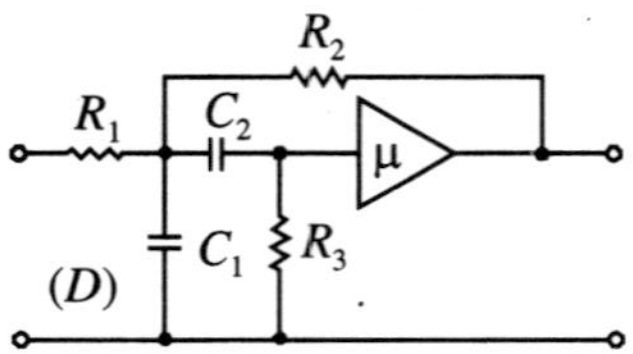 (D)	$K = \frac{\mu}{R_1 C_1}, f(s) = s$ $a_0 = \frac{R_1 + R_2}{R_1 R_2 R_3 C_1 C_2}$ $a_1 = \frac{1}{R_1 C_1} + \frac{1}{R_3}\left(\frac{1}{C_2} + \frac{1}{C_1}\right) + \frac{1-\mu}{R_2 C_1}$ $\omega_0 = \sqrt{\frac{R_1 + R_2}{R_1 R_2 R_3 C_1 C_2}}$ $Q = \frac{\sqrt{\frac{R_2 C_1 (R_1 + R_2)}{R_1 R_3 R_2}}}{1 + \frac{R_2}{R_1} + \frac{R_2}{R_3}\left(1 + \frac{C_1}{C_2}\right) - \mu}$	$S_{R_1}^{\omega_0} = S_{R_2}^{\omega_0} = -\frac{1}{4}$ $S_{R_3}^{\omega_0} = S_{C_1}^{\omega_0} = S_{C_2}^{\omega_0} = -\frac{1}{2}$ $S_{\mu}^{\omega} = 0$ $S_{R_1}^{Q} = -\frac{1}{4} + \frac{Q}{\sqrt{2}}$ $S_{R_2}^{Q} = \frac{3}{4} - \frac{3Q}{\sqrt{2}}$ $S_{R_3}^{Q} = -\frac{1}{2} + \sqrt{2Q}$ $S_{C_1}^{Q} = \frac{1}{2} - \frac{Q}{\sqrt{2}} = -S_{C_2}^{Q}$ $S_{\mu}^{Q} = \frac{4Q}{\sqrt{2}} - 1$	$R_1 = R_2 = R_3 = R$ $C_1 = C_2 = C$ $\mu = 4 - \frac{\sqrt{2}}{Q}$ $RC = \frac{\sqrt{2}}{\omega_0}$

Table 12.1 *(Continued)*

Structure	*Transfer function*	*Sensitivities*	*Design equations*
5. BP: Negative 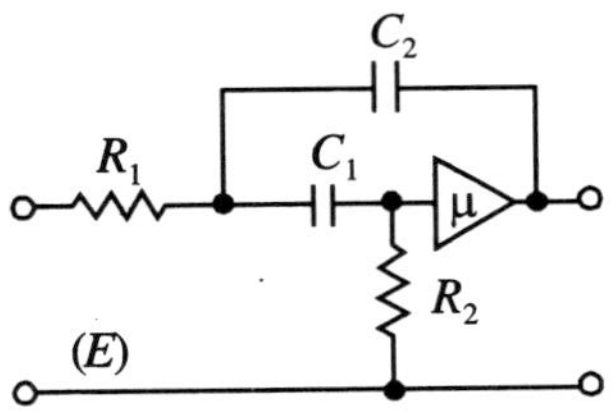	$K=-\frac{\mu}{(1+\mu)R_2C_2},\ f(s)=s$ $a_0=\frac{1}{(1+\mu)R_1R_2C_1C_2}$ $a_1=\frac{1}{1+\mu}\left[\frac{1}{R_1C_2}+\frac{1}{R_2C_1}+\frac{1}{R_2C_1}\right]$ $\omega_0=\frac{1}{\sqrt{(1+\mu)R_1R_2C_1C_2}}$ $Q=\frac{\sqrt{1+\mu}}{\sqrt{\frac{R_1C_2}{R_2C_1}}+\sqrt{\frac{R_1C_1}{R_2C_2}}+\sqrt{\frac{R_2C_1}{R_1C_2}}}$	$S_{R_1}^{\omega_0}=S_{R_2}^{\omega_0}=S_{C_1}^{\omega_0}=S_{C_2}^{\omega_0}=-\frac{1}{2}$ $S_{\mu}^{\omega_0}=\frac{1}{2}\left(\frac{1}{9Q^2}-1\right)$ $S_{R_1}^{Q}=-S_{R_2}^{Q}=S_{C_1}^{Q}=-S_{C_2}^{Q}=-\frac{1}{6}$ $S_{\mu}^{Q}=\frac{1}{2}\left(1-\frac{1}{9Q^2}\right)$	$R_1=R_2=R$ $C_1=C_2=C$ $\mu=9Q^2-1,\ RC=\frac{1}{3Q\omega_0}$

Table 12.1 *(Continued)*

Structure	*Transfer function*	*Sensitivities*	*Design equations*
6. BP: Infinite gain 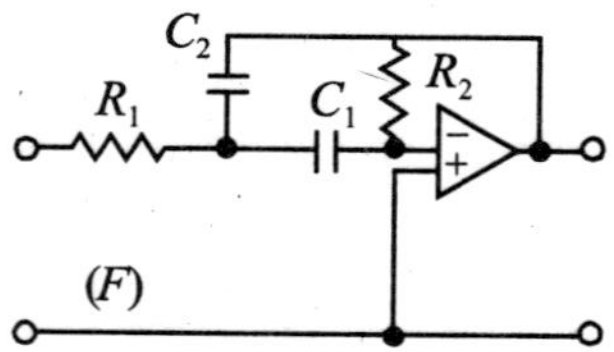	$K = -\frac{1}{R_1C_2}$, $f(s) = s$ $a_0 = \frac{1}{R_1R_2C_1C_2}$ $a_1 = \frac{1}{R_2}\left(\frac{1}{C_1} + \frac{1}{C_2}\right)$ $\omega_0 = \frac{1}{\sqrt{R_1R_2C_1C_2}}$ $Q = \frac{\sqrt{R_2/R_1}}{\sqrt{\frac{C_1}{C_2}} + \sqrt{\frac{C_2}{C_1}}}$	$S_{R_1}^{\omega_0} = S_{R_2}^{\omega_0} = S_{C_1}^{\omega_0} = S_{C_2}^{\omega_0} = -\frac{1}{2}$ $S_{R_1}^{Q} = -S_{R_2}^{Q} = -\frac{1}{2}$ $S_{C_1}^{Q} = S_{C_2}^{Q} = 0$	$C_1 = C_2 = C$ $\sqrt{\frac{R_2}{R_1}} = 2Q$ $C\sqrt{R_1R_2} = \frac{1}{\omega_0}$

Table 12.2

Structure	Transfer function	Design equation
1. HP Notch 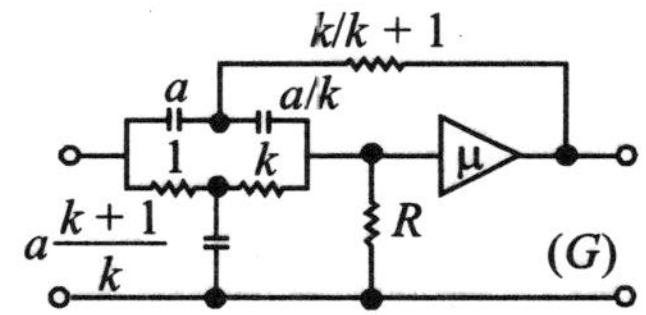	$K = \mu,\ b = \frac{1}{a^2}$ $a_0 = \left[1 + \frac{k+1}{R}\right]\frac{1}{a^2}$ $a_1 = \frac{k+1}{a}\left[\frac{1}{2} + \frac{2-\mu}{R}\right]$	$a = \frac{1}{\alpha},\ R = \frac{k+1}{\frac{\omega_0^2}{\alpha^2} - 1}$ $\mu = 2 + \frac{k}{k+1}\left[\frac{\omega_0^2}{\alpha^2} - 1 - \frac{\omega_0}{Q\alpha}\right]$
2. LP Notch 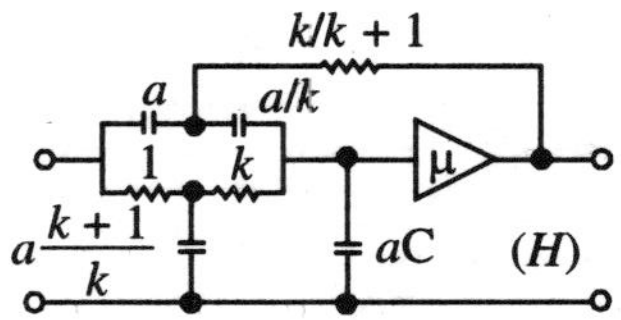	$K = \frac{\mu}{(k+1)C+1},\ b = \frac{1}{a^2}$ $a_0 = \frac{1}{a^2[(k+1)C+1]}$ $a_1 = \frac{(k+1)[C+(2-\mu)/k]}{a[(k+1)C+1]}$	$a = \frac{1}{\alpha},\ C = \left(\frac{\alpha^2}{\omega_0^2} - 1\right)\Big/(k+1)$ $\mu = 2 + \frac{k}{k+1}\left[\frac{\omega_0^2}{\alpha_0^2} - 1 - \frac{\omega_0}{\omega_0 Q}\right]$

(Contd.)

Table 12.2 *(Continued)*

Structure	*Transfer function*	*Design equation*
3. Active Notch 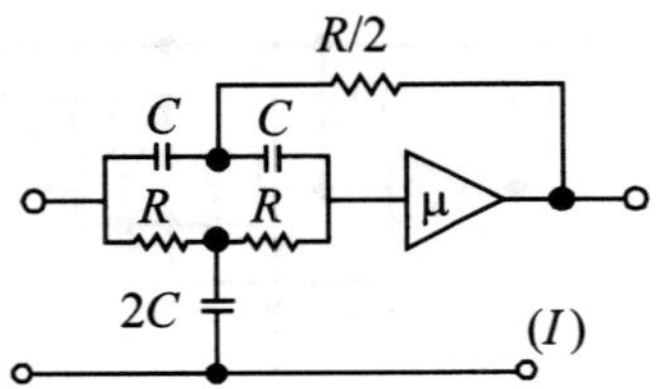	$K = \mu,\ b = \frac{1}{R^2C^2}$ $a_0 = b,\ a_1 = \frac{2}{RC}(2-\mu)$	$RC = \frac{1}{\omega_0}$, $\mu = 2 - \frac{1}{2Q}$.

Example 12.6. Consider the LP notch transfer function

$$T(s) = \frac{K(s^2 + 1.5)}{s^2 + 0.2s + 1}$$

From the design equations we have

$$a = \frac{1}{\sqrt{1.5}}, \; C = \left(\frac{1.5}{1} - 1\right)\frac{1}{k+1}$$

and
$$\mu = 2 + \frac{k}{k+1}\left(\frac{1.5}{1} - 1 - \sqrt{1.5} \times 0.2\right)$$

If we select k as 1, then

$$a = 0.816, \; C = 0.250, \; \mu = 2.128$$

The finite gain realisations considered here were the earliest to be introduced and are known as *Sallen-Key structures.* The band elimination structures using twin-T networks are known as *Kerwin structures.* There are many other structures realizing these standard filters.

12.5 STATE-VARIABLE REALIZATION

In the previous section, we discussed a few one-amplifier structures. Many multi-amplifier structures have also appeared in the literature of electrical engineering. One of the first three amplifier structures to appear is the "state-variable filter". The basic configuration is obtained by analog simulation of the prescribed biquadratic transfer function.

Let the transfer function be

$$T(s) = \frac{a_2 s^2 + a_1 s + a_0}{s^2 + b_1 s + b_0}.$$

A signal flow graph of this transfer function is shown in Fig. 12.33.

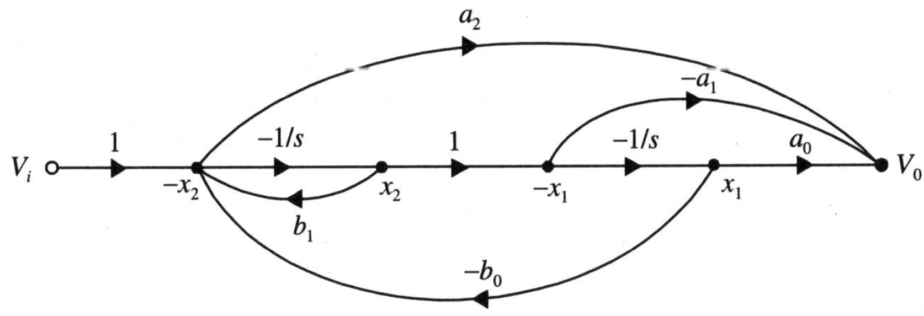

Fig. 12.33

From the figure we can write

$$\dot{x}_1 = -x_2$$

$$\dot{x}_2 = b_0 x_1 - b_1 x_2 - V_i$$

and

$$V_0 = a_0 x_1 + a_1 \dot{x}_1 - a_2 \dot{x}_2$$
$$= (a_0 - a_2 b_0)x_1 + (a_2 b_1 - a_1)x_2 + a_2 V_i$$

i.e.,

$$\begin{bmatrix} \dot{x}_1 \\ \dot{x}_2 \end{bmatrix} = \begin{bmatrix} 0 & -1 \\ b_0 & -b_1 \end{bmatrix} \begin{bmatrix} x_1 \\ x_2 \end{bmatrix} + \begin{bmatrix} 0 \\ -1 \end{bmatrix} V_i$$

$$V_0 = \left[(a_0 - a_2 b_0) \ (a_2 b_1 - a_1) \right] \begin{bmatrix} x_{1i} \\ x_2 \end{bmatrix} + a_2 V_i$$

which is of the form

$$\dot{x} = A\mathbf{x} + Bu$$
$$y = C\mathbf{x} + Du,$$

where x_1 and x_2 are the states (hence the name state-variable filter). The transfer-function of the system is (see Chap. 8)

$$C(sI - A)^{-1} B + D = T(s)$$

The analog simulation diagram of the signal flow graph of Fig. 12.33 is shown in Fig. 12.34. The structure uses two inverting integrators and two differential input summers.

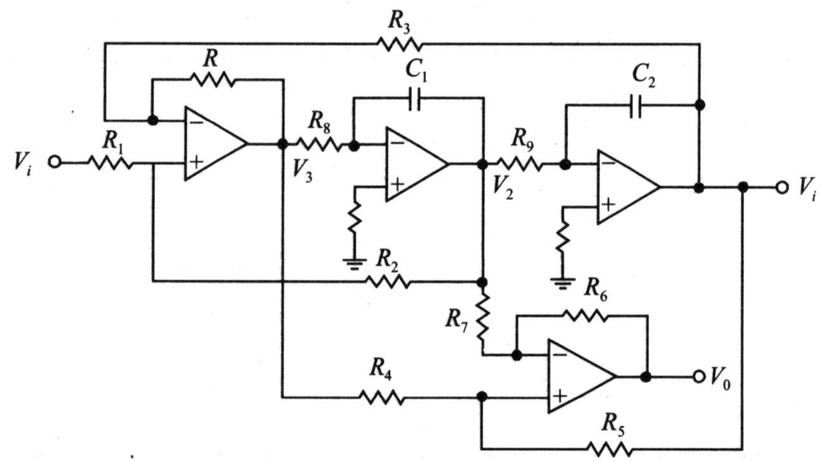

Fig. 12.34

The various transfer functions are given by

$$\frac{V_0}{V_i} = K\frac{N}{D}, \ \frac{V_1}{V_i} = K_1\frac{N_1}{D}, \ \frac{V_2}{V_i} = K_2\frac{N_2}{D} \text{ and } \frac{V_3}{V_i} = K_3\frac{N_3}{D}$$

where

$$D = R_8 R_9 C_1 C_2 s^2 + \frac{R_1(R + R_3)}{(R_1 + R_2)R_3} R_9 C_2 s + \frac{R}{R_3},$$

$$K = \frac{R_5 R_2 (R + R_3)(R_6 + R_7)}{(R_1 + R_2)(R_4 + R_5) R_3 R_7}$$

$$N = R_8 R_9 C_1 C_2 s^2 + \frac{(R_4 + R_5) R_6}{R_5 (R_6 + R_7)} R_9 C_2 s + \frac{R_4}{R_5},$$

$$K_1 = \frac{R_2 (R + R_3)}{(R_1 + R_2) R_3}, \; N_1 = 1$$

$$K_2 = -K_1 \qquad N_2 = R_9 C_2 s$$

$$K_3 = K_1 \qquad N_3 = R_8 R_9 C_1 C_2 s^2.$$

As can be observed from these equations, V_1 is a low-pass output, V_2 a band-pass output and V_3 a high-pass output. The centre frequency of the band-pass function (or the resonant frequency of the LP and the HP) and the Q factor are

$$\omega_0 = \frac{R}{\sqrt{R_3 R_8 R_9 C_1 C_2}}, \; Q = \sqrt{\frac{R R_3 R_8 C_1}{R_9 C_2}} \; \frac{R_1 + R_2}{R_1 (R + R_3)}$$

and the gain at this frequency is

$$T_{BP}(j\omega_0) = -\frac{R_2}{R_1}$$

The centre frequency can be adjusted (or tuned) by varying R and the Q by varying R_1 or R_2. The ω_0 and the Q sensitivities are

$$S_{R_3}^{\omega_0} = S_{R_8}^{\omega_0} = S_{R_9}^{\omega_0} = S_{C_1}^{\omega_0} = S_{C_2}^{\omega_0} = -S_{R_9}^{\omega_0} = -1/2,$$

$$S_{C_1}^{Q} = -S_{C_2}^{Q} = \frac{1}{2}, \; S_{R_8}^{Q} = -S_{R_9}^{Q} = \frac{1}{2}$$

If we let $R_1 = R_3 = R_5 = R_6 = R_8 C_1 = R_9 C_2 = 1$, we have

$$T_{BP}(s) = -\frac{R_2 (1 + R)}{1 + R_2} \; \frac{s}{s^2 + (1 + R)s/(1 + R_2) + R}$$

Hence, $\omega_0 = \sqrt{R}, \; T_{BP}(j\omega_0) = -R_2$.

The state variable filter structure is isotopic in nature in that it generates all three basic filter characteristics (LP, HP and BP) in the same realization. A general biquadratic function can be realised by a final summer amplifier.

This state-variable filter also goes under the name Kerwin-Huelsman-Newcomb (KHN) structure. This structure has found use in low frequency and low Q filters. Several of the commercially available modular filters are variations of the KHN structure.

12.6 ALL-PASS AND NOTCH-FILTER

In Sec. 9.6, we introduced the concept of all-pass networks. These find use as phase equilizers. All-pass filters and notch filters (see Sec. 12.4.2) can be designed in a number of ways. One method of design is by subtracting a band-pass filter transfer function from unity. By adjusting the mid-band gain of the BP function we can either realize an all-pass or a notch function.

Consider a second order band-pass function

$$T(s) = \frac{Ks}{s^2 + \frac{\omega_0}{Q}s + \omega_0^2}$$

If we subtract the function from unity we have

$$H(s) = 1 - T(s) = \frac{s^2 + \left(\frac{\omega_0}{Q} - K\right)s + \omega_0^2}{s^2 + \frac{\omega_0}{Q}s + \omega_0^2}$$

If we select $K = 2\omega_0/Q$, we have

$$H(s) = \frac{s^2 - \frac{\omega_0}{Q}s + \omega_0^2}{s^2 + \frac{\omega_0}{Q}s + \omega_0^2}$$

which is a biquadratic all-pass transfer function. On the other hand, if we select $K = \omega_0/Q$, we have

$$H(s) = \frac{s^2 + \omega_0^2}{s^2 + \frac{\omega_0}{Q}s + \omega_0^2}$$

yielding a notch filter.

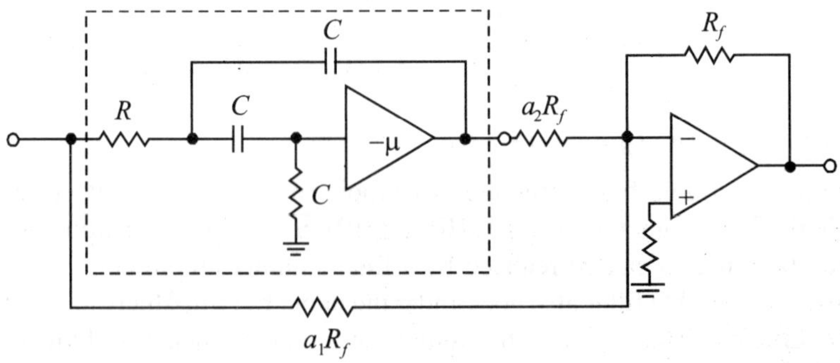

Fig. 12.35

Figure 12.35 shows a realization of an all-pass or a notch-filter. The network shown in the box is a band-pass filter with the transfer function

$$T(s) = -\frac{\omega_0\left(3Q - \frac{1}{3Q}\right)s}{s^2 + \frac{\omega_0}{Q}s + \omega_0^2}$$

(Here we have used the design relations $RC = 1/3Q\omega_0$ and $\mu = 9Q^2 - 1$). The overall transfer function of this network is

$$H(s) = \frac{s^2 + s\left[\frac{1}{Q} - \frac{a_1}{a_2}\left(3Q - \frac{1}{3Q}\right)\right]\omega_0 + \omega_0^2}{s^2 + \frac{\omega_0}{Q}s + \omega_0^2}$$

In order to realize an all-pass filter

$$\frac{1}{Q} - \frac{a_1}{a_2}\left(3Q - \frac{1}{3Q}\right) = -\frac{1}{Q}$$

i.e.,

$$\frac{a_1}{a_2} = \frac{6}{9Q^2 - 1}$$

For a notch filter

$$\frac{1}{Q} - \frac{a_1}{a_2}\left(3Q - \frac{1}{3Q}\right) = 0$$

i.e.,

$$\frac{a_1}{a_2} = \frac{3}{9Q^2 - 1} = \frac{3}{\mu}$$

Second-order all-pass and notch-filters can also be realized by embedding a differential input op. amp in a passive RC structure.

12.7 HIGHER ORDER FILTERS

In the previous sections, we have only realized a second-order transfer function. A standard method of realizing a higher order transfer function is to express it as a product of second-order transfer functions and realize each second-order transfer function as a single amplifier or state-variable filter. The final filter is obtained by cascading the individual second-order filter blocks.

Consider the transfer function $T(s)$ expressed as a product of second order factors.

i.e.,

$$T(s) = \frac{Kp(s)}{q(s)} = \prod_{i=1}^{r} \frac{K_i f_i(s)}{\left(s^2 + \frac{\omega_{0_i}}{Q_i}s + \omega_{0_i}^2\right)}$$

or
$$T(s) = \prod_{i=1}^{r} T_i(s) \text{ with } T_i(s) = \frac{K_i f_i(s)}{s^2 + \dfrac{\omega_{0_i}}{Q_i} s + \omega_{0_i}^2}$$

Here it is assumed that the transfer function is of even order. If $T(s)$ is of odd order, one of the $T_i s$ will be a first order function. The final realization is shown in Fig. 12.36. As the output of each of the amplifier is taken from an op. amp the effect of the loading of the succeeding section is not a design consideration in contrast to the cascade realization of passive structures.

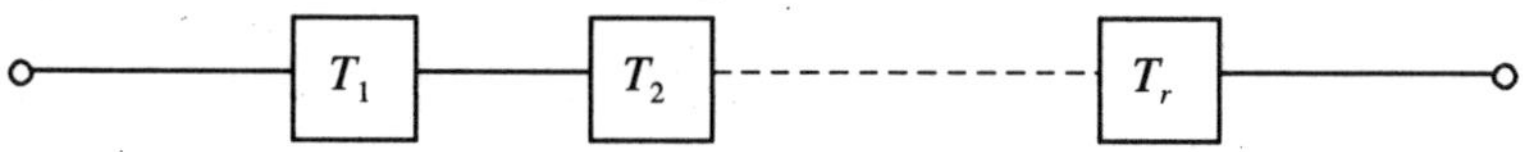

Fig. 12.36

A cascade realization of higher order filter is preferred to a direct form realization. In general, the former has lower sensitivity as compared to the latter.

Example 12.7. The transfer function of a sixth-order band-pass function is given by

$$T(s) = \frac{s(s^2 + 0.25)(s^2 + 2.25)}{(s^2 + 0.28s + 1.02)(s^2 + 0.9s + 0.83)(s^3 + 0.1s + 1.18)}$$

The centre frequency of the filter is normalized to 1 rad/sec. The filter has finite zeros of transmission at $\pm j\,0.5$ and $\pm j1.5$ and the band-width is 0.2 rad/sec. (This transfer function yields an asymmetrical band-pass filter and, as such, cannot be obtained by frequency transformation technique.)

For cascade realization the transfer function is expressed as a product of three second-order functions, viz.

$$T(s) = \prod_{i=1}^{3} T_i(s)$$

where
$$T_1(s) = \frac{s}{s^2 + 0.9s + 0.83}, \quad T_2(s) = \frac{s^2 + 0.25}{s^2 + 0.1s + 1.18}$$

and
$$T_3(s) = \frac{s^2 + 2.25}{s^2 + 0.2s + 1.02}$$

Of course, several other pairings of the poles and zeros are possible. $T_1(s)$ is a standard band-pass filter, $T_2(s)$ a high-pass notch filter and $T_3(s)$ a low-pass notch filter.

Realization of $T_1(s)$:

We realize $T_1(s)$ as a positive gain BP structure. From Table 12.1 (filter No. D)

$$R_1 = R_2 = R_3 = R, \; C_1 = C_2 = C$$

$$\omega_0 = \sqrt{0.83},\ Q = 1.012$$

Hence,

$$RC = \frac{\sqrt{2}}{\omega_0} = 1.55, \text{ and } \mu = 4 - \frac{\sqrt{2}}{Q} = 2.6$$

If we select $R = 1$, we have

$$R = 1,\ C = 1.55,\ \mu = 2.6.$$

The structure is shown in Fig. 12.37.

Realization of $T_2(s)$:

From Table 12.2 (filter No. G) we have

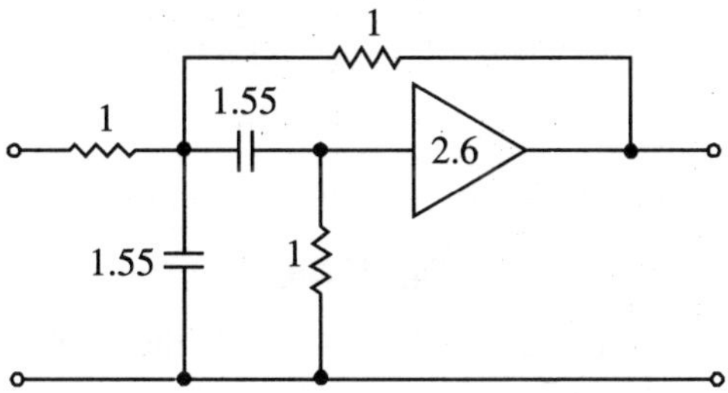

Fig. 12.37

$$\omega_0 = 1.18,\ Q = 10.86$$

and

$$\alpha = \sqrt{0.25} = 0.5$$

Hence,

$$a = \frac{1}{\alpha} = 2,\ R = \frac{k+1}{\dfrac{\omega_0^2}{a^2} - 1} = 0.27(k+1),$$

$$\mu = 2 + 3.52\frac{k}{k+1}.$$

If we select $k = 1$

$$a = 2,\ R = 0.54,\ \mu = 3.76.$$

The network is shown is Fig. 12.38.

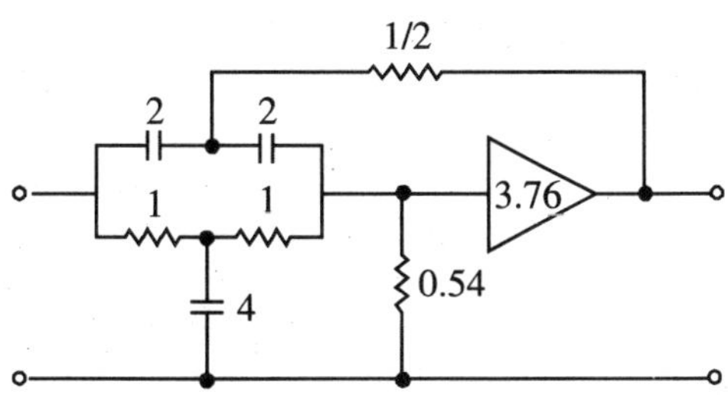

Fig. 12.38

Realization of $T_3(s)$:

From Table 10.2 (filter No. H) we have

$$\omega_0 = \sqrt{1.02} = 1.01,\ \alpha = \sqrt{2.25} = 1.5 \text{ and } Q = 5.05$$

$$a = \frac{1}{\alpha} = 0.67,\ C = \frac{1.21}{k+1} \text{ and } \mu = 2 + \frac{0.92k}{k+1}$$

If we select k as 1

$$a = 0.67,\ C = 0.61 \text{ and } \mu = 2.46.$$

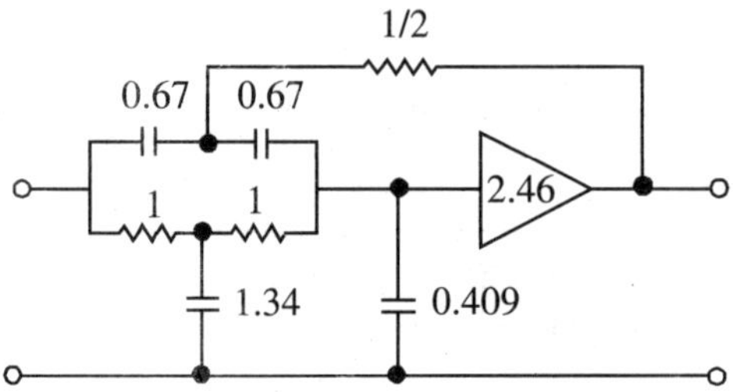

Fig. 12.39

The network is shown in Fig. 12.39. If we scale the impedance level to 10k Ω and the centre frequency to 5k rad/sec, we have the final filter as in Fig. 12.40. (This is obtained by multiplying all resistors by 10^4 and dividing all capacitors by 5×10^7).

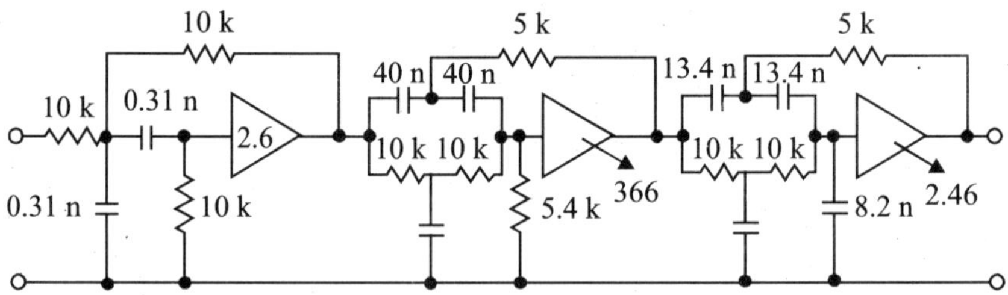

Fig. 12.40

In cascade structures the pairing of poles and zeros, the order of cascading the second order section and the distribution of gain in various sections determine the overall performance of the filters.

A higher order active filter can also be realised by first designing a LC ladder network terminated in a resistor and then transforming it to an active network. Though some of these transformation methods are applicable for general RLC networks, the zero first order sensitivity of a resistively terminated LC ladder to variations of the reactive elements at frequencies in the pass-band where maximum available power is delivered to the load makes it the prime candidate for such transformations. These methods can be, in general, classified as "Ladder Simulation" methods.

These methods mainly fall into two categories: 'element simulation' and 'simulated ladders'. In the former, the inductance is replaced by a 'simulated inductance' by either a GIC-based structure or by a capacitor loaded gyrator. This method is suitable for high-pass filters and other filters with grounded inductors. The realization of floating inductors poses a serious problem. Each floating inductors (or a group of inductors) can be simulated by gyrators or GICs connected back to back. This requires critical matching of gyration conductances. Floating inductance using GIC suffers a further disadvantage that two capacitors are used to simulate an inductor. A second method of element simulation is to transform the LC ladder to a RC-FDNR network by Bruton's FDNR transformation.

Simulated ladder structures fall into one of the following categories—coupled biquad or leap-frog structure. Wave active filters, SFG filters fall in the latter category.

In what follows we consider only FDNR transformation and coupled Bi-quads.

12.7.1 FDNR Transformation

Bruton has proposed a transformation which converts a passive RLC network to an active RC-FDNR network.

For example, consider a simple parallel RLC network of Fig. 12.41(*a*). The admittance of this network is

$$Y_1 = \frac{1}{R} + sC + \frac{1}{sL}$$

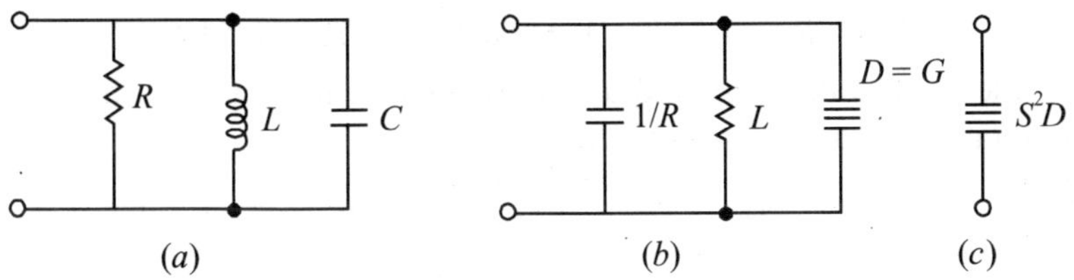

Fig. 12.41

To obtain an RC-FDNR realization Y_1 is replaced by sY_1 (this is often called *FDNR admittance transformation*).

i.e.,
$$Y_2 = sY_1 = \frac{s}{R} + s^2C + \frac{1}{L}$$

The corresponding network is shown in Fig. 12.41(*b*). The transformation converts a resistor to a capacitor (value-1/R), an inductor to a resistor (value = *L*) and a capacitor to an FDNR element. The symbolic representation of an FDNR is shown in Fig. 12.41(*c*). The RC-FDNR network is also called a *DCR network*. As indicated earlier, FDNR can be realized by two op. amp structures shown in Fig. 12.13.

The realization of a DCR network is obtained by transforming the passive RLC filter, element by element, by the FDNR transformation. An example illustrates the method.

Example 12.8. The transfer function of an L.P. elliptic filter, with the following specifications: pass-band ripple 1.25 dB, pass-band edge 1 rad/sec and stop-band edge 1.2 rad/sec stop-band attenuation 40 dB, is

$$T(s) = \frac{(s^2 + 1.2402)(s^2 + 1.7368)}{(s^2 + 0.4029s + 0.5903)(s^2 + 0.0906s + 0.9913)(s + 0.3569)}$$

The corresponding doubly terminated normalized ladder is given in Fig. 12.42 ($\omega_0 = 1$ rad/sec, $R = 1\Omega$).

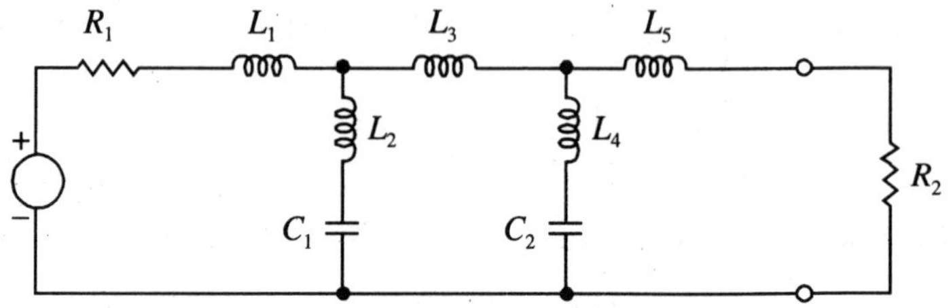

Fig. 12.42

$R_1 = R_2 = 1$, $L_1 = 2$, $L_1 = 0.3992$, $L_3 = 2.2477$, $L_4 = 1.1889$,
$L_5 = 1.5378$, $C_1 = 0.8305$, $C_2 = 0.5469$

The filter obtained by FDNR transformation is shown in Fig. 12.43 with
$C_1 = C_2 = 1$, $R_1 = 2$, $R_2 = 0.3992$, $R_3 = 2.2477$,
$R_4 = 1.1889$, $R_5 = 1.5378$, $D_1 = 0.8305$ and $D_2 = 0.5469$.

The filter is now denormalized to, for the sake of illustration, a frequency of 20.106 k rad/sec and an impedance level of 2260.72 Ω.

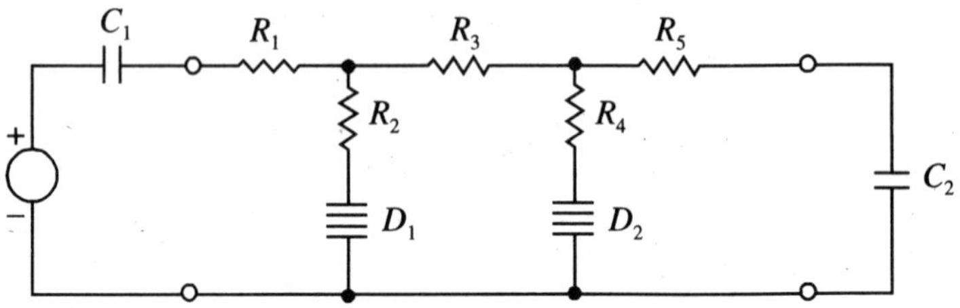

Fig. 12.43

The values of the elements are

$C_1 = C_2 = 22000$ pf, $R_1 = 4521.44$, $R_2 = 902.502$, $R_3 = 5081.42$, $R_4 = 2687.54$ $R_5 = 3476.60$, $D_1 = 0.9087 \times 10^{-12}$ and $D_2 = 0.5984 \times 10^{-12}$. (Here the impedance scaling factor is chosen to yield practical values for C_1 and C_2).

The two FDNR elements are realized by the circuit shown is Fig. 12.13 with
D_1: $C = C_1 = C_5 = 0.01827$ μF, $R_2 = R_3 = R_4 = R = 2722.28$ Ω
D_2: $C = C_1 = C_5 = 0.01203$ μF, $R_2 = R_3 = R_4 = R = 4133.70$ Ω

(observe that $D = C^2R$). The final network is shown in Fig. 12.44. Here, two unity gain voltage followers are added, one at the input end and the other at the output end for the purposes of impedance matching. In the figure 1M, resistors across the terminating capacitors have been introduced to provide the input bias current of the operational amplifiers of the GICs.

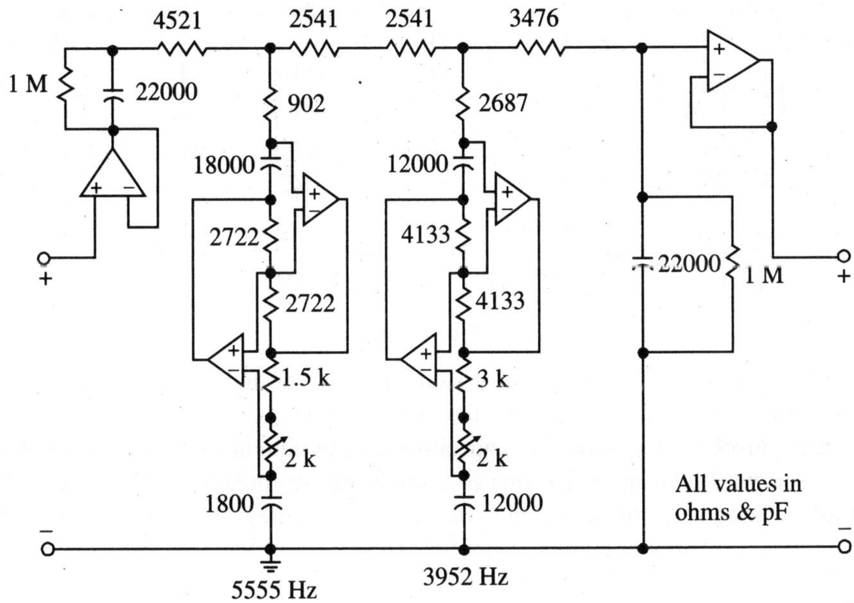

Fig. 12.44

12.7.2 Coupled Biquad Structure

Consider the low-pass ladder shown in Fig. 12.45. For this ladder, the following current-voltage relationships can be written:

$$I_i = \frac{V_i - V_1}{R_3}, \quad V_1 = \frac{I_i - I_2}{sC_1}$$

$$I_2 = \frac{V_1 - V_3}{L_2}, \quad V_3 = \frac{I_2 - I_4}{sC_3}$$

$$I_4 = \frac{V_3 - V_5}{sL_4}, \quad V_5 = \frac{I_4 - I_0}{sC_5}$$

$$I_0 = \frac{V_5}{R_L}$$

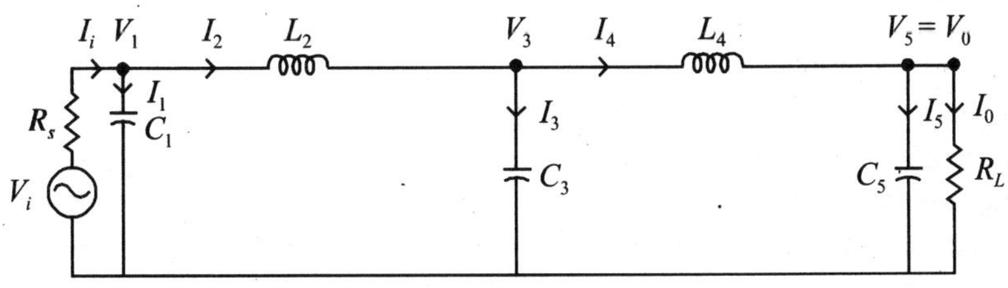

Fig. 12.45

Coupled-biquad structure is obtained by simulation of these current-voltage relationships. Fig. 12.46 gives a block diagram simulation of these equations. As

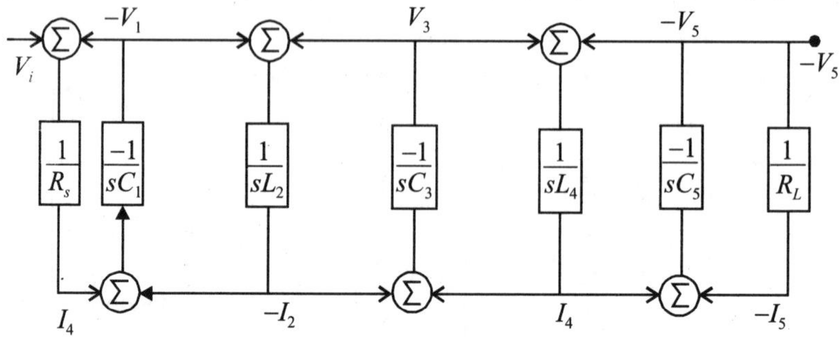

Fig. 12.46

can be seen from the figure sign, inversion is introduced in the structure to simulate the equations. It can also be seen that the shunt branches, and the reactive component branches, involve integration. The capacitive reactance branch indicates an inverting integrator while the inductive one is a noninverting integrator. To arrive at the coupled-biquad or the leapfrog structure, these integrators are replaced by appropriate operational amplifier circuits. This result in the final structure of Fig. 12.47.

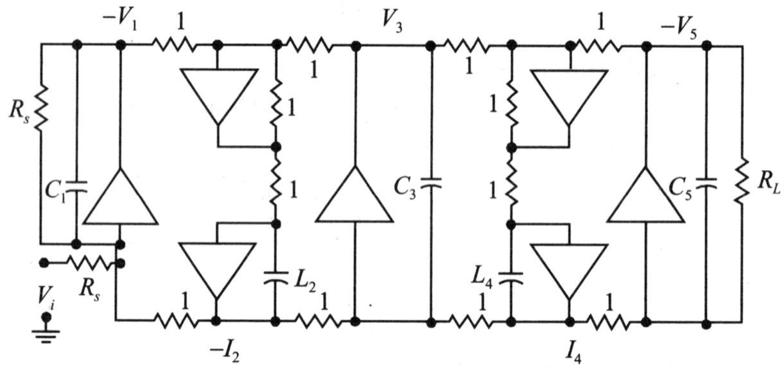

Fig. 12.47

As in the case of FDNR based structure, first a resistively terminated LC low-pass ladder satisfying the specification of a LP filter is designed. This is then transformed to a coupled-biquad structure by simulating the capacitive and inductive branches by inverting and noninverting integrators respectively, as shown. The salient feature of this structure is the one to one correspondence of the integrators to the reactive elements. Here we have discussed the realization of an LP filter and the realization of a BP filter can be found elsewhere.

The advantages often attributed to the design of higher order filters using cascaded biquadratic sections is the ability to tune the ω_0 and Q of each section without any interaction with others. This independent tunability achieved by isolation between the sections itself becomes a pitfall in higher order filter design. The effect of mistuning the pole frequency of one of the sections is significantly greater in the pass-band than the effect of an equivalent LC product change in the passive design. The pass-band response of a passive LC design, or its simulated versions, is much less sensitive to mistuning than that of the isolated cascaded realization. This reduced sensitivity in the pass-band is a result of coupling between the resonators. The stop-band sensitivity of the cascaded and ladder simulation methods are comparable.

The coupling between the resonators need not be obtained by one to one simulation of individual elements of the LC ladder. It is possible to duplicate the effect of coupling using certain other topologies. The structure shown in Fig. 12.48 is also a coupled biquad in which each block is a second order section. This method attempts to obtain the low sensitivity of the ladder while retaining the building block approach with independent tunability of the different section (see notes).

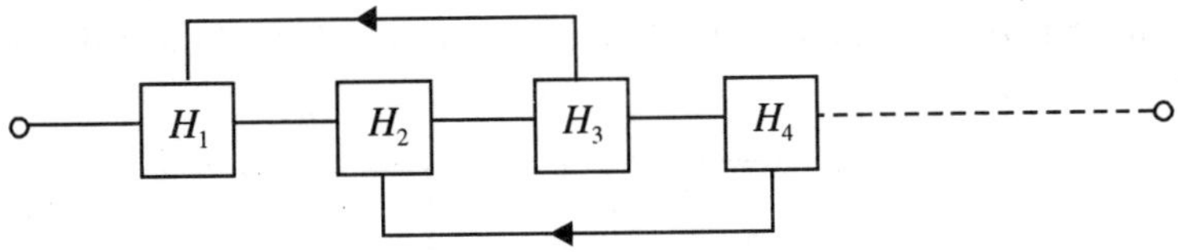

Fig. 12.48

12.8 NON-IDEAL OP. AMP

The filter design techniques developed so far in this chapter consider the op. amp as an ideal voltage amplifier. An ideal op. amp has infinite gain, infinite bandwidth, infinite input impedance and zero output impedance. A practical op. amp naturally deviates considerably from the ideal and this deviation manifests itself as a shift in the centre frequency and the quality factor of the designed filter. The overall frequency response of the filter transfer function may not only show considerable deviation from its design specification but may also become unstable. Before a filter circuit can be adopted for applications, it should be checked for these deviations and oscillations.

An operational amplifier, in general, has the main characteristics shown in Fig. 12.49(*a*). The 0 dB crossover takes, place at a rate of 18 dB/octave or more and any feedback around the op. amp may make the overall circuit unstable. To alleviate this problem, commercially available op. amps are internally compensated to yield a *single pole roll-off model* as shown in Fig. 12.49(*b*). The gain of the op. amp can now be written as

$$A = \frac{A_0 \omega_a}{s + \omega_a}$$

where A_0 is the dc gain and ω_a the corner frequency. The factor $A_0\omega_a$ is the *gain bandwidth product* (*GB*) of the amplifier and is generally also the 0 dB crossover

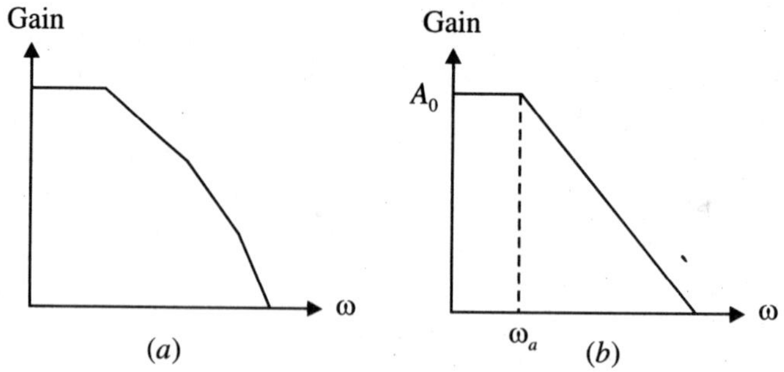

Fig. 12.49

frequency. For the commercially available 741 op. amps A_0 is of the order 100 dB and GB varies from 0.5 MHz to 1.2 MHz.

Apart from this frequency-dependent gain commercial 741s have an input impedance of the order of MΩ and an output impedance of the order of 50 Ω. If the op. amp has FET input then its input impedance can be of the order of 10^{11}Ω.

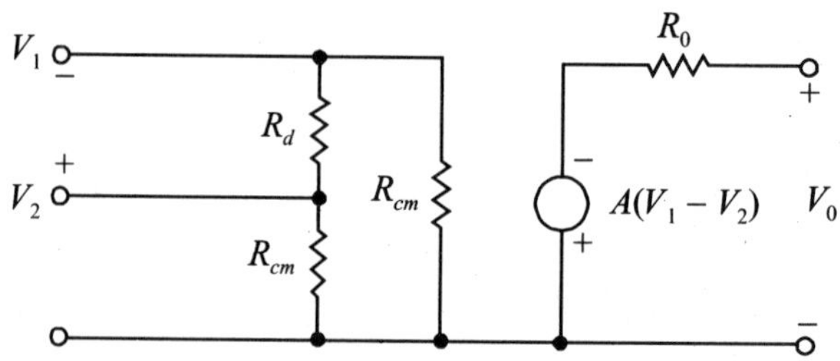

Fig. 12.50

A more accurate equivalent circuit of a differential input op. amp is shown in Fig. 12.50. R_{cm}, the common mode impedance is generally much higher than R_d, the differential impedance. The nullor model of a non-ideal op. amp is shown in Fig. 12.51 where $A_0 = R_2/R_1$ and $\omega_a = 1/R_2C$. This model does not show the input and output impedances.

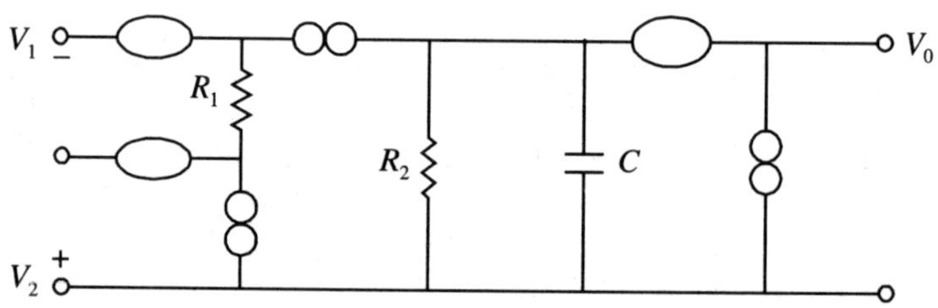

Fig. 12.51

The single pole role-off of an op. amp introduces error in the designed ω_0 and Q of a filter. The method of analysis of the filter with actual amplifier, though is conceptually the same as that with an ideal op. amp, it is algebraically much more involved. Due to the pole of the op. amp a second order filter transfer function becomes a third order function. In the standard method of analysis of the filter structure the actual op. amp is replaced by the equivalent circuit shown in Fig. 12.50 with both R_d and R_{cm} equal to infinity. (In practice it is found that the finite R_d does not affect the filter performance to any appreciable extent). The resulting third-order transfer function is now analysed to yield the actually realized $\hat{\omega}_0$ (say ω_0) and $\hat{Q}$ (say Q). The error due to finite GB and non-zero R_0 can be expressed as

$$\Delta\omega_0 = \hat{\omega}_0 - \omega_0$$

$$\Delta Q = \hat{Q} - Q$$

where ω_0 and Q are the centre frequency and quality factor with an ideal amplifier. The deviations in ω_0 and Q depend upon the actual filter structure. The exact analysis is rather involved and we summarize the effect of finite GB and non-zero R_0. These conclusions are applicable to a wide class of op. amp filters, including higher order realizations.

(*i*) The centre frequency of the filter ω_0 is reduced.

(*ii*) There is Q enhancement in the filter.

(*iii*) R_0 does not affect ω_0 significantly, but usually causes additional Q-enhancement.

(*iv*) R_0 may also add finite zeros to the transfer function and generally leads to a non-minimum phase transfer function.

12.9 DISCUSSIONS

The selection of an active filter structure depends mainly on the sensitivity of the transfer function to the variation in the passive elements and the active element parameters. A standard method of designing a higher order filter is to realize it as a cascade of second-order filters. But, of late, it has been shown that a coupled biquad structure has lower sensitivity than the cascase structures.

The deviation of ω_0 and Q of the realized structure can be reduced using an op. amp of higher gain bandwidth product. But the higher cost of these 'better' op. amp have limited the use of such amplifiers in bulk production of active filters and 741*s* are still being used in active filters design. The task of the filter designer is to select a structure which has the lowest sensitivity to A_0 (the highly variable and sensitive parameter of the op. amp) and GB. There have been many structures specifically designed to reduce the sensitivity. Some of the recently developed structures have a built-in compensation to reduce this effect. Even from this angle *GIC*-based and coupled-biquad structures seem to be highly competitive.

The effect of sensitivity of the transfer function to the passive parameters (resistances and capacitances) can be minimized by using precision components. Passive components can be designed, using thin film techniques with 0.1% accuracy or less. In hybrid integrated-circuit form, the component resistors and capacitors can be designed to have equal and opposite tracking and thus the total deviation of the transfer function due to temperature is minimized.

Another consideration in selecting a filter structure is the spread of the passive elements. It is a common practice to limit the spread of the capacitors to a ratio of 1:10 and those of resistors to 1 : 100. The value of the total capacitance determines the substrate area required for the fabrication of the filter in hybrid integrated-circuit form. Attempts have also been made to reduce the total capacitance of the filter structure.

The deviation of the filter performance due to the non-idealities of the op. amp is, as indicated, an important factor. Though commercially available op. amps are compensated to yield a single poll roll-off, op. amps can be designed to yield stable two-pole roll-off. The filters with such op. amps can be designed with lower deviation in ω_0 and Q. Of late, a new method of designing active filters without using external capacitors has been advanced. In this method, the single poll roll-off (or the internal compensating capacitance) is used to generate the required filter function by embedding the op. amp in a resistive structure. A two op. amp active R structure is capable of realizing an LP function and a BP function simultaneously. An HP active filter is, to a great extent, a misnomer. With the band limiting the op. amp it is not possible to design a real HP filter, or for that matter, an all-pass filter.

NOTES AND REFERENCES

Active network and filter design is so vast a topic to be covered as a chapter in a book. Here a minimum amount of material is covered to introduce the students to the concepts and the methods of filter design. The IEEE book edited by Mitra[1] provides exhaustive set of references.

The most useful active elements like controlled sources, op. amps, gyrators have been introduced here. Several other active element description can be found in literature. In addition to introducing such active elemetns Mitra's book (3, Chap. 2) discusses the interrelationship between the active elements and their basic characterisation in the AD-BC plane. As indicated, the nullor models are useful in developing transistor op. amp circuits for the active elements. Further details may be obtained from Mitra (3, Chap. 2) and Moschytz (4 vol. 1, Chap. 5) books.

Sensitivity analysis is of fundamental importance in active filter design. Sensitivity of a network to elemental variation depends on network topology also. A detailed study of sensitivity can be found in references[3] (Chap. 5) and[4] (vol. 1. Chap. 4) and the material may be supplemented by many excellent references cited in Ref. 1 and 2.

Most active filters are designed by coefficient matching technique. Here a few single and multiamplifier techniques are introduced. Lindiquist's book[5] discusses, in detail, the design of various filters. The treatment in this book is only for cascaded design.

Of late, GIC based structures, gyrators and FDNRs included, are becoming more popular for they yield low sensitivity structures. In addition, simulated ladder structures also yield low sensitivity structures. Bruton[6], Sedra and Brokett[7] present a detailed study of such structures. The realisation of band-pass filters are also considered in these references.

References 4 and 7 also provide the details of cascaded design, the pairing and ordering of sections.

Active filter design by formulae, tables and charts is presented by William's, book.[8] Many hardware considerations are presented in refs. 4 and 5.

The finite gain, bandwidth, input and output impedances of an operational amplifier deteriorate the performance of a filter. This particular aspect has received the attention of several researchers. Ref. 9 provides a good review of the effects of a non ideal op. amp.

Any discussion on active filter design without an accompanying lab is considered to be ineffective. As such, it is suggested that the following experiments be considered as an integral part of this chapter.

(*i*) Sallen-Key second order LP and BP filters.

(*ii*) KHN state-variable filter.

(*iii*) Cascade realisation of higher order filter (atleast 5th order).

(*iv*) Ladder simulation by FDNRs.

1. S.K. Mitra, *Active Inductorless Filters*, IEEE Press, 1971.
2. R. Schaumann, M. Soderstrand and K. Laker, *Modern Active Filter Design*, IEEE Press, 1981.
3. S.K. Mitra, *Analysis and Synthesis of Linear Active Networks*, Wiley, 1969.
4. G.S. Moschytz, *Linear Integrated Networks,* Vols. 1 and 2, Van. Nostrand Reinhold, 1974.
5. C.S. Lindiquist, *Active Network Design with Signal Filtering Applications*, Steward and Sons 1977.
6. L.T. Bruton RC, *Active Circuits Theory and Design*, Prentice Hall, Euglewood Cliffs, 1980.
7. A.S. Sedra and P.O. Brokett, *Filtering Theory and Design—Active and Passive*, Matrix, 1978.
8. A.B. Williams, *Active Filter Design*, Artech House, 1975.
9. K. Ramakrishna, K. Soundarajan and V.K. Aatre, *Effect of Amplifier Imperfections on Active Network, IEEE Transactions on Circuits and Systems*, Vol. CAS 26, No. 11, 1979, pp. 922–931.

APPENDIX A

MATRIX ALGEBRA

A matrix is a rectangular array of numbers representing a transformation. For example, the matrix Q_{11} of eqn. 2.6 transforms a set of chord currents I_c to a set of branch currents I_b of a given network with respect to a selected tree.

A matrix 'A' with m rows and n columns is represented as

$$A = [a_{ij}]_{m \times n}$$

where a_{ij} is the element in the ith row and jth column of A. (Here A has no bearing on the incidence matrix of a graph). The order of the matrix is $(m \times n)$. If $m = n$, then the matrix is called a 'square matrix'. The matrix obtained by interchanging the rows and columns of a matrix is called the 'transpose' of matrix A and is represented by A^t

$$A^t = [a_{ji}]_{n \times m}$$

with A as defined earlier. If the elements a_{ij} of a matrix are complex, the conjugate of the transpose is called the 'Hermitian transpose' and is represented by A^H.

Two matrices A and B of the same order defined by

$$A = [a_{ij}]_{m \times n} \text{ and } B = [b_{ij}]_{m \times n}$$

are said to be equal, if $a_{ij} = b_{ij}$ for all $i = 1, 2, ...m$, and $j = 1, 2, ...n$. If in a square matrix A, $a_{ij} = 0$ for all $i \neq j$ then the matrix is a 'diagonal' matrix. If all the diagonal entries of a diagonal matrix are 1, then it is called a 'unit or identity' matrix and is represented by U. If in a square matrix A, $a_{ij} = 0$ for $i > j$ (*i.e.*, all the entries below the diagonal) then such a matrix is an upper triangular matrix. On the other hand, if $a_{ij} = 0$ for $j > i$, it is a lower triangular matrix. If a matrix has only one row it is called a row matrix or row vector while a matrix with only one column is called a column matrix or a vector.

If a matrix A is such that $A = A^t$, then A is called a symmetric matrix. If $A = A^H$, then A is a Hermitian matrix. If $A = -A^t$ $(-A^H)$, then it is called a skew (skew-Hermitian) matrix. It can be easily seen that the diagonal entries of a skew-matrix are all zero.

A.1 MATRIX OPERATIONS

Addition: Let A and B be the two matrices defined by $A = [a_{ij}]_{m \times n}$ $B = [b_{ij}]_{m \times n}$ then the sum of the two matrices is another matrix of the same order, $(m \times n)$, defined by

$$C = [c_{ij}]_{m \times n}$$

where $c_{ij} = a_{ij} + b_{ij}$. Properties of this operation are

(*a*) $A + B = B + A$ (commutative)

(*b*) $A + (B + C) = (A + B) + C$ (associative)

(*c*) $(A + B)^t = A^t + B^t$

Scalar multiplication: The scalar multiplication of a matrix A by a scalar a is defined as

$$\alpha A = (\alpha a_{ij})_{m \times n}$$

Multiplication: Let A and B be two $(m \times n)$ and $(n \times p)$ matrices respectively. The product of the two matrices, AB, is another matrix of order $m \times p$, defined by

$$C = AB = [c_{ij}]_{m \times p}$$

where

$$c_{ij} = \sum_{k=1}^{n} a_{ik} b_{kj}$$

For a matrix A to be multiplied by a matrix B on the right, the number of columns of A must be equal to the number of rows of B. Two such matrices are said to be conformable for multiplication. If $m \neq p$, then BA is not defined. Even if AB and BA exist, AB is not generally equivalent to BA.

The properties of the equation of matrix multiplication are:

(*a*) $A(B + C) = AB + AC$ (distributive)
(*b*) $A(BC) = (AB)C$ (associative)
(*c*) If $AB = BA$, B and A are said to be commutative
(*d*) $AB = 0$ does not imply that either $A = 0$ or $B = 0$
(*e*) $(AB)^t = B^t A^t$.

Sometimes it may be convenient to write a matrix in partitioned form. For example, $a(4 \times 3)$, matrix can be written in partitioned form as

$$A = \left[\begin{array}{cc|c} a_{11} & a_{12} & a_{13} \\ a_{21} & a_{22} & a_{23} \\ \hline a_{31} & a_{32} & a_{33} \\ a_{41} & a_{42} & a_{43} \end{array}\right] = \begin{bmatrix} A_{11} & A_{12} \\ A_{21} & A_{32} \end{bmatrix}$$

where

$$A_{11} = \begin{bmatrix} a_{11} & a_{12} \\ a_{21} & a_{22} \end{bmatrix}, \quad A_{12} = \begin{bmatrix} a_{13} \\ a_{23} \end{bmatrix}$$

$$A_{21} = \begin{bmatrix} a_{31} & a_{32} \\ a_{41} & a_{42} \end{bmatrix}, \quad A_{22} = \begin{bmatrix} a_{33} \\ a_{43} \end{bmatrix}$$

Each submatrix A_{ij} is treated as an element of A. Let B be another partitioned matrix given by

$$B = \left[\begin{array}{cc|c} b_{11} & b_{12} & b_{13} \\ b_{21} & b_{22} & b_{23} \\ \hline b_{31} & b_{32} & b_{33} \end{array}\right] = \begin{bmatrix} B_{12} & B_{12} \\ B_{21} & B_{22} \end{bmatrix}$$

then

$$AB = \left[\begin{array}{c|c} A_{11}B_{11} + A_{12}B_{21} & A_{11}B_{12} + A_{12}B_{22} \\ \hline A_{21}B_{11} + A_{22}B_{21} & A_{21}B_{12} + A_{22}B_{22} \end{array}\right]$$

To perform the required multiplication the matrices should be conformably partitioned.

Inversion: Let A be a square matrix of order n, then an $(n \times n)$ matrix B is called the inverse of A if,

$$AB = BA = U_{n \times n}$$

B is represented by A^{-1}, *i.e.*, $B = A^{-1}$.

A matrix has an inverse, if it is non-singular (*i.e.*, the determinant of the matrix is non-zero). If an inverse exists, it is unique. The inverse of a matrix A is given by

$$A^{-1} = \frac{1}{\det A}\text{Adj}(A)$$

where $\det A$ is the determinant of A and Adj $A = [A_{ij}]^t$, where A_{ij} is the cofactor of the element a_{ij}. The cofactor, A_{ij} is defined by

$$A_{ij} = (-1)^{i+j} M_{ij}$$

where M_{ij} is the minor of the element a_{ij} and is the determinant obtained after deleting row i and column j.

It can be easily shown that

(*a*) $(AB)^{-1} = B^{-1}A^{-1}$

(*b*) $(A^t)^{-1} = (A^{-1})^t$

(*c*) If $AB = 0$ and A is non-singular then B is a null matrix

(*d*) $\det A = \dfrac{1}{\det A^{-1}}$

Let A be a non-singular matrix with real entries. If $A^t = A^{-1}$ then A is called an orthogonal matrix. If A is complex and $A^{-1} = A^H$, A is called unitary. The determinant of an orthogonal (unitary) matrix is 1. The matrix

$$A = \begin{bmatrix} \cos\theta & -\sin\theta \\ \sin\theta & \cos\theta \end{bmatrix}$$

is an orthogonal matrix.

Let a matrix A be partitioned as

$$A = \left[\begin{array}{c|c} \overbrace{A_{11}}^{m} & \overbrace{A_{12}}^{r} \\ \hline A_{21} & A_{22} \end{array}\right]_{n \times n} \begin{array}{l} \}m \\ \}r \end{array}$$

where A_{11} and A_{22} are non-singular, then it can be shown that

$$A^{-1} = \left[\begin{array}{c|c} (A_{11} - A_{12}A_{22}^{-1}A_{21})^{-1} & -(A_{11} - A_{12}A_{22}^{-1}A_{21})^{-1}A_{12}A_{22}^{-1} \\ \hline -(A_{22} - A_{21}A_{11}^{-1}A_{12})^{-1}A_{21}A_{11}^{-1} & (A_{22} - A_{21}A_{11}^{-1}A_{12})^{-1} \end{array}\right]$$

if the required inverses exist.

A.2 RANK OF A MATRIX AND ELEMENTARY TRANSFORMATION

The rank, $r(A)$ of a matrix A is the number of linearly independent rows (or columns, whichever is the minimum) of A. $r(A)$ is the order of the maximum order non-vanishing determinant that can be formed from the rows and columns of A. For example, consider the matrix

$$A = \begin{bmatrix} 1 & 2 & 3 \\ 4 & 5 & 6 \\ 7 & 8 & 9 \end{bmatrix}$$

The rank of this matrix is two. The rows (columns) are linearly dependent and we have

$$-r_1 + 2r_2 - r_3 = 0$$

where r_i corresponds to row i or column i of the matrix. A(2 × 2) determinant,

$$\begin{vmatrix} 1 & 2 \\ 4 & 5 \end{vmatrix}$$

is non-zero and hence the rank is 2.

Let $A = [a_{ij}]_{n\times n}$. If $r(A) = n$, then A is obviously non-singular. If A is a rectangular matrix of order $m \times n$ or (or $n \times m$) with $n > m$, and if $r(A) = m$, then the matrix is a maximum rank matrix. If A and B are conformable for multiplication, then

$$r(AB) \leq \min\,[r(A), r(B)]$$

If matrix A is non-singular and if the products AB and CA exist, then

$$r(AB) = r(B)$$

$$r(CA) = r(C)$$

i.e., pre- or post-multiplication by a non-singular matrix does not change the rank of a matrix.

A matrix can be transformed to its canonic form by elementary transformation. Elementary operations on a matrix are:

(*i*) Interchange of rows or columns,
(*ii*) Multiplication of a row (column) by a scalar.
(*iii*) Addition of scalar multiple of one or more rows (columns) to other rows (columns).

The elementary operation can be achieved by pre- or post-multiplying the matrix by a non-singular matrix. For example, pre-multiplication by a matrix of the form,

$$\begin{bmatrix} 0 & 1 & 0 & \cdots\cdots & 0 \\ 1 & 0 & 0 & \cdots\cdots & 0 \\ 0 & 0 & 1 & \cdots\cdots & 0 \\ \cdots & \cdots & \cdots & \cdots\cdots & \cdots \\ 0 & \cdots & \cdots & \cdots\cdots & 1 \end{bmatrix}$$

interchanges the first and second rows whereas post-multiplication by such a matrix interchanges columns 1 and 2. Pre-multiplication by a matrix of the form

$$\begin{bmatrix} 1 & k & 0 & . & . & . & . & . & . & . & 0 \\ 0 & 1 & 0 & . & . & . & . & . & . & . & . \\ 0 & 0 & 1 & & & & & & & & \\ \vdots & & & & & & & & & & \\ 0 & & & & & \cdots\cdots & & & \cdots\cdots & & \end{bmatrix}$$

adds k times the second row to the first row whereas post-multiplication by a matrix of the form

$$\begin{bmatrix} 1 & 0 & 0 & . & . & . & . & . & 0 \\ k & 1 & 0 & . & . & . & . & . & . \\ 0 & 0 & 1 & . & . & . & . & . & . \\ 1 & & & & & & & & \\ \vdots & & & & & & & & \\ 1 & & & & & & & & \\ 0 & \cdots\cdots & & \cdots\cdots & & & & & 1 \end{bmatrix}$$

adds k times the second columns to the first column.

The non-singular matrices used for elementary operations are called elementary matrices. The rank of a matrix A is invariant under elementary transformations. Every matrix A can be transformed to its canonic form by elementary transformation.

The canonic form of a $(m \times n)$ matrix A of rank r is

$$A_c = \left[\begin{array}{c|c} U_{r\times r} & 0 \\ \hline 0 & 0 \end{array}\right]$$

where $U_{r\times r}$ is a unit matrix of order $r(A_c = PAQ$, where P and Q are non-singular).

Two matrices A and B are said to be equivalent ($A \sim B$) if one can be transformed to the other by elementary transformations. Two matrices are equivalent if they have the same canonic form and this implies that two equivalent matrices have the same rank. Equivalence of matrices is an equivalence relation.

i.e., $A \sim A$ (reflexive)

$A \sim B$ then $B \sim A$ (symmetry)

$A \sim B$ and $B \sim C$ then $A \sim C$ (transitive)

Any relationship satisfying the three properties—reflexive, symmetry and transitive—is called an equivalence relationship. It can be shown that every non-singular matrix is a product of elementary matrices.

A.3 SYSTEM OF LINEAR EQUATION

Consider a system of linear simultaneous equation

$$A\mathbf{X} = \mathbf{Y}$$

where A is an $(m \times n)$ coefficient matrix, and X and Y are column vectors. If A is a square matrix (*i.e.*, $m = n$) then a non-trivial solution for X exists, if, A is non-singular. In this case, the solution is

$$\mathbf{X} = A^{-1}\mathbf{Y}$$

For a general A matrix, *i.e.*, for a system of m non-homogeneous system of equations in n unknowns, a solution exists, if, the rank of the augmented matrix is the same as the coefficient matrix, *i.e.*,

$$r[A:Y] = r(A)$$

(If $Y = 0$, the system of equations are said to be homogeneous).

A.4 EIGENVALUES AND EIGENVECTORS

An algebraic equation that often appears in the analysis of networks and systems is

$$A\mathbf{X} = \lambda\mathbf{X}$$

where the coefficient matrix A is a square matrix of order n, $\mathbf{X}$ is a n vector and λ is a scalar constant which may be complex. The preceding algebraic equation can be rewritten as

$$(A - \lambda U)\mathbf{X} - 0$$

and a non-trivial solution for $\mathbf{X}$ exists if $(A - \lambda U)$ is a singular matrix. The n scalar constants which render the matrix $(A - \lambda U)$ singular are called the eigenvalues of A (latent roots or characteristic roots) and the non-trivial solution vectors are called the 'eigenvectors' of A. The study of such a problem is classified as 'Eigenvalue problem'.

The n eigenvalues of A can be determined by the equation.

$$\det(A - \lambda U) = 0$$

The determinant can be written as

$$\lambda^n + \alpha_{n-1}\lambda^{n-1} + \alpha_{2n-2}\lambda^{n-2} + \ldots + \alpha_1\lambda + \alpha_0 = 0$$

and this is called the 'characteristic' equation of the matrix. The n roots of this equation λ_i, $i = 1, 2, \ldots, n$ are the eigenvalues of A. For each eigenvalue of the matrix we can form an equation.

$$A\mathbf{X}_i = \lambda_i \mathbf{X}_i$$

The solution vector X_i is the eigenvector corresponding to the eigenvalue λ_i.

Example. Let $A = \begin{bmatrix} 2 & -1 & 1 \\ -1 & 2 & 1 \\ -1 & 1 & 2 \end{bmatrix}$

Then the characteristic equation is

$$|A - \lambda U| = \det\left\{\begin{bmatrix} 2 & -1 & 1 \\ -1 & 2 & 1 \\ -1 & 1 & 2 \end{bmatrix} - \begin{bmatrix} \lambda & 0 & 0 \\ 0 & \lambda & 0 \\ 0 & 0 & \lambda \end{bmatrix}\right\} = 0$$

i.e.,
$$\det\begin{bmatrix} 2-\lambda & -1 & 1 \\ -1 & 2-\lambda & 1 \\ -1 & 1 & 2-\lambda \end{bmatrix} = 0$$

i.e.,
$$\lambda^3 - 6\lambda^2 + 11\lambda - 6 = 0$$

The eigenvalues are 1, 2 and 3.

i.e., $(\lambda - 1)(\lambda - 2)(\lambda - 3) = \lambda^3 - 6\lambda^2 + 11\lambda - 6 = 0$

(Observe that the product of the eigenvalues is the determinant of the matrix and the sum of the eigenvalues is the trace of the matrix. A trace of a matrix is the sum of its diagonal entries.) To find the eigenvectors we have to solve the equation

$$\begin{bmatrix} 2 & -1 & 1 \\ -1 & 2 & 1 \\ -1 & 1 & 2 \end{bmatrix}\begin{bmatrix} x_1 \\ x_2 \\ x_3 \end{bmatrix} = \lambda_i \begin{bmatrix} x_1 \\ x_2 \\ x_3 \end{bmatrix}$$

For
$$\lambda_1 = 1,\ [x_1\ x_2\ x_3]^t = [1\ 1\ 0]$$
$$\lambda_2 = 2,\ [x_1\ x_2\ x_3]^t = [1\ 1\ 1]$$
$$\lambda_3 = 3,\ [x_1\ x_2\ x_3]^t = [0\ 1\ 0]$$

The following results can be easily derived:

(*i*) The eigenvalues of a real symmetric matrix are all real.

(*ii*) A skew symmetric matrix has imaginary eigenvalues.

(*iii*) If a matrix is singular then it has a zero eigenvalue.

(*iv*) The eigenvectors associated with distinct eigenvalues of a matrix are linearly independent.

(*v*) The eigenvectors associated with distinct eigenvalues of a real symmetric matrix are orthogonal.

A matrix whose eigenvalues are all positive (negative) is called a positive definite (negative definite) matrix. Otherwise, it is said to be indefinite. If a matrix has a zero eigenvalue, it can only be a semidefinite matrix.

Two matrices A and B of the same order are said to be similar if there exists a non-singular matrix P so that

$$B = P^{-1}AP$$

Similarly, relationship is an equivalence relationship. Similar matrices have the same determinant, same characteristic equation and hence, the same eigenvalues, but not necessarily the same eigenvectors. If a matrix has distinct eigenvalues, it is similar to a diagonal matrix, the entries of which are the eigenvalues of the original matrix, *i.e.*, if A has distinct eigenvalues then,

$$S^{-1}AS = \Delta$$

where $\Delta = \text{dia}\,(\lambda_1, \lambda_2, \lambda_3, \ldots\lambda_n)$. This similarity in relationship can be rewritten as,

$$AS = S\Delta$$

If C_i are the columns of S, $i = 1, 2, \ldots n$, then we have

$$A\mathbf{C}_i = \lambda_i\mathbf{C}_i$$

indicating that C_i is the eigenvector of A with respect to the eigenvalue λ_i. Hence, we conclude that "an nth order matrix with distinct eigenvalues is similar to a diagonal matrix." The transformation matrix for this is the eigenvector matrix of the n independent eigenvectors of A.

For the example considered earlier

$$S = \begin{bmatrix} 1 & 1 & 0 \\ 1 & 1 & 1 \\ 0 & 1 & 1 \end{bmatrix} \text{ and } S^{-1} = \begin{bmatrix} 0 & 1 & -1 \\ 1 & -1 & 1 \\ -1 & 1 & 0 \end{bmatrix}$$

and

$$S^{-1}AS = \begin{bmatrix} 1 & 0 & 0 \\ 0 & 2 & 0 \\ 0 & 0 & 3 \end{bmatrix}$$

If A does not have the distinct eigenvalues then in, general, it cannot be diagonalized by similarity transformation. In such a case, the matrix can be reduced to its Jordan canonic form by similarity transformation.

i.e.,
$$p^{-1}AP = J$$

where

$$J = \begin{bmatrix} J_{r_1}(\lambda_1) & 0 & 0\ldots\ldots\ldots 0 \\ 0 & J_{r_2}(\lambda_2) & \\ \vdots & & \\ \vdots & & \\ 0 & \ldots\ldots\ldots & J_{r_k}(\lambda_k) \end{bmatrix}$$

λ_i, $i = 1, 2, \ldots, k$ are the eigenvalues of A and λ_i repeats r_i times, *i.e.*, $r_1 + r_2 + \ldots + r_k = n$.

The Jordan blocks $J_{r_i}(\lambda_i)$ are of the form

$$J_{r_i}(\lambda_i) = \begin{bmatrix} \lambda_i & 1 & 0 \dots\dots\dots 0 \\ 0 & \lambda_i & 1 \dots\dots\dots 0 \\ \vdots & & \\ \vdots & & 1 \\ 0 & \dots\dots\dots\dots\dots & \lambda_i \end{bmatrix}_{i \times r_i}$$

The proofs of many of the results considered here and the other aspects of matrix analysis have been dealt with in great detail in R. Bellman's book: *Introduction to Matrix Analysis*, McGraw-Hill.

APPENDIX B

LAPLACE TRANSFORMS

The time-domain network equations in general are integro-differential equations. We can convert these integro-differential equations to higher order differential equations and solve these differential equations to obtain the time-domain network response. A method of solving the network equations of a linear time-invariant network is to resort to Laplace transform technique. Laplace transformation of an integro-differential equation with constant coefficients leads to an algebraic equation in the Laplace variable 's'. The algebraic equation is solved in the Laplace domain and the time-domain response is obtained by inverse Laplace transforming the s-domain solution. (Compare this with the procedure of solving an equation $x^a = y$ by logarithms).

The one-sided Laplace transform of a time function $f(t)$, $0 < t < \infty$, represented by $F(s) = \mathcal{L}\,[f(t)]$, is defined by the integral

$$F(s) = \int_0^\infty f(t)e^{-st}\,dt \qquad \text{...(B.1)}$$

If the lower limit of integration is $-\infty$, then $F(s)$ is a two-sided Laplace transform. We have no occasion to use the latter.

The Laplace transform $F(s)$ of $f(t)$ is a function of the complex variable $s(s = \sigma + j\omega)$. The Laplace transform of a function exists if, the integral defined in (B.1) converges. The integral converges, in general, for some values of s in the right half plane.

A sufficient condition for the existence of the Laplace transform is given by the Theorem B.1.

Theorem B.1: Let $f(t)$ be a function which is piecewise continuous on every finite interval in the range $t \geq 0$ and satisfying

$$|f(t)| \leq M\,e^{\sigma t} \qquad \text{...(B.2)}$$

for some constants σ and M. Then the Laplace integral converges absolutely for all Re $s > \sigma$.

A function satisfying the inequality (B.2) is said to be of exponential order. The order of the function is the smallest number σ_0 that satisfies (B.2). We say that the Laplace integral converges absolutely for all $\sigma > \sigma_0$. σ_0 is called the *abscissa of convergence* and the region $c > \sigma_0$ is the *region of convergence*.

Example B.1. (*i*) $f(t) = u(t)$

$$F(s) = \int_0^\infty e^{-st}dt = -\frac{1}{s}e^{-st}\Bigg|_0^\infty$$

i.e.,
$$F(s) = \frac{1}{s}, \sigma > 0.$$

(ii) $f(t) = e^{\alpha t} u(t)$

$$F(s) = \int_0^\infty e^{\alpha t} e^{-st} dt$$

$$= \int_0^\infty e^{(\alpha - s)t} dt = \frac{1}{\alpha - s} e^{(\alpha - s)t} \Big|_0^\infty$$

i.e.,
$$F(s) = \frac{1}{s - \alpha}, \sigma > \alpha.$$

(iii) $f(t) = \delta(t)$

$$F(s) = \int_0^\infty \delta(t) e^{-st} dt$$

i.e., $F(s) = 1$ for all s.

(iv) $f(t) = t\, u(t)$

$$F(s) = \int_0^\infty t e^{-st} dt$$

$$= \frac{te^{-st}}{s}\Big|_0^\infty + \int_0^\infty -\frac{e^{-st}}{s} dt$$

$$= \frac{e^{-st}}{s}\Big|_0^\infty$$

i.e.,
$$F(s) = \frac{1}{s^2}, \sigma > 0$$

(v) $f(t) = \sin \omega t\, u(t)$

$$F(s) = \int_0^\infty \sin \omega t\, e^{-st} dt$$

$$= \int_0^\infty \left[\frac{e^{j\omega t} - e^{-j\omega t}}{2j}\right] e^{-st} dt$$

$$= \frac{1}{2j}\left[\frac{e^{(-s+j\omega)t}}{-s+j\omega}\Big|_0^\infty - \frac{e^{-(s+j\omega)t}}{-(s+j\omega)}\Big|_0^\infty\right]$$

$$= \frac{1}{2j}\left[\frac{1}{s - j\omega} - \frac{1}{s + j\omega}\right]$$

i.e.,

$$F(s) = \frac{\omega}{s^2 + \omega^2}, \; \sigma > 0.$$

Table B.1 gives some elementary time functions and the corresponding Laplace transforms. All time functions are defined for $t \geq 0$.

Table B.1

$f(t)$		$f(t)$	$T(s)$
$\delta(t)$	1	$\sin \omega t$	$\frac{\omega}{s^2 + \omega^2}$
$u(t)$	$\frac{1}{s}$	$\cos \omega t$	$\frac{s}{s^2 + \omega^2}$
e^{-at}	$\frac{1}{s + a}$	$e^{-at} \sin \omega t$	$\frac{\omega}{(s + a)^2 + \omega^2}$
	$\frac{1}{s^2}$	$e^{-at} \cos \omega t$	$\frac{s + a}{(s + a)^2 + \omega^2}$
t^n	$\frac{n!}{s^{n+1}}$	$\cosh \omega t$	$\frac{s}{s^2 - \omega^2}$
$t^n e^{-at}$	$\frac{n!}{(s + a)^{n+1}}$	$\sinh \omega t$	$\frac{\omega}{s^2 - \omega^2}$

Some of the important properties of Laplace transforms are:

1. *Linearity:* Let $f_1(t)$ and $f_2(t)$ be two time functions with the corresponding Laplace transforms $F_1(s)$ and $F_2(s)$, then

$$\mathcal{L}\,[\alpha_1 f_1(t) + a_2 f_2(t)] = \alpha_1 F_1(s) + \alpha_2 F_2(s) \qquad \text{...(B.3)}$$

   where α_1 and α_2 are two constants.

2. *Shifting theorem*: If $\mathcal{L}\,[f(t)] = F(s)$, then the Laplace transforms of $f(t - a)$ and $e^{at} f(t)$ are $e^{-as} F(s)$ and $F(s - a)$, respectively.

3. *Transform of derivatives*: If $\mathcal{L}\,[f(t)] = F(s)$, then

$$\mathcal{L}\left[\frac{df(t)}{dt}\right] = sF(s) - f(0)$$

   and $\mathcal{L}\left[f^{(n)}(t)\right] = s^n F(s) - s^{n-1} f(0) - s^{n-2} f'(0) \ldots - f^{(n-1)}(0)$

   where $f^{(n)}(t)$ is the nth derivative of $f(t)$ with respect to t.

4. *Transform of integrals:* If $\mathcal{L}[f(t)] = F(s)$, then

$$\mathcal{L}\left[\int_{-\infty}^{t} f(\tau)\,d\tau\right] = \frac{F(s)}{s} + \frac{f^{-1}(0_-)}{s}$$

where $f^{-1}(0_-) = \int_{-\infty}^{0} f(\tau)\,d\tau$

5. *Initial and final values:* If $\mathcal{L}[f(t)] = F(s)$, then

$$f(0) = \lim_{s\to\infty} sF(s)$$

and
$$\lim_{t\to\infty} f(t) = \lim_{s\to\infty} sF(s)$$

6. *Convolution:* Let $\mathcal{L}[f_1(t)] = F_1(s)$ and $\mathcal{L}[f_2(t)] = F_2(s)$, then

$$\mathcal{L}[f_1(t)^* f_2(t)] = F_1(s)F_2(s)$$

and
$$\mathcal{L}[f_1(t)\cdot f_2(t)] = F_1(s)^* F_2(s)$$

7. *Time scaling*: If $\mathcal{L}[f(t)] = F(s)$, then

$$\mathcal{L}[f(at)] = \frac{1}{a}F(s/a)$$

8. *Multiplication by t*: If $\mathcal{L}[f(t)] = F(s)$, then

$$\mathcal{L}[t^n f(t)] = (-1)^n \frac{d^n F(s)}{ds^n}$$

9. *Periodic function*: Let $f(t) = f(t + nT)$ where n is an integer, then

$$F(s) = \frac{1}{1-e^{-sT}} F_1(s)$$

where
$$F_1(s) = \int_0^T f(t)e^{-st}\,dt.$$

The inverse Laplace transform $\mathcal{L}^{-1}(.)$ of $F(s)$ is given by the complex contour integral

$$f(t) = \frac{1}{2\pi j}\int_{c-j\infty}^{c+j\infty} F(s)e^{st}\,ds \qquad \text{...(B.4)}$$

where the path of integration (Bromwich path) is along a vertical line $s = c$, from $-j^{\infty} + j^{\infty}$. The constant c is selected so that $c \geq \sigma_0$, where σ_0 is the abscissa of convergence of $F(s)$. When $F(s)$ alone is given, the path of integration is selected as a vertical line (in the complex s-plane) to the right of all the singularities of $F(s)$. [$F(s)$ is regular for all $\sigma > \sigma_0$]. Direct evaluation of the integral in eqn. B.4.

is rather complicated. We can resort to Cauchy's residue theorem to evaluate $f(t)$. If $F(s) \to 0$ as $s \to \infty$, then

$$f(t) = \sum \text{(residues of } F(s)e^{st} \text{ at the finite singularities of } F(s))$$

If $F(s)$, the Laplace transform of $f(t)$, is a rational function of s then $F(s)$ must be a proper fraction (the degree of the numerator is less than the degree of the denominator) in order to satisfy the condition that $F(s) \to 0$ as $s \to \infty$. The inversion by residues is always applicable to proper fractions.

Partial fraction expansion, and thus inversion of $F(s)$, is a very convenient method of determining $f(t)$. This method of inversion (by evaluating the residues) was discussed in Chapter 4.

Most of the books on Network Theory discuss the use and application of Laplace Transforms.

APPENDIX C

NATURAL FREQUENCIES

The natural response of a lumped linear network can be written as a sum of complex exponentials, *i.e.*,

$$y(t) = \sum_{i=1}^{n} k_i \exp(p_i t) \qquad \ldots(C.1)$$

where p_i and k_i are in general complex. The constants p_i which depend on the elements and the geometry of the network are called the natural frequencies (Chap. 6). A decomposition of the natural response into a series of exponentials places in evidence the *natural frequencies* (often called the mode of oscillations). Hence, a method of measuring natural frequencies is to express the observed response as a sum of complex exponentials. But, sometimes, not all the natural frequencies of the network will be reflected in the overall response of the network. For example, consider the network function.

$$F(s) = \frac{s^3 + 7s^2 + 14s + 8}{(s+3)(s+4)(s+5)(s+6)}$$

The partial fraction expansion of this is given by

$$F(s) = \frac{1/3}{s+3} + \frac{0}{s+4} + \frac{-6}{s+5} + \frac{20/3}{s+6}$$

Hence,

$$y(t) = \frac{1}{3}e^{-3t} - 6e^{-5t} + \frac{20}{3}e^{-6t}$$

Here we see that the natural frequency at $s = -4$ is not reflected in the output. Though the network function apparently seems to have four poles $p_1 = -3$, $p_2 = -5$ and $p_4 = -6$, there are only three complex exponentials in the output and we see that the residue corresponding to the pole at $s = -4$ is zero. This is due to the fact that $(s + 4)$ is a common factor in both the numerator and denominator. The actual network function is

$$F(s) = \frac{(s+1)(s+2)}{(s+3)(s+5)(s+6)}$$

Hence, though $s = -4$ may be an internal natural frequency of the network (*i.e.*, the natural frequency of some current or voltage internal to the network) so far as the terminal behaviour is concerned we can discard this natural frequency.

The cancellation of the common factor in the network function can be expressed in terms of *controllability* and *observability* of the particular natural frequency. In this example, we can say that the particular natural frequency at $s = -4$ is decoupled from the input (*i.e.*, not excitable from the prescribed terminals or uncontrollable) or decoupled from the output (*i.e.*, not measurable at the specified output terminals or unobservable) or decoupled from both the input and the output. The terms controllability and observability are commonly used in modern control theory. From the point of network synthesis we only design the network function after canceling the common factors between the numerator and denominator. If not, the realized network may have redundant or excess elements.

The constant k_1 in eqn. C.1 depend not only on the elements and geometry of the network but also on the initial conditions. The initial conditions may be so chosen that a particular natural frequency is not observed in a specified response. We demonstrate this with an example.

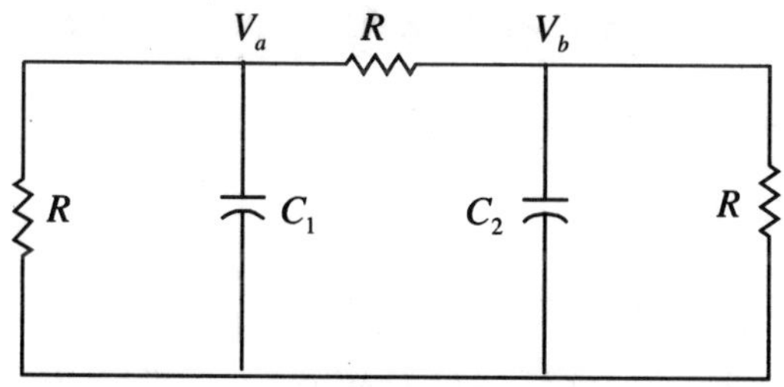

$C_1 = C_2 = C$

Fig. C.1

Consider the network shown in Fig. C.1. The capacitors C_1 and C_2 are initially charged to voltage $V_{C_1}(0)$ and $V_{C_2}(0)$. We are required to find $V_a(t)$ and $V_b(t)$.

The network under consideration has two energy storage elements (two capacitors) and is not degenerate. Hence, there are two natural frequencies. The node equation at nodes a and b is given by

$$\begin{bmatrix} \frac{2}{R}+sc & -\frac{1}{R} \\ -\frac{1}{R} & \frac{2}{R}+sc \end{bmatrix} \begin{bmatrix} V_a \\ V_b \end{bmatrix} - C \begin{bmatrix} V_{C_1}(0) \\ V_{C_2}(0) \end{bmatrix}$$

Solving for V_a and V_b we have

$$V_a = \frac{RC\left[(2+scR)V_{C_1}(0)+V_{C_2}(0)\right]}{(scR+3)(scR+1)}$$

and
$$V_b = \frac{RC\left[V_{C_1}(0)+(2+scR)V_{C_2}(0)\right]}{(scR+3)(scR+1)}$$

The natural frequencies are $-3/RC$ and $-1/RC$. We consider three different cases.

(*i*) $V_{C_1}(0) = V_{C_2}(0) = V(0)$

$$V_a = \frac{RCV(0) + [2 + scR + 1]}{(scR + 3)(scR + 1)} = \frac{V(0)}{s + (1/RC)} = V_b$$

i.e., $V_a(t) = V_b(t) = V(0)\, e^{-t/RC}$

(*ii*) $V_{C_1}(0) = -V_{C_2}(0) = V(0)$

$$V_a = \frac{RCV(0) + [2 + scR - 1]}{(scR + 3)(scR + 1)} = \frac{V(0)}{s + (2/RC)}$$

$$V_b = \frac{RCV(0) + [1 - (2 + scR)]}{(scR + 3)(scR + 1)} = -\frac{V(0)}{s + (3/RC)}$$

i.e., $V_a(t) = -V_b(t) = V(0)e^{-(3RC)t}$

(*iii*) $|V_{C_1}(0)| \neq |V_{C_2}(0)|$

$$V_a(t) = k_1 e^{-t/RC} + k_2 e^{(3/RC)t}$$

$$V_b(t) = k_1 e^{-t/RC} - k_2 e^{(3/RC)t}$$

where $$k_1 = \frac{V_{C_1}(0) + V_{C_2}(0)}{2}, \quad k_2 = \frac{V_{C_1}(0) - V_{C_2}(0)}{2}$$

In case (*i*) only one of the natural frequencies is observed in the response (observe the cancellation of the factor [*scR* + 3] in the calculations). In case (*ii*) again, only one natural frequency is observed. Here, the factor (*scR* + 1) cancels out because of the proper selection of the initial conditions. In case of (*iii*), both the natural frequencies observed in (*i*) and (*ii*) appear.

The study of the modes of oscillation of a network is certainly more involved than that is demonstrated by this simple example. The concept and algebraic characterization of the controllability and observability are fundamental to the study of optimal control systems.

APPENDIX D

BODE PLOTS

The specifications of a filter are given in the frequency domain. One of the methods of representing these specifications is by the *Bode Plot*.

Let a transfer function $H(j\omega)$ be represented as

$$H(j\omega) = |\, H(j\omega)\,| e^{j\phi(\omega)}$$

The Bode plots for this transfer function consist of

(*i*) 20 log $|\, H(j\omega)\,|$ vs log ω. The unit for 20 log $|\, H(j\omega)\,|$ is decibel (dB), and

(*ii*) $\phi(\omega)$ vs log ω.

A typical transfer function can be written in the factored form as

$$H(j\omega) = \frac{K\prod_{i=1}^{m_1}(1+j\omega\tau_{z_i})\prod_{i=1}^{m_2}\left[1+j2\xi_{z_i}\frac{\omega}{\omega_{z_i}}+\left(j\frac{\omega}{\omega_{z_i}}\right)^2\right]}{(j\omega)^r\prod_{i=1}^{n_1}(1+j\omega\tau_{p_i})\prod_{i=1}^{n_2}\left[1+j2\xi_{p_i}\frac{\omega}{\omega_{p_i}}+\left(j\frac{\omega}{\omega_{p_i}}\right)^2\right]} \quad \text{...(D.1)}$$

The poles and zeros of the transfer function are

zeros: $\quad -\frac{1}{\tau_{s_i}}, \qquad i = 1, 2, ..., m_1$

$\quad -\xi_{z_i}\omega_{z_i} + j\omega_{z_i}\sqrt{1-\omega_{z_i}^2}, \qquad i =1, 2, ..., m_2$

poles: $\quad -\frac{1}{\tau_{p_i}}, \qquad i = 1, 2, ..., n_1$

$\quad -\xi_{p_i}\omega_{p_i} + j\omega_{p_i}\sqrt{1-\xi_{p_i}^2}, \qquad i = 1, 2, ..., n_2$

*r*th order pole at the origin.

For example, if

$$H(s) = \frac{(s+a)(s^2+bs+d)}{s^2(s+\beta)(s^2+\alpha s+\delta)}$$

then $H(j\omega)$ can be written as

$$H(j\omega) = \frac{ad}{\beta\delta}\,\frac{\left(1+j\frac{\omega}{a}\right)\left[1+j\frac{\omega b}{d}+\left(i\frac{\omega}{d}\right)^2\right]}{(j\omega)^2\left(1+j\frac{\omega}{\beta}\right)\left[1+j\frac{\omega\alpha}{\delta}+\left(j\frac{\omega}{\delta}\right)^2\right]}$$

which is of the form of eqn. D.1.

From eqn. D.1, we have

$$20\log|H(j\omega)| = 20\log K + 20\sum_{i=1}^{m_1}\log\left|1 + j\omega\tau_{z_i}\right|$$

$$+20\sum_{i=1}^{m_2}\log\left|1 + j2\omega_{z_i}\frac{\omega}{\omega_{z_i}} + \left(j\frac{\omega}{\omega_{z_i}}\right)^2\right|$$

$$-20r\log\omega - 20\sum_{i=1}^{n_1}\log\left|1 + j\omega\tau_{p_i}\right|$$

$$-\sum_{i=1}^{n_2}20\log\left|1 + j2\xi_{p_i}\frac{\omega}{\omega_{p_i}} + \left(j\frac{\omega}{\omega_{p_i}}\right)^2\right|$$

and

$$\phi(\omega) = \sum_{i=1}^{m_1}\tan^{-1}\omega\tau_{z_i} + \sum_{i=1}^{m_2}\tan^{-1}\left(\frac{2\xi_{z_i}\omega_{z_i}\omega}{\omega_{z_i}^2 - \omega^2}\right)$$

$$-r90° - \sum_{i=1}^{n_1}\tan^{-1}\omega\tau_{p_i} - \sum_{i=1}^{n_2}\tan^{-1}\left(2\frac{\xi_{p_i}\omega_{p_i}\omega}{\omega_{p_i}^2 - \omega^2}\right).$$

The factors appearing in the Bode magnitude plots (*i.e.*, the plot of 20 log | $H(j\omega)$ | vs log ω) are of the form:

$20\log K$

$20\log\omega$

$20\log|1 + j\omega\tau|$

$20\log\left|1 + j2\xi\frac{\omega}{\omega_n} - \frac{\omega^2}{\omega_n^2}\right|$

(It is customary to consider the first two factors together). The terms appearing in the phase functions are of the form.

$\tan^{-1}\omega\tau$

$\tan^{-1}\left(\frac{2\xi\omega_n\omega}{\omega_n^2 - \omega^2}\right)$

$r\,90°$

(*i*) *Factor* $\frac{K}{(j\omega)^r}$: For this factor we have

$$20\log\frac{K}{(j\omega)^r} = 20\log K - r20\log\omega. \quad ...(D.2)$$

Consider two frequencies, ω_1 and ω_2, differing by a decade, *i.e.*, $\omega_2 = 10\omega_1$. Then

$$20 \log \omega_2 = 20 \log (10\, \omega_1)$$
$$= 20 + 20 \log \omega_1$$

Hence, we conclude the term 20 log ω has a slope of 20 dB/decade or 6 dB/octave. The plot of eqn. D.2 is a straight line with a slope of –20 dB/decade. At $\omega = 1$, this line passes through the point 20 log K. This straight line crosses the 0 dB axis at $\omega = K^{1/r}$. The angle contribution of this factor, is –90° r for all frequencies:

(*ii*) *Factor* $|1 + j\omega\tau|$. For this factor we have

$$20 \log |1 + j\omega\tau| = 20 \log [1 + \omega^2\tau^2]^{1/2}$$
$$= 10 \log [1 + \omega^2\tau^2] \qquad \text{...(D.3)}$$

For values of ω such that $\omega\tau << 1$, the plot is the 0 db axis. For values of $\omega\tau >> 1$ the plot is a line with a slope of 20 dB/decade. The latter crosses the 0 dB axis at $\omega = 1/\tau$. The plot of eqn. D.3 can be represented by two straight lines: a straight line along 0 dB axis up to $\omega = 1/\tau$ and beyond this point a straight line with a slope of 20 dB/decade. This representation is called the *asymptotic approximation*. The frequency $1/\tau$ is called the *corner frequency*. The actual value at $\omega = 1/\tau$ is 10 log (1 + 1) ≈ 3 dB.

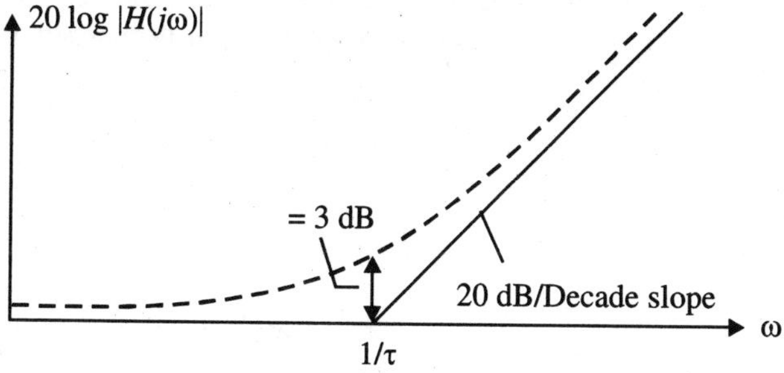

Fig. D.1

The plot is shown in Fig. D.1. If it is a pole factor, the plot will have a slope of –20 dB/decade and, at the corner frequency, its value would be –3 dB.

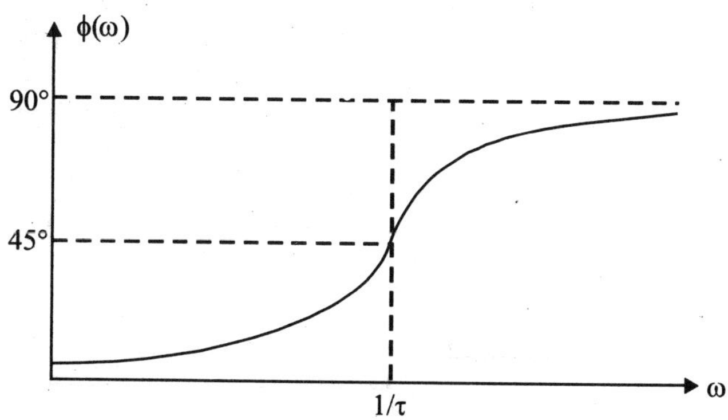

Fig. D.2

The phase corresponding to this factor is

$$\phi(\omega) = \tan^{-1}\omega\tau \qquad \text{...(D.4)}$$

and the plot is shown in Fig. D.2. For a pole factor the curve will be a mirror image of the one in Fig. D.2.; the phase varying from 0 to –90°

(*iii*) *Factor* $1 + j2\xi\frac{\omega}{\omega_n} - \frac{\omega^2}{\omega_n^2}$: For this factor, we have

$$20\log\left|1 + j2\xi\frac{\omega}{\omega_n^n} - \frac{\omega^2}{\omega_n^2}\right| = 20\log\left[(1-x^2)^2 + (2\xi x)^2\right]^{1/2}$$

$$= 10\log\left[(1-x^2)^2 + 4\xi^2x^2\right]$$

where $x = \omega/\omega_n$. For $x \le 1$ (*i.e.,* $\omega \le \omega_n$) the value is 0 dB and for $x >> 1$ (*i.e.,* $\omega >> \omega_n$) the value is $40 \log x$. Hence, the low and high frequency asymptotes are the 0 dB axis and a line with slope 40 dB/decade respectively. The corner frequency is at $\omega = \omega_n$ (*i.e.,* $x = 1$). The error between the actual Bode magnitude plot and its asymptotic approximation depends on the value of ξ.

If it is a pole factor, the high frequency asymptote is a straight line with a slope of – 40 dB/decade. Figure D.3 shows the plots for a pole factor.

The phase function for this factor is

$$\phi(\omega) = \tan^{-1}\left(\frac{2\xi x}{1-x^2}\right)$$

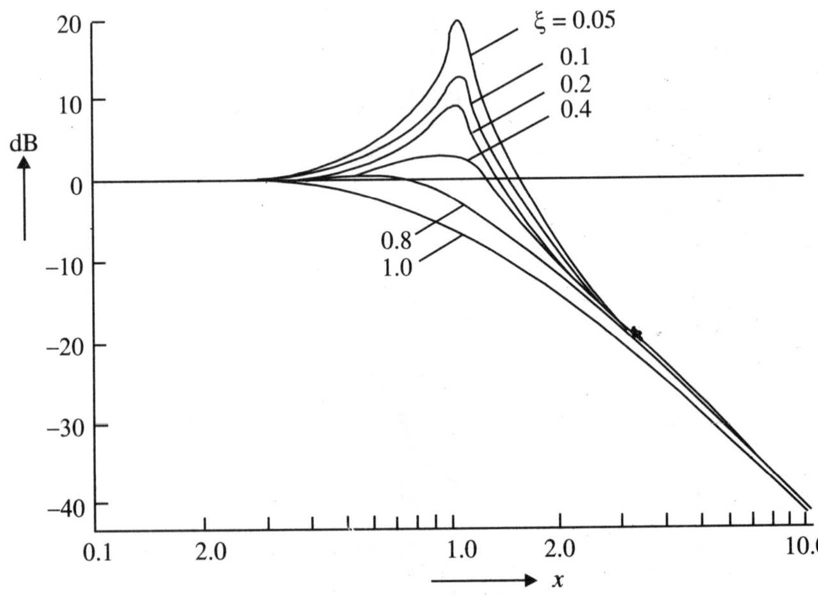

Fig. D.3

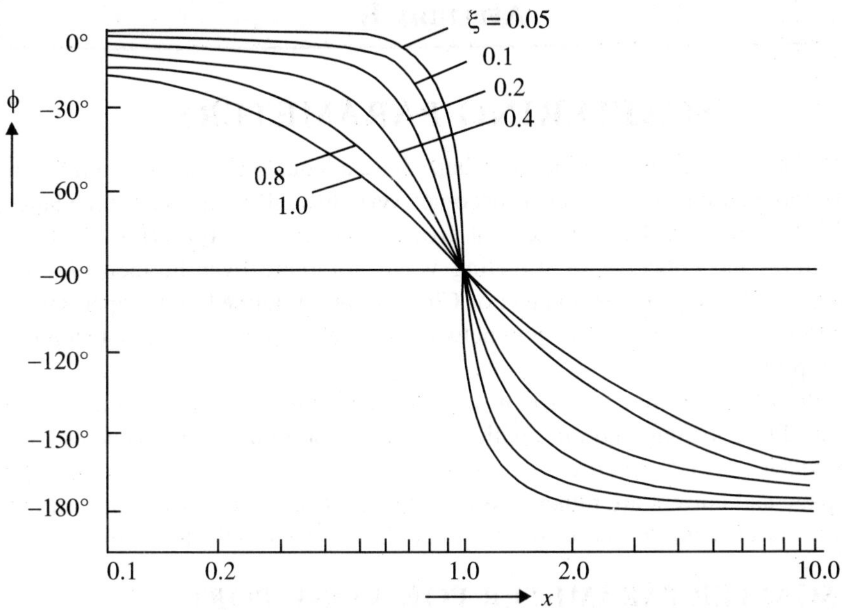

Fig. D.4

For $x << 1$, $\phi(\omega) = 0$ and $x << 1$, $\phi(\omega) = 180°$. If it is a pole factor, the latter is $-180°$. The phase plots for a pole factor are shown in Fig. D.4.

The overall Bode magnitude and phase plots are obtained by point-by-point summation of the individual plots of the various factors.

Most of the books on central systems discuss the frequency response through Bode plots in greater detail.

APPENDIX E

SCATTERING PARAMETERS

The two-port parameters defined in Chap. 7 are adequate to represent a network at low and medium frequencies. It becomes very difficult to measure impedance, admittance or hybrid parameters at frequencies above 100 MHz. At these frequencies the concepts of traveling wave dominate over lumped element techniques. Besides, the parameters of Chap. 7 are all defined under open circuit or shortcircuit condition. A two-port network generally operates with finite loading at both ports.

Scattering parameters defined in analogy with the transmission line theory consider finite loading and are applicable for a wide range of frequencies from low frequencies to the microwave ranges. At the high frequency end, the concepts of voltage and current are rather artificial and, as such, scatter parameter utilizes the concept of power in the form of the incident and the reflected waves.

E.1 SCATTER PARAMETER FOR A ONE-PORT

Let a one-port be driven by a source with an internal impedance Z_s (Fig. E.1).

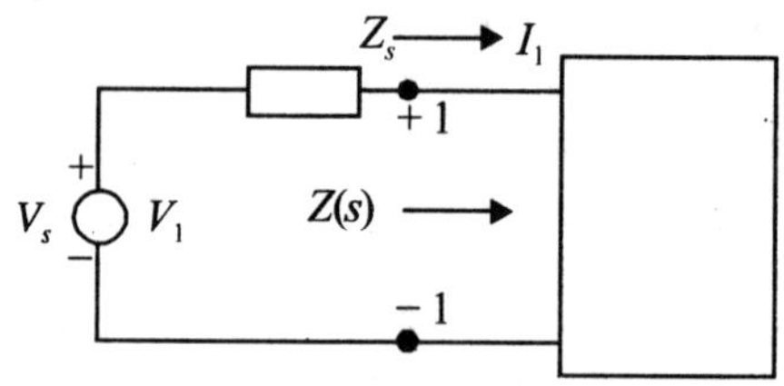

Fig. E.1

The network is matched to the source for maximum power transfer, when

$$Z(s) = Z_s(-s)$$

i.e., $Z(j\omega) = -Z_s(-j\omega)$, (see Sec. 6.6). If the source impedance is resistive, matching occurs when $Z = r_s$. Under matched condition we say that the network does not reflect any power. In general, we can write

$$V = V_i + V_r$$

$$I = I_i - I_r \qquad \text{...(E.1)}$$

where the subscript *i* refers to 'incident' and *r* to 'reflected' respectively. These two quantities are related by

$$\frac{V_i}{I_i} = \frac{V_r}{I_r} = r \qquad \text{...(E.2)}$$

where 'r' is called the characteristic impedance of the system. From (E.1) and (E.2) we have

$$\left.\begin{aligned} V_i &= \frac{1}{2}(V + rI), \quad I_i = \frac{1}{2}\left(\frac{1}{r}V + I\right) \\ V_r &= \frac{1}{2}(V - rI), \quad I_r = \frac{1}{2}\left(\frac{1}{r}V - I\right) \end{aligned}\right\} \qquad \text{...(E.3)}$$

The ratio of V_r to V_i is called the reflection coefficient ρ and is given by

$$\rho = \frac{V_r}{V_i} = \frac{I_r}{I_i} = \frac{Z - r}{Z + r} \qquad \text{...(E.4)}$$

Under matched condition $Z = r$, $\rho = 0$ and the network absorbs all the power from the source. If the source impedance is complex, the voltage reflection coefficient ($= V_r/V_i$) and current reflection coefficient ($= I_r/I_i$) are different.

In general, we can define

$$V_i = a_{11}V + a_{21}I, \; V_r = a_{21}V + a_{22}I$$

Under matched condition with a resistive source impedance, *i.e.*, $Z_s = Z(s) = r$, we have $V = rI$.

This leads to

$$V_i = V, \; V_r = 0, \; I_i = I, \; I_r = 0, \;\; \rho = 0$$

When the network is not matched to the source V_r and I_r represent the deviation of V and I from their value under matched condition.

i.e.,

$$V_r = V - V_i$$

$$I_r = I_i - I$$

which is the same as (E.1).

The asymmetry in the definition of the deviation can be justified by observing $I = \dfrac{V_s - V}{r}$ and $V_i = \dfrac{V_s}{2}$ under matched condition.

Hence,

$$I = \frac{2V_i - V}{r} = \left[\frac{V_i + (V_i - V)}{r}\right] = \frac{V_i}{r} - \frac{V_r}{r} = (I_i - I_r).$$

The normalized incident and reflected waves are defined as

$$\begin{aligned} a(s) &= \sqrt{r}\, I_i = V_i/\sqrt{r} \\ b(s) &= \sqrt{r}\, I_r = V_r/\sqrt{r} \end{aligned} \qquad \text{...(E.5)}$$

The normalized variables $a(s)$ and $b(s)$ are called 'scatter variables.' Scatter variables can also be written in terms of V and I. From eqns. (E.5) and (E.3)

$$a(s) = \frac{1}{\sqrt{r}}\left[\frac{1}{2}(V + rI)\right] = \frac{1}{2}[V_n + I_n]$$
$$b(s) = \frac{1}{\sqrt{r}}\left[\frac{1}{2}(V - rI)\right] = \frac{1}{2}[V_n - I_n] \qquad \ldots(\text{E.6})$$

where the normalized voltage and current variables are

$$V_n = \frac{V}{\sqrt{r}},\ I_n = \sqrt{r}I$$

The two scatter variables are related by the reflection coefficient. From (E.6) we have

$$\frac{b}{a} = \frac{V - rI}{V + rI} = \frac{IZ - rI}{IZ + rI} \qquad (V = IZ)$$

$$= \frac{Z - r}{Z + r} = \frac{Z_n - 1}{Z_n + 1} \qquad (Z_n \text{ is the normalized input impedance})$$

From (E.4), we have

$$b = Sa \qquad \ldots(\text{E.7})$$

where the scatter parameter, S, of the one-port network is the reflection coefficient (observe that normalization does not change the reflection coefficient). To physically interpret the scatter variables, consider power transfer through the network. The maximum power available from the source is given by

$$P_{\max} = \frac{|V_s|^2}{4r} = \frac{|2V_i|^2}{4r} = \left|\frac{V_i}{\sqrt{r}}\right|^2 = |a|^2$$

If the network is matched to the source, this maximum power is absorbed by the network and hence we can say that the reflected power (P_r) is zero. Under mismatch of the network and the source, the power absorbed by the network is

$$P = P_i - P_r$$

where P_i (= $P_{\max}$) is the incident power and P_r is the reflected power.

$$P_r^2 = \frac{V_r^2}{r} = |b|^2$$

Now, we can write the power balance equation as

$$\frac{P}{P_i} = 1 - \frac{P_r}{P_i} = 1 - |\rho(j\omega)|^2 = 1 - |S|^2 \qquad \ldots(\text{E.8})$$

For a passive one-port network $P/P_i \le 1$, hence $|\ \rho(j\omega)\ | \le 1$.

From the preceding argument it is evident that the power absorbed by the network is

$$P = |a|^2 - |b|^2 \qquad \ldots(\text{E.9})$$

and that a and b have the dimension of $\sqrt{\text{power}}$, say $\sqrt{\text{watt}}$.

Example. Consider the network shown in Fig. E.2

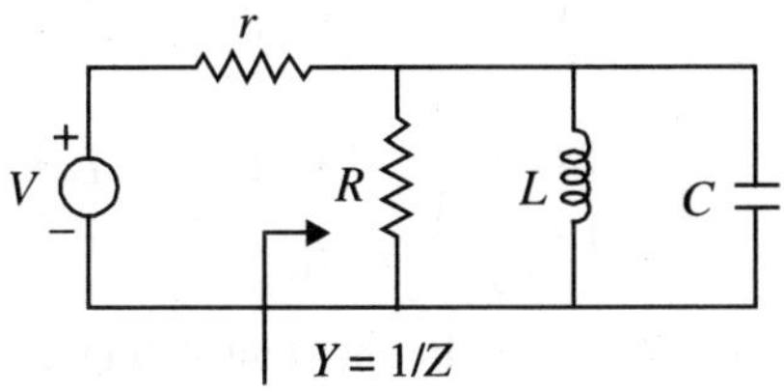

Fig. E.2

For this network

$$Z = \frac{j\omega L}{(1-\omega^2 LC) + j\omega LG}$$

and from the definition of scatter parameter

$$S = \frac{-r(1-\omega^2 LC) + j\omega L(1-rG)}{r(1-\omega^2 LC) + j\omega L(1+rG)}$$

From this equation it can be seen that $|S| \le 1$ and if $G \to 0$ *i.e.*, the network is an *LC* network, then

$$S = \frac{-r(1-\omega^2 LC) + j\omega L}{r(1-\omega^2 LC) + j\omega L}$$

and $|S| = 1$ for all ω.

The scatter parameter can also be related to the input admittance instead of the input impedance.

$$S = \rho = \frac{Z-r}{Z+r} = 1 - \frac{2r}{Z+r} = 1 - 2rY_a$$

where Y_a is the admittance of the one-port including the source resistance r.

E.2 SCATTERING MATRIX FOR A TWO-PORT

Figure E.3 shows the general configuration of the two-port network to be considered here.

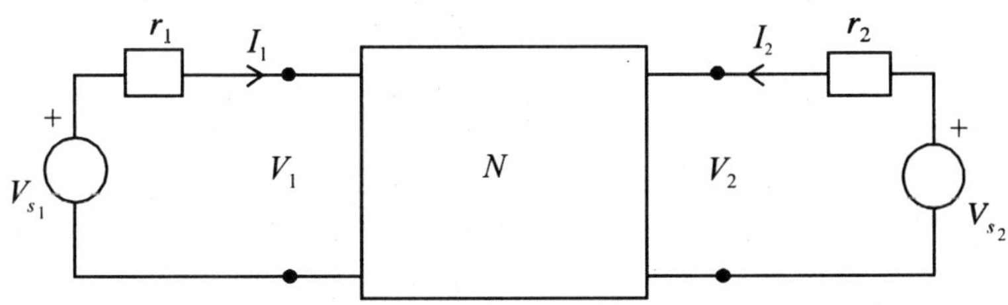

Fig. E.3

In analogy with one-port network description we define

$$\mathbf{V}_r = \mathbf{V} - \mathbf{V}_i \text{ and } \mathbf{I}_r = \mathbf{I}_i - \mathbf{I}$$

where
$$\mathbf{V}_r = \begin{bmatrix} V_{r_1} \\ V_{r_2} \end{bmatrix}, \mathbf{V} = \begin{bmatrix} V_{i_1} \\ V_{i_2} \end{bmatrix}, \mathbf{V} = \begin{bmatrix} V_1 \\ V_2 \end{bmatrix}$$

$$\mathbf{I}_r = \begin{bmatrix} I_{r_1} \\ I_{r_2} \end{bmatrix}, \mathbf{I}_i = \begin{bmatrix} I_{i_1} \\ I_{i_2} \end{bmatrix}, \mathbf{I} = \begin{bmatrix} I_1 \\ I_2 \end{bmatrix}$$

and V_{ik}, V_{rk}, I_{ik} and I_{rk} ($k = 1, 2$) are the incident voltage, the reflected voltage, the incident current and the reflected current waves respectively at the two ports. Under matched conditions (*i.e.*, at each port the network is matched to the source resistance at that port)

$$\mathbf{V}_i = V - R\mathbf{I} = R\mathbf{I}_i = \frac{1}{2}V_s$$

where
$$R = \begin{bmatrix} r_1 & 0 \\ 0 & r_2 \end{bmatrix} \text{ and } \mathbf{V}_s = \begin{bmatrix} V_{s_1} \\ V_{s_2} \end{bmatrix}$$

The incident and the reflected voltage and current waves can be written as

$$\mathbf{V}_i = \frac{1}{2}[\mathbf{V} + R\mathbf{I}], \mathbf{V}_r = \frac{1}{2}[\mathbf{V} - R\mathbf{I}]$$

$$\mathbf{I}_i = \frac{1}{2}R^{-1}[\mathbf{V} + R\mathbf{I}], \mathbf{I}_r = \frac{1}{2}R^{-1}[\mathbf{V} - R\mathbf{I}] \qquad \text{...(E.10)}$$

(compare these with E.3)

As in the case of one-port network, the normalized incident and the reflected waves are defined as

$$\mathbf{a} = R^{1/2}\,\mathbf{I}_i = R^{-1/2}\,\mathbf{V}_i$$
$$\mathbf{b} = R^{1/2}\,\mathbf{I}_r = R^{-1/2}\,\mathbf{V}_r$$

where
$$\mathbf{a} = \begin{bmatrix} a_1 \\ a_2 \end{bmatrix}, \mathbf{b} = \begin{bmatrix} b_1 \\ b_2 \end{bmatrix}.$$

Hence,
$$a_1 = \frac{1}{2\sqrt{r_1}}[V_1 + r_1 I_1], b_1 = \frac{1}{2\sqrt{r_1}}[V_1 - r_1 I_1]$$
$$a_2 = \frac{1}{2\sqrt{r_1}}[V_2 + r_1 I_2], b_2 = \frac{1}{2\sqrt{r_2}}[V_2 - r_2 I_2] \qquad \text{...(E.11)}$$

If we define the normalized voltage and the normalized current vectors as

$$\mathbf{V}_n = R^{-1/2}\mathbf{V}, \mathbf{I}_n = R^{1/2}\mathbf{I}$$

we have
$$\mathbf{V}_n = \mathbf{a} + \mathbf{b} \text{ and } \mathbf{I}_n = \mathbf{a} - \mathbf{b}$$

The scattering matrix of a two-port is defined by

$$\boldsymbol{b} = S\,\boldsymbol{a}$$

i.e.,
$$\begin{bmatrix} b_1 \\ b_2 \end{bmatrix} = \begin{bmatrix} S_{11} & S_{12} \\ S_{21} & S_{22} \end{bmatrix} \begin{bmatrix} a_1 \\ a_2 \end{bmatrix} \quad \text{...(E.12)}$$

where S is the scattering matrix. The scatter parameters are defined as

$$S_{11} = \left.\frac{b_1}{a_1}\right|_{a_2=0}, \quad S_{12} = \left.\frac{b_1}{a_2}\right|_{a=0}$$
$$S_{21} = \left.\frac{b_2}{a_1}\right|_{a_2=0}, \quad S_{22} = \left.\frac{b_2}{a_2}\right|_{a_1=0} \quad \text{...(E.13)}$$

The four scatter parameters can be related to the reflection coefficients at port 1 and port 2 and the transmission coefficient of the network.

(*i*) S_{11} is defined under the condition that the incident wave at port 2 is zero. From (E.10) we have $V_2 = -r_2I_2$ and there is no voltage source at port 2. In this case, port 2 is matched. From (E.13) and (E.11), we have

$$S_{11} = \left.\frac{b_1}{a_1}\right|_{a_2=0} = \frac{V_1 - rI_1}{V_1 + rI_1} = \frac{Z_1 - r_1}{Z_1 + r_1} = \rho_1$$

i.e., S_{11} is the reflection coefficient at port 1, with port 2 matched.

(*ii*) Similar to (*i*) it can be seen that

$$S_{22} = \frac{Z_2 - r_2}{Z_2 + r_2} = \rho_2$$

i.e., S_{22} is the reflection coefficient at port 2 with port 1 matched.

(*iii*) $S_{21} = \left.\frac{b_2}{a_1}\right|_{a_2=0}$

With $a_2 = 0$, (E.10) yields $V_2 = -r_2 I_2$

and $b_2 = \frac{1}{2\sqrt{r_2}}[2V_2] = \frac{V_2}{\sqrt{r_2}}$

(see eqns. 11.13 and 11.16)

(*iv*) Similar to (*iii*) it can be shown that

$$S_{21} = 2\sqrt{\frac{r_2}{r_1}\frac{V_1}{V_{s_2}}} \quad \text{...(E.14)}$$

As in the case of a one-port network, the scatter parameters can be related to the power transfer through the network. From (E.14) we have

$$|S_{21}(j\omega)|^2 = 4\frac{r_1}{r_2}\left|\frac{r_2(j\omega)}{V_{s_1}(j\omega)}\right|^2$$

and hence from eqn. 12.13 we see that

$$|S_{21}(j\omega)|^2 = \frac{P_2}{P_{2\max}}$$

From (E.9) we have the power absorbed by the network at port i as

$$P_i = |a_i|^2 - |b_i|^2$$

Hence, the total power absorbed by the two-port network is

$$P_T = P_1 + P_2 = |a_1|^2 + |a_2|^2 - |b_1|^2 - |b_2|^2$$

or $$P_T = a^H a - b^H b$$

where a^H and b^H are the Hermitian transposes of a and b respectively.

In view of (E.12), this can be written as

$$P_T = a^H (u - S^H S) a$$

which relates power to the scattering matrix.

For a lossless network P_T is zero and consequently

$$U - S^H S = 0$$

i.e., $$S^H S = U$$

This yields $$|S_{11}|^2 + |S_{21}|^2 = 1$$

$$|S_{12}|^2 + |S_{22}|^2 = 1$$

(If the network is reciprocal, observe that $S_{12} = S_{21}$). For a passive network P_T being non-negative, the matrix $(U - S^H S)$ must be positive—semi-definite.

The scattering matrix S can be related to the impedance and admittance matrices of the two-port. It can be easily shown that

$$S = U - 2Y_{an}$$

$$= (Z_n - U)(Z_n + U)^{-1}$$

$$= (U - Y_n)(U + Y_n)^{-1}$$

where $$Y_{an} = R^{1/2} Y_a R^{1/2},\ Z_n = R^{-1/2} Z_{oc} R^{-1/2}$$

$$Y_n = R^{1/2}\ Y_{sc} R^{1/2}$$

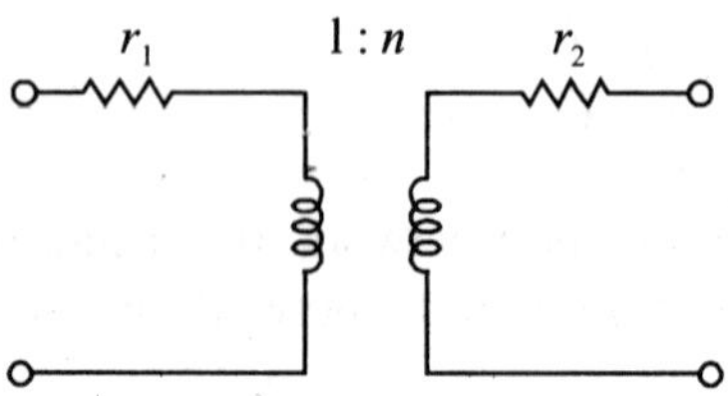

Fig. E.4

and Y_a is the admittance matrix of the two-port including the two source resistances. Z_{oc} and Y_{sc} are the open circuit impedance and the short-circuit admittance matrices respectively of the two-port as defined in Chap. 7.

For example, for the ideal transformer of turns ratio $1:n$

$$Y_a = \frac{1}{r_2 + n^2 r_1}\begin{bmatrix} n^2 & -n \\ -n & 1 \end{bmatrix}, \; Y_{an} = \frac{1}{r_2 + n^2 r_1}\begin{bmatrix} n^2 r_1 & -n\sqrt{r_1 r_2} \\ -n\sqrt{r_1 r_2} & r_2 \end{bmatrix}$$

and finally,

$$S = \frac{1}{r_2 + n^2 r_1}\begin{bmatrix} r_2 - n^2 r_1 & 2n\sqrt{r_1 r_2} \\ 2n\sqrt{r_1 r_2} & n^2 r_1 - r_2 \end{bmatrix}$$

It is worth noting that an ideal transformer has neither impedance nor admittance characterization.

A detailed description of the scattering parameters and their application is available in *Theory of Linear Active Networks* by E.S. Kuh and R.A. Rohrer, Holden-Day, San Francisco, 1967.

IMAGE PARAMETER FILTER DESIGN

F.1 IMAGE PARAMETERS

In the classical filter design the filter networks used are *symmetric T* or π networks. Consider the symmetric T network shown in Fig. F. 1. Let the two-port be terminated in a load Z_L. Then the input impedance Z_{in} of the network is

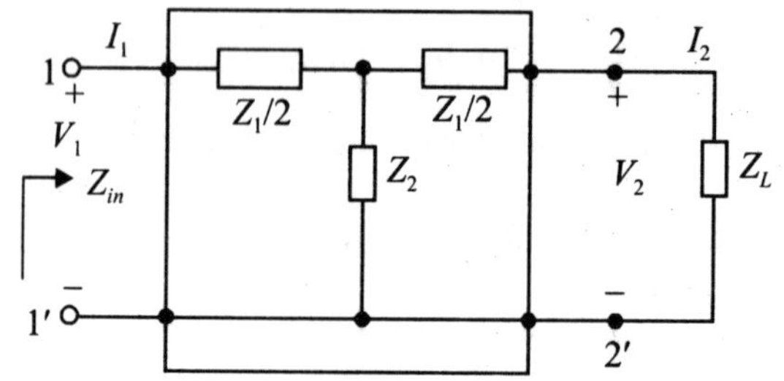

Fig. F.1

$$Z_{in} = \frac{Z_1}{2} + \frac{Z_2\left(\frac{Z_1}{2} + Z_L\right)}{Z_2 + \frac{Z_1}{2} + Z_L}$$

i.e.,

$$Z_{in} = \frac{Z_1}{2} + \frac{Z_2(Z_1 + 2Z_L)}{Z_1 + 2(Z_2 + Z_1)} \quad \text{...(F.1)}$$

The *image* (*iterative, characteristic*) *impedance* Z_I of the network is defined as the load or terminating impedance chosen so that

$$Z_{in} = Z_L$$

Substituting $Z_{in} = Z_L = Z_I$ in eqn. F. 1 and solving for Z_I, we have

$$Z_I = \sqrt{Z_1Z_2 + \frac{Z_1^2}{4}}$$

or

$$Z_I = \sqrt{Z_1Z_2\left(1 + \frac{Z_1}{4Z_2}\right)} \quad \text{...(F.2)}$$

If we consider π-network as shown in Fig. F.2 instead we have

$$Z_{in} = \frac{2Z_2\left(Z_1 + \frac{2Z_2Z_L}{2Z_2 + Z_L}\right)}{2Z_2 + Z_1 + \frac{2Z_2Z}{2Z_2 + Z_L}}$$

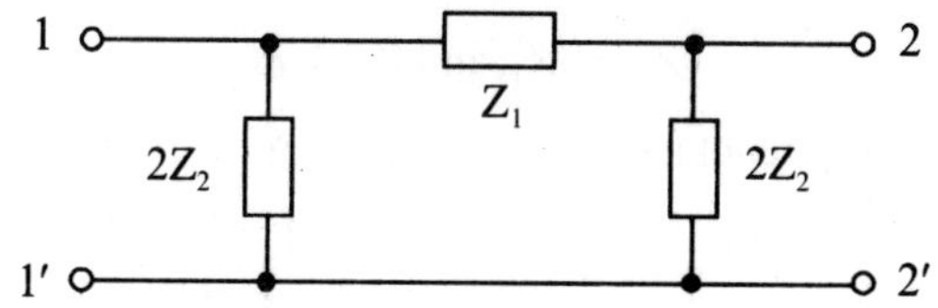

Fig. F.2

The image impedance then is

$$Z_I = \sqrt{\frac{Z_1 Z_2}{1+\frac{Z_1}{4Z_2}}} \qquad \text{...(F.3)}$$

The image impedances, in general, are irrational functions of *s*. The image impedances can be obtained by measuring the open-circuit and short-circuit impedances of the network.

From Fig. F.1 the open-circuit impedance (*i.e.*, the impedance Z_{in} with terminals 22′ open) Z_{OC} is

$$Z_{OC} = Z_2 + \frac{Z_1}{2} \qquad \text{...(F.4)}$$

and the short-circuit admittance (*i.e.*, the admittance Y_{in} with terminals 22′ short-circuited) Y_{SC} is

$$Y_{SC} = \frac{1}{\frac{Z_1}{2} + \frac{Z_2 Z_1/2}{Z_1 + \frac{Z_1}{2}}} \qquad \text{...(F.5)}$$

From eqns. F.2, F.4 and F.5, we see that

$$Z_I = \sqrt{\left(\frac{Z_{OC}}{Y_{SC}}\right)} \qquad \text{...(F.6)}$$

From the symmetric π network

$$Z_{OC} = \frac{1}{1/2z_2 + 1/(Z_1 + 2Z_2)}$$

$$Y_{SC} = \frac{1}{2Z_2} + \frac{1}{Z_1}$$

from which it can be shown that eqn. F.6 once again holds good.

F.2 PASS AND STOP BANDS

In general the *coupling network* (the symmetric *T* or π) is a reactive network (LC network) and the termination is a resistance. In this case, let

$$Z_1 = jX_1 \text{ and } Z_2 = jX_2$$

where the signs of the reactances are absorbed in X_1 and X_2.

Then

$$Z_1Z_2 = -X_1X_2, \frac{Z_1^2}{4} = \frac{-X_1^2}{4} \qquad \text{...(F.7)}$$

The filter pass and stop bands are obtained when the image impedance is purely resistive or reactive respectively. When Z_I is purely resistive, the network accepts power from the source and delivers the same power to load. This is only possible because the coupling network is reactive and hence lossless. Thus, purely resistive Z_I indicates pass band condition. In this case, there is *image matching*. When Z_I is purely reactive, no power is absorbed from the source and hence no power is delivered to the load. Purely reactive Z_I indicates stop band conditions. There is no image matching in this case.

From eqn. F.2 and F.7 we have

$$Z_I = \sqrt{-X_1X_2\left(1+\frac{X_1}{4X_2}\right)}$$

Two possibilities exist, depending on the reactance types X_1 and X_2.

(*i*) X_1 and X_2 are of opposite reactance types: For this case $-X_1X_2$ is positive and Z_I depends on $1 + (X_1/4X_2)$. If $1 + (X_1/4X_2)$ is positive, Z_I is purely real and if $1 + (X_1/4X_2)$ is negative, Z_I is purely imaginary. X_1 and X_2 being of the opposite reactance types $X_1/4X_2$ is negative. Hence, pass band (Z_I resistive) exists for those frequencies which satisfy the equation

$$-1 < \frac{X_1}{4X_2} < 0 \qquad \text{...(F.8)}$$

and stop band (reactive Z_I) exists for those frequencies which satisfy

$$\frac{X_1}{4X_2} < -1 \qquad \text{...(F.9)}$$

(*ii*) X_1 and X_2 are of the same reactance type: Here X_1/X_2 is positive and $-X_1X_2$ negative making Z_I always purely imaginary. Hence, there is no possibility of a pass band. (The network acts like a potential divider).

From the foregoing discussion we can conclude that X_1 and X_2 must be of the opposite reactance types for pass band and stop band to exist (bounds on pass band and stop band frequencies are given by eqn. F.8 and F.9 respectively).

From Fig. F.1

$$V_1 = Z_I\mathbf{I}_1, V_2 = -V_1I_2$$

i.e.,

$$\frac{V_1}{V_2} = \frac{-I_2}{I_1} = e^{\tau}$$

where τ the propagation constant is

$$\tau = a + j\beta.$$

Here α is the attenuation constant and β is the phase constant. From simple analysis it can be shown that

$$e^{\tau} = 1 + \frac{Z_1}{2Z_2} + \sqrt{\frac{Z_1}{Z_2} + \left(\frac{Z_1}{2Z_2}\right)^2}$$

i.e.,

$$\tau = \ln\left\{1 + \frac{Z_1}{2Z_2} + \sqrt{\frac{Z_1}{Z_2} + \left(\frac{Z_1}{2Z_2}\right)^2}\right\} \qquad \ldots(F.10)$$

Filter performance of different filter structures can be compared on the basis of their propagation constants.

F.3 CONSTANT *k* FILTERS

From eqn. F.8 we conclude that for a pass band to exist X_1 and X_2 must be opposite type reactances. When X_1 and X_2 are combinations of inductors and capacitors, the overall reactance must be of the opposite type to generate pass band frequencies. The image parameter filters depend on the product $Z_1Z_2 = k^2$ and hence such filters are called *constant k* filters.

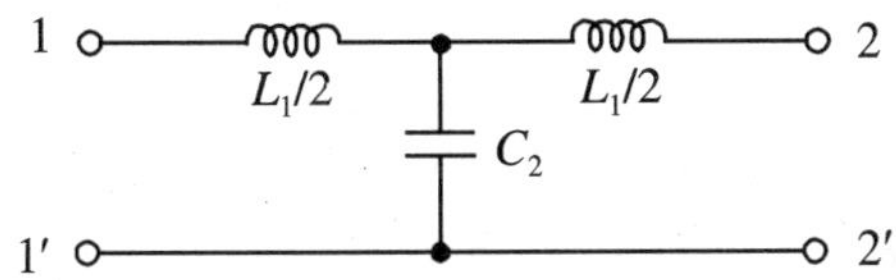

Fig. F.3

Consider the *T* network shown in Fig. F.3. From physical consideration we can conclude that it is a low-pass network. For this network $Z_1 = sL_1$ and $Z_2 = \frac{1}{sC_2}$ and $Z_1Z_2 = \frac{L_1}{C_2}$. From eqn. F.8 pass-band exist for those frequencies which satisfy the relation

$$-1 < \frac{-s^2 L_1 C_2}{4} < 0$$

i.e., pass-band extends from $\omega = 0$ to $\omega = \frac{2}{\sqrt{L_1C_2}}$. Hence, the filter is a *low-pass filter.* The *cut-off frequency* of the filter is

$$f_C = \frac{1}{\pi\sqrt{L_1C_2}} \qquad \ldots(F.11)$$

The ideal filter response is shown in Fig. F.4. After some simple algebra, eqn. F.10 yields

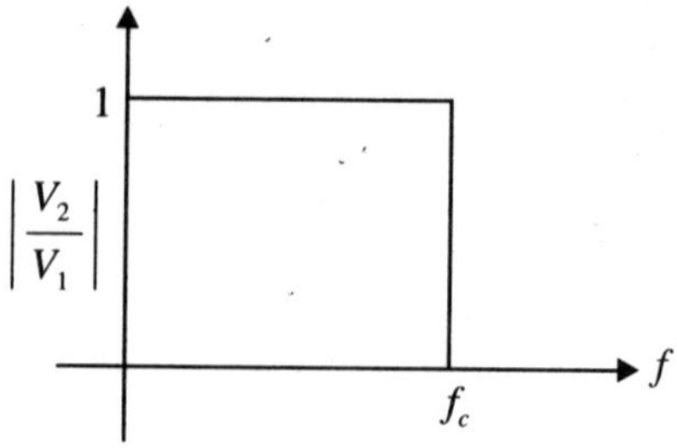

Fig. F.4

$$e^{\tau} = 1 - 2\frac{f^2}{f_c^2} + 2\frac{f}{f_c}\sqrt{\frac{f^2}{f_c^2} - 1} \qquad \text{...(F. 12)}$$

For pass-band frequencies, *i.e.*, $f < f_c$, eqn. F. 12 yields

$$e^{\tau} = 1 - 2\frac{f^2}{f_c^2} + j2\frac{f}{f_c}\sqrt{1 - \frac{f^2}{f_c^2}}$$

and the attenuation is

$$\alpha = \ln\left|e^{\tau}\right| = \ln\left[\left(1 - 2\frac{f^2}{f_c^2}\right) + 4\frac{f^2}{f_c^2}\left(1 - \frac{f^2}{f_c^2}\right)\right]^{1/2} = 0$$

indicating that attenuation in the pass-band is zero. The phase β is given by

$$\beta = \cos^{-1}\frac{A}{\sqrt{A^2 + B^2}}$$

where A and B are the real and imaginary parts of eqn. F. 12. On simplifying we have

$$\beta = \cos^{-1}\left[1 - \frac{2f^2}{f_c^2}\right]$$

Hence, the phase changes from 0 at $f = 0$ to π radians at $f = f_c$.

For stop-band frequencies, *i.e.*, $f > f_c$, eqn. F. 12 yields

$$\alpha = \ln\left[1 - \frac{2f^2}{f_c^2} + \frac{2f}{f_c}\left(\frac{f^2}{f_c^2} - 1\right)\right]^{1/2}$$

and $\beta = \pi$ radians. The attenuation increases with frequency.

The image impedance of the filter, from eqn. F. 2. is

$$Z_I = \sqrt{\frac{L_1}{C_2}\left(1 - \frac{f^2}{f_c^2}\right)} \qquad \text{...(F. 13)}$$

from which it is observed that Z_I is not constant over the entire pass band and, as such, there is image matching at only one frequency $f = 0$. This is one of the major disadvantages of the constant k filter. The design equations for the filter can be easily obtained from eqns. F.11 and F.13. The termination resistance R_L is the same as Z_I at $f = 0$, *i.e.*,

$$R_L = \sqrt{\frac{L_1}{C_2}} \quad \text{...(F.14}a\text{)}$$

But $$f_c = \frac{1}{\pi\sqrt{L_1C_2}} = \frac{1}{\pi R_L C_2}$$

Hence, $$C_2 = \frac{1}{\pi f_c R_L} \quad \text{...(F.14}b\text{)}$$

and $$L_1 = \frac{R_L}{\pi f_c} \quad \text{...(F.14}c\text{)}$$

If we *normalize* (see Sec. 4.9)

$$\omega_c = 1 \text{ rad/sec}$$

and $$R_L = 1\ \Omega$$

eqn. F.14 yields

$$L_1 = 2H,\ C_2 = 2F$$

We can *denormalize* this normalized filter to any desired cut-off frequency and termination. Here, our discussion is limited to T networks. One can easily obtain the design equations for a π network.

Example F.1. We are required to realize a constant k, low-pass filter to cut-off at 2 KHz with a terminating resistance of 600 Ω.

Let γ_n and Ω_n be the impedance and the frequency scaling factors, respectively. Then

$$\gamma_n = \frac{600}{1} = 600$$

$$\Omega_n = \frac{2\pi(2\times 10^3)}{1} = 4\pi\times 10^3$$

The low-pass prototype (cut-off at 1 rad/sec and operating into a load resistance of 1 Ω) has $L_1 = 2H$, $C_2 = 2F$. The parameters of the required filter are

$$R_L = 600$$

$$L = \frac{\gamma_n L_1}{\Omega_n} = \frac{600\times 2}{4\pi\times 10^3} = 95.493 \text{ mH}$$

and $$C = \frac{C_1}{\gamma_n \Omega_n} = \frac{2}{600 \times 4\pi \times 10^3} = 265 \text{ nf}$$

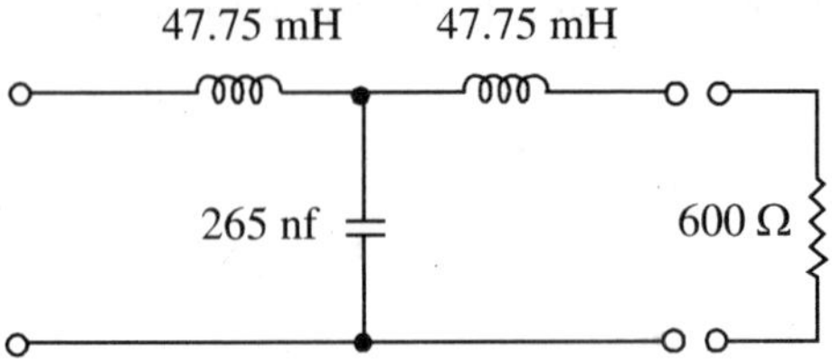

Fig. F.5

The final filter is shown in Fig. F.5.

F.4 *m*-DERIVED FILTERS

The image impedance of the low-pass constant *k* filter is given by eqn. F.13. The load resistance of the filter is

$$R_L = \sqrt{\frac{L_1}{C_2}}$$

Hence, we have

$$\frac{Z_1}{R_L} = \sqrt{1 - \frac{f^2}{f_c^2}} \qquad \text{...(F.15)}$$

Image matching exists when this ratio is 1, *i.e.*, $Z_I = R_L$. From eqn. F.15 it is obvious that matching exists only at $f = 0$. As f increases Z_I decreases and at $f = f_c$, $Z_I = 0$. Hence, the image impedance does not approximate a constant value very well over the pass band as is required by the design.

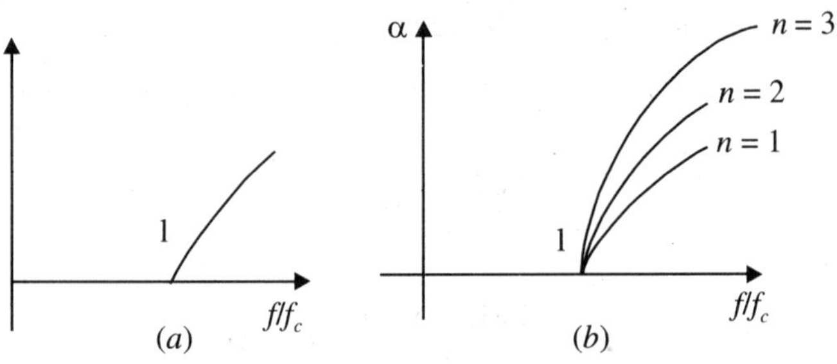

Fig. F.6

Figure F.6(*a*) shows the attenuation of a constant *k* filter. It can be seen that the attenuation does not increase rapidly beyond the cut-off point. This can be improved by cascading a larger number of sections (Fig. F.6(*b*)). This procedure has the disadvantage of increasing the cost of filters (a large number of components). The sharpness of cut-off is not very much improved. Besides, the image impedance is still not constant over the pass band.

In order to alleviate the problems of a constant k filter we have to modify the network, so that the image matching occurs over a larger band of frequencies. The modification proposed by Zobel is to scale the series arm impedances of constant k T-section and select the shunt arm impedance so that the modified and the original prototype have the same image impedance. The latter allows us to cascade the prototype and the modified structure. In the derivation of the modified structure a new parameter m is introduced and the resulting filter is called an *m-derived* filter.

Let $Z_{m_1}(Z_1)$ and $Z_{m_2}(Z_2)$ be the series and shunt arm impedances respectively of the m-derived (prototype) section. The series arm impedance is chosen as

$$Z_{m_1} = mZ_1 \qquad \text{...(F.16)}$$

In order that the two sections have the same image impedance (eqn. F.2)

$$Z_1Z_2 + \frac{Z_1^2}{4} = Z_{m_1}Z_{m_2} + \frac{Z_{m_1}^2}{4}$$

i.e.,

$$Z_{m_2} = \frac{1}{Z_{m_1}}\left\{Z_1Z_2 + \frac{Z_1^2}{4} - \frac{Z_{m_1}^2}{4}\right\}$$

Hence, from eqn. F.16 we have

$$Z_{m_2} = \frac{Z_2}{m} + \frac{1-m^2}{4m}Z_1 \qquad \text{...(F.17)}$$

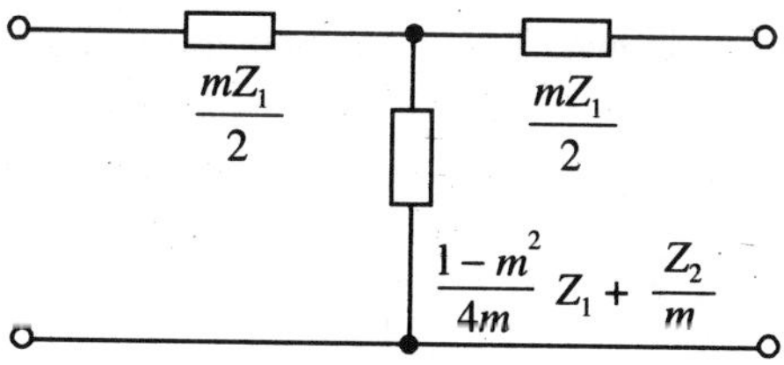

Fig. F.7

The m-derived section is shown in Fig. F.7. The one corresponding to the low-pass prototype of Fig. F.3 is shown in Fig. F.8. In order that all the elements be finite and positive, m must satisfy the relation

$$0 < m < 1$$

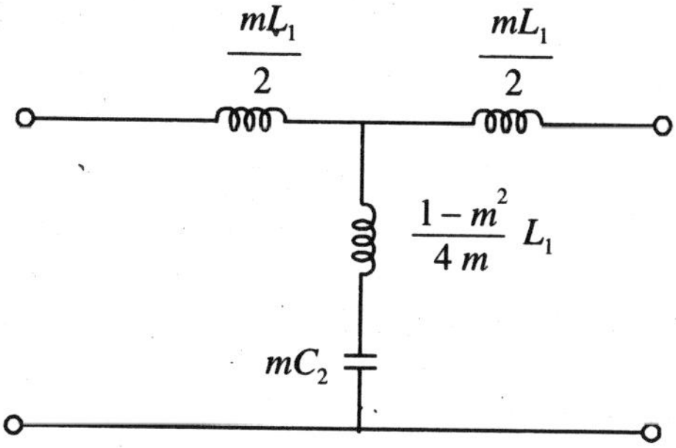

Fig. F.8

The m-derived section has a resonant shunt arm. At the resonant frequency of this arm the transmission from the input to the output is prevented and thus we have created a finite transmission zero (the constant k has a transmission zero at infinite frequency). The resonant frequency of the shunt arm is

$$\omega_r = \frac{1}{\sqrt{\frac{(1-m^2)}{4m} L_1 (m_c^2)}}$$

i.e.,

$$\omega_r = \omega_c \Big/ \sqrt{1-m^2}$$

If the finite transmission zero is specified, the value of m is

$$m = \sqrt{1 - \frac{\omega_c^2}{\omega_r^2}} \qquad \text{...(F. 18)}$$

It can be observed from eqn. F. 16 and F. 17 that

$$Z_{m_1} Z_{m_2} \neq k^2$$

and thus m-derived section is not of constant k type. The m derived section is identical to the constant k type when $m = 1$.

The m-derived filter by itself is of no direct use in filter design. A cascade combination of m-derived and constant k sections is the final form of the filter generally used in the design. Such a filter is said to be of the *composite type.* Consider the half section m-derived filter, as shown in Fig. F. 9. Let Z_{I_1} and Z_{I_2} be the image impedances, as shown. These impedances can be calculated by using eqn. F. 6.

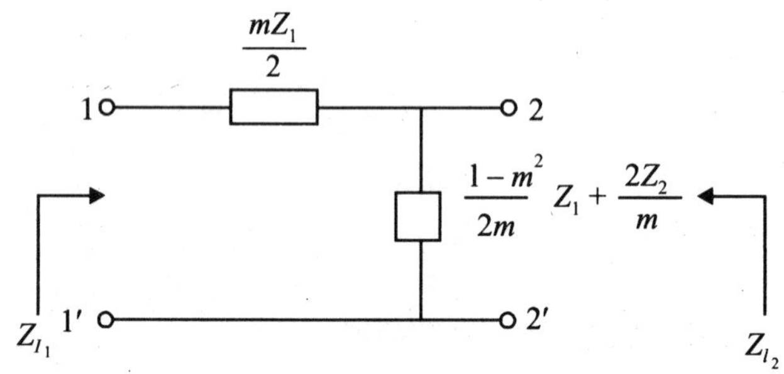

Fig. F.9

$$Z_{I_1} = \sqrt{\frac{Z_{oc_1}}{y_{sc_1}}} = \frac{\frac{mZ_1}{2} + \frac{1-m^2}{2m} Z_1 + \frac{2Z_2}{m}}{2/mZ_1}$$

$$= \sqrt{Z_1 Z_2 + \left(Z_1^2/4\right)}$$

Hence, $Z_{I_1} = Z_I$ of the prototype.

$$Z_{I_2} = \sqrt{\frac{Z_{oc_2}}{Y_{sc_2}}}$$

$$= \sqrt{\frac{\{(1-m^2/2m)Z_1 + (2Z_2/m)\}\{(1-m^2/2m)Z_1 + 2Z_2/m)\}mZ_1/2}{(Z_1/2m)+(2Z_2/m)}}$$

$$= \frac{(1-m^2/2)Z_1 + 2Z_2}{\sqrt{1+(4Z_2/Z_1)}} \qquad \text{...(F. 19)}$$

For the low-pass section $Z_1 = j\omega L_1$ and $Z_2 = \dfrac{1}{j\omega C_2}$ and eqn. F.1.

$$Z_{I_2} = \frac{\dfrac{1-m^2}{2} j\omega L_1 + \dfrac{2}{j\omega C_2}}{\sqrt{1+4\dfrac{1}{j\omega C_2}\dfrac{1}{j\omega L_1}}} = \frac{\dfrac{2}{j\omega C_2}\left\{1-\dfrac{1-m^2}{4}\omega^2 L_1 C_2\right\}}{\dfrac{2}{\omega\sqrt{L_1 C_2}}\sqrt{\dfrac{\omega^2 L_1 C_2}{4}-1}}$$

i.e., $$Z_{I_2} = \sqrt{\frac{L_1}{C_2}}\frac{\{1-(1-m^2)2(L_1C_2/4)\}}{\sqrt{1-\omega^2\dfrac{L_1C_2}{4}}}$$

From eqns. F. 11 and F. 14 (*a*), we have

$$R_{i_2} = \frac{R_L\left[1-(1-m^2)f_c^2\right]}{\sqrt{1-f^2/f_c^2}} \qquad \text{...(F. 20)}$$

Figure F. 10 shows a plot of Z_{I_2} for different values of m. From the figure it can be observed that for $m = 0.6$ image matching occurs for nearly 85% of the pass band, *i.e.,* Z_{I_3}/R_L is unity over 85% of the pass band. Thus, by selecting the parameter m we can achieve better matching, which results in sharper cut-off of the filter.

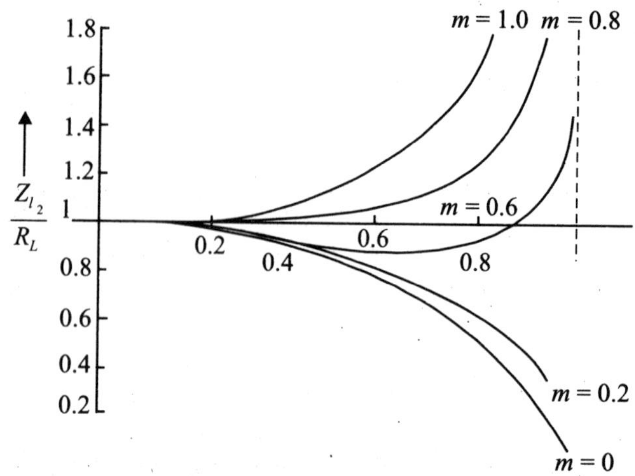

Fig. F.10

Figure F.11 shows the final form of the composite low-pass filter. The half sections of m-derived structure at the end gives the required impedance of eqn. F.20. The sharpest cut-off is obtained by making $m = 0.6$ for the end sections. The cut-off frequency is, of course, determined by L_1 and C_2.

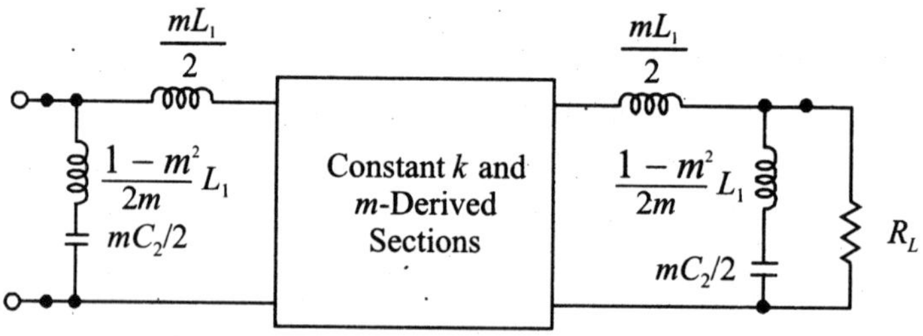

Fig. F.11

Example F.2. A composite low-pass filter is to satisfy the following specifications:

Cut-off frequency 20 k rad/sec

$$R_L = 600\ \Omega$$

Infinite attenuation at

$$\omega_{r_1} = 22k \text{ rad/sec}$$

$$\omega_{r_2} = 40k \text{ rad/sec}$$

$$\omega_{r_3} = \infty$$

The filter must have three sections. Two m-derived sections to produce transmission zeros at ω_{r_1} and ω_{r_2} and a constant k section to produce the basic transmission characteristic and the transmission zero at ω_{r_2}. The end sections are half sections of m-derived structures with $m = 0.6$.

Let the cut-off frequency be normalized to 1 rad/sec and the load resistance to 1 Ω. Then from eqn. F. 14

$$L_1 = 2h \text{ and } C_2 = 2f$$

The normalized transmission zeros are

$$\omega_{\gamma_{1n}} = \frac{22\times10^3}{20\times10^3} = 1.1$$

$$\omega_{\gamma_{2n}} = \frac{40\times10^3}{20\times10^3} = 2.0$$

The value of m for the corresponding m-derived sections are, from eqn. F. 18,

$$m_1 = \sqrt{1-\frac{1}{(1.1)^2}} = 0.417$$

and

$$m_2 = \sqrt{1-\frac{1}{(2)^2}} = 0.866$$

The normalized filter is shown in Fig. F. 12, and F. 13 shows the final normalized network. To denormalize the network elements have to be properly scaled.

$$\Omega_n = 20 \times 10^3$$

$$\gamma_n = 600$$

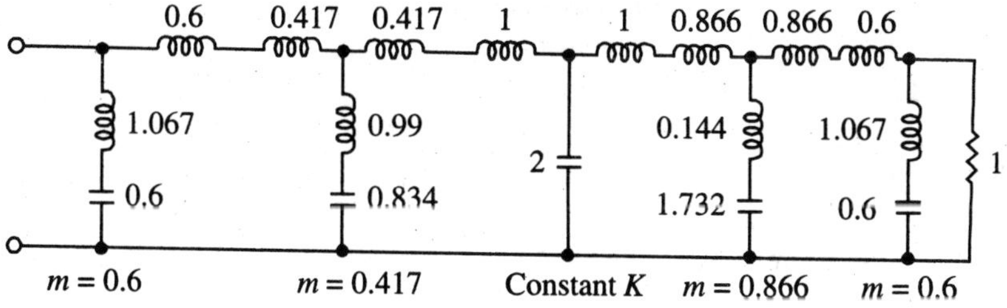

Fig. F. 12

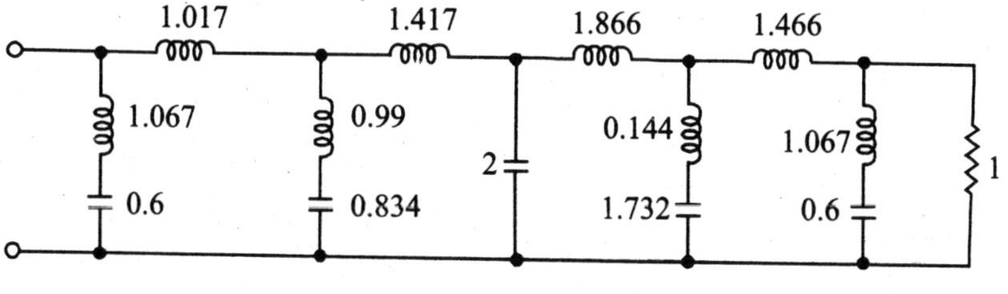

Fig. F. 13

The denormalized values of inductors and capacitors are obtained as (Sec. 4.9)

$$L_d = \frac{\gamma_n L_n}{\Omega_n}$$

$$C_d = \frac{C_n}{\gamma_n \Omega_n}$$

where L_b and C_d are the denormalized, and L_n and C_n the normalized values of inductor and capacitor respectively. The denormalized network is shown in Fig. F.14. Of course, we can obbain alternate filter structures by changing the order of cascading of m-derived and constant k-sections.

The design of an image-parameter low-pass filter requires the specification of the cut-off frequency and frequencies of infinite attenuation (transmission zeros).

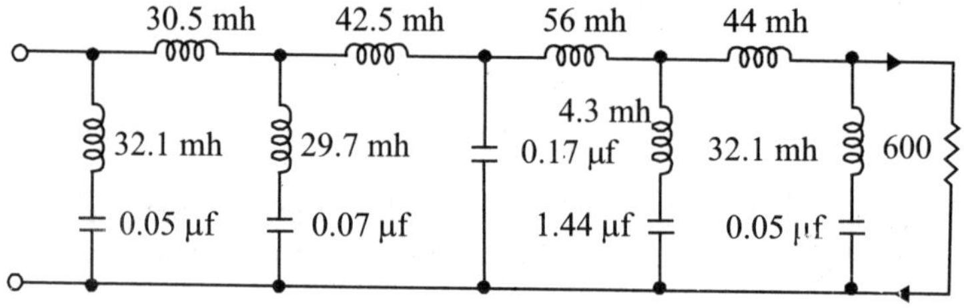

Fig. F.14

The basic filtering characteristic is provided by the k-section and the finite transmission zeros by the m-derived sections. The end sections, half-sections of m-derived section with m = 0.6, provide the required image matching. As demonstrated by example F. 2, the design of even a very high order image parameter filter is quite simple.

Though we have restricted our discussion to T section, the design of composite filter with π section can be similarly developed. The design of lattice image parameter is little more involved and is not discussed in order to keep the discussion simple.

Index

A

Active
 elements 329
 networks 12, 329
 notch 354, 362, 366
Adjoint
 network 140, 344
Admittance 52, 54, 173, 237, 371
 parameters 173
All-pass
 filter 367
 function 292
All-pole transfer function 297, 303
Approximation 276
 Bessel 305
 Butterworth 279, 319
 Chebyshev 286, 319
 elliptic 304, 372
 inverse Chebyshev 303
Asymptotically stable system 105
Attenuation function 279, 317

B

Band elimination filter 299, 354
Band-pass filter 111, 297, 354, 358, 367
 approximation 300
 geometrically symmetric 301
Bandwidth 109, 298, 375
Bessel 308
 approximation 305
Bode plots 164, 397
Branch 21, 219
 transformation 42, 57
Bruton transformation 371
Butterworth
 approximation 279
 circle 283, 289
 filter 313, 319
 polynomial 284
 transfer function 291

C

Capacitance 6
 incremental 6
Capacitor 73
 charge-controlled 6
 linear 7
 non-linear 6
 voltage controlled 6
Cascade connection 186
Cauer form I 239, 248, 254, 257
Cauer form II 239, 249, 254, 259
Centre frequency 298, 300, 368
Characteristic function 316
Chebyshev
 approximation 285, 319
 filter 289, 309
 filter poles 289
 polynomial 286
 transfer function 291
Chord 20, 219
 transformation 43, 134
Coefficient matching technique 311, 350

Coefficient of coupling 65
Common base transistor 172
Complete graph 19
Complex frequency 150
Composite filter 420
Constant-*k* filter 413
Continued fractions 152, 230, 239, 249, 320
Controlled sources 5, 68, 330
Convolution 89, 223
Coupled biquad 374
Coupled coils 73
Coupling network 411
Cut-set 21
  admittance matrix 53, 58
  equations 57, 63
  matrix 28

D

Darlington
  pair 15
  realization 318
Degenerate
  circuit 100, 213
  cut-set 103, 213
  networks 100
Delta function 90
Discrimination parameter 280
Dominant matrix 81
Driving point
  admittance 150, 173
  immittances 227, 235, 255
  impedance 150, 172
Duals 24, 112, 254, 320

E

Edge 18
  admittance matrix 53, 82
  capacitance matrix 82
  impedance matrix 47
Elastance 46
Energy 7
  functions 234
Equivalent
  network 186
  source network 213
E shift 50
Euclid's algorithm 242
Excess elements 215

F

FDNR transformation 371
Feldtkellers condition 316
  f-circuit 20, 59, 219
  f-circuit matrix 22, 219
  f-cut-set 26, 219
  f-cut-set matrix 28, 219
Foster form I 237, 246
Foster form II 238, 248
Fourier series 123
Frequency dependent
  negative conductance 337
  negative resistance 337, 371
Frequency normalizing factor 114
Frequency scaling 113
Frequency transformation 294, 300, 325

G

Gain bandwidth product 375
Generalised
  element 45
  impedance converter 336
Geometrically symmetric filter 301
Graph connected 19
  linear 18
  complete 19
  spanning 18
  sub 18
Gyrator 68, 334

H

Half-power frequencies 110
High-pass filter 296, 371
High-pass network 159, 259
High-pass notch 354, 361
Hurwitz polynomial 228
Hybrid parameters 176, 177
Hysteresis 8

I

Ideal filter 277
Image
  impedance 410, 414, 416
  matching 412
  parameters 410
Immittance 227, 236, 255
Impedance 46
  denormalizing factors 322

normalizing factors 114
parameters 171, 178
Impulse response 89, 97, 223
Incidence matrix 26
Indefinite admittance matrix 189
Inductance 8
incremental 8
mutual 63
Inductor
current-controlled 8
flux-controlled 8
linear 7
non-linear 8
Initial conditions 46, 90, 208
Insertion
loss 314, 318
phase 315
power ratio 315
voltage ratio 315
Inverse
hybrid parameters 177
Laplace transforms 83, 392
transmission parameters 174
I-shift 50

K

Kerwin structure 363
KHN structure 365
Kirchhoff's current law 9, 27, 133
Kirchhoff's voltage law 9, 31, 133

L

Laddcr
similated 370
Ladder network 181, 255, 370
Lagrangian tree 58
Laplace transforms 46, 389
LC ladders 266, 314
Linear graph 18
Linear network 10, 97, 122
Linear phase filter 305
Loop
equations 59, 62, 64
impedance matrix 60, 64
Low-pass
filter 257, 277, 321, 350
network 159
notch 354, 361

Lumped circuit elements 2
Lumped network 12

M

Matrix 380
Maximally flat response 279
Maximum power transfer 141, 315
m-derived filter 416
Mesh 33
equations 54, 64
impedance matrix 55, 64
matrix 34, 55
transformation 43
Minimum-phase network 272
Mutual inductance 63

N

Natural frequency 146, 394
Natural response 104, 147
Negative impedance converter
current inversion 333, 340
voltage inversion 333, 340
Negative impedance inverter 336, 342
Network
active 12, 329
equation 40
linear 10, 97, 122
lumped 2, 13
non-linear 10
one-port 12, 170, 402
passive 12, 136, 235
theory 1
time invariant 11, 97
time variant 11
topology 17
two-port 12, 170, 405
Network functions 150, 195
Node 18
admittance matrix 52, 64
equations 50, 61, 64
transformation 41, 134
voltage 51
Non-casual system 277
Non-minimum phase function 273, 377
Norator 13, 337
Norton's
equivalent 50, 127, 128
theorem 127

Notch filter 354, 366
Nullator 13, 337
Nullor 337

O

One-port 13, 170, 402
Open circuit natural frequency 155
Open circuit stability 158
Operational amplifier 182, 197, 331, 350
  non-ideal 375
Operational impedance 45
Order of complexity 149, 224
Oriented graph 24, 134

P

Partial fractions 83, 237
Passive network 12, 136, 235
Path 19
Phase response 292
Pivotal condensation 192
Planar graphs 24, 33, 34
Poles 84, 152, 282, 289
Pole removal 239, 249
  partial 262
Pole-zero plot 154
Pole weakening 265
Positive real function 227, 231
Principle of conservation of charge 100
Principle of continuity of flux
  linkages 103
Proper tree 219
  modified 219
Propagation constant 413

Q

Q-factor 109, 292, 377
Quadratral symmetry 231, 319

R

RC
  function 244
  ladders 256
  network 350
Reactance function 235, 294
Reciprocal network 136, 173
Reciprocity theorem 136
Reflection coefficient 316, 403
Residue 84, 236, 245
Resistance 2
  incremental 3
Resistor
  current-controlled 3
  linear 2
  non-linear 2, 12
  voltage-controlled 3
Resonance 107
Resonant-frequency 107, 418
RL network 148, 254
Routh-Hurwitz array 228

S

Sallen-key structures 363
Scatter parameters 171, 402
Selectivity parameter 279
Sensitivity 130, 140, 342, 370
Sensitivity
  adjoint network 140
  gain 343
  phase 343
  root 344
Short circuit
  natural frequencies 155
  stability 158
Sinusoidal steady-state response 105
Source
  controlled 5, 330
  current 4, 49
  dependent 5
  independent 4
  voltage 4, 50
  transformation 49
State
  equations 208
  transition matrix 223
  variable filter 365
  variables 208, 363
State equations 208
  equivalent source method 213
  graph theoretic method 218
Step response 91
Strurm's
  functions 233
  test 232
Substitution theorem 123, 128
Superposition theorem 122, 128

T

Taylor-series 278
Tellegen's theorem 133, 141
Thevenin
  equivalent 50, 127
  theorem 127
Transducer function 316
Transfer function 89, 97, 151, 222, 255, 350, 351
Transform networks 71
Transmission
  coefficient 315
  parameters 174, 181
  zeros 158, 256, 257, 266, 418
Tree 19
  modified proper 219
  proper 219
Two-port
  functions 255
  networks 12, 170, 312, 405
  parameters 171

U

Unit step 91

V

Vertex 18
Voltage transfer ratio 152, 255
Vratsanov's theorem 202

Z

Zero input response 80, 97
Zero state response 80, 97
Zeros 84, 152, 228, 316
Zero shirting 262